Lecture Notes in Mathematics

Edited by A. Dold and B. Eckmann

1320

H. Jürgensen G. Lallement
H. J. Weinert (Eds.)

Semigroups Theory and Applications

Proceedings of a Conference held in
Oberwolfach, FRG, Feb. 23 – Mar. 1, 1986

Springer-Verlag
Berlin Heidelberg New York London Paris Tokyo

Editors

Helmut Jürgensen
Department of Computer Science, The University of Western Ontario
London, Ontario, Canada, N6A 5B7

Gérard Lallement
Department of Mathematics, Pennsylvania State University
University Park, PA 16802, USA

Hanns Joachim Weinert
Institut für Mathematik, Technische Universität Clausthal
3392 Clausthal Zellerfeld, Federal Republic of Germany

Mathematics Subject Classification (1980): 20L05, 20M05, 20M07, 20M10, 20M17, 20M18, 20M20, 20M25, 20M35, 20M50, 68Q45, 68Q50, 68T15, 94B45, 94B60

ISBN 3-540-19347-2 Springer-Verlag Berlin Heidelberg New York
ISBN 0-387-19347-2 Springer-Verlag New York Berlin Heidelberg

Library of Congress Cataloging-in-Publication Data. Semigroups: theory and applications: Proceedings of a conference held in Oberwolfach, FRG, Feb. 23 – Mar. 1, 1986 / H. Jürgensen, G. Lallement, H.J. Weinert, eds. p. cm.–(Lecture notes in mathematics; 1320) Bibliography: p. ISBN 0-387-19347-2 (U.S.)
1. Semigroups–Congresses. I. Jürgensen, Helmut. II. Lallement, Gerard, 1935-. III. Weinert, Hanns-Joachim. IV. Series: Lectures notes in mathematics (Springer-Verlag); 1320.
QA3.L28 no. 1320 [QA171] 512'.2–dc 19 88-14699

Printing and binding: Druckhaus Beltz, Hemsbach/Bergstr.
2146/3140-543210

PREFACE

During the week of February 23rd to March 1st, 1986, a conference on semigroups was held at Oberwolfach, Germany, at the Mathematisches Forschungsinstitut. It was organized by H. Jürgensen (The University of Western Ontario), G. Lallement (Pennsylvania State University), and H. J. Weinert (Technische Universität Clausthal). It was the third conference on semigroups held at Oberwolfach, this time with an emphasis on combinatorial semigroups and their applications. The previous ones were held in 1978 and 1981. Their proceedings have been published as volumes 855 and 998 of these Lecture Notes in Mathematics.

The conference was attended by 53 participants from 15 countries: 11 from Germany; 25 from the countries of Czechoslovakia, Finland, France, Hungary, the Netherlands, Poland, Portugal, the Soviet Union, the United Kingdom, and Yugoslavia; 15 from Canada and the United States; 1 from each of Australia and Taiwan. The conference program included 42 lectures, most of which are presented in this volume.

The organizers would like to express their gratitude to the staff at Oberwolfach for creating excellent conditions for the meeting, and to the editors of the Lecture Notes in Mathematics for publishing these proceedings. They also thank all authors and the referees for the work they contributed to the publication of this volume. Special thanks are due to Dr. U. Hebisch (Technische Universität Clausthal) for his continued and indispensable assistance in the preparation of the conference itself and of this volume.

H. Jürgensen, G. Lallement, H. J. Weinert
London (Ontario), University Park (Pennsylvania), and Clausthal-Zellerfeld,
December 1987.

INTRODUCTION

The papers gathered in this volume reflect various trends of research activity over the past several years in pure algebraic semigroup theory, in some areas of theoretical computer science related to semigroup theory (languages, automata, rewriting rules, systems of equations), and in areas of ring theory, universal algebras, and category theory where the objects of interests do have some direct connections with semigroups.

The following brief analysis of the papers regroups them under somewhat artificial headings. This is essentially intended to help the reader gain a better understanding of the general aims of researchers in the various fields mentioned above.

1. Congruences

Unlike in group theory or ring theory, congruences on a semigroup are somewhat difficult to apprehend. In general, subobjects replacing the kernels are not available. Inverse and regular semigroups have proven to offer the best grounds of approach, and the paper by *B. P. Alimpić and D. N. Krgović*, where some classes of congruences on regular semigroups are studied, illustrates perfectly this point.

In the sixties the work of Rhodes on complexity of finite semigroups led him to consider sequences of morphisms collapsing a semigroup to a singleton, each individual morphism of the sequence collapsing as little as possible. The corresponding notion is that of minimal congruence. This is the object of the article by *M. Demlová and V. Koubek* which provides a classification of minimal congruences, and studies their relationship to the extension problem. In the same context, subdirectly irreducible semigroups (i. e. semigroups with a finest congruence distinct from equality) are of interest. An example of structural investigation of this kind for a special class of semigroups is provided by *A. Nagy's* article.

Structural properties of the lattice of all congruences have also been studied. It is well-known, for example, that the lattice of congruences of a completely simple semigroup is semimodular. Here *P. R. Jones* determines almost all varieties of semigroups having a semimodular lattice of congruences and his paper contains results relevant to both congruences and varieties.

2. Varieties and pseudovarieties

Besides the paper by *P. R. Jones* mentioned above, another one by *P. G. Trotter* concentrates on varieties of completely regular semigroups (formerly called unions of groups). These varieties have been vigorously investigated in recent years, e. g. by Petrich, Gerhardt, Jones, and Pollák. Here *P. G. Trotter* determines the injective objects ('injective' means that any morphism $S \to I$ extends to $T \to I$ where T is an extension of S) in several completely regular varieties.

Pseudo-varieties of finite semigroups and monoids are classes closed under sub, quotient, and finite direct products (while for varieties there are no finiteness restrictions). Following Eilenberg's correspondence theorem between varieties of rational languages and pseudo-varieties of monoids, a wealth of activity has been devoted to make this correspondence more precise in special cases. Talks illustrating this were given at the conference by J. Sakarovitch and by H. Straubing and D. Thérien. In the same vein the paper by *J. Almeida* deals with the problem of the connection between a pseudo-variety V of semigroups and the pseudo-variety MV generated by the monoids S^1 for all S in V.

3. Languages

The relationship between star-free languages and first order logic was established by McNaughton in 1971 (see Counterfree Automata, MIT Press). The connection has been investigated further more recently, especially when similarities were detected between the dot-depth hierarchy of Brzozowski and Knast, and the quantifier alternating depth of first order sentences. The paper by *D. Lippert and W. Thomas*, which clarifies the differences between the dot operation in languages and the existential quantifier in first order formulas, is a contribution to this line of work.

In recent years the Western Ontario school has produced many new results on languages and free semigroups dealing with properties of disjunctive languages, various conditions on codes, and properties of partial orders on free semigroups. The papers by *M. Petrich and G. Thierrin* and by *M. Katsura and H. J. Shyr* illustrate this original approach to the study of languages.

The paper by *G. Pollák* dealing with infima in the power set of a free monoid is more set theoretically oriented but it can also be viewed as a contribution to language theory. I should also mention an interesting lecture by D. Perrin (not reported here) where he uses classical semigroup theory results to investigate properties of infinite words.

4. Presentations, equations in free monoids

R. V. Book gave an overview of results on presentations of semigroups and monoids with the so-called Church-Rosser property. The paper by *K. Madlener and F. Otto* contains numerous results on groups having such presentations. In my own paper I survey most of the known results on the decidability of the word problem for one-relator semigroups, concentrating mostly on results of the Russian school.

The paper by *K. Culik II and J. Karhumäki* deals with a problem related to the Ehrenfeucht conjecture proved in 1985 (Each system of equations over a free monoid A^*, A finite, with finitely many variables, is equivalent to a finite subsystem). The question they consider here is when such a finite subsystem can effectively be found. In another paper on equations, *J.-C. Spehner* uses an earlier result of his on presentations of submonoids of free monoids, to give a classification of certain systems of equations in three variables.

Other important recent developments were presented at the Conference but are not reported in this volume: The plactic monoid and its connections with Young tableaux by M. P. Schützenberger; the study of presentations of inverse semigroups by S. W. Margolis and J. C. Meakin.

5. Inverse semigroups and generalizations

The papers by *N. R. Reilly* and by *G. A. Freiman and B. M. Schein* present problems of interest either directly in the area of inverse semigroups or inspired by inverse semigroups. In her paper, *M. B. Szendrei* studies certain classes of semigroups with involutions and shows that the free objects in these classes admit descriptions that are quite similar to the well-known descriptions of free inverse semigroups e. g. by Scheiblich and Munn. Similarly, *J. Fountain* studies certain free right adequate semigroups (S is right adequate if each $\mathcal{L}^*$-class has an idempotent, where $a\mathcal{L}^*b$ iff $a\mathcal{L}b$ in an oversemigroup, and the idempotents commute). Again the free objects Fountain considers do have descriptions extending those of free inverse semigroups.

6. Semigroups of endomorphisms

V. Fleischer and U. Knauer prove that the endomorphism monoid of an act (i. e. of a monoid acting on a set) has a nice representation as a wreath-product of a monoid and a small category. *S. M. Goberstein* studies more generally correspondences. A correspondence on a universal algebra A is simply a subalgebra of $A \times A$. A survey of known results on correspondences on universal algebras and groups is made, and new results on semigroup correspondences are announced.

7. Semigroups and other algebraic structures

a) In the theory of partial semigroups an extension of $(S_1, \circ_1)$, where $\circ_1$ denotes the partial operation on S_1, is defined as $(S_2, \circ_2)$ such that $S_1 \subseteq S_2$ and $a \circ_1 b = c$ implies $a \circ_2 b = c$. In his paper *E. S. Ljapin* develops a number of conditions for the existence of a semigroup extension for a partial semigroup.

b) A typical example of a "transfer" theorem in the theory of semigroup rings is as follows: The monoid ring $R[M]$ is Artinian if and only if the ring R is Artinian and M is a finite monoid (Zelmanov). *J. Okniński* studies here similar types of transfer theorems with respect to the Krull dimensions of rings.

Based on semimodules over semirings *H. J. Weinert* extends the notion of (generalized) algebras over rings by introducing (generalized) semialgebras over semirings including those where infinite sums are used.

c) A semiring is said to be a weak p. o. semiring if it has a partial order compatible with its addition only. The paper by *U. Hebisch and L. C. A. van Leeuwen* contains results on embeddings, and on weak p. o. semirings S such that $(S, +)$ or $(S, \cdot)$ are idempotent semigroups.

d) *K. D. Schmidt* introduces a new class of partially ordered semigroups called minimal clans, and shows how their properties allow to retrieve properties of both Boolean rings and lattice-ordered groups, thereby solving a problem posed about 20 years ago by Birkhoff.

e) A category is called universal if it contains the category of graphs as a full subcategory. *P. Goralčik and V. Koubek* prove here the following interesting result: The category of all extensions of a semigroup S is universal if and only if S has no idempotents.

f) The object of the paper by *W. Lex* are acts in the general meaning of semi-automata, especially lattices of torsion theories of acts as proposed by him and Wiegandt. In this context a new characterization of the non- trivial abelian groups is obtained.

g) Is it possible to get machines to prove theorems for you? Not quite. The machines still need assistance from the operator, as shown in *R. B. McFadden's* paper, using several problems in the theory of semigroups, the last of which I liked particularly.

As these short analyses show, a large variety of topics have been the object of lectures at the Conference. It is a clear sign that the algebraic theory of semigroups is steadily growing over the years, both in strength and in depth. It also appears that semigroups are increasingly connected to more and more distinct areas of Mathematics. This is perhaps the most important warrant of the future vitality of the field.

Gerard Lallement
University Park (Pennsylvania), November 1987

TABLE OF CONTENTS

SOME CONGRUENCES ON REGULAR SEMIGROUPS

Branka P. Alimpić
Dragica N. Krgović

Prirodno-matematički fakultet
Studentski trg 16
YU 11000 Beograd

Matematički institut
Knez Mihailova 35
YU 11000 Beograd

A congruence ρ on a regular semigroup S is uniquely determined by its kernel $\ker \rho = \{x \in S \mid (\exists e \in E) x \rho e\}$ and trace $\mathrm{tr}\, \rho = \rho|_{E(S)}$ [2] . Let Con S be the congruence lattice of S, K and T equivalences on Con S defined by $\rho\, K\, \xi \iff \ker \rho = \ker \xi$ and $\rho\, T\, \xi \iff \mathrm{tr}\, \rho = \mathrm{tr}\, \xi$. It is known that K-classes $[\rho_K, \rho^K]$ and T-classes $[\rho_T, \rho^T]$ are intervals on Con S ([13],[15]). In this paper K-classes with $\mathrm{tr}\, \rho^K = \omega_E$ and T-classes with $\ker \rho^T = S$ are considered. It turns out that such a K-class consists exactly of E-unitary congruences on S, and such a T-class consists exactly of band of groups congruences on S. Similarly, K-classes for which ρ^K is a Clifford congruence consist of E-reflexive congruences. These results generalize corresponding results for inverse semigroups [14].

Throughout this paper, S stands for an arbitrary regular semigroup. For $X \subseteq S$, E(X) denotes the set of idempotents of X. If ρ is a relation on S, then ρ^* denotes the least congruence on S which contains ρ. If ρ is an equivalence on S then ρ^0 denotes the greatest congruence on S contained in ρ. Let $X \subseteq S$. A congruence ρ on S saturates X if for any $a \in X$, the ρ-class $a\rho$ is contained in X. In particular, a congruence ρ on S is idempotent pure if ρ saturates E(S). If Θ_X is the equivalence on S induced by the partition $\{X, S \setminus X\}$ of S, then Θ_X^0 is the greatest congruence which saturates X. We write τ instead of Θ_E^0. If ρ is a congruence on S and α is an equivalence on S/ρ, then the equivalence $\bar{\alpha}$ on S is defined by

$$a\, \bar{\alpha}\, b \iff (a\rho)\ \alpha\ (b\rho) \qquad (a,b \in S).$$

Obviously, $\bar{\alpha}$ is a congruence on S if and only if α is a congruence on S/ρ.

For undefined notations or terminology see [3] or [14].

RESULT 1. [9]. For any congruences ρ and ξ on S,

$$\rho \; T \; \xi \iff \bar{\mathcal{H}}_{S/\rho} = \bar{\mathcal{H}}_{S/\xi}.$$

COROLLARY 1. Let ρ and ξ be congruences on S such that $\rho \; T \; \xi$. Then $\mathcal{H}_{S/\rho} \in \mathrm{Con}(S/\rho) \iff \mathcal{H}_{S/\xi} \in \mathrm{Con}(S/\xi)$. Moreover, if $\mathcal{V}$ is a variety of bands then $\mathcal{H}_{S/\rho}$ is a $\mathcal{V}$-congruence if and only if $\mathcal{H}_{S/\xi}$ is such one.

RESULT 2. ([15],[13],[9]). Let ρ and ξ be congruences on S. Then

(i) $\rho T = [\rho_T, \rho^T]$ and $\rho K = [\rho_K, \rho^K]$ are intervals of Con S.

(ii) $\rho_T = (\mathrm{tr}\rho)^*$, $\rho^T = \bar{\mathcal{H}}^0_{S/\rho}$,

$\rho_K = \{(x,x^2) \mid x \in \ker \rho\}^*$, $\rho^K = \theta^0_{\ker \rho}$.

(iii) $\mathrm{tr}\ \rho \subseteq \mathrm{tr}\ \xi \Rightarrow \rho_T \subseteq \xi_T$ and $\rho^T \subseteq \xi^T$,

$\ker \rho \subseteq \ker \xi \Rightarrow \rho_K \subseteq \xi_K$.

(iv) $\ker \rho \subseteq \ker \xi$ and $\mathrm{tr}\ \rho \subseteq \mathrm{tr}\ \xi \Rightarrow \rho \subseteq \xi$.

Using this result and Theorem [19] it is easy to prove the following lemma.

LEMMA 1. Let be a nonempty family of congruences on S. Then

$$\bigcap_{\rho \in \mathcal{F}} \rho^T = (\bigcap_{\rho \in \mathcal{F}} \rho)^T \quad \text{and} \quad \bigvee_{\rho \in \mathcal{F}} \rho_T = (\bigvee_{\rho \in \mathcal{F}} \rho)_T.$$

REMARK. This result is a part of Theorem 4.13[10].

RESULT 3. [18]. For $X \subseteq S$,

$$a \; \theta^0_X \; b \iff (\forall x,y \in S^1)(xay \in X \iff xby \in X) \qquad (a,b \in S).$$

COROLLARY 2. Let ρ be a congruence on S and let $\tau_{S/\rho}$ be the greatest idempotent pure congruence on S/ρ. Then $\rho^K = \bar{\tau}_{S/\rho}$. Consequently

$$\rho \; K \; \xi \iff \bar{\tau}_{S/\rho} = \bar{\tau}_{S/\xi} \qquad (\rho,\xi \in \mathrm{Con}\ S).$$

Proof. $a \; \bar{\tau}_{S/\rho} \; b \iff (a\rho)\tau_{S/\rho}(b\rho)$

$\iff (\forall x,y \in S^1)((xay)\rho \in E(S/\rho) \iff (xby)\rho \in E(S/\rho))$

$\iff (\forall x,y \in S^1)(xay \in \ker \rho \iff xby \in \ker \rho)$

$\iff a \; \theta^0_{\ker \rho} \; b$ (by Result 3)

$\iff a \; \rho^K \; b$ (by Result 2).

If ω denotes the universal congruence on S then $\sigma = \omega_T$ $[\beta = \omega_K]$ is the least group [band] congruence on S. Similarly, if ε denotes the equality on S then $\mu = \varepsilon^T [\tau = \varepsilon^K]$ is the greatest idempotent separating [idempotent pure] congruence on S.

Using Result 2 we obtain

PROPOSITION 1. The following inclusions are valid for any congruence $\rho \in$ Con S.

(*i*) $\rho \cap \tau \subseteq \rho_T \subseteq \rho \cap \sigma$

(*ii*) $\rho \cap \mu \subseteq \rho_K \subseteq \rho \cap \beta$

(*iii*) $\rho \vee \mu \subseteq \rho^T \subseteq \rho \vee \beta$

(*iv*) $\rho^K \subseteq \rho \vee \sigma$

Proof. (*i*) Since $\rho \subseteq \omega$ implies $\rho_T \subseteq \omega_T = \sigma$ it follows $\rho_T \subseteq \rho \cap \sigma$. From $\mathrm{tr}(\rho \cap \tau) \subseteq \mathrm{tr}\,\rho = \mathrm{tr}\,\rho_T$ and $\ker(\rho \cap \tau) = E \subseteq \ker \rho_T$ we have $\rho \cap \tau \subseteq \rho_T$.

(*ii*) The argument here is similar to that in the proof of (*i*) and is omitted.

(*iii*) Since $\varepsilon \subseteq \rho$ implies $\mu = \varepsilon^T \subseteq \rho^T$ it follows $\rho \vee \mu \subseteq \rho^T$. From $\ker \rho^T \subseteq S = \ker(\rho \vee \beta)$ and $\mathrm{tr}\,\rho^T = \mathrm{tr}\,\rho \subseteq \mathrm{tr}(\rho \vee \beta)$ we have $\rho^T \subseteq \rho \vee \beta$.

(*iv*) From $\mathrm{tr}\,\rho^K \subseteq \omega_E = \mathrm{tr}(\rho \vee \sigma)$ and $\ker \rho^K = \ker \rho \subseteq \ker(\rho \vee \sigma)$ we have $\rho^K \subseteq \rho \vee \sigma$.

The following example shows that the analoque of the first inclusion of (*iii*), i.e. the inclusion $\rho \vee \tau \subseteq \rho^K$, does not hold in general.

EXAMPLE 1. [14;III.4.11]. Let S be a semilattice of two groups G and H of order 2 determined by an isomorphism $\varphi : G \to H$. Let ρ be the Rees congruence on S relative to H. Then $\rho \vee \tau = \omega$ and $\rho^K = \rho \neq \omega$.

By [14] a semigroup in which $\mathcal{H}$ is a congruence is cryptic. A completely regular cryptic semigroup (i.e. a band of groups) is a cryptogroup.

The next theorem characterizes T-classes with $\ker \rho^T = S$.

THEOREM 1. The following statements concerning a congruence ρ on S are equivalent.

(*i*) ρ is a cryptogroup congruence.

(*ii*) ρ^T is a band congruence.

(*iii*) $\rho^T = \rho \vee \beta$.

(*iv*) $\mathrm{tr}\,\rho = \mathrm{tr}\,(\rho \vee \beta)$.

Proof. (*i*) => (*ii*). Since S/ρ is a cryptogroup, $\mathcal{H}_{S/\rho}$ is a congruence on S/ρ which together with Result 2 (*ii*), shows that $\rho^T = \overline{\mathcal{H}}_{S/\rho}$. Since $\mathcal{H}$-classes of S/ρ are groups we have

$$(\forall a \in S)(\exists e \in E(S))\ a\rho\ \mathcal{H}_{S/\rho}\ e\rho \qquad \text{(by Lallement's Lemma)}$$

$$\iff (\forall a \in S)(\exists e \in E(S))\ a\ \overline{\mathcal{H}}_{S/\rho}\ e$$

$\iff \rho^T$ is a band congruence.

(*ii*) => (*iii*). The hypothesis implies that $\rho^T \supseteq \rho \vee \beta$, and thus by Proposition 1 (*iii*), we have $\rho^T = \rho \vee \beta$.

(*iii*) => (*iv*). This is obvious.

(*iv*) => (*i*). This is immediate from Corollary 1.

COROLLARY 3. On a regular semigroup S the following conditions are equivalent.

(*i*) S is a cryptogroup.

(*ii*) $\mu = \beta$.

(*iii*) For every $\rho \in \text{Con } S$, $\rho^T = \rho \vee \beta$.

(*iv*) For every $\rho \in \text{Con } S$, $\rho_K = \rho \wedge \mu$.

Proof. (*i*) <=> (*ii*) is a consequence of Theorem 1.

(*ii*) => (*iii*) and (*ii*) => (*iv*) follow immediately from Proposition 1.

(*iii*) => (*ii*). $\mu = \varepsilon^T = \varepsilon \vee \beta = \beta$.

(*iv*) => (*ii*). $\beta = \omega_K = \omega \wedge \mu = \mu$.

REMARK. Equivalences (*i*) <=> (*ii*) <=> (*iii*) are implicitly in [17] and [12].

The next simple result describes the least cryptogroup congruence.

PROPOSITION 2. The congruence $\kappa = \beta_T$ is the least cryptogroup congruence on S.

Proof. Since $\text{tr}(\beta_T) = \text{tr } \beta$, Theorem 1 implies that β_T is a cryptogroup congruence on S. If ρ is any cryptogroup congruence on S, ρ^T is a band congruence, so $\beta \subseteq \rho^T$. Hence by Result 2 (*iii*), $\beta_T \subseteq (\rho^T)_T = \rho_T \subseteq \rho$.

Let η denotes the least semilattice congruence on S. Similarly one proves the following series of results.

THEOREM 2. *The following statements concerning a congruence* ρ *on* S *are equivalent.*

(*i*) ρ *is a Clifford congruence.*

(*ii*) ρ^T *is a semilattice congruence.*

(*iii*) $\rho^T = \rho \vee \eta$.

(*iv*) $\mathrm{tr}\,\rho = \mathrm{tr}(\rho \vee \eta)$.

COROLLARY 4. *On a regular semigroup* S *the following conditions are equivalent.*

(*i*) S *is a Clifford semigroup.*

(*ii*) $\mu = \eta$.

(*iii*) For every $\rho \in \mathrm{Con}\, S$, $\rho^T = \rho \vee \eta$.

PROPOSITION 3. *The congruence* $\eta_T = \nu$ *is the least Clifford congruence on* S.

Following R.Feigenbaum [1], for any non-empty subset H of S the *closure* $H\omega$ of H is defined by $H\omega \overset{def}{=} \{x \in S \mid (\exists h \in H) hx \in H\}$. H is *closed* if $H\omega \subseteq H$. If H is a subsemigroup of S or if it is full ($E(S) \subseteq H$), then $H \subseteq H\omega$.

A regular semigroup S is E-*unitary* if the set E(S) is closed.Any E-unitary semigroup is orthodox [4].

A subset H of S is called *self-conjugate* if $x'Tx \subseteq T$ for every x of S and every inverse x' of x. Let U be the least full self-conjugate subsemigroup of S, and let σ be the least group congruence on S. According to [1], $\ker \sigma = U\omega$. If the semigroup S is orthodox, $U = E(S)$.

For a subset H of S, and any congruence ρ on S, let
$H\rho = \{x \in S \mid (\exists h \in H) x \,\rho\, h\}$.

RESULT 4 [6]. *For any congruence* ρ *on* S

$$\ker(\rho \vee \sigma) = (U\rho)\omega.$$

RESULT 5 [6]. *Let* S *and* T *be regular semigroups and* $\Phi: S \to T$ *a homomorphism of* S *onto* T. *If* U *is the least self-conjugate full subsemigroup of* S, $U\Phi$ *is the least such subsemigroup of* T.

Now we shall consider K-classes of Con S with $\mathrm{tr}\,\rho^K = \omega_E$ where ω_E is the universal congruence on E(S).

THEOREM 3. *The following statements for a congruence* ρ *on* S *are equivalent.*

(*i*) ρ *is* E-*unitary*.

(*ii*) ker ρ *is closed*.

(*iii*) $\ker \rho = \ker(\rho \vee \sigma)$.

(*iv*) $\rho^K = \rho \vee \sigma$.

(*v*) ρ^K *is a group congruence*.

Proof. (*i*) <=> (*ii*). ρ is E-unitary

$\iff (\forall a,h \in S)((ha)\rho,\ h\rho \in E(S/\rho) \Rightarrow a\rho \in E(S/\rho))$

$\iff (\forall a,h \in S)(ha,h \in \ker \rho \Rightarrow a \in \ker \rho)$

<=> ker ρ is closed.

(*i*) => (*iii*). Let $x \in S$. Then

$x \in \ker(\rho \vee \sigma) \iff x \in (U\rho)\omega$ (by Result 4)

$\iff (\exists s \in S)(s \in U\rho$ and $sx \in U\rho)$

$\Rightarrow (\exists s \in S)(s\rho \in U(S/\rho)$ and $(sx)\rho \in U(S/\rho))$ (by Result 5)

$\Rightarrow (\exists s \in S)(s\rho \in E(S/\rho)$ and $(sx)\rho \in E(S/\rho))$

(since S/ρ is orthodox)

$\Rightarrow x\rho \in E(S/\rho)$ (since S/ρ is E-unitary)

$\Rightarrow x \in \ker \rho$.

Thus $\ker(\rho \vee \sigma) \subseteq \ker \rho$. Since the opposite inclusion is obvious, (iii) follows.

(*iii*) => (*iv*). From $\ker \rho^K = \ker \rho = \ker(\rho \vee \sigma)$ it follows $\rho^K \supseteq \rho \vee \sigma$. By Proposition 1 (*iv*) we have $\rho^K = \rho \vee \sigma$.

(*iv*) => (*v*). This is obvious.

(*v*) => (*i*). The hypothesis implies that ρ^K is E-unitary and by (*i*) <=> (*ii*) it follows $\ker \rho = \ker \rho^K$ is closed. Thus ρ is E-unitary.

COROLLARY 5. *On a regular semigroup* S, *the following conditions are equivalent*

(*i*) S *is* E-*unitary*.

(*ii*) $\sigma = \tau$.

(*iii*) *For every* $\rho \in \operatorname{Con} S$, $\rho_T = \rho \cap \tau$.

(*iv*) *Every idempotent pure congruence on* S *is* E-*unitary*.

(*v*) *There exists an idempotent pure* E-*unitary congruence on* S.

REMARK. Equivalence (*i*) <=> (*ii*) is proved also in [16].

The proof of the following proposition is similar to the proof of the Proposition 2.

PROPOSITION 4. *The congruence* $\pi = \sigma_K$ *is the least* E-*unitary congruence on* S.

Using the Corollary 5 and Lemma 1 one can prove that the following holds.

PROPOSITION 5. *Let* S *be an* E-*unitary regular semigroup. The mapping*

$$\varphi: \rho \to \rho \cap \tau$$

is a complete lattice homomorphism of Con S *onto the lattice of idempotent pure congruences on* S.

Let S be an orthodox semigroup and let Y be the least inverse congruence on S. Then we have

PROPOSITION 6. *For an orthodox semigroup* S *the following conditions are equivalent.*

(*i*) S *is* E-*unitary*.

(*ii*) Y *is* E-*unitary*.

(*iii*) ba Y a => b ∈ E (a,b ∈ S).

Proof. (*i*) <=> (*ii*) follows from Corollary 5.

(ii) <=> S/Y is E-unitary
<=> (baYa ⇒ bY ∈ E(S/Y)) (by Proposition III 7.2.[14])
<=> (iii) (since Y is idempotent pure).

REMARK. The equivalence (*i*) <=> (*ii*) is also proved in [8] and [11].

In the remainder of the paper we consider K-classes which consist of E-reflexive congruences. A semigroup S is E-*reflexive* if exy ∈ E(S) => eyx ∈ E(S) for every x,y ∈ S and e ∈ E(S). We observe that every E-unitary semigroup is E-reflexive [4].

RESULT 6 [7]. *On a regular semigroup the following conditions are equivalent*

(*i*) $\nu \subseteq \tau$.

(*ii*) *Every* η-*class of* S *is* E-*unitary*.

(*iii*) S *is* E-*reflexive*.

We can now prove an analogue of Theorem 3.

THEOREM 4. *The following statements concerning a congruence* ρ *on* S *are equivalent*.

(i) ρ *is* E-reflexive.

(ii) ker $\rho \cap N$ *is closed in* N *for every* η-*class* N *of* S.

(iii) ρ^K *is a Clifford congruence*.

(iv) ker ρ = ker($\rho \vee \nu$).

Proof. (i) => (ii). Let N be an η-class of S and let a $\in$ N. Then we have

a $\in$ (ker $\rho \cap N)\omega_N$ => ($\exists$ x)(xa $\in$ ker $\rho \cap N$ and x $\in$ ker $\rho \cap N$)

=> ($\exists$ x)(xa,x $\in$ ker ρ and a η x)

=> ($\exists$ x)((xρ),(xa)$\rho \in$ E(S/ρ) and (aρ) $\eta_{S/\rho}$ (xρ))

=> a$\rho \in$ E(S/ρ)   (by Result 6)

=> a $\in$ ker ρ.

(ii) => (i). ($\forall$ N) ker $\rho \cap N$ is closed in N

=> ($\forall$ N)$\rho|_N$ is an E-unitary congruence on N

=> S/$\rho \cap \eta$ is a semilattice of E-unitary semigroups

=> S/$\rho \cap \eta$ is E-reflexive   (by Result 6)

=> S/ρ is E-reflexive   (since ker ρ = ker($\rho \cap \eta$)).

=> ρ is an E-reflexive congruence.

(i) <=> (iii). S/ρ is E-reflexive <=> $\tau_{S/\rho}$ is a Clifford congruence
                                              (by Result 6)

                              <=> ρ^K is a Clifford congruence
                                              (by Corollary 2)

(iii) => (iv). Since ρ^K is a Clifford congruence we have $\nu \subseteq \rho^K$ which yields $\rho \vee \nu \subseteq \rho^K$, so ker($\rho \vee \nu$) $\subseteq$ ker ρ^K = ker ρ. But ker $\rho \subseteq$ ker($\rho \vee \nu$) and therefore ker ρ = ker($\rho \vee \nu$).

(iv) => (iii). From ker ρ^K = ker ρ = ker($\rho \vee \nu$) it follows that $\rho^K \supseteq \rho \vee \nu \supseteq \nu$, hence ρ^K is a Clifford congruence on S.

The following proposition is an analogue of Proposition 4.

PROPOSITION 7. *The congruence* $\lambda = \nu_K$ *is the least* E-reflexive *congruence on* S.

One may ask whether the equivalence (i) <=> (ii) of the Result 6 would remain true if ν and η were replaced by κ and β respectively. It can be proved that $\kappa \subseteq \tau$ implies that E(N) is closed in N for every β-class N of S. The following counter-example given us by P.R.Jones, shows that the opposite implication does not remain true.

EXAMPLE 2 [5]. Let $D = \mathcal{M}(G;I,I,P)$ where $I = \{0,1\}$, $G = \{e,a\}$ and $P = \begin{pmatrix} e & e \\ e & a \end{pmatrix}$. Let S be the ideal extension of D by the group $\{1,x\}$, where 1 is the identity for S and

$$(i,g,\lambda)x = (i,g,\lambda+1 \pmod 2)$$

$$x(i,g,\lambda) = \begin{cases} (i,g,\lambda) & \text{if} \quad i = 0 \\ (i,ag,\lambda) & \text{if} \quad i = 1 \end{cases}$$

for $g \in G$, $i,\lambda \in I$.

The semigroup S is a band of E-unitary semigroups, and it is not a band of groups. On the other hand, the equality is the only idempotent pure congruence on S, so $\kappa \subsetneq \tau$ is not true.

Acknowledgment.

It is our pleasure to record here our thanks to P.R.Jones and P.G.Trotter for useful discussions and comments.

REFERENCES

1. R.Feigenbaum, Kernels of regular semigroup homomorphisms, Doctoral dissertation, University of South Carolina, 1975.

2. R.Feigenbaum, Regular semigroup congruences, Semigroup Forum, 17 (1979), 373-377.

3. J.M.Howie, An Introduction to Semigroup Theory, Academic Press, London, 1976.

4. J.M.Howie and G.Lallement, Certain fundamental congruences on a regular semigroup, Proc. Glasgow Math. Assoc., 7 (1966), 145-159.

5. P.R.Jones, Mal'cev products of varieties of completely regular semigroups, J.Austral.Math.Soc. T.42(1987),227-246.

6. D.R.LaTorre, Group congruences on regular semigroups, Semigroup Forum 24 (1982), 327-340.

7. D.R.LaTorre, The least semilattice of groups congruence on a regular semigroup, Semigroup Forum 27 (1983), 319-329.

8. F.E.Masat, Proper regular semigroups, Proc.Amer.Math.Soc. 71 (1978), 189-192.

9. F.Pastijn, Congruences on regular semigroups - A Survey, Proc. Marquette Conf. on Semigroups (1984), 159-175.

10. F.Pastijn and M.Petrich, Congruences on regular semigroups, Trans. Amer.Math.Soc. 295 (1986), 607-633.

11. F.Pastijn and M.Petrich, Regular semigroups as extensions, Pitman Advanced Publishing Program, Boston, 1985.

12. F.Pastijn and M.Petrich, The congruence lattice of a regular semigroup, Preprint.

13. F.Pastijn and P.G.Trotter, Lattices of completely regular semigroup varieties, Pac.J.Math. 119 (1985), 191-214.

14. M.Petrich, Inverse semigroups, Wiley, New York, 1984.

15. N.R.Reilly and K.E.Scheiblich, Congruences on regular semigroups, Pac.J.Math. 23 (1967), 349-360.

16. T.Saitô, Ordered regular proper semigroups, J.Algebra 8 (1968), 450-477.

17. E.Spitznagel, The lattice of congruences on a band of groups, Glasgow Math.J. 14 (1973), 189-197.

18. M.Teissier, Sur les equivalences reguliere dans les demi-groups, C.R.Acad.Sci.Paris 232 (1951), 1987-1989.

19. P.G.Trotter, On a problem of Pastijn and Petrich, Semigroup Forum, 34(1986), 249-252.

ON PSEUDOVARIETIES OF MONOIDS

Jorge Almeida
Centro de Matemática
Universidade do Minho
Braga, Portugal

1. INTRODUCTION

In connection with his study of power semigroups, Pin [5] introduced the operator $\underline{M}$ on pseudovarieties of semigroups defined as follows. For a semigroup S, the monoid S^1 coincides with S if S is a monoid; otherwise, S^1 is obtained from S by adjoining a neutral element. For a pseudovariety $\underline{V}$ of semigroups, $\underline{MV}$ denotes the pseudovariety of monoids generated by all monoids of the form S^1 with $S \in \underline{V}$.

In this note, we begin a systematic study of the operator $\underline{M}$. Our approach involves examining the links with the operator $\mathcal{M}$ defined analogously at the level of varieties. This allows us to obtain results on $\underline{M}$ by dealing with identities. Our main lemma describes a finite basis of identities for $\mathcal{MV}$ whenever $\mathcal{V}$ is a variety defined by a single identity of the form

$$x_1\cdots x_p y_1\cdots y_m z_1\cdots z_q = x_1\cdots x_p y'_1\cdots y'_{m+1}\, z_1\cdots z_q .$$

We are then able to give direct unified proofs of some known results [8,6,5] and also solve a problem proposed in [5] which asks for a characterization of $\underline{M}\ \underline{LI}$ where $\underline{LI}$ denotes the pseudovariety of all locally trivial finite semigroups.

2. PRELIMINARIES

The main references we adopt on semigroups and pseudovarieties are Eilenberg [3] and Lallement [4]. For the interplay between varieties in Birkhoff's sense and pseudovarieties as defined by Eilenberg and Schützenberger, see Ash [2]. Also, see Reiterman [7] for the notion of pseudoidentity.

For a variety $\mathcal{V}$ of semigroups, let $\mathcal{MV}$ denote the variety of monoids generated by all monoids of the form S^1 with $S \in \mathcal{V}$. Recall a generalized variety is a union of a directed family of varieties. For a generalized variety W of semigroups, MW denotes the generalized variety of monoids generated by all S^1 with $S \in W$. Since $W_1 \subseteq W_2$ implies $MW_1 \subseteq MW_2$, if $W = \cup_{i\in I} W_i$ is a union of a directed family of (generalized) varieties of semigroups, then $MW = \cup_{i\in I} MW_i$. In particular, if W is a variety, the two definitions of MW given above agree.

For a class C of algebras of a given type, C^F denotes the class of all finite members of C. Recall every pseudovariety is of the form W^F for some generalized variety W.

<u>LEMMA</u> 2.1. If W is a generalized variety of semigroups generated by its finite members then $\underline{M}(W^F) = (MW)^F$.

<u>PROOF</u>. From the definitions, we have $\underline{M}(W^F) \subseteq (MW)^F$ in general. Suppose now $M \in (MW)^F$. Then, since W is generated by its finite members, there is a finite set I and, for each $i \in I$, a finite member S_i of W and a (not necessarily finite) exponent e_i, and there is a submonoid T of $\Pi_{i\in I}(S_i{}^1)^{e_i}$ such that M is a homomorphic image of T. Since M is finite, we may assume that T is finitely generated. Since I is finite and each S_i is finite, we may then assume that each exponent e_i is also finite. Then we conclude that T is finite, whence $M \in \underline{M}(W^F)$.

For a variety V of monoids, let V_S denote the variety of semigroups generated by the elements of V regarded as semigroups. Similar definitions can be made at the levels of generalized varieties and pseudovarieties and the analogues of the compatibility results given above for M and $\underline{M}$ also hold for these new operators.

<u>LEMMA</u> 2.2. For any variety V of monoids, $M(V_S) = V$.

<u>PROOF</u>. Since $V \subseteq V_S$ as classes, we have $V \subseteq M(V_S)$. Now, suppose $S \in V_S$. Then, there exists $M \in V$ and a subsemigroup T of M of which S is a homomorphic image. But, then S^1 is a homomorphic image of a submonoid of M. Hence $S^1 \in V$. It follows that $M(V_S) \subseteq V$.

<u>COROLLARY</u> 2.3. The operator M is onto, $V \mapsto V_S$ is one-to-one and V_S is the largest variety W of semigroups such that $MW \subseteq V$. Moreover, $V \mapsto V_S$

defines a lattice embedding of the lattice of varieties of monoids into the lattice of all varieties of semigroups.

Let $X = \{x,y,z,t,x_1,x_2,\ldots,y_1,y_2,\ldots\}$ denote the free monoid on X and $X^+ = X^* \setminus \{1\}$. For $w \in X^*$, let $|w|$ denote the length of w and $c(w)$ represent the set of all variables which occur in w. If $Y \subseteq X$, w_Y denotes the word obtained from w by deletion of all elements of $X \setminus Y$. We also write $|w|_Y$ for $|w_Y|$, w_α for $w_{\{\alpha\}}$ and $w_{\alpha,\beta}$ for $w_{\{\alpha,\beta\}}$ where $\alpha, \beta \in X$. An identity of monoids (resp. semigroups) is an expression $w = v$ with $w,v \in X^*$ (resp. $w,v \in X^+$). Since a monoid has a unique neutral element, an identity $w = 1$ with $w \in X^+$ and $y \notin c(w)$ is equivalent to $wy = y = yw$ in the sense that the monoids satisfying the first or the last identities are the same. Thus, when convenient, we may use only semigroup identities even when working with monoids. For a set Σ of identities, $[\Sigma]$ denotes, according to the context, either the class of all semigroups or the class of all monoids satisfying Σ. For a set Π of pseudoidentities, $[\![\Pi]\!]$ (resp. $\mathbf{[}\Pi\mathbf{]}$) denotes the pseudovariety of all semigroups (resp. monoids) satisfying Π.

3. MAIN LEMMA

For non-negative integers p,m,q not all zero, let $\lambda_{p,m,q}$ denote the identity

$$x_1\ldots x_p y_1\ldots y_m z_1\ldots z_q = x_1\ldots x_p t_1\ldots t_{m+1} z_1\ldots z_q.$$

<u>LEMMA</u> 3.1. Let $V_{p,m,q}$ denote the class of all monoids M satisfying the following conditions:

(i) $|u|_x \geq p+m+1$, $|v|_x \geq q$ imply $M \models uv = u'v$ where u' is obtained from u by removing the last occurrence of x;

(ii) $|u|_{x,y} \geq p$, $|w|_{x,y} \geq q$, $|uw|_{x,y} \geq p+q+\max\{m-2,0\}$ imply $M \models uxyw = uyxw$.

Then $V_{p,m,q} = M[\lambda_{p,m,q}]$.

<u>PROOF</u>. The inclusion $M[\lambda_{p,m,q}] \subseteq V_{p,m,q}$ is immediate. For the reverse inclusion, we show that $V_{p,m,q}$ satisfies every identity which holds in $M[\lambda_{p,m,q}]$.

For a word w, let $\overline{w}$ be obtained from w by deleting all occurrences of any variable between p+m occurrences of it to the left and q occurrences to the right. Clearly $V_{p,m,q} \models w = \overline{w}$. Suppose $M[\lambda_{p,m,q}] \models u = v$. $M[\lambda_{p,m,q}] \subseteq V_{p,m,q}$, it follows that, to establish $V_{p,m,q} \models u = v$, we may assume $u = \overline{u}$ and $v = \overline{v}$. Then, $|u|_\alpha = |v|_\alpha \leq p+m+q$ for all $\alpha \in X$, since $\lambda_{p,m,q} \vdash u_\alpha = v_\alpha$.

Suppose $u = u_1xu_2$, $v = v_1yv_2$ with $(u_1)_{x,y} = (v_1)_{x,y}$. Since $\lambda_{p,m,q} \vdash u_{x,y} = v_{x,y}$, we must have $|u_1|_{x,y} \geq p$ and also $|u_2|_{x,y} = |v_2|_{x,y} \geq q$, $|u|_{x,y} = |v|_{x,y} \geq p+m+q$.

Suppose now that $u \neq v$ and write $u = \varphi\, x\, \chi\, y\, \psi\, \omega$, $v = \varphi\, y\, \pi\, \omega$ where $y \notin c(\chi)$ and the last variables in ψ and π are not the same. Consider $f(u,v) = (|x\chi y\psi|, |\chi|)$ and suppose that f(u,v) is lexicographically minimal for $V_{p,m,q} \not\models u = v$, $M[\lambda_{p,m,q}] \models u = v$, $u = \overline{u}$, $v = \overline{v}$. If $\chi = 1$, then $V_{p,m,q} \models u = u'$ where $u' = \varphi y x \psi \omega$ since $|\varphi|_{x,y} \geq p$, $|\psi\omega|_{x,y} \geq q$, $|u|_{x,y} \geq p+m+q$ by the preceding paragraph. If $\chi = \chi' z$, let $u' = \varphi x \chi' y z \psi \omega$ so that $V_{p,m,q} \models u = u'$ just as before. In any case, the parameter $f(u',v)$ strictly precedes f(u,v) in the lexicographical order and so $V_{p,m,q} \models u' = v$. Hence $V_{p,m,q} \models u = v$, a contradiciction. This completes the proof of the inclusion $V_{p,m,q} \subseteq M[\lambda_{p,m,q}]$.

From the Lemma we conclude in particular that $V_{p,m,q}$ is a variety. Actually, it is easy to see that $V_{p,m,q}$ admits a finite basis of identities. Namely, M satisfies condition (i) if and only if it satisfies the following identity

$$xy_1 \ldots xy_{p+m} x z_1 x \ldots z_q x = xy_1 \ldots xy_{p+m} z_1 x \ldots z_q x$$

while condition (ii) can be replaced by the identities

$$uzwytw = uzyxtw$$

where $p+m-2+q \leq |uw|_{x,y} \leq p+m+q$, $|u|_{x,y} \geq p$, $|w|_{x,y} \geq q$ and u and w are obtained from $u_{x,y}$ and $w_{x,y}$, respectively, by inserting between any two consecutive letters a different previously non-occurring variable.

4. APPLICATIONS

Consider the following pseudovarieties of semigroups:

$\underline{N} = \cup_n [\![\lambda_{0,n,0}]\!] = [\![x^\omega = 0]\!]$

$$\underline{K} = \cup_n [\![\lambda_{n,0,0}]\!] = [\![x^\omega y = x^\omega]\!],\ \underline{K}_1 = [\![\lambda_{1,0,0}]\!]$$

$$\underline{LI} = \cup_n [\![\lambda_{n,0,n}]\!] = [\![x^\omega y x^\omega = x^\omega]\!]$$

Here, $x \mapsto x^\omega$ denotes the implicit unary operation which associates with an element s of a finite semigroup the idempotent in the subsemigroup generated by s. Hence, the second equalities on each line follow from the observation that, for a finite semigroup S with n elements and set of idempotents E, $S^n = SES$ (cf. [3], Proposition III. 9.2).

Before applying the main lemma to the calculation of $\underline{M}$ on the above pseudovarieties, we need a preliminary result whose proof is inspired by Eilenberg's proof of the result cited above.

LEMMA 4.1. Let M be a monoid with n elements satisfying the pseudoidentity $x^\omega yxzx^\omega = x^\omega yzx^\omega$ and let $s \in M$. Then $(MsM)^n = Ms^nM$.

PROOF. The inclusion $Ms^nM \subseteq (MsM)^n$ is obvious. For the reverse inclusion, let $t_0, t_1, \ldots, t_n \in M$ and consider $a_k = t_0st_1s\ldots t_{k-1}st_k$ $(k = 1, \ldots, n)$. If the elements a_k $(k = 1, \ldots, n)$ of M are all distinct, then $a_k = s^n$ for some k and so $a_n = 1s^n(st_{k+1}\ldots st_n)$, as desired. Otherwise, say $a_k = a_j$ with $k < j$. Let $e = (st_{k+1}\ldots st_j)^\omega$ and let $b = t_{k+1}st_{k+2}\ldots st_j$. Then

$e = esbe$	since $e = esb$ because $M \models x^\omega = x^{\omega+1}$
$= esebe$	since $M \models x^\omega yxzx^\omega = x^\omega yzx^\omega$ (*)
$= esesb^2e$	since $e = esb$
$= es^2b^2e$	as in (*)
$= \ldots = es^nb^ne$	proceeding in a similar way
$= es^nes^nb^ne$	as in (*) using $s^n = s^\omega$
$= es^ne$	using the above

Hence, $a_k = a_j = a_ke = a_kes^ne$ so that $a_n \in Ms^nM$, as claimed.

Let $\mathcal{W} = \cup_{p,m,q}[\lambda_{p,m,q}] = \cup_n[\lambda_{n,0,n}]$. Since every finitely generated member of $\mathcal{W}$ is finite, every generalized variety contained in $\mathcal{W}$ is generated by its finite members. Clearly $\mathcal{W}^F = \underline{LI} \supseteq \underline{K} \supseteq \underline{N}$. Further, the pseudoidentity $x^\omega yxzx^\omega = x^\omega yzx^\omega$ is easily seen to hold in every monoid of the form S^1 with $S \in \underline{LI}$.

THEOREM 4.2 (Straubing). $\underline{MN} = [\![x^\omega = x^{\omega+1},\ x^\omega y = yx^\omega]\!]$.

PROOF. By the results of section 2 and Lemma 3.1, $\underline{MN} = \cup_n \mathcal{V}_{0,n,0}^F$. Now, if

$$\mathcal{W}_n = [\, xy_1 \ldots xy_n x = xy_1 \ldots xy_n = y_1 x \ldots y_n x \,],$$

it is easy to see that $\mathcal{V}_{0,n,0} \subseteq \mathcal{W}_n \subseteq \mathcal{V}_{0,2n,0}$ whence $\cup_n \mathcal{V}_{0,n,0}^F = \cup_n \mathcal{W}_n^F$. Finally, using Lemma 4.1, we obtain $\cup_n \mathcal{W}_n^F = [\![x^\omega = x^{\omega+1},\ x^\omega y = yx^\omega]\!]$: if $M \models \{x^\omega = x^{\omega+1},\ x^\omega y = yx^\omega\}$ and M has n elements, let $s, t_1, \ldots, t_n \in M$; then $st_1 \ldots st_n = as^n b = cs^n$ and so $st_1 \ldots st_n s = cs^{n+1} = cs^n = st_1 \ldots st_n$, whence $M \models xy_1 \ldots xy_n x = xy_1 \ldots xy_n$ while the identity $xy_1 \ldots xy_n x = y_1 x \ldots y_n x$ is verified in an analogous manner; hence $M \in W_n$.

THEOREM 4.3 (Pin, Straubing and Thérien). $\underline{MK}_1 = [\![xyx = xy]\!]$.

PROOF. We have $\underline{MK}_1 = \mathcal{V}_{1,0,0}^F = [\![xyx = xy,\ xxy = xyx]\!] = [\![xyx = xy]\!]$.

THEOREM 4.4 (Pin). $\underline{MK} = [\![x^\omega yx = x^\omega y]\!]$.

PROOF. Here, we have

$$\begin{aligned} \underline{MK} &= \cup_n \mathcal{V}_{n,0,0}^F \\ &= \cup_n [\![xy_1 \ldots xy_n x = xy_1 \ldots xy_n]\!] \\ &= [\![x^\omega yx^\omega = x^\omega y]\!] \end{aligned}$$

the last two equalities being obtained just as in the proof of Theorem 4.1.

THEOREM 4.5. $\underline{M}\ \underline{LI} = [\![x^\omega yxzx^\omega = x^\omega yzx^\omega\ ,\ x^\omega yxztz^\omega = x^\omega yzxtz^\omega]\!]$.

PROOF. As usually,

$$\begin{aligned} \underline{M}\ \underline{LI} &= \cup_n \mathcal{V}_{n,0,n}^F \\ &= \cup_n [\![xy_1 \ldots xy_n xz_1 x \ldots z_n x = xy_1 \ldots xy_n z_1 x \ldots z_n x, \\ &\qquad xy_1 \ldots xy_n xzt_1 z \ldots t_n z = xy_1 \ldots xy_n zxt_1 z \ldots t_n z]\!] \\ &= [\![x^\omega yxzx^\omega = x^\omega yzx^\omega,\ x^\omega yxztz^\omega = x^\omega yzxtz^\omega]\!], \end{aligned}$$

where the last two equalities are obtained in a routine way.

5. CONCLUSION

It would be interesting to find a systematic way of calculating $\mathcal{MV}$ given a basis of identities for $\mathcal{V}$. This appears to be quite hard for general $\mathcal{V}$, but it should be accessible in case we restrict our attention to permutative $\mathcal{V}$ (i.e, each member of $\mathcal{V}$ satisfies some nontrivial permutation identity).

Let $\underline{\text{Perm}}$ denote the class of all finite permutative semigroups. As a particular instance of the above program, in view of Theorem 4.5 and the equality $\underline{\text{Perm}} = \underline{\text{LI}} \vee \underline{\text{Com}}$ where $\underline{\text{Com}}$ denotes the class of all finite commutative semigroups (cf. Almeida [1], Corollary 3.4), it appears natural to conjecture that

$$\underline{M}\ \underline{\text{Perm}} = [\![x^{\omega}yx^{\omega}zx^{\omega} = x^{\omega}yzx^{\omega},\ x^{\omega}yxztz^{\omega} = x^{\omega}yzxtz^{\omega}]\!].$$

Finally, we would like to point out that the previously known proofs of the results of section 4 involved complicated language-theoretic arguments.

REFERENCES

1. J. Almeida, Power pseudovarieties of semigroups I, Semigroup Forum 33 (1986) 357-373.
2. C. Ash, Pseudovarieties, generalized varieties and similary described classes, J. Algebra 92 (1985) 104-115.
3. S. Eilenberg, Automata, Languages and Machines, Vol. B, Academic Press, New York, 1986.
4. G. Lallement, Semigroups and Combinatorial Applications, Wiley-Interscience, New York, 1979.
5. J.E. Pin, Semigroupe des parties et relations de Green, Can. J. Math. 36 (1984) 327-343.
6. J.E. Pin, H. Straubing and D. Thérien, Small varieties of finite semigroups and extensions, J. Austral. Math. Soc. (Series A) 37 (1984) 269-281.
7. J. Reiterman, The Birkhoff theorem for finite algebras, Algebra Universalis 14 (1982) 1-10.
8. H. Straubing, The variety generated by finite nilpotent monoids, Semigroup Forum 24 (1982) 25-38.

SYSTEMS OF EQUATIONS OVER A FINITELY GENERATED FREE MONOID HAVING AN EFFECTIVELY FINDABLE EQUIVALENT FINITE SUBSYSTEM*

K. Culik II
Department of Computer Science
University of Waterloo
Ontario, Canada

J. Karhumäki
Department of Mathematics
University of Turku
Finland

Abstract. It has been proved recently, cf, [AL], that each system of equations over a finitely generated free monoid having only a finite number of variables has an equivalent finite subsystem. We discuss the problem when such a finite subsystem can be effectively found. We show that this is the case when the system is defined by finite, algebraic or deterministic two-way transducers.

1. *Introduction*

Throughout the history of mathematics compactness results, that is results stating that something which is specified by an infinite way is actually specified by a finite subpart of this infinite specification, have been eagerly looked for. In recent years a remarkable compactness property of free monoids has been revealed. More precisely, it has been shown in [AL] and [Gu] that each system of equations over a finitely generated free monoid and having a finite number of variables is equivalent to a finite subsystem.

This compactness result is closely related to the *Ehrenfeucht Conjecture*, cf. [K], which is as follows: For each subset L of a finitely generated free monoid Σ^* there exists a finite subset F of L such that for any two morphisms h and g from Σ^* into another free monoid the equation $h(x) = g(x)$ holds for all x in L if and only if it holds for all x in F. F is called a *test set* for L. It is straightforward to conclude that the Ehrenfeucht Conjecture follows directly from the above compactness property of systems of equations, which, hence, could be called the *Generalized Ehrenfeucht Conjecture*. It was shown in [CK1], as a first step towards the solution of the Ehrenfeucht Conjecture, that these two statements are in fact equivalent.

After knowing that each system of equations possesses an equivalent finite subsystem a natural question to be asked is "under which conditions can such a finite subsystem be found effectively." This is the topic of this note.

* This work was supported by the Natural Sciences and Engineering Research Council of Canada under Grant A-2403

We first recall from [CK1] a connection between the Ehrenfeucht Conjecture and its generalized version showing that the conjecture holds effectively for certain types of subsets of Σ^* if and only if systems of equations of the "corresponding" type possess effectively equivalent finite subsystems. Then we start to consider systems of equations defined by different kinds of transducers, that is automata with outputs. Such devices suit very well to describe infinite systems of equations - for each successful computation the input word defines the lefthand side of an equation and the corresponding output word defines the righthand side of the same equation.

We consider three types of transducers: finite transducers, pushdown transducers and determinisitic two-way transducers. We show that in each of these cases the corresponding systems of equations possess effectively equivalent finite subsystems. In the first two cases proofs are based on pumping properties of sets of words, and the results are proved already in [CK1] and [ACK]. In the third case the detailed proof is much more complicated as is shown here.

The weakest type of equations for which it is known that the equivalent finite subsystems cannot be found effectively are equations defined by linear bounded automata ("context sensitive equations"). This follows from the undecidability of morphic equivalence on context sensitive languges [CS].

2. *Preliminaries*

We assume that the reader is familiar with the basic facts of formal language theory, cf. e.g. [H], as well as those of free monoids. Consequently, we define here in details only a few most infrequently used notions as well as our special terminology, while some other notions are described only informally.

Let Σ be a finite alphabet and $N = \{x_1,..,x_n\}$ a finite set of *variables* such that $\Sigma \cap N = \emptyset$. An *equation* with n variables (or unknowns) over the free monoid Σ^* generated by Σ is of the form

(1) $$u = v \quad with \quad u,v \in N^* .$$

A *system of equations* is any collection of equations. A *solution* of a system of equations over Σ^* is a morphism $h : N^* \to \Sigma^*$ satisfying $h(u) = h(v)$ for all equations $u = v$ in the system. Thus, a solution can be identified with an n-tuple of words. Two systems of equations are called *equivalent* if they have exactly the same solutions.

Observe that in defining equations we did not allow constants, i.e., u and v in (1) were in N^* rather than in $(N \cup \Sigma)^*$. This was done only for the sake of convenience, since without affecting our considerations constants in equations can be eliminated by introducing for each symbol a in Σ a new variable X_a and replacing each occurrence of a by X_a and adding a finite set of new equations $X_a = a$.

Following [CK1] we next introduce our special notions. In what follows we identify an equation $u = v$ with the pair (u,v). Consequently, a system S of equations with unknowns N can be viewed as a binary relation over N, i.e., $S \subseteq N^* \times N^*$. Now, let L be a family of languages (over the same alphabet) and R a family of binary relations over N. We say that **R** is *morphically characterized* by **L** if the following holds: A binary relation R is in **R** if and only if there exist a language L in **L** and two morphisms h and g such that $R = \{(h(w), g(w)) \mid w \varepsilon L\}$. Finally, we say that a system of equations (that is a binary relation) is of *type* **L** if it belongs to the family of relations morphically characterized by **L**.

A connection between the Ehrenfeucht Conjecture for a family **L** of languages and its generalized version for systems of equations of type **L** (for definitions cf. Introduction) can now be obtained, as is shown in [CK1]:

Theorem 1. For any family **L** of languages the following statements are equivalent :

(i) For each effectively given L in **L** a test set can be effectively found,

(ii) For each effectively given system S of equations of type **L** a finite equivalent subsystem can effectively be found.

A natural way (at least for computer scientists) to define infinite systems of equations is to use transducers, that is to say automata with outputs. In this paper we shall be considering three types of transducers which are informally described in the following lines (for more details cf. [H]). A *finite transducer* is a finite (nondeterministic) automaton provided with an output structure, that is for each transition a (possibly empty) output is produced. Similarly, a *pushdown transducer* is an ordinary pushdown automaton provided with an output structure. Finally, a *deterministic two-way transducer* is obtained from a deterministic two-way automaton by adding a single output to each transition rule.

Let T be an arbitrary transducer of any of the above types. Then if N denotes the input alphabet (that is the alphabet of the underlying automaton) and M denotes the output alphabet then T defines via successful computations a binary relation $S_T \subseteq N^* \times M^*$. Consequently, each transducer defines a system of equation with $N \cup M$ as the set of variables.

Next we argue in favour of our above special notion by using some known results from the theory of transducers. Let *Reg* and *CF* denote the families of regular (or rational) and context free (or algebraic) languages, respectively. We said that a system of equations is of type *Reg* iff it is morphically characterized by the family *Reg*, which, in turn, means by the well known Nivat Theorem, cf. [B], that the system is defined by a finite transducer. We call such systems of equations *rational*. Similarly, a system of equations is of type *CF* iff it is defined by a pushdown transducer, cf. [CC]; hence, we call these relations *algebraic*.

Finally, it is clear that the family of arbitrary binary relations is of type "the family of all languages".

We proceed by giving two examples of systems of equations.

Example 1. Let $L \subseteq N^*$ be a regular language. Then the system of equations defined by

$$S = \{x = x^R \mid x \in L\},$$

when x^R denotes the reverse of the word x, is algebraic, since it is obvious how to construct a pushdown transducer for S.

Example 2. Let $d : N^* \to N^*$ be a morphism defined by $d(a) = aa$ for each a in N. Then the relation defined by

$$S = \{d(x) = xx^R \mid x \in \Sigma^*\}$$

can be realized by a deterministic two-way transducer. The same conclusion holds if x ranges over an arbitrary given regular language instead of Σ^*.

In order to be able to express relations defined by deterministic two-way transducers in terms of type **L** for some family **L** of language we shall need the following definitions. Let w be a word in the alphabet Σ and $h_1, \ldots, h_k$, for $k \geq 1$, be a set of endomorphisms of Σ^*.

Define

$$L_o = \{w\}$$

$$L_{i+1} = L_i \cup \bigcup_{j=1}^{k} h_j(L_i) \quad \text{for} \quad i \geq 0$$

and

$$L = \bigcup_{i=0}^{\infty} L_i .$$

Languages **L** thus defined are called *DTOL Languages*. Further a language L is called an *HDTOL Language* iff it is a morphic image of a DTOL language. The family of all HDTOL languages is denoted by *HDTOL*. More about these and related language families can be found from [RS].

The family *HDTOL* has the following properties. Firstly, it contains all regular or even all linear context-free languages as is easy to see. Secondly, it is incomparable with the family of context-free languages, cf. [RS]. Finally, the most important property of HDTOL languages from the point of view of this note is that these languages are "purely morphically defined". As an illustration of the power of HDTOL languages we give the following example.

Example 3. The language

$$L' = \{x\, x^R x \mid x \in \{a,b\}^*\}$$

is an HDTOL language. Indeed, $L' = h(\bigcup_{i=o}^{\infty} L_i)$, where

$$L_o = \{w\}$$

$$L_{i+1} = L_i \cup h_a(L_i) \cup h_b(L_i), \quad \text{for} \quad i \geq o,$$

and the morphisms $h_a, h_b : \{w, a, b, A, \bar{A}, B, \bar{B}\}^* \to \{a, b, A, \bar{A}, B, \bar{B}\}^*$ and the morphism $h : \{w, a, b, A, \bar{A}, B, \bar{B},\}^* \to \{a, b\}^*$ are defined as follows:

$$\begin{array}{llllll}
h_a: & w \to A\bar{A}A & h_b: & w \to B\bar{B}B & h: & w \to \varepsilon \\
 & A \to aA & & A \to bA & & A \to a \\
 & \bar{A} \to \bar{A}a & & \bar{A} \to \bar{A}b & & \bar{A} \to a \\
 & B \to aB & & B \to bB & & B \to b, \\
 & \bar{B} \to \bar{B}a & & \bar{B} \to \bar{B}b & & \bar{B} \to b \\
 & a \to a & & a \to a & & a \to a \\
 & b \to b & & b \to b & & b \to b
\end{array}$$

where ε denotes the empty word.

3. *Results*

In this section we consider systems of equations defined by the above three types of transducers, and conclude that in each case an equivalent finite subsystem can be effectively found.

Theorem 2. For each rational system S of equations (given by a finite transducer) an equivalent finite subsystem S' can be effectively found.

Outline of the proof. A straightforward consequence of pumping properties of regular languages and of the following implication, cf. [ACK] or [K] : For any words $x, y, u, v, \bar{x}, \bar{y}, \bar{u}$ and $\bar{v}$ we have

$$\left.\begin{array}{l} xy = \bar{x}\bar{y} \\ xuy = \bar{x}\bar{u}\bar{y} \\ xvy = \bar{x}\bar{v}\bar{y} \end{array}\right\} \Rightarrow xuvy = \bar{x}\bar{u}\bar{v}\bar{y}$$

□

It follows from the proof of Theorem 2 that not only an equivalent finite subsystem S' can be found but it can also be strongly bounded. Indeed, assume without loss of generality that S is given by a *normalized* finite transducer (that is to say that inputs read and outputs produced in single transition steps are of the length at most 1). Then the S' can be chosen to contain only those equations in which the words (in unknowns) are shorter than two times the cardinality of the state set of the finite transducer.

The proof of Theorem 2 used pumping properties of regular languages. Similarly we can use pumping properties of context-free languages to establish the Ehrenfeucht Conjecture for this family. However, in this case the detailed proof is quite lengthy, cf. [ACK], but since everything can be done effectively we conclude by Theorem 1 the following.

Theorem 3. For each algebraic system of equations (given by a pushdown transducer) there effectively exists an equivalent finite subsystem.

Next we turn to consider systems of equations defined by deterministic two-way transducers. In order to establish the above compactness property also in this case we need a different approach. In this case the systems of equations are not characterized by any family of languages (cf. discussion after Theorem 5), however, the family of HDTOL languages plays an important role. For this family we have:

Theorem 4. Each system of equations of the type *HDTOL* possesses effectively an equivalent finite subsystem.

Proof: By Theorem 1 it is enough to show that the Ehrenfeucht Conjecture holds effectively for HDTOL languages. This, in turn, was shown in [CK2], cf. also [CK1], using the (noneffective) validity of the Ehrenfeucht Conjecture, cf. [AL], and a decidability result of Makanin, cf. [Mak], stating that it can be tested whether a given equation over a free monoid has a solution.

□

From Theorem 4 we obtain

Theorem 5. Each system S of equations defined by a deterministic two-way transducer possesses effectively an equivalent finite subsystem.

Proof. Let S be defined by a deterministic two-way transducer T which means that

$$(u,v) \in S \quad \textit{iff} \quad v = T(u).$$

Without loss of generality we may assume that the input and output alphabets of T coincide, say are equal to N. Since we can allow endmarkers in our transducers it is easy to construct from T another deterministic two-way transducer, say T_1, such that

$$T_1(u) = \bar{u}\ T(u) \quad \text{for } \textit{all} \quad u \in N^*$$

where $\bar{u}$ is the barred copy of u.

Next we define the language

$$L = \{T_1(u) \mid u \in N^*\}. \tag{2}$$

Then, clearly

$$S = \{(h(u), \bar{h}(u)) \mid u \in L\}$$

where the morphisms $h, \bar{h} : (N \cup \bar{N})^* \to (N \cup \bar{N})^*$ are defined by

$$h(a) = \varepsilon \text{ and } \bar{h}(a) = a \quad \text{for all} \quad a \text{ in } N$$
$$h(\bar{a}) = a \text{ and } \bar{h}(\bar{a}) = \varepsilon \quad \text{for all} \quad \bar{a} \text{ in } \bar{N}.$$

So, by Theorem 4, it remains to be shown that L is an HDTOL language.

In order to see this we first note that the domain of T is regular, cf. [H]. Secondly, it was shown in [ERS] that the image of an EDTOL language, which is, by definition, of the form $K \cap \Sigma^*$ where K is a DTOL language and Σ is an alphabet, under a deterministic two-way transducer is an EDTOL language, too. Finally, it is known that the families of EDTOL and HDTOL languages coincide, cf. [NRSS], and so by the fact that each regular language is an HDTOL language we conclude that L in (2) is an HDTOL language. Furthermore, by the above references, it can be effectively constructed from T completing the proof of Theorem 5.

□

By the proof of Theorem 5, each system of equations defined by a deterministic two-way transducer is of type *HDTOL*. The converse is not true. In fact, the family of systems of equations defined by deterministic two-way transducers cannot be morphically characterized by any family of languages, since, for example the domains and the images of these transducers determine different families of languages, as was seen in the proof of Theorem 5. It also follows from the proof of Theorem 5 that systems of equations of the form

(3) $\{(u, T(u)) \mid u \in L\}$ with $L \in HDTOL$ and T a deterministic two-way transducer

or even

(3′) $\{(T_1(u), T_2(u) \mid u \in L\}$ with $L \in HDTOL$ and T_1 and T_2 determininstic two-way transducers.

are of type *HDTOL*, yielding the following strengthing of Theorem 5:

Theorem 6. For each system of equations of the form (3) or even (3′) there effectively exists an equivalent finite subsystem.

The fact that deterministic two-way transducers are single-valued implies that (3) does not give all systems of equations of type *HDTOL* either, while (3′) clearly characterizes the family of equations of type *HDTOL*

In this section we have considered two incomparable extensions of rational systems of equations, namely algebraic and *HDTOL* systems. In both cases finite equivalent subsystems can be effectively found. Obviously, these results can still be extended slightly: for each union of algebraic and *HDTOL* systems, which need not be of either of the types, an equivalent finite subsystem can be effectively found.

4. *Applications and concluding remarks*

We start this final section by pointing out a couple of applications of our previous results. We hope (and believe) that more will be found in the future.

Application 1. Let X be a finite set of words over an alphabet Σ. We consider the semigroup X^+ generated by X, and we are particularly interested in the set of all identities of X^+ in Σ^*. It is straightforward to see, cf. e.g. [Mar], that this set of identities forms, in our terms, a rational system of equations with X as the set of variables. Consequently, by Theorem 2, it has a finite equivalent subsystem which, moreover, can be effectively found. This means that all the identities of X^+ are actually implied by a finite effectively findable set of identities of X^+, cf. also [HK] and [S] for a more general result. As a conclusion we have found a short proof for the following result:

Corollary 1. It is decidable whether two finitely generated subsemigroups of a free semigroup are isomorphic.

Application 2. Let us call a word x palindromic if $x = x^R$. Now we raise the question of deciding whether a given language is a subset of the set of all palindromic words. For regular languages the problem can be settled by Example 1 and Theorem 3. Indeed, let L be a regular language. Then the relation $\{(x, x^R) \mid x \in L\}$ is algebraic and hence equivalent with a finite relation $\{(x, x^R) \mid x \in F\}$, where $F \subseteq L$ and can be effectively found. Now, the result follows since L is palindromic iff the relation $\{(x, x^R) \mid x \in L\}$ holds.

A similar argumentation can be used to solve the problem for HDTOL languages, since the relation $\{(x, x^R) \mid x \in L\}$, where $L \in$ *HDTOL*,, is of type *HDTOL*, cf. Example 3 and the proof of Theorem 5. More about these and similar problems can be found in [HKK].

As a concluding remark we want to compare our results to some related results. We first observe, cf. also [CK1] and [ACK]:

Corollary 2. The equivalence problem for rational (resp. algebraic or of type *HDTOL*) systems of equations is decidable.

Proof. By our theorems in Section 3, in each case systems of equations can be replaced by finite systems of equations. Hence, the result follows since the equivalence of two finite systems of equations can be tested as was shown in [CK1].

By Corollary 2 we can decide whether two finite transducers define equivalent systems of equations. On the other hand it is a well-known result cf. [Gr] or [B] that it is undecidable whether two finite transducers are equivalent, that is whether they define the same relation.

References

[ACK] Albert, J., Culik II, K. and Karhumäki, J., Test sets for context-free languages and algebraic systems of equations, Inform. Control 52 (1982) 172-186.

[AL] Albert, M.H. and Lawrence, J., A proof of Ehrenfeucht's Conjecture, Theoret. Comput. Sci. 41 (1985) 121-123.

[B] Berstel, J., Transductions and Context-Free Languages (Teubner, Stuttgrard, 1979).

[CC] Culik II, K., and Choffrut, C., Properties of finite and pushdown transducers, SIAM, J. Comput. 12 (1983) 300-315.

[CK1] Culik II, K., and Karhumäki, J., Systems of equations over a free monoid and Ehrenfeucht's Conjecture, Discrete Mathematics 43 (1983) 139-153.

[CK2] Culik II, K., and Karhumäki, J., The decidability of the DTOL sequence equivalence problem and related decision problems, University of Waterloo, Department of Computer Science, Research Report CS-85-05 (1985).

[CS] Culik II, K., and Salomaa, A., On the Decidability of Homomorphism Equivalence for Languages, *J. Comput. System Sci. 17* (1978) 163-175.

[ERS] Engelfriet, J., Rozenberg, G., and Stutzki, G., Tree transducers, *L* systems and two-way machines, J. Comput. Systems Sci. 20 (1980) 150-202.

[Gr] Griffiths, T., The unsolvability of the equivalence problem for ε-free nondeterministic generalized machines, J. Assoc. Comput. Mach. 15 (1968) 409-413.

[Gu] Guba, V.S., personal communication (1985).

[H] Harrison, M.A., Introduction to Formal Language Theory (Addison-Wesley, Reading MA, 1982).

[HK] Harju, T., and Karhumäki, J., On the defect theorem and simplifiability, semigroup Forum 33 (1986) 199-217.

[HKK] Horvâth, S., Karhumäki, J., and Kleijn, H.C.M., Decidability and characterization results concerning palindromicity, EIK, to appear.

[K] Karhumäki, J., The Ehrenfeucht Conjecture: A compactness claim for finitely generated free monoids, Theoret. Comput. Sci. 29 (1984) 285-308.

[Mak] Makanin, G.S., The Problem of solvability of equations in a free semigroup, Mat. Sb. 103 (1977) 147-236 (English transl. in : Math USSR Sb. 32 (1977) 129-198).

[Mar] Markov, Al. A., On finitely generated subsemigroups of a free semigroup, Semigroup Forum 3 (1971) 251-258.

[NRSS] Nielsen, M., Rozenberg, G., Salomaa, A. and Skyum S., Nonterminals, homomorphisms and codings in different variations of OL-systems, I. Deterministic systems, Acta Informatica 4 (1974) 87-106.

[S] Spehner,J.-C., Tout sous-monoide finiment engendré d'un monoide libre admet une présentation de Malcev finie, C.R. Acad. Sc. Paris, 301, Série I, no. 18 (1985).

[RS] Rozenberg, G., and Salomaa, A., The Mathematical Theory of *L* Systems (Academic Press, New York, 1980).

MINIMAL CONGRUENCES AND COEXTENSIONS IN SEMIGROUPS

Marie Demlová
Department of Mathematics
Faculty of Electrical Engineering ČVUT
166 27 Praha 6, Czechoslovakia

Václav Koubek
Computing Center of Charles University
Faculty of Mathematics and Physics
118 00 Praha 1, Czechoslovakia

Introduction

Congruences are basic tools for the investigation of algebraic structures. For example, the lattice theoretic properties of the congruence lattice give a good deal of information on the structure of terms - see [10]. Also the structure theory of algebraic objects is based on congruences - e.g. an algebra is subdirectly irreducible if it has a finest non-identical congruence. The semigroup structure theory - see [14] - is developed by means of semigroup congruences; the theory of syntactic semigroups - see [13] - is based on the notion of syntactic congruence. These facts lead to the detailed investigation of semigroup congruences.

This paper focuses on the minimal semigroup congruences. A congruence is minimal if it is non-identical and only the identical congruence is finer than it. We continue papers [9,15]. Rhodes [15] characterized and classified the minimal congruences of finite semigroups. In the paper [9] the minimal congruences of finite semigroups in which the Green relations coincide were described in detail. The minimal congruences are divided into two classes - either every non-singleton class meets two distinct $\mathcal{J}$-classes - such a congruence is called an outer congruence or every non-singleton class is contained in a singleton $\mathcal{J}$-class - such a congruence is called an inner congruence. We completely describe the outer minimal congruences in Chapters V and VI. The inner minimal congruences are described only for a special class of semigroups - the so called Green semigroups in

Chapters VII, VIII, and IX. Chapter II is devoted to Green semigroups. We can say that the Green semigroups form the greatest class of semigroups such that the Green relations have the structure of a box-product. Chapter I contains some basic semigroup notions, Chapter III is devoted to some basic facts about transformation semigroups.

The counterpart to a congruence is a coextension. The first papers studying semigroup coextensions generalized the Schreier group extensions. The second impulse for developing the coextensions was given by the structural semigroup theory - the coextensions were described by bitranslations - see [14]. A general description of coextensions was suggested by Grillet [11]. In this paper we study coextensions in connection with the minimal congruences. Our way toward a description of coextensions is a combination of the both methods and it is contained in Chapter IV. In Chapters V-IX the coextensions corresponding to the minimal congruences are characterized.

I. Semigroup notions

The aim of this chapter is to recall the semigroup notions which we use in the subsequent chapters.

If X is a proper subset of Y then we shall write $X \subset Y$ (i.e. $X \neq Y$) and $X \subseteq Y$ denotes that $X \subset Y$ or $X = Y$.

For a semigroup S denote by S^1 the semigroup which is obtained from S by adjoining a unity if S has none, otherwise $S^1 = S$. Denote by $\mathcal{J}(S)$ (or $\mathcal{D}(S)$, or $\mathcal{L}(S)$, or $\mathcal{R}(S)$, or $\mathcal{H}(S)$) the set of all $\mathcal{J}$-classes (or $\mathcal{D}$-classes, or $\mathcal{L}$-classes, or $\mathcal{R}$-classes, or $\mathcal{H}$-classes) of S. For $x \in S$, $J(x)$ (or $D(x)$, or $L(x)$, or $R(x)$, or $H(x)$) is the $\mathcal{J}$-class (or $\mathcal{D}$-class, or $\mathcal{L}$-class, or $\mathcal{R}$-class, or $\mathcal{H}$-class) of S containing x. For a subset $X \subseteq S$ define $J(X) = \bigcup\{J(x);\ x \in X\}$, analogously for $D(X)$, $L(X)$, $R(X)$, $H(X)$. Further denote by $W(X)$ the greatest two-sided ideal with $X \cap W(X) = 0$ (clearly, $W(X)$ can be empty). If $A = \{x\}$ then we write $W(x)$ instead of $W(\{x\})$.

If I is an ideal of S, then S/I is the *Rees quotient* of S by I, and $S-I$ is called a *co-ideal*. If $I = 0$ then $S/I = S$.

It is well-known that equivalences on a set S form a complete lattice (with the *identical equivalence* = the smallest equivalence =

$= \Delta$ and the *trivial equivalence* = the biggest equivalence = ∇). For an equivalence ε denote by $Car(\varepsilon)$ the union of all non-singleton classes of ε, i.e. $Car(\varepsilon) = \{x;\ \exists y \neq x,\ (x,y)\in\varepsilon\}$. Further εx is the class of ε containing x, i.e. $\{\varepsilon x;\ x\in X\}$ is the decomposition of X induced by ε. The set of all congruences on a semigroup S is a complete sublattice of the lattice of all equivalences. For a set A denote by

$\varepsilon(A)$ - the smallest congruence on S such that $(x,y)\in\varepsilon(A)$ for every $x,y\in A$,

$\eta(A)$ - the biggest congruence on S with $Car(\eta(A))\subseteq A$.

A congruence τ is called *elementary* if $\tau = \varepsilon(A)$ for some two-element subset of S. If $A = \{x,y\}$ then we shall write $\varepsilon(x,y)$ instead of $\varepsilon(\{x,y\})$. A congruence τ is *minimal* if $\tau \neq \Delta$ and Δ is the only congruence smaller than τ.

Clearly

Proposition 1.1: A congruence τ on a semigroup S is minimal if and only if $\tau = \varepsilon(x,y)$ for every $x \neq y$ with $(x,y)\in\tau$.■

Minimal congruences were studied by Rhodes [15] for finite semigroups. As an easy extension of his method we obtain:

Proposition 1.2: If τ is a minimal congruence on a semigroup S then either $Car(\tau)\subseteq J$ for some $\mathcal{J}$-class J of S or there exist $J_0,J_1\in\mathcal{J}(S)$ such that $Car(\tau)\subseteq J_0\cup J_1$ and every non-singleton class of τ has a non-empty meet with both J_0 and J_1 and if there exist two distinct τ-equivalent elements $x,y\in J_i$ then $J_0\cup J_1\subseteq S^1J_iS^1$ (i.e. $J_0\cup J_1$ is contained in the two-sided ideal generated by J_i).

Proof: For every non-singleton class A of τ we have $\tau\cap\varepsilon(A) = \tau$ and $\tau\cap\varepsilon(W(A)) = \Delta$. Thus if $A\cap J \neq 0$ for some $\mathcal{J}$-class J of S then for every non-singleton class B of τ we have $B\cap J \neq 0$. Assume that there exist three $\mathcal{J}$-classes J_i, $i\in 3$ with $A\cap J_i \neq 0$ for every $i\in 3$. Then for every $i,j\in 3$, $i \neq j$ we have $\tau\cap\varepsilon(J_i\cup J_j) = \tau$. This contradicts to the fact that there exist $k\in 3$ such that $J_k\cap Car(\varepsilon(\cup\{J_i;\ i\in 3\setminus\{k\}\})) = 0$. Hence either there exists a $\mathcal{J}$-class J with $A\subseteq J$, and in this case every non-singleton class B of τ fulfils $B\subseteq J$, or there exist two $\mathcal{J}$-classes J_i, $i\in 2$ such that

$A\subseteq J_0\cup J_1$, and then every non-singleton class B of τ fulfils $B\subseteq J_0\cup J_1$ and $J_i\cap B \neq 0$ for $i\in 2$. In the former case $Car(\tau)\subseteq J$ in the latter one $Car(\tau)\subseteq J_0\cup J_1$. Finally, if $x \neq y$, $(x,y)\in\tau$ then $\tau\cap\varepsilon(x,y) = \tau$ whence if $x,y\in J_i$ then $J_0\cup J_1\subseteq S^1J_iS^1$.∎

Proposition 1.2 is a basis of our classification of minimal congruences. A minimal congruence τ is called *inner* if $Car(\tau)\subseteq J$ for some *J*-class J of S otherwise τ is *outer*. Moreover, in the following we use a more detailed classification. A minimal congruence τ on S

is *of type 1* (or a *1-minimal congruence*) if τ is outer, $Car(\tau)\subseteq J_0\cup J_1$ for *J*-classes J_0 and J_1, and $S^1J_0S^1$ and $S^1J_1S^1$ are incomparable (with respect to the inclusion);

is *of type 2* (or a *2-minimal congruence*) if τ is outer, $Car(\tau)\subseteq J_0\cup J_1$ for *J*-classes J_0 and J_1, and either $S^1J_0S^1\subseteq S^1J_1S^1$ or $S^1J_1S^1\subseteq S^1J_0S^1$;

is *of type 3* (or a *3-minimal congruence*) if $\tau\subseteq H$;

is *of type 4* (or a *4-minimal congruence*) if $\tau\subseteq L$ or $\tau\subseteq R$ and $\tau\not\subseteq H$;

is *of type 5* (or a *5-minimal congruence*) if $\tau\subseteq D$ and neither $\tau\subseteq L$ nor $\tau\subseteq R$;

is *of type 6* (or a *6-minimal congruence*) if $\tau\subseteq J$ and $\tau\not\subseteq D$.

Note that by Proposition 1.2 minimal congruences of types 3-6 are always inner.

We describe 1-minimal congruences in Chapter V, 2-minimal congruences in Chapter VI. We do not know any description of inner minimal congruences in the class of all semigroups, therefore we restrict ourselves to a special class of semigroups - the so called Green semigroups (see Chapter II). By Lemma 2.1, $D = J$ for every Green semigroup. Thus no 6-minimal congruence exists in a Green semigroup. Chapters VII, VIII, and IX are devoted to a description of 3-minimal, 4-minimal, and 5-minimal congruences in the class of all Green semigroups.

Rhodes [15] defined an analogous classification of minimal congruences for finite semigroups. If we compare both classifications we obtain that

1-minimal congruences = Class IV of the Rhodes classification,
2-minimal congruences = Class III of the Rhodes classification,

3-minimal congruences = Class I of the Rhodes classification, 4-minimal and 5-minimal congruences = Class II of the Rhodes classification.

The Rhodes classification applies only to finite semigroups; it does not obtain 6-minimal congruences because of $\mathcal{J} = \mathcal{D}$. We get a different classification by subdividing his Class II, while we do not describe minimal congruences of types 3-6 for a general semigroup.

Finally, for a semigroup S and an element $x \in S$, denote by f_x the inner left translation of the element x, i.e. $f_x(y) = xy$ for every $y \in S$, and g_x the inner right translation of the element x, i.e. $g_x(y) = yx$ for every $y \in S$. A $\mathcal{J}$-class J is called *non-regular* if $JJ \cap J = 0$, otherwise J is *regular*. It is well-known that J is regular if and only if J contains an idempotent.

A normal subgroup N of a group G is *minimal* if $N \neq \{1\}$ and every normal subgroup H of G with $H \subseteq N$ fulfils either $H = N$ or $H = \{1\}$.

II. Green semigroups

In this chapter we define and study a special class of semigroups. A semigroup S fulfils the *Green Theorem* if the following hold:

(*) for $a,b \in S$, $x,y,v \in S^1$ with $a = xb$ and $b = yav$ there exists $z \in S^1$ with $b = za$;

(**) for $a,b \in S$, $x,y,v \in S^1$ with $a = bx$ and $b = yav$ there exists $z \in S^1$ with $b = az$.

We say that S is a *Green semigroup* if S fulfils the Green Theorem.

Green semigroups play a substantial role in Chapters VII, VIII, and IX where minimal inner congruences for Green semigroups are described. We investigate properties of Green semigroups.

First we show

Lemma 2.1: If S is a Green semigroup then $\mathcal{J} = \mathcal{D}$.

Proof: If $(a,b) \in \mathcal{J}$ then there exist $u,v \in S^1$ with $a = ubv$. Hence $(a,ub),(b,ub) \in \mathcal{J}$ and because S is a Green semigroup we conclude that $(b,ub)\in\mathcal{L}$ and $(ub,ubv = a)\in\mathcal{R}$. Thus $\mathcal{J} = \mathcal{L}\circ\mathcal{R} = \mathcal{D}$.■

The bicyclic semigroup is a witness of the fact that Green semigroups are proper subclass of the class of the semigroups fulfilling $\mathcal{J} = \mathcal{D}$.

Theorem 2.2: Let S be a semigroup. Then the following are equivalent:

(i) S is a Green semigroup;

(ii) for every $a,b\in S$, if $(a,ba)\in\mathcal{J}$ then $(a,ba)\in\mathcal{L}$ and if $(a,ab)\in\mathcal{J}$ then $(a,ab)\in\mathcal{R}$;

(iii) for every $x\in S$ we have $S^1x\subseteq L(x)\cup W(x)$ and $xS^1\subseteq R(x)\cup W(x)$.

Proof: (i)⇒(ii): Let $a,b\in S$ with $(a,ba)\in\mathcal{J}$ then there exist $u,v\in S^1$ with $a = ubav$ and by (*) we have $(a,ba)\in\mathcal{L}$. Analogously, $(a,ab)\in\mathcal{J}$ implies $(a,ab)\in\mathcal{R}$.
(ii)⇒(i): If $a = xb$ and $(a,b)\in\mathcal{J}$ then $(a = xb,b)\in\mathcal{L}$ by (ii) and (*) holds. Analogously, by the dual argument (**) holds.
(ii)⇒(iii): If $x\in S$, $a\in S^1$ then either $ax = x$ or $(x,ax)\in\mathcal{L}$ or $(x,ax)\notin\mathcal{J}$ thus either $ax = x$ or $ax\in L(x)$ or $ax\in W(x)$. Analogously for xS^1.
(iii)⇒(ii): If $a,b\in S$ then $ba\in L(a)\cup W(a)$ thus either $(a,ba)\in\mathcal{L}$ or $(a,ba)\notin\mathcal{J}$. Analogously, the second assertion.■

Corollary 2.3: Every commutative semigroup is a Green semigroup.

Proof: If S is a commutative semigroup then $\mathcal{J} = \mathcal{H}$ in S and hence (iii) of Theorem 2.2 holds.■

Proposition 2.4: If S is a Green semigroup then the following hold:

1) for every $L\in\mathcal{L}(S)$ and for every $s\in S$ either g_s is a bijection from L onto $Ls\in\mathcal{L}(S)$ or $Ls\subseteq W(L)$;

2) for every $R\in\mathcal{R}(S)$ and for every $s\in S$ either f_s is a bijection from R onto $sR\in\mathcal{R}(S)$ or $sR\subseteq W(R)$;

3) if $a,b,x\in S^1$ such that $(axb,x)\notin\mathcal{J}$ then either $(ax,x)\notin\mathcal{J}$ or $(xb,x)\notin\mathcal{J}$.

Proof: Let $L \in \mathcal{L}(S)$, then $L \cup W(L)$ is a left ideal by (iii) of Theorem 2.2. Hence for every $s \in S$, $(L \cup W(L))s$ is a left ideal. If $Ls \cap J(L) \neq 0$ then $Ls \cap J(L) \in \mathcal{L}(S)$. Moreover, there exists $t \in S$ with $Lst \subseteq L \cup W(L)$ and for some $x \in L$, $xst = x$ because, according to Lemma 2.1, we have $\mathcal{J} = \mathcal{D}$. For every $y \in L$ there exists $u \in S^1$ with $y = ux$; hence we conclude that $yst = y$. Thus $Ls \subseteq J(L)$ and $Ls \in \mathcal{L}(S)$, and g_s is a bijection from L onto Ls.

The proof of 2) is dual.

To prove 3) assume that $(ax,x) \in \mathcal{J}$. Then for a suitable $c \in S^1$, we have $cax = x$. Hence $caxb = xb$ and $(xb,x) \in \mathcal{J}$ implies $(axb,x) \in \mathcal{J}$ - a contradiction.∎

Corollary 2.5: For a Green semigroup S, if $J \in \mathcal{J}(S)$ then $\eta(J) \cap \mathcal{L}$, $\eta(J) \cap \mathcal{R}$, $\eta(J) \cap \mathcal{H}$ are congruences.

Proof: Since $\mathcal{L}$ is a right congruence, we have that $\eta(J) \cap \mathcal{L}$ is a right congruence. If $(x,y) \in \eta(J) \cap \mathcal{L}$ then for $z \in S^1$ we have that $(zx,zy) \in \eta(J)$. Thus either $zx = zy$ or $zx, zy \in J$. Since S is a Green semigroup we obtain from Theorem 2.2 (ii) that $(x,zx),(y,zy) \in \mathcal{L}$. Then $(x,y) \in \mathcal{L}$ implies that $(zx,zy) \in \mathcal{L}$ - thus $\eta(J) \cap \mathcal{L}$ is a congruence. By the dual argument $\eta(J) \cap \mathcal{R}$ is a congruence and $\eta(J) \cap \mathcal{H} = (\eta(J) \cap \mathcal{L}) \cap (\eta(J) \cap \mathcal{R})$ is also a congruence.∎

It is well-known that every finite semigroup, moreover every periodical semigroup is a Green semigroup because it fulfils the Green Theorem. A generalization of periodical semigroups - the so called quasiperiodical semigroups - was defined in the paper [4]. We say that a semigroup S is *quasiperiodical* if for every element $a \in S$ there exists a natural number n with $a^n S = a^{n+1} S$. If we apply results from [6,7,12] we obtain that our definition coincides with the original definition of quasiperiodical semigroups given in [4]. Moreover:

Proposition 2.6: For a semigroup S the following are equivalent:

(i) S is a quasiperiodical semigroup;

(ii) for every $a \in S$ there exists a natural number n such that $Sa^n = Sa^{n+1}$;

(iii) for every $a \in S$ there exists $b \in S$ and a natural number n such that $ba^{n+i} = a^{n+i}b = a^{n+i-1}$ for every natural number i.

Moreover, if S is quasiperiodical then for every $a \in S$ and a natural number n with $a^nS = a^{n+1}S$ we have that f_a is a bijection of a^nS onto itself and g_a is a bijection of Sa^n onto itself.

Proof: (iii)⇒(i) and (ii): If $ba^{n+i} = a^{n+i}b = a^{n+i-1}$ for every $i = 0,1,\ldots$, then for every $x \in S$, $a^nbx = a^{n-1}x$ and $xba^n = xa^{n-1}$, thus $a^nS = a^{n-1}S$ and $Sa^n = Sa^{n-1}$.
(i)⇒(iii): Since $a^nS = a^{n+1}S$ we obtain by [6,7,12] that the translation f_a has a kernel (i.e. $f_a^n(S) = f_a^{n+1}(S)$). If f_a has an increasing kernel then by [6] and [7] there exists $b \in S$ such that f_b has no kernel, i.e. $b^nS \neq b^{n+1}S$ for every natural number n - this is a contradiction, thus f_a has to have a bijection kernel and by [12] there exists $b \in S$ such that for a suitable n, $a^{n+i}b = ba^{n+i} = a^{n+i-1}$ for every natural number i.
(ii)⇒(iii) follows by the dual arguments.∎

The next statement follows from [12].

Theorem 2.7: Every quasiperiodical semigroup is a Green semigroup.

Proof: We prove (*) and (**). Let $a,b \in S$, $x,y,v \in S^1$ with $a = xb$, $b = yav$. By Proposition 2.6, there exist $w \in S$ and a natural number n such that $w(yx)^{n+i} = (yx)^{n+i}w = (yx)^{n+i-1}$ for every natural number i. Thus $b = yxbv = (yx)^nbv^n = w(yx)^{n+1}bv^n = wyxb = wya$ - thus it suffices to set $z = wy$ and (*) holds. By the dual argument we obtain (**).∎

A free semigroup over a non-empty set is a witness of the fact that quasiperiodical semigroups are a proper subclass of Green semigroups.

Consider the following semigroup: $S = \{a_i, b_i, c_i;\ i$ is an integer$\} \cup \{0\}$ where

$a_i \cdot a_j = a_{i+j}$, $b_i \cdot b_j = b_{i+j}$, $a_i \cdot c_j = c_{i+j}$, $c_i \cdot b_j = c_{i+j}$, for every i,j, otherwise the multiplication is 0.

By a direct inspection S is a quasiperiodical semigroup such that $\mathcal{H} = \mathcal{J}$ and $\mathcal{H}(S) = \{A,B,C,\{0\}\}$ where $A = \{a_i;\ i$ is an integer$\}$, $B = \{b_i;\ i$ is an integer$\}$, $C = \{c_i;\ i$ is an integer$\}$. Set $T = S-\{a_i;\ i\leq 0\}$. Then T is a subsemigroup, $\mathcal{R} = \mathcal{D} = \mathcal{J}$ and $\mathcal{R}(T) = \{\{a_i\};\ i>0\}\cup\{B,C,\{0\}\}$. Since for every integer i, $T^1c_i = \{c_j;\ j\geq i\}\cup\{0\}$ we have by Theorem 2.2 (iii) that T is not a Green semigroup. Notice that T fulfils the statements of Proposition 2.4 (the statements 1) and 2) are hereditary). Thus we have

<u>Theorem 2.8: The class of Green semigroups and quasiperiodical semigroups are not closed under subsemigroups. The class of Green semigroups is not closed under quotients, the class of quasiperiodical semigroups is closed under quotients.</u>

Proof: The first statement follows from the example above Theorem 2.8. Since by Theorem 2.2 every free semigroup is a Green semigroup, clearly Green semigroups are not closed under quotients. The last statement immediately follows from the definition of quasiperiodicity.∎

Finally, we give an example of a Green semigroup which is not a subsemigroup of a quasiperiodical semigroup. Consider the semigroup $S = \{a_i,b_i;\ i\geq 0,\ i$ is an integer$\}\cup\{0\}$ where

$a_i\cdot a_j = a_{i+j}$, $b_j\cdot a_i = a_i\cdot b_j = b_k$ where $k = \min\{0,j-i\}$ for every i,j, and otherwise the multiplication is 0.

By a direct inspection, S is a commutative semigroup and hence S is a Green semigroup - see Corollary 2.3. To show that S is not a subsemigroup of a quasiperiodical semigroup it suffices to note that $a_0(B\cup\{0\}) = B\cup\{0\}$ where $B = \{b_i;\ i\geq 0$ is an integer$\}$ and that f_{a_0} is not injective on $B\cup\{0\}$.

<u>III. Transformation semigroups</u>

Here we recall some basic notions concerning transformation semigroups and connections with the algebraic semigroups.

For a mapping $f:X\to Y$, denote by $\operatorname{Ker} f$ the equivalence on X such that $(x,y)\in\operatorname{Ker} f$ if and only if $f(x) = f(y)$, and by $\operatorname{Im} f$ a subset of Y with $\mathrm{Im}\, f = \{f(x);\ x\in X\}$.

A pair (X,Φ) where X is a set and Φ is a set of mappings of X into itself which is closed under composition is called a *transformation semigroup*. An equivalence τ on X is called a *congruence* of (X,Φ) if $(x,y)\in\tau$ implies $(f(x),f(y))\in\tau$ for every $f\in\Phi$. Again, the set of all congruences of (X,Φ) is a complete sublattice of the lattice of all equivalences on X. A congruence τ is said to be *minimal* if only the identical equivalence is finer than $\tau \neq \Delta$. We say that transformation semigroups (X,Φ), (Y,Ψ) are *isomorphic* if there exists a bijection $\varphi:X\to Y$ such that for every $f\in\Phi$ there exists $g\in\Psi$ with $\varphi\circ f = g\circ\varphi$ and for every $g\in\Psi$ there exists $f\in\Phi$ with $\varphi\circ f = g\circ\varphi$. Then φ is called an *isomorphism* from (X,Φ) onto (Y,Ψ). If (X,Φ) is a transformation semigroup and τ is a congruence, then $(X,\Phi)/\tau$ is the transformation semigroup (Y,Ψ) where Y is the set of all classes of τ and $\Psi = \{g:Y\to Y;\ \exists\ f\in\Phi$, for every class A of τ, $g(A)\supseteq f(A)\}$. A transformation semigroup (X,Φ) is *transitive* if for every order pair (x,y) of elements of X there exists $f\in\Phi$ with $f(x) = y$. A transformation semigroup (X,Φ) is *a-transitive* where $a\in X$ if for every order pair (x,y) of elements of $X-\{a\}$ there exists $f\in\Phi$ with $f(x) = y$ and for every $f\in\Phi$, $f(a) = a$.

If S is a semigroup, $H\in\mathcal{H}(S)$ then the *Schutzenberger group* $\Psi(H)$ is the pair (H,Ψ) where $\Psi = \{g:H\to H;\ \exists\ a\in S$ with $Ha = H$ and $g(x) = xa$ for every $x\in H\}$. Define $\Psi^d(H) = (H,\Psi^d)$ where $\Psi^d =$ $= \{g:H\to H;\ \exists\ a\in S$ with $aH = H$ and $g(x) = ax$ for every $x\in H\}$. We recall that every $g\in\Psi\cup\Psi^d$ is a bijection of H and $\Psi(H)$ and $\Psi^d(H)$ are transitive, moreover, every $g\in\Psi$ commutes with every $f\in\Psi^d$ - see e.g. [1]. Furthermore, if $H_0,H_1\in\mathcal{H}(S)$ belong to the same $\mathcal{D}$-class then $\Psi(H_0)$ and $\Psi(H_1)$ are isomorphic and also $\Psi^d(H_0)$ and $\Psi^d(H_1)$ are isomorphic.

Let S be a Green semigroup, $J\in\mathcal{J}(S)$. Denote $\mathcal{L}(J) = \{L;\ L\in\mathcal{L}(S),\ L\subseteq J\}$ and $\mathcal{R}(J) = \{R;\ R\in\mathcal{R}(S),\ R\subseteq J\}$. Define transformation semigroups $\mathcal{L}(S,J)$ and $\mathcal{R}(S,J)$ as follows:
$\mathcal{L}(S,J) = (X,\Phi)$ where $X = \mathcal{L}(J)\cup\{0\}$, $\Phi = \{f:X\to X;\ \exists\ a\in S$ such that for $L\in\mathcal{L}(J)$, $f(L) = La$ if $La\subseteq J$, $f(L) = 0$ else, $f(0) = 0\}$;
$\mathcal{R}(S,J) = (Y,\Psi)$ where $Y = \mathcal{R}(J)\cup\{0\}$, $\Psi = \{g:Y\to Y;\ \exists\ a\in S$ such that for $R\in\mathcal{R}(J)$, $g(R) = aR$ if $aR\subseteq J$, $g(R) = 0$ else, $g(0) = 0\}$.
Then clearly

<u>Proposition 3.1: If</u> S <u>is a Green semigroup,</u> $J\in\mathcal{J}(S)$ <u>then</u> $\mathcal{L}(S,J)$ <u>and</u> $\mathcal{R}(S,J)$ <u>are 0-transitive transformation semigroups.</u>

Proof is straightforward.∎

If τ is a congruence on S then we can define relations τ_L^J and τ_R^J on $\mathcal{L}(S,J)$ and $\mathcal{R}(S,J)$:

$(L_0,L_1)\in\tau_L^J$ if and only if there exist $x,y\in S$ with $(x,y)\in\tau$ and $x\in L_0$, $y\in L_1$,

$(L,0)\in\tau_L^J$ if and only if there exist $x,y\in S$ with $(x,y)\in\tau$, $x\in L$ and $y\in W(x)\cap S^1xS^1$,

$(L_0,L_1,L\in\mathcal{L}(J))$.

Analogously

$(R_0,R_1)\in\tau_R^J$ if and only if there exist $x,y\in S$ with $(x,y)\in\tau$ and $x\in R_0$, $y\in R_1$,

$(R,0)\in\tau_R^J$ if and only if there exist $x,y\in S$ with $(x,y)\in\tau$, $x\in R$ and $y\in W(x)\cap S^1xS^1$,

$(R_0,R_1,R\in\mathcal{R}(J))$.

Then we have

<u>Proposition 3.2: If</u> τ <u>is a congruence on a Green semigroup</u> S <u>and</u> $J\in\mathcal{J}(S)$ <u>then</u> τ_L^J <u>is a congruence on</u> $\mathcal{L}(S,J)$, τ_R^J <u>is a congruence on</u> $\mathcal{R}(S,J)$.

Proof is obtained by a direct inspection.∎

Let S be a semigroup and $H\in\mathcal{H}(S)$. Recall that, for every $x\in H$, $\Psi(H)$ (or $\Psi^d(H)$) uniquely determines a group G on the set H such that x is the unity of G and $\Psi(H)$ (or $\Psi^d(H)$) is the set of all right (or left, respectively) inner translations of G. Moreover, if x is an idempotent of S then G is a subsemigroup of S. We use these observations for a description of left and right inner translations of a Green semigroup S, with respect to a $\mathcal{J}$-class J of S.

For a group G and sets X,Y denote by $\mathcal{M}(G,X,Y)$ a triple (U,Φ,Ψ) where $U = G\times X\times Y$, $\Phi = \{f_{a,x,z};\ a\in G,\ x,z\in X\}$ where $f_{a,x,z}$ is a mapping from $G\times\{x\}\times Y$ to $G\times\{z\}\times Y$, $f_{a,x,z}(b,x,y) = (ab,z,y)$; $\Psi = \{g_{a,y,v};\ a\in G,\ y,v\in Y\}$ where $g_{a,y,v}$ is a mapping from $G\times X\times\{y\}$ to

$G \times X \times \{v\}$, $g_{a,y,v}(b,x,y) = (ba,x,v)$.
Denote by $\mathcal{M}_1(G,X,Y) = U$, $\mathcal{M}_2(G,X,Y) = \Phi$, $\mathcal{M}_3(G,X,Y) = \Psi$. Then the following hold:

(+) for every $a,b \in G$, $x,z \in X$, $y \in Y$ there exists exactly one $f \in \Phi$ with $f(a,x,y) = (b,z,y)$ (set $f = f_{ba^{-1},x,z}$);

(++) for every $a,b \in G$, $x \in X$, $y,v \in Y$ there exists exactly one $g \in \Psi$ with $g(a,x,y) = (b,x,v)$ (set $g = g_{a^{-1}b,y,v}$);

(+++) every $f \in \Phi$ or $g \in \Psi$ is injective and Φ and Ψ are closed under composition;

(++++) for every $a,b,c \in G$, $x,z \in X$, $y,v \in Y$ we have $g_{b,y,v} \cdot f_{c,x,z}(a,x,y) = f_{c,x,z} \cdot g_{b,y,v}(a,x,y)$.

Proposition 3.3: Let S be a Green semigroup, $J \in \mathcal{J}(S)$. There exist a group G, sets X,Y, and a bijection $\varphi: J \to G \times X \times Y$ such that:

(i) for every $R \in \mathcal{R}(J)$ there exists $x \in X$ with $\varphi(R) = G \times \{x\} \times Y$;

(ii) for every $L \in \mathcal{L}(J)$ there exists $y \in Y$ with $\varphi(L) = G \times X \times \{y\}$;

(iii) $\{$the domain-range restriction of f_a onto R and aR; $a \in S$, $R, aR \in \mathcal{R}(J)\} = \{\varphi^{-1} \cdot f \cdot \varphi;\ f \in \mathcal{M}_2(G,X,Y)\}$;

(iv) $\{$the domain-range restriction of g_a onto L and La; $a \in S$, $L, La \in \mathcal{L}(J)\} = \{\varphi^{-1} \cdot g \cdot \varphi;\ g \in \mathcal{M}_3(G,X,Y)\}$.

Proof: Choose $H \in \mathcal{H}(S)$ with $H \subseteq J$. Then $\Psi(H)$ determines a group G (see a note above Proposition 3.3). Set $X = \mathcal{R}(J)$, $Y = \mathcal{L}(J)$. We shall define a bijection φ. Choose $L^0 \in \mathcal{L}(J)$, $R^0 \in \mathcal{R}(J)$, $H^0 \in \mathcal{H}(J)$ such that $L^0 \cap R^0 = H^0$, and if J contains an idempotent then there exists an idempotent $h \in H^0$, otherwise choose an arbitrary element $h \in H^0$. For every $L \in \mathcal{L}(J)$ choose $a_L, b_L \in S^1$ such that $La_L = L^0$, $L^0 b_L = L$ and $xb_L a_L = x$, $ya_L b_L = y$ for every $x \in L^0$, $y \in L$; for every $R \in \mathcal{R}(J)$ choose $c_R, d_R \in S^1$ such that $c_R R = R^0$, $d_R R^0 = R$ and $d_R c_R x = x$, $c_R d_R y = y$ for every $y \in R^0$, $x \in R$. Without loss of generality we can assume that $a_{L^0} = b_{L^0} = c_{R^0} = d_{R^0}$ is the unity of S^1, further assume that G is a group on H^0 with h as the unity. Let $x \in J$, assume that $x \in R \cap L$ where $R \in \mathcal{R}(J)$, $L \in \mathcal{L}(J)$ and let $g \in G$ such that $h \cdot g = c_R x a_L$ (where the multiplication on the left side is in the group G and on the right side is in the semigroup S). Then define $\varphi(x) = (g,R,L)$. Since f_{c_R} or g_{a_L} is injective on R

or on L we obtain for $x,y \in R \cap L$, $x \neq y$, that $c_R x a_L \neq c_R y a_L$, hence φ is bijective. Moreover, clearly (i) and (ii) are fulfilled. From the property of G, $\Psi(H^0)$ and $\Psi^d(H^0)$ we clearly obtain that the domain-range restriction of f_a (or g_a) onto R and aR (or L and La) such that $a \in S$, $R, aR \in \mathcal{R}(J)$ (or $L, La \in \mathcal{L}(J)$) belongs to $\{\varphi^{-1} \circ f \circ \varphi;\ f \in \mathcal{M}_2(G,X,Y)\}$ (or $\{\varphi^{-1} \circ g \circ \varphi;\ g \in \mathcal{M}_3(G,X,Y)\}$, respectively). Since the domain-range restriction of f_a and g_a fulfils (+), (++), (+++), and (++++) we obtain (iii) and (iv) by an easy calculation.∎

Proposition 3.3 shows how to coordinatize every $\mathcal{J}$-class in a Green semigroup.

Definition 3.4: Let S be a Green semigroup, J be a $\mathcal{J}$-class of S. Choose $x \in J$ such that x is an idempotent if J contains some idempotent. Denote by $\Psi(H(x)) = (H, \Psi)$ the Schutzenberger group of $H(x)$ in S. For every $\mathcal{L}$-class L of S with $L \subseteq J$ choose $b_L \in S^1$ such that $x b_L \in L$, for every $\mathcal{R}$-class R of S with $R \subseteq J$ choose $a_R \in S^1$ such that $a_R x \in R$ and assume that $x = x b_{L(x)} = a_{R(x)} x$. Then every element z of J is identified with a triple $(g, R(z), L(z))$ such that $z = a_{R(z)} g(x) b_{L(z)}$. We shall denote $G(z) = g$. If we consider Ψ as a group with the operation composition then we have:

if $b \in S$, $R \in \mathcal{R}(J)$ and $bR \subseteq J$ then the domain-range restriction of f_b on R and bR is in $\mathcal{M}_2(\Psi, \mathcal{R}(J), \mathcal{L}(J))$;

if $b \in S$, $L \in \mathcal{L}(J)$ and $Lb \subseteq J$ then the domain-range restriction of g_b on L and Lb is in $\mathcal{M}_3(\Psi, \mathcal{R}(J), \mathcal{L}(J))$.

Then $x, \{a_R;\ R \in \mathcal{R}(J)\}, \{b_L;\ L \in \mathcal{L}(J)\}$ is an *initialization of coordinates* on J in S and the *coordinate of an element* z is the triple $(G(z), R(z), L(z))$.∎

IV. Coextension scheme

A *coextension* of S, in the maximum generality of the term, is just an arbitrary surjective homomorphism $f: T \rightarrow S$. In this chapter we describe general coextensions in terms of a certain family of mappings which we call a coextension family. By imposing various conditions on the coextension family we obtain in the subsequent chapters coextensions with desired properties.

Two coextensions $f:T\to S$, $g:U\to S$ are *isomorphic* if there exists an isomorphism $h:T\to U$ with $g\circ h = f$. If T and U are isomorphic semigroups and $f:T\to S$, $g:U\to S$ are coextensions then f and g need not be isomorphic coextensions, see the following examples: S - is the left zero semigroup on the set $\{a,b\}$. The semigroup T is given by the multiplication table

T	a	b	c
a	a	a	a
b	b	b	b
c	b	b	b

and define $f:T\to S$, $f(a) = a$, $f(b) = f(c) = b$, and $g:T\to S$, $g(a) = b$, $g(b) = g(c) = a$, then f and g are not isomorphic.

We say that a coextension f is *of type* i, or shortly an *i-coextension* if $\operatorname{Ker} f$ is an i-minimal congruence. Let A be a subset of S. The pair (S,A) is *coextendable* if there exists a coextension $f:T\to S$ such that $f(\operatorname{Car}(\operatorname{Ker} f))\subseteq A$. In this case f is said to be a *coextension* of (S,A). If moreover, f is an i-coextension then (S,A) is *i-coextendable*.

For a subset A of a semigroup S and an element $a\in A$ let $V_{A,a} = \{(u,v);\ u,v\in S-A,\ uv = a\}$ and denote by $\sigma_{A,a}$ the equivalence on $V_{A,a}$ such that $((u,tv),(ut,v))\in\sigma_{A,a}$ for $(u,tv),(ut,v)\in V_{A,a}$.

A subset A of a semigroup S is called *multiply-closed* if for every $a\in A$, $b\in S$ either $ab\in A$ or $A\cap S^1abS^1 = 0$ and simultaneously either $ba\in A$ or $A\cap S^1baS^1 = 0$. Clearly, S is multiply-closed, every $\mathcal{J}$-class is multiply-closed, and for every minimal congruence τ, $J(\operatorname{Car}(\tau))$ is multiply-closed.

<u>Construction 4.1 (Coextension scheme):</u>

Let S be a semigroup and A a multiply-closed subset of S. Assume that there are given

a) for every $a\in A$ a non-empty set T_a and a mapping $h_a:V_{A,a}\to T_a$;

b) for every $b\in S-A$ and every $a\in A$ with $ba\in A$ (or $ab\in A$) a mapping $f_b^a:T_a\to T_{ba}$ (or $g_b^a:T_a\to T_{ab}$, respectively);

c) for every $x\in T_a$, $a\in A$ and every $b\in A$ with $ab\in A$ (or $ba\in A$) a mapping $f_x^b:T_b\to T_{ab}$ (or $g_x^b:T_b\to T_{ba}$, respectively);

and that the following conditions hold:

d) for every $a\in A$ we have $T_a\cap S = 0$ and $\{T_a;\ a\in A\}$ is a set of pairwise disjoint sets;

e) for every $a,b\in A$ with $ab\in A$ and for every $x\in T_a$, $y\in T_b$ we have

$f_x^b(y) = g_y^a(x)$;

f) for every $a\in A$, $\sigma_{A,a}$ is finer than $\operatorname{Ker} h_a$;

g) for every $a,b,c\in S-A$ with $abc\in A$ we have

if $ab\in A$, $bc\notin A$ then $g_c^{ab}(h_{ab}(a,b)) = h_{abc}(a,bc)$,

if $ab\notin A$, $bc\in A$ then $h_{abc}(ab,c) = f_a^{bc}(h_{bc}(b,c))$,

if $ab,bc\in A$ then $g_c^{ab}(h_{ab}(a,b)) = f_a^{bc}(h_{bc}(b,c))$;

h) for every $b,c\in S-A$, $a\in A$ we have

if $bac\in A$ then $g_c^{ba}\circ f_b^a = f_b^{ac}\circ g_c^a$,

if $bca\in A$ then $f_u^a = f_b^{ca}\circ f_c^a$,

if $abc\in A$ then $g_u^a = g_c^{ab}\circ g_b^a$

where $u = bc$ if $bc\notin A$, $u = h_{bc}(b,c)$ if $bc\in A$;

i) for every $c\in S$, $a,b\in A$ with $acb\in A$ and for every $x\in T_a$, $y\in T_b$ we have

if $c\notin A$ then $f_x^{cb}\circ f_c^b(y) = g_y^{ac}\circ g_c^a(x)$,

if $c\in A$ then $f_x^{cb}\circ g_y^c = g_y^{ac}\circ f_x^c$;

j) for every $c\in S-A$, $a,b\in A$ and $x\in T_a$ we have

if $abc\in A$ then $g_c^{ab}\circ f_x^b = f_x^{bc}\circ g_c^b$,

if $cba\in A$ then $f_c^{ab}\circ g_x^b = g_x^{cb}\circ f_c^b$.

The family $\Gamma = \{\Gamma_i;\ i\in 6\}$ where $\Gamma_0 = \{T_a;\ a\in A\}$, $\Gamma_1 = \{h_a:V_{A,a}\to T_a;\ a\in A\}$, $\Gamma_2 = \{f_b^a:T_a\to T_{ba};\ b\in S-A,\ a\in A$ with $ba\in A\}$, $\Gamma_3 = \{g_b^a:T_a\to T_{ab};\ b\in S-A,\ a\in A$ with $ab\in A\}$, $\Gamma_4 = \{f_x^a:T_a\to T_{ba};\ x\in T_b,\ a,b\in A$ with $ba\in A\}$, $\Gamma_5 = \{g_x^a:T_a\to T_{ab};\ x\in T_b,\ a,b\in A$ with $ab\in A\}$ is called a *coextension family* of (S,A).

Given a coextension family Γ of (S,A) define a mapping $k = \operatorname{Coex}(\Gamma)$ of a groupoid $(T,\circledast)$ into S by

$T = (S-A)\cup(\cup\{T_a;\ a\in A\})$;

$$k(t) = \begin{cases} t & \text{for } t\in S-A,\\ a & \text{for } t\in T_a;\end{cases}$$

for $t,s\in T$ define

$$t\circledast s = \begin{cases} k(t)\cdot k(s) & \text{if } k(t)\cdot k(s)\notin A\\ h_{ts}(t,s) & \text{if } k(t),k(s)\notin A \text{ and } k(t)\cdot k(s)\in A\\ f_t^a(s) & \text{if } k(s) = a\in A \text{ and } k(t)\cdot k(s)\in A\\ g_s^a(t) & \text{if } k(t) = a\in A \text{ and } k(t)\cdot k(s)\in A.\blacksquare\end{cases}$$

Then we have

<u>Proposition 4.2:</u> <u>$\circledast$ and k are well-defined.</u> <u>$(T,\circledast)$ is a semigroup, $k:T\to S$ is a coextension of (S,A).</u>

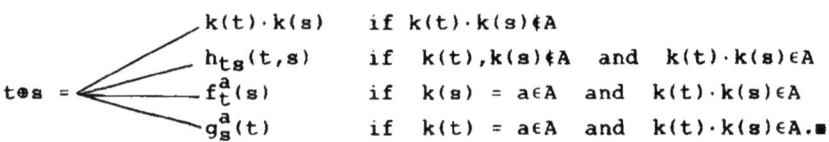

Proof: By d) k is a mapping, and by e) the two apparent possibilities for defining $t\circledast s$ coincide.
Since A is multiply-closed we obtain for $a\in A$:

if $abc\in A$ or $cab\in A$ then $ab\in A$,

if $bac\in A$ or $cba\in A$ then $ba\in A$.

Hence we immediately see that all mappings and their compositions which enter conditions h), i), and j) exist.

Since $k(t) = t$ for $t\in S-A$ and $k(t) = a$ for $t\in T_a$ we have by a) that k is surjective. If $k(t)\cdot k(s)\in A$ then we have:
$t,s\in S-A$ imply $t\circledast s = h_a(t,s)$ where $a = k(t)\cdot k(s) = t\cdot s$, thus

$k(t\circledast s) = k(t)\cdot k(s)$;

$k(s)\in A$ implies $t\circledast s = f_t^{k(s)}(s)\in T_{k(t)\cdot k(s)}$, thus $k(t\circledast s) = k(t)\cdot k(s)$;
$k(t)\in A$ implies $t\circledast s = g_s^{k(t)}(t)\in T_{k(t)\cdot k(s)}$, thus $k(t\circledast s) = k(t)\cdot k(s)$.

Therefore k is a homomorphism from $(T,\circledast)$ to $(S,\cdot)$. We have to show $(T,\circledast)$ is a semigroup, i.e. that $\circledast$ is an associative operation. Let $t,u,v\in T$. First $k((t\circledast u)\circledast v) = (k(t)\cdot k(u))\cdot k(v) =$
$= k(t)\cdot(k(u)\cdot k(v)) = k((t\circledast(u\circledast v))$ because $(S,\cdot)$ is a semigroup. Since for $x\in S-A$, $k^{-1}(x)$ is a singleton we have $(t\circledast u)\circledast v = t\circledast(u\circledast v)$ if $k((t\circledast u)\circledast v)\in S-A$.

Assume that $k((t\circledast u)\circledast v)\in A$. If $k(t),k(u),k(v)\notin A$ then one of the following hold:

$k(t)\cdot k(u),k(u)\cdot k(v)\notin A$ then f) implies $(t\circledast u)\circledast v = (t\cdot u)\circledast v =$
$= h_{tuv}(t\cdot u,v) = h_{tuv}(t,u\cdot v) = t\circledast(u\circledast v)$;

$k(t)\cdot k(u)\in A$, $k(u)\cdot k(v)\notin A$ then g) implies $(t\circledast u)\circledast v = h_{tu}(t,u)\circledast v =$
$= g_v^{tu}(h_{tu}(t,u)) = h_{tuv}(t,u\cdot v) = t\circledast(u\circledast v)$;

$k(t)\cdot k(u)\notin A$, $k(u)\cdot k(v)\in A$ then by g), $(t\circledast u)\circledast v = h_{tuv}(t\cdot u,v) =$
$= f_t^{uv}(h_{uv}(u,v)) = t\circledast h_{uv}(u,v) = t\circledast(u\circledast v)$;

$k(t)\cdot k(u),k(u)\cdot k(v)\in A$ then by g), $(t\circledast u)\circledast v = h_{tu}(t,u)\circledast v =$
$= g_v^{tu}(h_{tu}(t,u)) = f_t^{uv}(h_{uv}(u,v)) = t\circledast h_{uv}(u,v) = t\circledast(u\circledast v)$.

If $k(t)\in A$, $k(u),k(v)\notin A$ then h) implies

$$(t\circledast u)\circledast v = g_u^{k(t)}(t)\circledast v = g_v^{k(t)\cdot k(u)}(g_u^{k(t)}(t)) = g_{u\circledast v}^{k(t)}(t) = t\circledast(u\circledast v).$$

If $k(v)\in A$, $k(t),k(u)\notin A$ then by h),

$$(t\circledast u)\circledast v = f_{t\circledast v}^{k(v)}(v) = f_t^{k(u)\cdot k(v)}(f_u^{k(v)}(v)) = t\circledast f_u^{k(v)}(v) = t\circledast(u\circledast v).$$

If $k(t),k(v)\in A$, $k(u)\notin A$ then i) implies

$$(t\circledast u)\circledast v = g_u^{k(t)}(t)\circledast v = g_v^{k(t)\cdot k(u)}(g_u^{k(t)}(t)) =$$
$$= f_t^{k(u)\cdot k(v)}(f_u^{k(v)}(v)) = t\circledast f_u^{k(v)}(v) = t\circledast(u\circledast v).$$

Then remaining cases fulfil $(t\circledast u)\circledast v = f_t^{k(u)}(u)\circledast v =$
$= g_v^{k(t)\cdot k(u)}(f_t^{k(u)}(u)) = f_t^{k(u)\cdot k(v)}(g_v^{k(u)}(u)) = t\circledast g_v^{k(u)}(u) = t\circledast(u\circledast v)$

where we use h) if $k(t),k(v)\notin A$, $k(u)\in A$, i) if $k(t),k(u),k(v)\in A$, j) if $k(u)\in A$ and either $k(t)\in A$, $k(v)\notin A$ or $k(t)\notin A$, $k(v)\in A$. Hence $(T,\circledast)$ is a semigroup and the proof is complete.■

Theorem 4.3: Let S be a semigroup, A be a multiply-closed subset of S. If $p:(U,*)\to(S,\cdot)$ is a coextension of (S,A) then there exists a coextension family Γ such that $p:(U,*)\to(S,\cdot)$ is isomorphic with $\mathrm{Coex}(\Gamma)$.

Proof: For every element $a\in A$ choose a set T_a of cardinality $\mathrm{card}\ p^{-1}(a)$ such that $\{T_a;\ a\in A\}$ fulfils d). Let φ_a be a bijection between T_a and $p^{-1}(a)$. For $x\in S-A$, take $x'\in U$ with $p(x')=x$; since $p^{-1}(x)$ is a singleton, x' is uniquely determined. For $a\in A$, define $h_a:V_{A,a}\to T_a$ such that for $(b,c)\in V_{A,a}$ we have $\varphi_a(h_a(b,c)) = b'*c'$. Since p is a homomorphism we have $p(b'*c') = p(b')\cdot p(c') = a$. Since $(U,*)$ is a semigroup we conclude that $\{h_a:V_{A,a}\to T_a;\ a\in A\}$ fulfils f). Set $T=(S-A)\cup(\cup\{T_a;\ a\in A\})$ and define $\varphi:T\to U$ such that $\varphi(c)=c'$ for $c\in S-A$, $\varphi(c)=\varphi_a(c)$ for $c\in T_a$. A mapping φ is a bijection because p is a coextension of (S,A). If $k:T\to S$ such that $k(c)=c$ for $c\in S-A$, $k(c)=a$ for $c\in T_a$ then $k=p\circ\varphi$. Denote by $\circledast$ the associative operation on T induced by φ and $*$. Then $k:(T,\circledast)\to(S,\cdot)$ is a coextension isomorphic to $p:(U,*)\to(S,\cdot)$. For $t\in T$, $a\in A$ with $k(t)\cdot a\in A$ (or $a\cdot k(t)\in A$) define $f_t^a:T_a\to T_{k(t)\cdot a}$ (or $g_t^a:T_a\to T_{a\cdot k(t)}$) such that $f_t^a(u)=t\circledast u$ (or $g_t^a(u)=u\circledast t$) for every $u\in T_a$. Since $(T,\circledast)$ is a semigroup we obtain by a direct inspection that $\Gamma_2=\{f_b^a:T_a\to T_{ba};\ b\in S-A,\ a\in A$ with $ba\in A\}$, $\Gamma_3=\{g_b^a:T_a\to T_{ab};$ $b\in S-A,\ a\in A$ with $ab\in A\}$, $\Gamma_4=\{f_x^a:T_a\to T_{ba};\ x\in T_b,\ a,b\in A$ with $ba\in A\}$, $\Gamma_5 = \{g_x^a:T_a\to T_{ab};\ x\in T_b,\ a,b\in A$ with $ab\in A\}$ fulfil the conditions e)-j). Thus if we set $\Gamma=\{\Gamma_0=\{T_a;\ a\in A\},\ \Gamma_1=\{h_a:V_{A,a}\to T_a;$ $a\in A\},\Gamma_2,\Gamma_3,\Gamma_4,\Gamma_5\}$ then Γ is a coextension family and obviously $\mathrm{Coex}(\Gamma)=k:(T,\circledast)\to(S,\cdot)$.■

Assume that Γ, Λ are two coextension families of (S,A) such that $\mathrm{Coex}(\Gamma)$ and $\mathrm{Coex}(\Lambda)$ are isomorphic coextensions. If $\Gamma_0=\{T_a;$ $a\in A\}$, $\Lambda_0=\{U_a;\ a\in A\}$ then for every $a\in A$ there exists a bijection $\mu_a:T_a\to U_a$ fulfilling:

k) if $\Gamma_1=\{h_a:V_{A,a}\to T_a;\ a\in A\}$, $\Lambda_1=\{h'_a:V_{A,a}\to U_a;\ a\in A\}$ then $\mu_a\circ h_a = h'_a$ for every $a\in A$;

ℓ) if $\Gamma_2=\{f_a^b:T_b\to T_{ab};\ b\in A,\ a\in S-A$ with $ab\in A\}$, $\Gamma_3=\{g_a^b:T_b\to T_{ba};$

$b\in A$, $a\in S-A$ with $ba\in A\}$, $\Lambda_2 = \{m_a^b:U_b \to U_{ab};\ b\in A,\ a\in S-A$ with $ab\in A\}$, $\Lambda_3 = \{n_a^b:U_b \to U_{ba};\ b\in A,\ a\in S-A$ with $ba\in A\}$ then for every $a'\in S-A$, $b'\in A$ with $a'b'\in A$ we have $\mu_{a'b'}\cdot f_{a'}^{b'} = m_{a'}^{b'}\cdot\mu_{b'}$ and for every $a'\in S-A$, $b'\in A$ with $b'a'\in A$ we have $\mu_{b'a'}\cdot g_{a'}^{b'} = n_{a'}^{b'}\cdot\mu_{b'}$;

m) if $\Gamma_4 = \{f_x^b:T_b \to T_{ab};\ x\in T_a,\ a,b\in A$ with $ab\in A\}$, $\Gamma_5 = \{g_x^b:T_b \to T_{ba};\ x\in T_a,\ a,b\in A$ with $ba\in A\}$, $\Lambda_4 = \{m_x^b:U_b \to U_{ab};\ x\in U_a,\ a,b\in A$ with $ab\in A\}$, $\Lambda_5 = \{n_x^b:U_b \to U_{ba};\ x\in U_a,\ a,b\in A$ with $ba\in A\}$ then for every $a',b'\in A$ with $a'b'\in A$ and for every $x\in T_{a'}$ we have $\mu_{a'b'}\cdot f_x^{b'} = m_{\mu(x)}^{b'}\cdot\mu_{b'}$ and for every $a',b'\in A$ with $b'a'\in A$ and for every $x\in T_{a'}$ we have $\mu_{a'b'}\cdot g_x^{b'} = n_{\mu(x)}^{b'}\cdot\mu_{b'}$

where μ is an isomorphism of $\mathrm{Coex}(\Gamma)$ onto $\mathrm{Coex}(\Lambda)$ satisfying $\mu(c) = c$ for $c\in S-A$, $\mu(c) = \mu_a(c)$ for $c\in T_a$, $a\in A$. On the other hand, if there exist bijections $\{\mu_a:T_a \to U_a;\ a\in A\}$ fulfilling k), ℓ), and m) then μ is an isomorphism from $\mathrm{Coex}(\Gamma)$ onto $\mathrm{Coex}(\Lambda)$. Thus we can summarize:

Theorem 4.4: Let S be a semigroup, A be a multiply-closed subset of S, Γ, Λ be two coextension families of (S,A). Then $\mathrm{Coex}(\Gamma)$ and $\mathrm{Coex}(\Lambda)$ are isomorphic coextensions if and only if there exists a family $\{\mu_a:T_a \to U_a;\ a\in A\}$ of bijections fulfilling k), ℓ), and m). Every isomorphism is uniquely determined by a family $\{\mu_a:T_a \to U_a;\ a\in A\}$ of bijections fulfilling k), ℓ), and m).∎

Convention: In the sequel, if a coextension family Γ of (S,A) is given and we do not write otherwise then we assume that $\Gamma_0 = \{T_a;\ a\in A\}$, $\Gamma_1 = \{h_a:V_{A,a} \to T_a;\ a\in A\}$, $\Gamma_2 = \{f_b^a:T_a \to T_{ba};\ b\in S-A,\ a\in A$ with $ba\in A\}$, $\Gamma_3 = \{g_b^a:T_a \to T_{ab};\ b\in S-A,\ a\in A$ with $ab\in A\}$, $\Gamma_4 = \{f_x^a:T_a \to T_{ba};\ x\in T_b,\ a,b\in A$ with $ba\in A\}$, $\Gamma_5 = \{g_x^a:T_a \to T_{ab};\ x\in T_b,\ a,b\in A$ with $ab\in A\}$.

Corollary 4.5: If A is a multiply-closed subset of a semigroup S with $AA\cap A = 0$ then every coextension k is determined by a coextension family Γ such that $\Gamma_4 = \Gamma_5 = 0$ and Γ fulfils a),b),d),f),g), and h). Moreover, in h) $u = bc$.∎

In this paper, if we use the coextension scheme then the multiply-closed subset A of S often satisfies the following conditions:

(*) there exists a decomposition $L(A) = \{A_i;\ i\in I\}$ of A such that for every $x\in S$, $i\in I$, we have $xA_i\cap A\subseteq A_i$ and either $A_ix\cap A = 0$ or $A_ix\subseteq A_{i'}$ for some $i'\in I$;

(**) there exists a decomposition $R(A) = \{A_j;\ j\in J\}$ of A such that for every $x\in S$, $j\in J$, we have $A_jx\cap A\subseteq A_j$ and either $xA_j\cap A = 0$ or $xA_j\subseteq A_{j'}$ for some $j'\in J$.

Then we say that A has a *cross decomposition*. If A is a multiply-closed subset of S having a cross decomposition then we can formulate conditions a)-j) from Construction 4.1 of a coextension family in a simpler way. Note that every $\mathcal{J}$-class in a Green semigroup is a multiply-closed subset having a cross decomposition (that is the decomposition into $\mathcal{L}$-classes and $\mathcal{R}$-classes).

Definition 4.6: Let A be a multiply-closed subset of a semigroup S having a cross decomposition. Denote by $L(A) = \{C_i;\ i\in I\}$, $R(A) = \{D_j;\ j\in J\}$ the decompositions from (*) and (**) in the definition of a cross decomposition. Assume that there are given:

a') a set T' disjoint with $S-A$;

b') a surjective mapping $k':T'\to A$ (for a subset $X\subseteq A$ denote by $\kappa(X) = (k')^{-1}(X))$;

c') for every $a\in A$, a mapping $h_a:V_{A,a}\to T'$ such that $\operatorname{Im} h_a\subseteq\kappa(a)$;

d') for every $a\in S-A$ and $j,j'\in J$ (or $i,i'\in I$) with $aD_j\subseteq D_{j'}$ (or $C_ia\subseteq C_{i'}$), a mapping $f_a^j:\kappa(D_j)\to\kappa(D_{j'})$ (or $g_a^i:\kappa(C_i)\to\kappa(C_{i'})$) such that for every $x\in\kappa(D_j)$ (or $x\in\kappa(C_i)$) if $x\in\kappa(b)$ for $b\in A$ then $f_a^j(x)\in\kappa(ab)$ (or if $x\in\kappa(b)$ for $b\in A$ then $g_a^i(x)\in\kappa(ba)$, respectively);

e') for every $a\in A$, $j,j'\in J$ (or $i,i'\in I$) with $aD_j\subseteq D_{j'}$ (or $C_ia\subseteq C_{i'}$) and for every $x\in\kappa(a)$, a mapping $f_x^j:\kappa(D_j)\to\kappa(D_{j'})$ (or $g_x^i:\kappa(C_i)\to\kappa(C_{i'})$) such that for every $y\in\kappa(D_j)$ (or $y\in\kappa(C_i)$) if $y\in\kappa(b)$ for $b\in A$ then $f_x^j(y)\in\kappa(ab)$ (or $g_x^i(y)\in\kappa(ba)$, respectively).

Further assume that the following conditions are fulfilled:

f') for every $a\in S^1-A$, $x\in\kappa(C_i)$, $y\in\kappa(D_j)$, with $C_iaD_j\subseteq A$ we have $f_x^k\cdot f_a^j(y) = g_y^l\cdot g_a^i(x)$ where $aD_j\subseteq D_k$, $C_ia\subseteq C_l$;

g') for every $a\in A$, $\operatorname{Ker} h_a\supseteq\sigma_{A,a}$;

h') for every $a,b,c\in S-A$ with $abc\in C_i\cap D_j$ for $i\in I$, $j\in J$ we have if $ab\in C_k$ for $k\in I$, $bc\notin A$ then $g_c^k(h_{ab}(a,b)) = h_{abc}(a,bc)$,

if $ab\notin A$, $bc\in D_l$ for $l\in J$ then $h_{abc}(ab,c) = f_a^l(h_{bc}(b,c))$,
if $ab\in C_k$ for $k\in I$, $bc\in D_l$ for $l\in J$ then $g_c^k(h_{ab}(a,b)) =$ $= f_a^l(h_{bc}(b,c))$;

i') for every $b,c\in S-A$
if for $j,l\in J$ we have $cD_j\subseteq D_l$, $bD_l\subseteq A$ then $f_a^j = f_b^l\cdot f_c^j$,
if for $i,k\in I$ we have $C_ib\subseteq C_k$, $C_kc\subseteq A$ then $g_a^i = g_c^k\cdot g_b^i$,
where $a = bc$ if $bc\notin A$, $a = h_{bc}(b,c)$ if $bc\in A$;

j') for every $f_x^j:\kappa(D_j)\to\kappa(D_{j'})$, $g_y^i:\kappa(C_i)\to\kappa(C_{i'})$ where $x,y\in T'\cup(S-A)$ we have $f_x^j\cdot g_y^i(z) = g_y^i\cdot f_x^j(z)$ for every $z\in C_i\cap D_j$.

Then $\Lambda = \{\Lambda_0 = \{T'\}, \Lambda_1 = \{k'\}, \Lambda_2 = \{h_a;\ a\in A\}, \Lambda_3 =$ $= \{f_a^j:\kappa(D_j)\to\kappa(D_{j'});\ a\in S-A$ with $aD_j\subseteq D_{j'},\ j,j'\in J\}, \Lambda_4 =$ $= \{g_a^i:\kappa(C_i)\to\kappa(C_{i'});\ a\in S-A$ with $C_ia\subseteq C_{i'},\ i,i'\in I\}, \Lambda_5 =$ $= \{f_x^j:\kappa(D_j)\to\kappa(D_{j'});\ x\in\kappa(a),\ a\in A$ with $aD_j\subseteq D_{j'},\ j,j'\in J\}, \Lambda_6 =$ $= \{g_x^i:\kappa(C_i)\to\kappa(C_{i'});\ x\in\kappa(a),\ a\in A$ with $C_ia\subseteq C_{i'},\ i,i'\in I\}\}$ is called a *special coextension family.*∎

Theorem 4.7: For a multiply-closed subset A of a semigroup S having a cross decomposition there exists a one-to-one correspondence between coextension families and special coextension families.

Proof: If Γ is a coextension family then we set $T' = \{T_a;$ $a\in A\}$; $k'(x) = a$ for $x\in T_a$, $a\in A$; for $x\in(S-A)\cup T'$, if f_x^j has to be defined, we set f_x^j to be the union of f_x^b, $b\in D_j$, analogously for g_x^j. Then by a direct inspection we obtain that a')-i') hold. To prove j') it suffices to use Proposition 4.2 - T is a semigroup hence $(xz)y = x(zy)$ - j') is a consequence of this fact.

If Λ is a special coextension family, then we set $T_a = \kappa(a)$, and f_x^a (or g_x^a) are the domain-range restrictions of f_x^j (or g_x^i) where $a\in D_j$ (or $a\in C_i$, respectively). By a straightforward calculation we obtain that a')-j') imply a)-j).∎

If Λ is a special coextension family and Γ is a coextension family corresponding to Λ then $\mathrm{Coex}(\Lambda) = \mathrm{Coex}(\Gamma)$.

Remark 4.8: Assume that we have a family Λ fulfilling a'),b'),c'),d'),e'),g'),h'),i'), and j') and that the conditions f") and f"') hold:

f") for every $i\in I$, $j\in J$ with $C_iD_j\subseteq C_i\cap D_j$ there exists $u\in\kappa(C_i\cap D_j)$ such that for every $x\in\kappa(C_i)$, $y\in\kappa(D_j)$ we have $x = f_x^j(u)$, $y = g_y^i(u)$;

f"') for every $a\in S-A$, $i\in I$, $j\in J$ with $C_iaD_j\subseteq C_i\cap D_j$ there exists $z\in\kappa(C_i\cap D_j)$ such that for every $y\in\kappa(D_j)$, $x\in\kappa(C_i)$ we have $f_x^k\cdot f_a^j = f_x^k\cdot f_z^j$ and $g_y^l\cdot g_a^i = g_y^l\cdot g_z^i$ where $aD_j\subseteq D_k$ for $k\in J$, $C_ia\subseteq C_l$ for $l\in I$.

Then f') also holds. Indeed, if c is the unity then f") and j') imply that $f_x^j(y) = g_y^i(x)$. If $a\in S-A$ then by f"') we obtain $f_x^k\cdot f_a^j(y) = f_x^k\cdot f_z^j(y) = f_x^k\cdot g_y^i(z) = g_y^l\cdot f_x^j(z) = g_y^l\cdot g_z^i(x) = g_y^l\cdot g_a^i(x)$.

The reason for introducing conditions f") and f"') is that for a J-class A, f') is equivalent with f") and f"') and conditions f") and f"') are more natural than f').

V. Congruences of type 1

This chapter is devoted to the study of the 1-minimal congruences and 1-coextensions.

Proposition 5.1: A congruence $\tau \neq \Delta$ of a semigroup S is a 1-minimal congruence if and only if the following hold:

(i) $\tau\cap J = \Delta$;

(ii) there exist two J-classes J_0, J_1 of S with $Car(\tau)\subseteq J_0\cup J_1$ such that $S^1J_0S^1$ and $S^1J_1S^1$ are incomparable sets (with respect to inclusion).

Moreover, if τ is a 1-minimal congruence then

A) $\tau\cap(J_0\times J_1)$ is a bijection from J_0 onto J_1;

B) J_0 and J_1 are non-regular J-classes of S;

C) $\{\tau x;\ x\in J_0\cup J_1\}$ is a J-class of S/τ.

Proof: Let τ be a 1-minimal congruence. Then by the definition (ii) holds. (i) follows from Proposition 1.2. If $(u,v)\in J$ then there exist $a,b\in S^1$ with $aub = v$ - hence by Proposition 1.2 we obtain A). If $x\in J_0$, $y\in J_1$ with $(x,y)\in\tau$ then $(xz,yz),(zx,zy)\in\tau$ for every $z\in J_0$ and by (ii) $yz,zy\notin J_0\cup J_1$, thus $xz = yz$, $zy = zx$ and $J_0J_0\cap J_0 = 0$. Analogously $J_1J_1\cap J_1 = 0$ and B) is proved. C) is a consequence of (ii) and A).

Assume that τ satisfies (i) and (ii). Assume that there exists a non-identical congruence σ of S finer than τ. Since $\sigma \neq \Delta$ there exists $(x,y)\in\sigma$, $x \neq y$. By (ii) $x,y\in J_0 \cup J_1$, by (i) either $x\in J_0$, $y\in J_1$ or $x\in J_1$, $y\in J_0$ because $(x,y)\in\tau$. Assume that $x\in J_0$, $y\in J_1$, then for every $a,b\in S^1$ with $axb\in J_0$ we have $(axb,ayb)\in\sigma\subseteq\tau$ and hence $ayb\in J_1$. If $(u,v)\in\tau$, $u \neq v$ then either $u\in J_0$, $v\in J_1$ or $v\in J_0$, $u\in J_1$ and $\tau u = \{u,v\}$ - hence $\sigma = \tau$.∎

<u>Theorem 5.2: Let S be a semigroup, and let x,y be different elements of S. Then $\varepsilon(x,y)$ is a 1-minimal congruence if and only if the following hold:</u>

<u>(i)</u> S^1xS^1 <u>and</u> S^1yS^1 <u>are incomparable;</u>

<u>(ii)</u> <u>for every</u> $a,b\in S^1$ <u>if</u> $(axb,x)\notin J$ <u>or</u> $(ayb,y)\notin J$ <u>then</u> $axb = ayb$;

<u>(iii)</u> <u>for every</u> $a,b,c,d\in S^1$ <u>we have</u> $axb = cxd$ <u>if and only if</u> $ayb = cyd$.

Proof: Assume that $\varepsilon(x,y)$ is a 1-minimal congruence. Then by Proposition 5.1, $J(x) \neq J(y)$, $Car(\varepsilon(x,y))\subseteq J(\{x,y\})$, and $S^1J(x)S^1$ and $S^1J(y)S^1$ are incomparable - thus (i) holds. Since $(axb,ayb)\in\varepsilon(x,y)$ then (ii) holds. To prove (iii) we note that $axb = cxd$ implies $(axb,ayb),(cyd,axb)\in\varepsilon(x,y)$ and Proposition 5.1 implies $ayb = cyd$.

Conversely, assume that x,y fulfil (i), (ii), and (iii). The congruence $\varepsilon(x,y)$ is the reflexive, symetric, and transitive closure of the relation $R = \{(axb,ayb);\ a,b\in S^1\}$. By (ii) either $axb = ayb$ or $(axb,x),(ayb,y)\in J$. By (iii) R is a union of non-adjacent edges. Hence $\varepsilon(x,y)\cap J = \Delta$ and $Car(\varepsilon(x,y))\subseteq J(x)\cup J(y)$. By (i), $S^1J(x)S^1$ and $S^1J(y)S^1$ are incomparable, thus by Proposition 5.1, $\varepsilon(x,y)$ is a 1-minimal congruence.∎

<u>Corollary 5.3: Let S be a Green semigroup. Then $\varepsilon(x,y)$ is a 1-minimal congruence if and only if the following hold:</u>

<u>(i)</u> S^1xS^1 <u>and</u> S^1yS^1 <u>are incomparable;</u>

<u>(ii)</u> <u>for every</u> $a\in S$ <u>if</u> $(ax,x)\notin J$ <u>or</u> $(ay,y)\notin J$ <u>then</u> $ax = ay$, <u>if</u> $(xa,x)\notin J$ <u>or</u> $(ya,y)\notin J$ <u>then</u> $xa = ya$;

<u>(iii)</u> <u>for every</u> $a,b\in S^1$ <u>we have</u>

$ax = bx$ <u>if and only if</u> $ay = by$,

$ax = xb$ <u>if and only if</u> $ay = yb$,

$xa = xb$ <u>if and only if</u> $ya = yb$.

Proof: If $\varepsilon(x,y)$ is a 1-minimal congruence then by Theorem 5.2, (i), (ii), and (iii) hold. Assume that x,y satisfy (i), (ii), and (iii). Then (i) of Theorem 5.2 holds. According to Proposition 2.4 3) and (ii) we obtain (ii) of Theorem 5.2. We prove (iii) of Theorem 5.2. Let $a,b,c,d \in S^1$ with $axb = cxd$. If $(axb,x) \notin \mathcal{J}$ then by (ii) and Proposition 2.4 we obtain $ayb = axb = cxd = cyd$. Thus we can assume that $axb \in J(x)$. Since $axb = cxd$ we conclude that $(ax,cx) \in \mathcal{H}$. Hence there exists $e \in S^1$ with $ax = ecx$. Since S is a Green semigroup there exists $c' \in S^1$ with $c'cx = x$. Then $(x,c'ecx) \in \mathcal{H}$ and therefore $xe' = c'ecx$ for some $e' \in S^1$. Thus by (iii) we have $ayb = ecyb = cc'ecyb = cye'b = cyd$ because $cc'ecx = ecx$ and $xe'b = xd$ (we have $cxe'b = cc'ecxb = ecxb = axb = cxd$ and hence $xe'b = c'cxe'b = c'cxd = xd$). Theorem 5.2 concludes the proof that $\varepsilon(x,y)$ is a 1-minimal congruence.■

Corollary 5.4: Let S be a Green semigroup with two $\mathcal{J}$-classes J_0, J_1 such that $S^1J_0S^1$ and $S^1J_1S^1$ are incomparable. Let ξ be the greatest congruence on $J_0 \cup J_1$ such that $(a,b) \in \xi$ if for every $x,y \in S^1$ we have:

if $(ax,a) \notin \mathcal{J}$ or $(bx,b) \notin \mathcal{J}$ then $ax = bx$;

if $(xa,a) \notin \mathcal{J}$ or $(xb,b) \notin \mathcal{J}$ then $xa = xb$;

$xa = ya$ if and only if $xb = yb$;

$xa = ay$ if and only if $xb = by$;

$ax = ay$ if and only if $bx = by$.

Choose $u \in J_0$. Then $\{\varepsilon(u,v);\ (u,v) \in \xi,\ v \in J_1\}$ is the set of all 1-minimal congruences τ with $\mathrm{Car}(\tau) \subseteq J_0 \cup J_1$.

Proof: By Corollary 5.3 if $x \in J_0$, $y \in J_1$ and $\varepsilon(x,y)$ is a 1-minimal congruence then $(x,y) \in \xi$. Now Proposition 5.1 completes the proof.■

Corollary 5.5: Let τ be a 1-minimal congruence on a semigroup S. Then S is a Green semigroup if and only if S/τ is a Green semigroup.

Proof: Apply Proposition 5.1 and Theorem 2.2.■

We now investigate 1-coextensions. Assume that S is a semigroup, J is a non-regular $\mathcal{J}$-class of S, and that $\varphi: T \to S$ is a 1-coextension of $(S,.J)$. We apply the preceding results of this chapter and the results of the chapter IV. If Γ is a coextension family with $\mathrm{Coex}(\Gamma) = \varphi$ then by Proposition 5.1, for every $a \in J$, T_a has exactly two elements. If we choose $a \in J$ and set $T_a = \{(a,i);\ i \in 2\}$ then for every $t,u,v,w \in S^1$ with $tau = vaw$ we have $f_t^{au} \cdot g_u^a(a,i) = f_t^{aw} \cdot g_w^a(a,i)$ for $i \in 2$. Thus we can set $T_x = \{(x,i);\ i \in 2\}$ for every $x \in J$ such that for every $t \in S^1$ with $tx \in J$ (or $xt \in J$) we have $f_t^x(x,i) = (tx,i)$ (or $g_t^x(x,i) = (xt,i)$, respectively) for $i \in 2$. Further define $h'_a : V_{J,a} \to \{0,1\}$ such that $h_a(b,c) = (bc, h'_a(b,c))$ for every $(b,c) \in V_{J,a}$, $a \in J$. By a straightforward calculation we obtain that the h'_a satisfy:

(+) for every $a,b,c \in S$ with $abc \in J$ we have
 if $ab, bc \notin J$ then $h'_{abc}(ab,c) = h'_{abc}(a,bc)$,
 if $ab \in J$, $bc \notin J$ then $h'_{ab}(a,b) = h'_{abc}(a,bc)$,
 if $ab \notin J$, $bc \in J$ then $h'_{abc}(ab,c) = h'_{bc}(b,c)$,
 if $ab, bc \in J$ then $h'_{ab}(a,b) = h'_{bc}(b,c)$.

On the other hand, if we have a family $\{h'_a : V_{J,a} \to \{0,1\};\ a \in J\}$ of mappings fulfilling (+) then we can define $T_a = \{(a,i);\ i \in 2\}$, $h_a : V_{J,a} \to T_a$, $h_a(b,c) = (a, h'_a(b,c))$ for every $a \in J$, $(b,c) \in V_{J,a}$. For $x \in J$, $t \in S$ with $tx \in J$ (or $xt \in J$) define $f_t^x(x,i) = (tx,i)$ (or $g_t^x(x,i) = (xt,i)$, respectively). Then we obtain a coextension family Γ and by Proposition 5.1 $\mathrm{Coex}(\Gamma)$ is a 1-minimal coextension. Thus we can summarize:

<u>Theorem 5.6: Let</u> S <u>be a semigroup,</u> J <u>be a non-regular</u> $\mathcal{J}$-<u>class of</u> S<u>. Then, up to isomorphism, every 1-coextension</u> φ <u>of</u> (S,J) <u>is determined by a family of mappings</u> $\{h'_a : V_{J,a} \to \{0,1\};\ a \in J\}$ <u>fulfilling (+). On the other hand, every family of mappings</u> $\{h'_a : V_{J,a} \to \{0,1\};\ a \in J\}$ <u>fulfilling (+) determines exactly one 1-coextension</u> φ <u>of</u> (S,J) <u>in a canonical way. Two families</u> $\{h'_a : V_{J,a} \to \{0,1\};\ a \in J\}$ <u>and</u> $\{h''_a : V_{J,a} \to \{0,1\};\ a \in J\}$ <u>fulfilling (+) determine isomorphic 1-coextension if and only if either</u> $h'_a = h''_a$ <u>for every</u> $a \in J$ <u>or</u> $h'_a = 1 - h''_a$ <u>for every</u> $a \in J$.

Proof: The last statement follows from Theorem 4.4.∎

<u>Theorem 5.7: Let</u> S <u>be a semigroup,</u> J <u>be a</u> $\mathcal{J}$-<u>class of</u> S<u>. Then</u> (S,J) <u>is 1-coextendable if and only if</u> J <u>is non-regular.</u>

Proof: If (S,J) is 1-coextendable then, according to Proposition 5.1, J is non-regular. If J is a non-regular $\mathcal{J}$-class then by Theorem 5.6 it suffices to take a family $\{h_a' : V_{J,a} \to \{0,1\};\ a \in J\}$ where h_a' is a constant mapping to 0 for every $a \in J$. Then (+) is fulfilled and hence (S,J) is 1-coextendable.∎

Finally consider the semigroup S determined by the following multiplication table:

S	a	b	c	d	e	f	g	0
a	0	g	0	0	0	0	0	0
b	0	0	0	0	0	0	0	0
c	0	0	0	g	0	0	0	0
d	0	0	0	0	0	0	0	0
e	0	0	0	0	0	g	0	0
f	0	0	0	0	0	0	0	0
g	0	0	0	0	0	0	0	0
0	0	0	0	0	0	0	0	0

Then $A = \{g\}$ is a non-regular $\mathcal{J}$-class, $V_{A,g} = \{(a,b),(c,d),(e,f)\}$ and $\sigma_{A,g} = \Delta$. Thus every mapping $h' : V_{A,g} \to \{0,1\}$ fulfils (+), and hence every mapping $h' : V_{A,g} \to \{0,1\}$ determines a 1-coextension. Consider h_0' is a constant mapping to 0, h_1', h_2' are mappings where $(h_1')^{-1}(1) = \{(a,b)\}$, $(h_2')^{-1}(1) = \{(c,d)\}$. Let $\varphi_i : T_i \to S$ be the 1-coextensions corresponding to h_i', $i \in 3$. Then $\varphi_0, \varphi_1, \varphi_2$ are non-isomorphic 1-coextensions, but T_1 and T_2 are isomorphic and T_0 and T_1 are not isomorphic.

VI. Congruences of type 2

<u>Proposition 6.1: A congruence $\tau \neq \Delta$ of a semigroup S is a 2-minimal congruence if and only if there exist elements $x,y \in S$ such that $S^1 y S^1 \subset S^1 x S^1$ and the following conditions hold:</u>

(i) $Car(\tau) \subseteq J(\{x,y\})$;

(ii) $\tau \cap \varepsilon(J(y)) = \Delta$, $\tau \cap \eta(J(x)) = \Delta$.

<u>Moreover, if τ is a 2-minimal congruence then</u>

A) <u>for every $u \in J(x)$ there exists exactly one $u_\tau \in J(y)$ with $(u,u_\tau) \in \tau$;</u>

B) <u>if $u,v \in J(x)$ with $(u,v) \in \tau$ then there exist $a,b,c,d \in S^1$</u>

with $aub = v$, $au_\tau b = u_\tau$, and $\{cud, cvd\} = \{u, u_\tau\}$;

C) $\{\tau z;\ z\in J(x)\cup J(y)\}$ is a J-class of S/τ.

Proof: Let τ be a 2-minimal congruence, then by definition there exist $x,y\in S$ such that $S^1yS^1\subset S^1xS^1$ and (i) holds. By Proposition 1.2 we obtain (ii). Hence A) is obvious. To show B) it suffices to note that if $aub = v$ then $(au_\tau b, v)\in\tau$ and $(u,v)\in\tau$ implies $u_\tau = v_\tau$. The existence of c,d follows from $\tau\cap\eta(J(x)) = \Delta$. C) is clear.

Conversely, assume that τ fulfils (i) and (ii). Let $\sigma \neq \Delta$ be a congruence finer than τ. Since $\sigma \neq \Delta$ there exist different $u,v\in S^1$ with $(u,v)\in\sigma$ (whence $(u,v)\in\tau$), thus $u,v\in J(\{x,y\})$ and $\{u,v\}\cap J(x) \neq 0$. Since $\sigma\cap\eta(J(x))$ is finer than $\tau\cap\eta(J(x)) = \Delta$ we can assume without loss of generality that $u\in J(x)$, $v\in J(y)$. Take $s,t\in S$ with $(s,t)\in\tau$. If $s = t$, then clearly $(s,t)\in\sigma$. Assume that $s \neq t$. Then $s,t\in J(\{x,y\})$ and $\{s,t\}\cap J(x) \neq 0$. We have two cases: either $\{s,t\}\cap J(y) \neq 0$ or $\{s,t\}\subseteq J(x)$. If $\{s,t\}\cap J(y) \neq 0$ then we can assume that $s\in J(x)$, $t\in J(y)$. There exist $a,b\in S^1$ with $aub = s$. Then $(aub, avb)\in\sigma$, whence $(aub,avb)\in\tau$, and, by (ii), $avb = t$, thus $(s,t)\in\sigma$. If $\{s,t\}\subseteq J(x)$ then there exist $a,b,c,d\in S^1$ with $aub = s$, $cud = t$. Then $(s,avb),(t,cvd)\in\sigma$. Whence $(s,avb),(t,cvd)\in\tau$. Since $(s,t)\in\tau$ we have, by (ii), that $avb = cvd$ and thus $(s,t)\in\sigma$. Therefore $\sigma = \tau$.∎

Theorem 6.2: Let S be a semigroup, x,y be elements of S with $S^1yS^1\subset S^1xS^1$. Then $\varepsilon(x,y)$ is a 2-minimal congruence if and only if the following hold:

(i) for every $a,b\in S^1$, if $(axb,x)\notin J$ or $(ayb,y)\notin J$ then $axb = ayb$;

(ii) for every $a,b,c,d\in S^1$, if $axb = cxd$ then $ayb = cyd$;

(iii) for every $a,b,c,d\in S^1$ with $ayb = cyd$ and $ayb \neq axb \neq cxd \neq cyd$ there exist $e,f\in S^1$ with $(eaxbf, ecxdf)\notin J$.

Proof: Let $\varepsilon(x,y)$ be a 2-minimal congruence. Since $(axb,ayb)\in\varepsilon(x,y)$ for every $a,b\in S^1$ we obtain (i) and (ii) by Proposition 6.1. To prove (iii) it suffices to note that $(eaxbf,ecxdf)\in J$ for every $e,f\in S^1$ implies $\varepsilon(axb,cxd)\cap\eta(J(x)) \neq \Delta$ (by (i) $axb,cxd\in J(x)$). Since $\varepsilon(axb,cxd) = \varepsilon(x,y)$ by Proposition 1.1 we obtain a contradiction with Proposition 6.1.

Conversely, if x,y fulfil (i), (ii), and (iii) then by (i) $Car(\varepsilon(x,y)) \subseteq J(\{x,y\})$, by (i) and (ii) $\varepsilon(x,y) \cap \varepsilon(J(y)) = \Delta$ and by (iii) $\varepsilon(x,y) \cap \eta(J(x)) = \Delta$. Thus Proposition 6.1 implies that $\varepsilon(x,y)$ is a 2-minimal congruence.■

<u>Proposition 6.3: Let</u> τ <u>be a 2-minimal congruence on a Green semigroup</u> S. <u>Then</u>

<u>A)</u> $\tau \cap H = \Delta$;

<u>B)</u> <u>if for</u> $u,v \in S$ <u>there exist</u> $u',v' \in S$ <u>such that</u> $S^1uS^1 \subset S^1u'S^1$, $S^1vS^1 \subset S^1v'S^1$ <u>and</u> $(u,u'),(v,v') \in \tau$ <u>then there exist</u> $x,y,x',y' \in S^1$ <u>such that</u> $S^1xS^1 \subset S^1x'S^1$, $S^1yS^1 \subset S^1y'S^1$, $(x,x'),(y,y') \in \tau$, <u>and</u> $(u,x),(y,v) \in L$, $(u,y),(x,v) \in R$.

Proof: Let $(x,y) \in \tau \cap H$. Then $\varepsilon(x,y)$ is the reflexive, symmetric, and transitive closure of the relation $\{(axb,ayb);\ a,b \in S^1\}$. Since S is a Green semigroup we obtain by Proposition 2.4 that $(axb,x) \notin J$ if and only if $(ayb,y) \notin J$. Since $\varepsilon(x,y)$ is finer than τ we conclude by Proposition 1.2 that $Car(\varepsilon(x,y)) \subseteq J(x)$. Thus either $x = y$ or $\varepsilon(x,y) \cap \eta(J(x)) \neq \Delta$, and by Proposition 6.1 we have that $x = y$, whence $\tau \cap H = \Delta$.

We prove B). Since τ is a 2-minimal congruence we have that $(u,v),(u',v') \in J$. From the fact that S is a Green semigroup we obtain elements $a,b \in S^1$ with $au'b = v'$. Then by Proposition 6.1, $aub = v$ and by Green Theorem $(au',u'),(au,u),(u'b,v'),(ub,v) \in L$, $(u'b,u'),(ub,u),(au',v'),(au,v) \in R$. Set $au=x$, $au' = x'$, $ub = y$, $u'b = y'$. Then $(x,x'),(y,y') \in \tau$ and the proof is complete.■

<u>Theorem 6.4: Let</u> S <u>be a Green semigroup,</u> x,y <u>be elements of</u> S <u>with</u> $S^1yS^1 \subset S^1xS^1$. <u>Then</u> $\varepsilon(x,y)$ <u>is a 2-minimal congruence if and only if the following hold:</u>

<u>(i)</u> <u>for every</u> $a \in S^1$ <u>if</u> $(ax,x) \notin J$ <u>or</u> $(ay,y) \notin J$ <u>then</u> $ax = ay$, <u>if</u> $(xa,x) \notin J$ <u>or</u> $(ya,y) \notin J$ <u>then</u> $xa = ya$;

<u>(ii)</u> <u>for every</u> $a,b \in S^1$ <u>if</u> $ax = bx$ <u>then</u> $ay = by$, <u>if</u> $xa = bx$ <u>then</u> $ya = by$, <u>if</u> $xa = xb$ <u>then</u> $ya = yb$;

<u>(iii)</u> $\varepsilon(x,y) \cap \eta(J(x)) = \Delta$.

Proof: By Proposition 6.1 and Theorem 6.2, if $\varepsilon(x,y)$ is a 2-minimal congruence then (i), (ii), and (iii) hold.

Conversely, assume that (i), (ii), and (iii) hold. Since S is a

Green semigroup we have by Proposition 2.4 for $a,b\in S^1$: if $(axb,x)\notin J$ then either $(ax,x)\notin J$ or $(xb,x)\notin J$; and if $(ayb,y)\notin J$ then either $(ay,y)\notin J$ or $(yb,y)\notin J$. Hence by (i) we obtain that for $a,b\in S^1$ if either $(axb,x)\notin J$ or $(ayb,y)\notin J$ then $axb = ayb$. As a consequence of this fact we have that $Car(\varepsilon(x,y))\subseteq J(\{x,y\})$ because $\varepsilon(x,y)$ is the reflective, symmetric, and transitive closure of the relation $R = \{(axb,ayb);\ a,b\in S^1\}$. Note if $(u,v)\in R$ then either $u = v$ or $u\in J(x)$, $v\in J(y)$. Thus to prove that $\varepsilon(x,y)\cap\varepsilon(J(y)) = \Delta$ it suffices to show that if $(u,v),(u,w)\in R$ then $v = w$. Assume that $a,b,c,d\in S^1$ with $axb = cxd$, we have to prove that $ayb = cyd$. If $axb\notin J(x)$ then $ayb = axb = cxd = cyd$ by the foregoing part of the proof. If $axb\in J(x)$ then analogously as in the proof of Theorem 5.3 there exist $e,c',e'\in S^1$ with $ax = ecx$, $c'cx = x$, $xe' = c'ecx$ (we recall that $(x,c'ecx)\in H$) then $cc'ecx = ecx$ and $cxe'b = cxd$ hence $xe'b = cc'ecxb = c'cxd = xd$. If we apply (ii) then we obtain $ayb = ecyb = cc'ecyb = cye'b = cyd$, this completes the proof of $\varepsilon(x,y)\cap\varepsilon(J(y)) = \Delta$. If we use (iii) and Proposition 6.1 we have that $\varepsilon(x,y)$ is a 2-minimal congruence.∎

Theorem 6.5: Let S be a semigroup, x,y be elements of S with $(x,y)\in J$. Then $\varepsilon(x,y)$ is a 2-minimal congruence if and only if there exists $z\in S$ with $S^1zS^1\subset S^1xS^1$ such that the following hold:

(i) $\varepsilon(x,z)$ is a 2-minimal congruence;

(ii) there exist $a,b\in S^1$ with $\{axb,ayb\} = \{x,z\}$;

(iii) there exist $a,b\in S^1$ with $axb = y$ and $azb = z$.

If S is a Green semigroup then (ii) can be formulated as follows:

(ii') there exists $a\in S^1$ such that either $ax = x$, $ay = z$, or $ax = z$, $ay = y$ or $xa = x$, $ya = z$ or $xa = z$, $ya = y$.

Proof: If $\varepsilon(x,y)$ is a 2-minimal congruence then by Proposition 6.1 there exists $z\in S$ with $S^1zS^1\subset S^1xS^1$ and $(x,z)\in\varepsilon(x,y)$. By Proposition 1.1 $\varepsilon(x,z)$ is a 2-minimal congruence. Thus if we apply Theorem 6.2 then we obtain (ii) and (iii).

Conversely, if (i), (ii), and (iii) hold then by (ii), $\varepsilon(x,z)$ is finer than $\varepsilon(x,y)$ and by (iii), $\varepsilon(x,y)$ is finer than $\varepsilon(x,z)$, thus (i) completes the proof. Moreover, (ii') implies (ii). Assume that S is a Green semigroup and (ii) holds, then by Proposition 2.4 we get (ii').∎

Corollary 6.6: Let τ be a 2-minimal congruence on a Green semigroup S, then S/τ is a Green semigroup.

Proof: Combine Theorem 2.2 and Proposition 6.1.■

We now investigate 2-coextensions. We use the Coextension scheme (see Construction 4.1). Assume that $\varphi:T \to S$ is a 2-coextension of (S,J) where J is a $\mathcal{J}$-class of S. Let Γ be a coextension family such that φ is isomorphic to $Coex(\Gamma)$. Then there exist two $\mathcal{J}$-classes J_0, J_1 of T with $S^1J_1S^1 \supset S^1J_0S^1$ and $Car(Ker\ \varphi) \subseteq J_0 \cup J_1$. Thus according to Proposition 6.1, for every $a \in J$, $T_a \cap J_0$ is a singleton, denote by $\{t_a\} = T_a \cap J_0$. Thus if we consider that J_0 and J_1 are $\mathcal{J}$-classes of T we obtain that the following conditions hold (for simplicity set $T_x = \{x\}$ for $x \in S-J$):

1) for every $a \in J$ the element $t_a \in T_a$ is given;
2) for every $a \in J$, $b \in S-J$ if $ba \in J$ then $f^a_b(t_a) = t_{ba}$, if $ab \in J$ then $g^a_b(t_a) = t_{ab}$;
3) for every $a,b \in J$, $x \in T_a$ if $ab \in J$ then $f^b_x(t_b) = t_{ab}$, if $ba \in J$ then $g^b_x(t_b) = t_{ba}$;
4) for every if $a,b \in J$, $ab \in J$ then both $f^b_{t_a}$, $g^a_{t_b}$ are constants to t_{ab};
5) for every $a,b \in J$, $x \in T_a-\{t_a\}$, $y \in T_b-\{t_b\}$ there exist $u,z \in S^1$, $v \in T_u$, $w \in T_z$ such that $uaz = b$ and $g^{ua}_w \circ f^a_v(x) = y$;
6) for every $a \in J$, $x \in T_a-\{t_a\}$ there exist $b,c \in S$, $y \in T_b$, $z \in T_c$ such that $bac \in J$ and $f^{ac}_y \circ g^c_z(x) = t_{bac}$;
7) for every $a \in J$, $x,y \in T_a-\{t_a\}$ with $x \neq y$ there exist $b,c \in S^1$, $v \in T_b$, $w \in T_c$ with $a = bac$ and $\{f^{ac}_v \circ g^a_w(x), f^{ac}_v \circ g^a_w(y)\} = \{x,t_a\}$.

Conditions 2),3), and 4) are equivalent with the fact that $J_0 = \{t_a;\ a \in J\}$ is a $\mathcal{J}$-class of T and that $T^1J_0T^1 \cap (\cup\{T_a;\ a \in J\}) = J_0$. Condition 5) is equivalent with the fact that $J_1 = \cup\{T_a-\{t_a\};\ a \in J\}$ is a $\mathcal{J}$-class of T and the condition 6) is equivalent with the fact that $J_0 \subset T^1J_1T^1$. Condition 7) is equivalent with the fact that $Ker(Coex(\Gamma)) \cap \eta(J_1) = \Delta$. Since by the definition of T, $Car(Ker(Coex(\Gamma))) \subseteq J_0 \cup J_1$ and $Ker(Coex(\Gamma)) \cap \varepsilon(J_0) = \Delta$ we obtain by Proposition 6.1 that $Ker(Coex(\Gamma))$ is a 2-minimal congruence if and only if Γ satisfies 1)-7). Thus we can summarize:

Theorem 6.7: Let S be a semigroup, J be a $\mathcal{J}$-class of S. Then every 2-coextension of (S,J) is determined by a coextension family Γ fulfilling 1)-7) and every coextension family Γ (where $A = J$) fulfilling 1)-7) determines a 2-coextension of (S,J).∎

Theorem 6.8: Let S be a semigroup with a $\mathcal{J}$-class J. Then (S,J) is 2-coextendable if and only if there exist $x\in J$, $a,b\in S^1$ with $axb = x$ and either $a\in S$ or $b\in S$. Hence if J has at least two elements or J contains an idempotent then (S,J) is 2-coextendable.

Proof: Assume that S and J fulfil the required condition. Choose $x\in J$, a point $u\notin S$. Let T be the inflation of x in S by u. If $\varphi:T\to S$ is the corresponding natural inflation homomorphism then φ is a minimal coextension of (S,J). Let Γ be a coextension family corresponding to φ, i.e. $\varphi = \mathrm{Coex}(\Gamma)$. Then $T_a = \{a\}$ for every $a\in J-\{x\}$, $T_x = \{x,u\}$. If we set $t_a = a$ for $a\in J$ then Γ evidently fulfils 1),2),3),4),5), and 7). 6) follows from the condition on J. Hence by Theorem 6.7, φ is a 2-coextension of (S,J).

Conversely, if $\varphi:T\to S$ is a 2-coextension and Γ is a coextension family such that φ and $\mathrm{Coex}(\Gamma)$ are isomorphic then Condition 6) of Theorem 6.7 implies that J satisfies the required condition. The rest is clear.∎

Finally, we investigate the condition implying that for a 2-coextension $\varphi:T\to S$ of a Green semigroup S we obtain that T is also a Green semigroup. For this reason we strengthen Conditions 6) and 7).

6') for every $a\in J$, $x\in T_a-\{t_a\}$ there exist $b\in S$, $y\in T_b$ such that either $ab\in J$ and $g_y^a(x) = t_{ab}$, or $ba\in J$ and $f_y^a(x) = t_{ba}$;

7') for every $a\in J$ and $x,y\in T_a-\{t_a\}$ with $x \neq y$ there exist $b\in S$, $z\in T_b$ such that either $ab = a$ and $\{g_z^a(x),g_z^a(y)\} = \{x,t_a\}$ or $ba = a$ and $\{f_z^a(x),f_z^a(y)\} = \{x,t_a\}$.

If T is a Green semigroup and $\varphi:T\to S$ is a 2-coextension of (S,J) and Γ is a coextension family such that φ and $\mathrm{Coex}(\Gamma)$ are isomorphic then by Proposition 2.4 we easily obtain that 6) and 6') and 7) and 7') are equivalent. Moreover, consider Condition 8):

8) if $x \in T_a - \{t_a\}$, $y \in T_b - \{t_b\}$ for $a,b \in J$ such that for some $c \in S$ and $v \in T_c$ we have $ca = b$ and $f_v^a(x) = y$ (or $ac = b$ and $g_v^a(x) = y$) then there exist $d \in S$, $w \in T_d$ with $db = a$ and $f_w^b(y) = x$ (or $bd = a$ and $g_w^b(y) = x$, respectively).

If T is a Green semigroup then validity of Green Theorem implies 8). On the other hand, if S is a Green semigroup then the Green Theorem is satisfied for $a,b \in T$ with $a,b \notin \cup\{T_x - \{t_x\};\ x \in J\}$ and 8) implies its validity also for the set $\cup\{T_x - \{t_x\};\ x \in J\}$. Thus if we combine Theorem 6.7 with these observations we have:

<u>Theorem 6.9: Let S be a Green semigroup, J be a *J*-class of S. If $\varphi: T \to S$ is a coextension and Γ is a coextension family such that φ and $\mathrm{Coex}(\Gamma)$ are isomorphic then φ is a 2-coextension and T is a Green semigroup if and only if Γ fulfils 1)-5), 6'), 7'), 8).</u>

Proof see above.∎

This chapter ends with an example:

Let $S = \{a_i, b_i;\ i$ is an integer, $i>0\} \cup \{d_i;\ i$ is an integer$\} \cup \{0\}$, $T = S \cup \{c_i;\ i$ is an integer$\}$ and define a multiplication such that

for all integers $i,j>0$, $a_i \cdot a_j = a_{i+j}$, $b_i \cdot b_j = b_{i+j}$,

for all integers i,j where $i>0$, $a_i \cdot c_j = c_{i+j}$, $c_j \cdot b_i = c_{j-i}$,

for all integers i, $d_i \cdot c_i = c_i$,

otherwise the multiplication is equal to 0.

Then the *J*-equivalence in S is equal to Δ, the *D*-equivalence in T is equal to Δ and the *J*-equivalence in T has exactly one non-singleton class $\{c_i;\ i$ is an integer$\}$. Thus T is not a Green semigroup and S is a Green semigroup. Let τ be a congruence on T with exactly one non-singleton class $\{c_i;\ i$ is an integer$\} \cup \{0\}$ and $\varphi: T \to S$ be a homomorphism such that $\varphi(x) = x$ for $x \in S$, $\varphi(c_i) = 0$ for every integer i. By direct inspection we obtain that τ is a congruence, φ is a coextension and $\tau = \mathrm{Ker}(\varphi)$. Since $d_i \cdot c_i = c_i$ and $d_i \cdot c_j = 0$ for $i \neq j$ we have that $\eta(\{c_i;\ i$ is an integer$\}) = \Delta$ and hence by Proposition 6.1, τ is a 2-minimal congruence. Thus Corollary 6.6 cannot be strengthened. Note also that φ is not an inflation.

VII. Congruences of type 3

In this and the following chapters we restrict ourselves to the class of Green semigroups. We investigate the inner minimal congruences only in the class of Green semigroups. Proposition 2.4 and Corollary 2.5 show the reason why we restrict to this class.

Proposition 7.1: Let S be a Green semigroup. Then a congruence $\tau \neq \Delta$ on S is a 3-minimal congruence if and only if the following hold:

(i) there exists a J-class J of S with $Car(\tau) \subseteq J$;

(ii) τ is finer than H;

(iii) for an H-class $H \subseteq J$ and for the Schutzenberger group $\Psi(H) = (H,\Psi)$ we have that $\Phi = \{g \in \Psi;$ for every $y \in H$, $(y,g(y)) \in \tau\}$ is a minimal normal subgroup of the group Ψ with the operation composition.

Proof: If τ is a 3-minimal congruence then clearly (i) and (ii) hold. To prove (iii) consider a congruence σ on S such that $Car(\sigma) \subseteq J \in J(S)$ and σ is finer than H. Denote by $\Phi = \{f \in \Psi;$ for every $y \in H$, $(y,f(y)) \in \sigma\}$. We show that Φ is a normal subgroup of Ψ. Let $f,g \in \Phi$. Since σ is symmetrical we have $f^{-1} \in \Phi$, since σ is transitive we obtain $f \circ g \in \Phi$. Thus Φ is a subgroup. Let $f \in \Phi$, $g \in \Psi$. Then there exist $a,b \in S^1$ with $g(y) = yb$, $g^{-1}(y) = ya$ for every $y \in H$. Since $f \in \Phi$ we have $(y,f(y)) \in \sigma$ for every $y \in H$ whence $(yb,f(yb)) \in \sigma$ for every $y \in H$ and thus $(y = yba, f(yb)a = g^{-1} \circ f \circ g(y)) \in \sigma$. Hence $g^{-1} \circ f \circ g \in \Phi$ and we showed that Φ is a normal subgroup of Ψ. Assume that $\Lambda \subseteq \Phi$ is also a normal subgroup of Ψ. Define a relation ν as follows: $(u,v) \in \nu$ if either $u = v$ or $(u,v) \in H$, $u,v \in J$ and there exist $a,b \in S^1$ with $aub \in H$ and for some $f \in \Lambda$ we have $f(aub) =$ $= avb$.

Clearly, ν is reflexive. Since $f \in \Lambda$ implies $f^{-1} \in \Lambda$ we have that ν is symmetric. If $f(aub) = avb$ for $(u,v) \in H$, $u,v \in J$, $aub \in H$ then for every $c \in S^1$ with $cub \in H$ we have $f(cub) = cvb$ because there exists $c' \in S^1$ with $c'ax = cx$ for every $x \in H(u)$ and f is a right inner translation. Consider $c \in S^1$ such that $auc \in H$. Then there exists $g\in\Psi$ with $g(axc) = axb$ for every $x\in H(u)$. Hence $g^{-1}\cdot f\cdot g(auc) = g^{-1}\cdot f(aub) = g^{-1}(avb) = avc$, and $g^{-1}\cdot f\cdot g\in\Lambda$ by the normality of Λ. Thus if $f(aub) = avb$ for $f\in\Lambda$ then for every $c,d\in S^1$, $cud\in H$ there exists $g\in\Lambda$ with $g(cud) = cvd$. Since Λ is a subgroup we conclude that ν is transitive - thus ν is an equivalence.

Assume that $(u,v)\in\nu$. Then either $u = v$ or $u,v\in J$, $(u,v)\in \mathcal{H}$ and there exist a,b,c,d such that $aub\in H$, $caubd = u$ (and also $cavbd = v$). Since $\Lambda\subseteq\Phi$ we have $(aub,avb)\in\tau$ whence $(u,v)\in\tau$. Let $z\in S$. If $(zu,u)\notin \mathcal{J}$ or $(zv,v)\notin \mathcal{J}$ then $zu = zv$ since $Car(\tau)\subseteq J$. Analogously, if $(uz,u)\notin \mathcal{J}$ or $(vz,v)\notin \mathcal{J}$ then $uz = vz$. If $(zu,u)\in \mathcal{J}$ then $(zv,v)\in \mathcal{J}$ and $(zu,zv)\in \mathcal{H}$, $zu,zv\in J$, analogously, if $(zv,v)\in \mathcal{J}$ then $(zu,zv)\in \mathcal{H}$, $zu,zv\in J$. If $(uz,u)\in \mathcal{J}$ then $uz,vz\in J$ and $(uz,vz)\in \mathcal{H}$, if $(vz,v)\in \mathcal{J}$ then $vz,uz\in J$, $(vz,uz)\in \mathcal{H}$. If $(uz,vz)\in \mathcal{H}$, $uz,vz\in J$ then there exist $a,b\in S^1$ with $auzb\in H$ and by the foregoing part of the proof there exists $g\in\Lambda$ with $g(auzb) = avzb$ - whence $(uz,vz)\in\nu$. By duality if $(zu,zv)\in \mathcal{H}$, $zu,zv\in J$ then $(zu,zv)\in\nu$ and thus ν is a congruence. Moreover, ν is finer than σ and $\nu = \sigma$ if and only if $\Phi = \Lambda$. A consequence of these facts is (iii).

Assume that (i), (ii), and (iii) hold. If σ is a finer congruence than τ then σ satisfies (i) and (ii), thus $\Lambda = \{g\in\Psi;$ for every $y\in H$, $(y,g(y))\in\sigma\}$ is a normal subgroup of Ψ. Since Φ is minimal we have that either $\Lambda = \Phi$ and $\sigma = \tau$ or $\Lambda = \{1\}$ and $\sigma = \Delta$ - thus τ is a minimal congruence and by (i) and (ii) τ is a 3-minimal congruence.∎

<u>Theorem 7.2:</u> <u>Let</u> S <u>be a Green semigroup,</u> x,y <u>be different elements of</u> S. <u>Then</u> $\varepsilon(x,y)$ <u>is a 3-minimal congruence if and only if the following hold:</u>

<u>(i)</u> $(x,y)\in \mathcal{H}$;

<u>(ii)</u> <u>for every</u> $a\in S^1$ <u>if</u> $(ax,x)\notin \mathcal{J}$ <u>or</u> $(ay,y)\notin \mathcal{J}$ <u>then</u> $ax = ay$, <u>if</u> $(xa,x)\notin \mathcal{J}$ <u>or</u> $(ya,y)\notin \mathcal{J}$ <u>then</u> $xa = ya$;

<u>(iii)</u> <u>if</u> $\Psi(H) = (H,\Psi)$ <u>is the Schutzenberger group of</u> $H(x)$ <u>then the smallest normal subgroup of</u> Ψ <u>(with the operation composition) containing</u> $f\in\Psi$ <u>with</u> $f(x) = y$ <u>is minimal.</u>

Proof: If $\varepsilon(x,y)$ is a 3-minimal congruence then (i) follows from the definition. Since $Car(\varepsilon(x,y))\subseteq J(x)$ we obtain (ii). Since f is a right inner translation we have $f(ax) = ay$ for every $a \in S^1$ with $aH(x) \subseteq H(x)$, thus Proposition 7.1 implies (iii).

Conversely, assume that x,y satisfy (i), (ii), and (iii). Then $\varepsilon(x,y)$ is the reflexive, symmetric, and transitive closure of the relation $R = \{(axb,ayb);\ a,b \in S^1\}$. By Proposition 2.4, (i) and (ii) we have that $Car(\varepsilon(x,y)) \subseteq J(x)$ and $\varepsilon(x,y)$ is finer than $\mathcal{H}$. We know that $\Phi = \{g \in \Psi;$ for every $u \in H(x),\ (g(u),u) \in \varepsilon(x,y)\}$ is a normal subgroup of Ψ containing f. If Λ is the smallest normal subgroup of Ψ containing f then $\Lambda \subseteq \Phi$ and we know that the relation ν is a congruence where $(u,v) \in \nu$ if and only if either $u = v$ or $(u,v) \in \mathcal{H}$, $u,v \in J(x)$ and there exist $a,b \in S^1$, $g \in \Lambda$ with $aub \in H(x)$, $g(aub) = avb$. Since $f \in \Lambda$ we have $(x,y) \in \nu$ and hence $\varepsilon(x,y)$ is finer than ν. But $\Lambda \subseteq \Phi$ implies the converse inclusion and thus $\nu = \varepsilon(x,y)$. Proposition 7.1 concludes the proof.■

From Theorem 2.2 and Proposition 7.1 we have:

<u>Corollary 7.3: Let</u> τ <u>be a 3-minimal congruence on a Green semigroup</u> S<u>. Then</u> S/τ <u>is also a Green semigroup.</u>■

We investigate 3-coextensions between Green semigroups. Let $\varphi: T \to S$ be a 3-coextension of (S,J) such that T and S are Green semigroups and J is a $\mathcal{J}$-class of S. Note that $\varphi^{-1}(J)$ has an idempotent if and only if J has an idempotent. Denote $J' = \varphi^{-1}(J)$, $\mathcal{R}(J') = \{R;\ R \in \mathcal{R}(T),\ R \subseteq J'\}$, $\mathcal{L}(J') = \{L;\ L \in \mathcal{L}(T),\ L \subseteq J'\}$, $\mathcal{R}(J) = \{R;\ R \in \mathcal{R}(S),\ R \subseteq J\}$, $\mathcal{L}(J) = \{L;\ L \in \mathcal{L}(S),\ L \subseteq J\}$. Then we have $\mathcal{R}(J) = \{\varphi(R);\ R \in \mathcal{R}(J')\}$ and $\mathcal{L}(J) = \{\varphi(L);\ L \in \mathcal{L}(J')\}$. Choose an initialization of coordinates y, $\{c_R;\ R \in \mathcal{R}(J')\}$, $\{d_L;\ L \in \mathcal{L}(J')\}$ of J' in T, then $x = \varphi(y)$, $\{a_R = \varphi(c_{\varphi^{-1}(R)});\ R \in \mathcal{R}(J)\}$, $\{b_L = \varphi(d_{\varphi^{-1}(L)});\ L \in \mathcal{L}(J)\}$ is an initialization of coordinates of J in S. If $\Psi(H(y)) = (H(y),\Psi)$ is the Schutzenberger group of $H(y)$ in T, $\Phi(H(x)) = (H(x),\Phi)$ is the Schutzenberger group of $H(x)$ in S then by Propositions 3.3 and 7.1 there exists a minimal normal subgroup Λ of Ψ (Ψ and Φ are groups where the operation is the composition) such that Ψ/Λ and Φ are isomorphic. Then φ induces a group homomorphism $\psi: \Psi \to \Phi$ such that $\psi^{-1}(1) = \Lambda$ and $\varphi(a,R,L) = (\psi(a),\varphi(R),\varphi(L))$. Thus for a special coextension family Ω such that φ and $Coex(\Omega)$ are isomorphic, we can assume that $T' = \Psi \times \mathcal{R}(J) \times \mathcal{L}(J)$, $k'(a,R,L) = (\psi(a),R,L)$. We redefine the coordinates on J' in T - instead of (a,R,L) for $a\in\Psi$, $R\in\mathcal{R}(J')$, $L\in\mathcal{L}(J')$ we set $(a,\varphi(R),\varphi(L))$ - it is possible because φ induces bijections from $\mathcal{R}(J')$ onto $\mathcal{R}(J)$ and from $\mathcal{L}(J')$ onto $\mathcal{L}(J)$. Further for every (a,R,L), if $f^{R'}_{(a,R,L)}$ is defined then $f^{R'}_{(a,R,L)}(b,R',L') = (ab,R,L')$, and if $g^{L'}_{(a,R,L)}$ is defined then $g^{L'}_{(a,R,L)}(b,R',L') = (ba,R',L)$. By Proposition 3.3 for every $b\in S-J$ and $R\in\mathcal{R}(J)$ with $bR\in\mathcal{R}(J)$ there exists $\mu(b,R)\in\Psi$ with $f^R_b = f_{(\mu(b,R),R,bR)}$ and for every $b\in S-J$ and $L\in\mathcal{L}(J)$ with $Lb\in\mathcal{L}(J)$ there exists $\nu(b,L)\in\Psi$ with $g^L_b = g_{(\nu(b,L),L,Lb)}$. Thus a special coextension family Ω (if an initialization of coordinates in J of S is given) is determined by a group Ψ, a family of mappings $\{h'_a:V_{J,a}\to\Psi;\ a\in J\}$ and mappings

$$\mu:\{(b,R);\ b\in S-J,\ R\in\mathcal{R}(J)\ \text{with}\ bR\in\mathcal{R}(J)\}\to\Psi$$
$$\nu:\{(b,L);\ b\in S-J,\ L\in\mathcal{L}(J)\ \text{with}\ Lb\in\mathcal{L}(J)\}\to\Psi$$

fulfilling:

1) there exists a minimal normal subgroup Λ of Ψ such that there exists a surjective homomorphism $\psi:\Psi\to\Phi$ with kernel Λ (i.e. $\psi^{-1}(1) = \Lambda$);
2) for every $a\in J$, $\sigma_{J,a}$ is finer than $\operatorname{Ker} h'_a$ and $\psi(h'_a(b,c)) = G(a)$ for every $(b,c)\in V_{J,a}$;
3) for every $b_1,b_2\in S-J$, $R_1,R_2\in\mathcal{R}(J)$ with $b_1R_1\subseteq R_2$, $b_2R_2\subseteq J$ we have
   if $b_2b_1\notin J$ then $\mu(b_2,R_2)\mu(b_1,R_1) = \mu(b_2b_1,R_1)$,
   if $b_2b_1\in J$ then $\mu(b_2,R_2)\mu(b_1,R_1) = h'_{b_2b_1}(b_2,b_1)$;
4) for every $b_1,b_2\in S-J$, $L_1,L_2\in\mathcal{L}(J)$ with $L_1b_1\subseteq L_2$, $L_2b_2\subseteq J$ we have
   if $b_1b_2\notin J$ then $\nu(b_1,L_1)\nu(b_2,L_2) = \nu(b_1b_2,L_1)$,
   if $b_1b_2\in J$ then $\nu(b_1,L_1)\nu(b_2,L_2) = h'_{b_1b_2}(b_1,b_2)$;
5) for $b\in S-J$, $R\in\mathcal{R}(J)$, $L\in\mathcal{L}(J)$ with $bR\cup Lb\subseteq J$ we have $\mu(b,R) = \nu(b,L)$;
6) for every $b\in S-J$, $(a,R,L)\in J$ with $b(a,R,L) = (a',R',L)$ (or $(a,R,L)b = (a',R,L')\in J$) and for every $c\in\psi^{-1}(a)$ we have $\psi(\mu(b,R)c) = a'$ (or $\psi(c\ \nu(b,L)) = a'$, respectively);
7) for every $a,b,c\in S-J$ with $abc\in J$ we have
   if $ab\in J$, $bc\notin J$ then $h'_{abc}(a,bc) = h'_{ab}(a,b)\nu(c,L(ab))$,
   if $ab\notin J$, $bc\in J$ then $h'_{abc}(ab,c) = \mu(a,R(bc))h'_{bc}(b,c)$,
   if $ab,bc\in J$ then $h'_{ab}(a,b)\cdot\nu(c,L(ab)) = \mu(a,R(bc))h'_{bc}(b,c)$;
8) either J is regular or for every $R,R'\in\mathcal{R}(J)$, $L,L'\in\mathcal{L}(J)$ we have $\{\mu(b,R);\ b\in S-J,\ bR = R'\} = \{\nu(b,L);\ b\in S-J,\ Lb = L'\} = \Psi$.

Define T', k, f_a^R, g_a^L as above and $h_a(b,c) = (h'_a(b,c),R(a),L(a))$. Then condition 6) is equivalent with the complementary conditions in d') and e'). g') is equivalent with the first part of 2), h') is equivalent with 7), i') is equivalent with 3) and 4). 5) is equivalent with f"') (we recall that for $A \in \mathcal{J}(S)$, f') is equivalent with f") and f"')). The second statement in 2) is equivalent with supplementary condition c'). Finally, 1) is implied by Proposition 7.1 and 8) is equivalent with the fact that $T' = J'$ is a $\mathcal{J}$-class of T. Therefore we say that a group Ψ, together with a family of mappings $\{h'_a : V_{J,a} \to \Psi;\ a \in J\}$ and mappings μ, ν fulfilling 1)-8), is called a *3-coextension family*. On the other hand, note that if Θ is a 3-coextension family which defines as above a special coextension family Λ then it fulfils f") and j') - it follows from Chapter III, conditions (+), (++), and (++++). We prove that then $\mathrm{Coex}(\Lambda)$ is a 3-coextension. Since by the definition of $M(\Psi,\mathcal{R}(J),\mathcal{L}(J))$ we have that for every $R \in \mathcal{R}(J)$, $\{(a,R,L);\ a \in \Psi, L \in \mathcal{L}(J)\}$ is an $\mathcal{R}$-class of T, for every $L \in \mathcal{L}(J)$, $\{(a,R,L);\ a \in \Psi, R \in \mathcal{R}(J)\}$ is an $\mathcal{L}$-class we conclude that $\mathrm{Ker}(\mathrm{Coex}(\Lambda))$ is finer than $\mathcal{H}$ and $\mathrm{Car}(\mathrm{Ker}(\mathrm{Coex}(\Lambda))) \subseteq J' = T'$ which is by 8) a $\mathcal{J}$-class of T. Thus 1) and Proposition 7.1 implies that $\mathrm{Coex}(\Lambda)$ is a 3-coextension. Thus we can summarize:

Theorem 7.4: Let S be a Green semigroup, J be a $\mathcal{J}$-class of S. Then every 3-coextension $\varphi: T \to S$ of (S,J) such that T is a Green semigroup is, up to isomorphism, determined by a 3-coextension family. On the other hand, every 3-coextension family of (S,J) determines a 3-coextension $\varphi: T \to S$ of (S,J) such that T is a Green semigroup.∎

As an application we obtain:

Theorem 7.5: Let S be a Green semigroup with a regular $\mathcal{J}$-class J. Then (S,J) is 3-coextendable.

Proof: Let H be an $\mathcal{H}$-class of S containing an idempotent x with $H \subseteq J$. Choose an initialization of coordinates x, $\{a_R;\ R \in \mathcal{R}(J)\}$, $\{b_L;\ L \in \mathcal{L}(J)\}$. Let $\Psi(H) = (H,\Psi)$ be the Schutzenberger group. Define $G = \Psi \times C_2$ (C_2 is a cyclic group of order 2). For $b,c \in S-J$ with $bc \in J$, define $h'_{bc}(b,c) = (bc,0)$. For $b \in S-J$, $R \in \mathcal{R}(J)$ with $bR \subseteq J$ set

$\mu(b,R) = (g,0)$ where $g\in\Phi$ fulfils $ba_R x = a_{bR}g(x)$. For $b\in S-J$, $L\in\mathcal{L}(J)$ with $Lb\subseteq J$ set $\nu(b,L) = (g,0)$ where $g\in\Phi$ and it is fulfilled $xb_L b = g'(x)b_{Lb}$ for $g'\in\Phi^d$ satisfying $g(x) = g'(x)$. The validity of 1),2),7), and 8) is clear. Since S is a Green semigroup we obtain by a direct inspection 3),4),5), and 6). Theorem 7.4 concludes the proof.∎

For a non-regular $\mathcal{J}$-class of a Green semigroup S the situation is more complicated. In this case the definition of a 3-coextension family simplifies - Condition 5) holds evidently, Conditions 3),4), and 7) are simpler, but Condition 8) need not be fulfilled. The following examples show that in this case (S,J) may, but need not, be 3-coextendable.

Consider the semigroups S and T given by the following multiplication tables:

S	a	b	c	0
a	a	b	c	0
b	b	a	c	0
c	c	c	0	0
0	0	0	0	0

T	a	b	c	d	0
a	a	b	c	d	0
b	b	a	d	c	0
c	c	d	0	0	0
d	d	c	0	0	0
0	0	0	0	0	0

Define $\varphi:T\to S$, $\varphi(x) = x$ for $x\in\{a,b,c,0\}$, $\varphi(d) = c$. Clearly, φ is a surjective homomorphism (coextension). If $J = \{c\}$ then by a direct inspection we obtain that J is a non-regular $\mathcal{J}$-class of S and φ is a 3-coextension of (S,J). Set $J' = \{c,d\}$ then J' is a non-regular $\mathcal{J}$-class of T. We show that (T,J') is not 3-coextendable. Indeed, assume that $\psi:U\to T$ is a 3-coextension of (T,J') then $\psi^{-1}(J') = J''$ is a non-regular $\mathcal{J}$-class of U with $|J''|>|J'|$. Since J' is also an $\mathcal{H}$-class of T, by 8) there exist $|J''|$ pairs (x,R) in T such that $x\in T-J'$, R is an $\mathcal{R}$-class of T and $xR\subseteq J'$. Since J' contains exactly one $\mathcal{R}$-class, and 0 is a zero of T, we obtain a contradiction because the number of such pairs is equal to $|J'|<|J''|$.

<u>Remark:</u> Note that the both semigroups are commutative. Thus it is impossible to extend Theorem 7.5 for non-regular $\mathcal{J}$-classes of commutative semigroups. We conjecture that Theorem 7.5 cannot be extended to non-regular $\mathcal{J}$-classes of any "reasonable" class of semigroups.

VIII. Congruence of type 4

Analogously as in the foregoing chapter we restrict ourselves to the class of Green semigroups. Then by Corollary 2.5 we obtain: if τ is a 4-minimal congruence, then $\tau \cap H = \Delta$ and either $\tau \cap R = \Delta$ or $\tau \cap L = \Delta$. If $\tau \cap R = \Delta$ then we say that τ is a 4L-*minimal congruence*, if $\tau \cap L = \Delta$ then τ is a 4R-*minimal congruence*. We prove and formulate results only for 4L-minimal congruences, the analogous results hold for 4R-minimal congruences but we omit them.

Proposition 8.1: Let S be a Green semigroup. Then a congruence τ on S is a 4L-minimal congruence if and only if the following hold:

(i) $\tau \cap L = \tau \neq \Delta$, $\tau \cap R = \Delta$;

(ii) there exist a J-class J of S such that $Car(\tau) \subseteq J$ and τ_R^J is a minimal congruence on $R(S,J)$ (see Chapter III).

Moreover, if τ fulfils (i) and $Car(\tau) \subseteq J$ for some $J \in J(S)$ then for every non-singleton class A of τ and for every $a \in S^1$ either Aa is a singleton or Aa is a class of τ and g_a is one-to-one on A.

Proof: Assume that τ is a 4L-minimal congruence. Then clearly (i) holds. Further $Car(\tau) \subseteq J$ for some $J \in J(S)$. By Proposition 3.2 τ_R^J is a congruence on $R(S,J)$, we prove that it is minimal. Let σ be a congruence on $R(S,J)$ finer than τ_R^J. Define a relation ν on S: $(a,b) \in \nu$ if and only if either $a = b$ or $a,b \in J$, $(a,b) \in \tau$ and $(R(a),R(b)) \in \sigma$. We show that ν is a congruence on S - clearly ν is reflexive, symmetric, and transitive, thus ν is an equivalence. Let $x \in S^1$, $(a,b) \in \nu$. Then $(a,b) \in \tau$ because ν is finer than τ. Thus $(xa,xb) \in \tau$ and hence either $xa = xb$ or $xa,xb \in J$. In the former case we have $(xa,xb) \in \nu$, in the latter one we have $xR(a) = R(xa)$, $xR(b) = R(xb)$ and hence $(R(xa),R(xb)) \in \sigma$ because σ is a congruence on $R(S,J)$, whence $(xa,xb) \in \nu$. Further $(ax,bx) \in \tau$ and again either $ax = bx$ then $(ax,bx) \in \nu$ or $ax,bx \in J$, then $R(a) = R(ax)$ and $R(b) = R(bx)$ by the Green Theorem and therefore $(ax,bx) \in \nu$. Thus ν is a congruence. Since $\nu = \Delta$ if and only if $\sigma = \Delta$ and $\tau = \nu$ if and only if $\tau_R^J = \sigma$ we conclude that τ_R^J is a minimal congruence on $R(S,J)$ and (ii) is proved.

If τ fulfils (i) and $Car(\tau) \subseteq J$ for some $J \in J(S)$ then every non-singleton class A of τ is a subset of J and thus either Aa is a singleton or $Aa \subseteq J$ and by Proposition 2.4 we have that g_a is one-to-one on A. Since there exists $b \in S^1$ with Aab = A we have that Aa is a class of τ.

If τ fulfils (i) and (ii) then every congruence ν on S finer than τ fulfils (i) and $Car(\nu) \subseteq J$. Moreover ν_R^J is finer than τ_R^J. Then we easily obtain that $\tau = \nu$ if and only if $\tau_R^J = \nu_R^J$ and $\nu = \Delta$ if and only if $\nu_R^J = \Delta$. Thus (ii) implies that τ is minimal. By (i) we obtain that τ is a 4L-minimal congruence.∎

Theorem 8.2: Let S be a Green semigroup. Then a congruence τ on S is a 4L-minimal congruence if and only if the following conditions hold:

(i) $\tau \cap L = \tau \neq \Delta$, $\tau \cap R = \Delta$;

(ii) there exists a J-class J of S with $Car(\tau) \subseteq J$;

(iii) for every non-singleton class A of τ and for every pair $\{a,b\}$ of different elements from A we have that A is a weak component of the relation $\{(xay,xby);\ x,y \in S^1\}$;

(iv) if A,B are two non-singleton classes of τ with L(A) = = L(B) then there exist different elements a,b of A and $c,d \in S^1$ such that cad,cbd are different elements of B.

Proof; Assume that τ is a 4L-minimal congruence. Then by Proposition 8.1, (i) and (ii) hold. If τ is a minimal congruence then by Proposition 1.1 $\varepsilon(a,b) = \tau$ thus A is a class of $\varepsilon(a,b)$. But $\varepsilon(a,b)$ is the reflexive, symmetric, and transitive closure of the relation $\{(xay,xby);\ x,y \in S^1\}$ - hence we have (iii) and (iv).

Conversely, assume that τ satisfies (i), (ii), (iii), and (iv). If A is a non-singleton class of τ then for every pair $a,b \in A$, $a \neq b$, we have by (iii) that A is a class of $\varepsilon(a,b)$. If B is another non-singleton class of τ then by (iv) there exist different elements $c,d \in B$ with $(c,d) \in \varepsilon(a,b)$. Thus $\varepsilon(a,b) = \tau$ and by Proposition 1.1 τ is minimal. (i) and (ii) complete the proof that τ is a 4L-minimal congruence.∎

Note that τ is a minimal congruence if and only if τ fulfils (iii) and (iv).

Theorem 8.3: Let S be a Green semigroup, x,y be different elements of S. Then $\varepsilon(x,y)$ is a 4L-minimal congruence if and only if the following conditions hold:

(i) $(x,y)\in L$ and $(x,y)\notin H$;

(ii) for every $a\in S$, if $(ax,x)\notin J$ or $(ay,a)\notin J$ then $ax = ay$, if $(xa,x)\notin J$ or $(ya,y)\notin J$ then $xa = ya$;

(iii) the smallest congruence σ on $R(S,J(x))$ such that $(R(x),R(y))\in\sigma$ is minimal;

(iv) if a,b are different elements of a weak component of the relation $\{(cxd,cyd);\ c,d\in S^1\}$ then $(a,b)\in H$ implies $a = b$.

Remark: (iii) is equivalent with the following fact: if c,d are different elements of a weak component of the relation $\{(xay,xby);\ x,y\in S^1\}$ then a,b are elements of the same weak component of the relation $\{(xcy,xdy);\ x,y\in S^1\}$.

Proof of Theorem 8.3: If $\varepsilon(x,y)$ is a 4L-minimal congruence then by the definition (i) holds. Condition (ii) follows from the fact that $Car(\varepsilon(x,y))\subseteq J(x)$. Proposition 8.1 implies (iii) and $\varepsilon(x,y)\cap H = \Delta$ implies (iv).

Conversely, assume that x,y satisfy (i), (ii), (iii), and (iv). Then $\varepsilon(x,y)$ is the reflexive, symmetric, and transitive closure of the relation $R = \{(axb,ayb);\ a,b\in S^1\}$. By (i), $R\cap L \neq 0$. By (ii) and Proposition 2.4, we obtain that $Car(\varepsilon(x,y))\subseteq J(x)$. By (iv), we can conclude that $\varepsilon(x,y)\cap H = \Delta$. Since $(x,y)\in L$ and S is a Green semigroup we obtain that for $a,b\in S^1$ either $(axb,ayb)\in L$ or $(axb,x)\notin J$ or $(ayb,y)\notin J$. Thus $\varepsilon(x,y)$ is finer than L. By a direct inspection we obtain that $\varepsilon(x,y)_R^{J(x)}$ is the smallest congruence on $R(S,J(x))$ which merges $R(x)$ and $R(y)$. Thus (i) and (ii) of Proposition 8.1 hold whence $\varepsilon(x,y)$ is a 4L-minimal congruence.∎

Proposition 8.4: Let S be a Green semigroup. Let A be a subset of S such that

1) *there exists an* $\mathcal{L}$*-class* L *of* S *with* $A \subseteq L$;

2) *if* $a,b \in A$ *such that* $(a,b) \in \mathcal{H}$ *then* $a = b$;

3) *if* $a \in A$ *and* $c,d \in S^1$ *such that* $cad \in A$ *then* $cAd \subseteq A$;

4) *if* $a \in A$ *and* $c \in S^1$ *such that* $(ca,a) \notin \mathcal{J}$ *(or* $(ac,a) \notin \mathcal{J}$*) then* cA *(or* Ac*, respectively) is a singleton.*

Then the smallest congruence τ *on* S *such that* $(a,b) \in \tau$ *whenever* $a,b \in A$ *fulfils:*

A) τ *is finer than* $\mathcal{L}$;

B) $Car(\tau) \subseteq J(A)$;

C) A *is a class of* τ;

D) $\tau \cap \mathcal{H} = \Delta$.

Proof: Clearly, τ is the reflexive, symmetric, and transitive closure of the relation $R = \{(cad,cbd);\ c,d \in S^1,\ a,b \in A\}$. By 4) we have that $Car(\tau) \subseteq J(A)$. By 3) we conclude that A is a class of τ. By 1) if $a,b \in A$ then $(a,b) \in \mathcal{L}$ therefore for every $c,d \in S^1$ with $(cad,cbd) \in \mathcal{J}$ we obtain by 4) and by the fact that S is a Green semigroup that $(cad,cbd) \in \mathcal{L}$, thus τ is finer than $\mathcal{L}$. Assume that $x,y \in S$ with $(x,y) \in \tau \cap \mathcal{H}$. Then either $x \in J(A)$ and by B) $x = y$ or there exist $c,d \in S^1$ and $a \in A$ such that $cxd = a$. In the second case since S is a Green semigroup we have by Proposition 2.4 that $(cxd,cyd) \in \mathcal{H}$, thus $cyd = cxd$ by 2) and C). Proposition 2.4 implies $x = y$. Thus $\tau \cap \mathcal{H} = \Delta$.■

Corollary 8.5: If S *is a Green semigroup and* τ *is a 4L-minimal congruence then* S/τ *is also a Green semigroup.*

Proof follows from Theorem 2.2 and Proposition 8.1.■

We shall investigate 4L-coextensions, i.e. coextensions $\varphi: T \to S$ such that $Ker\ \varphi$ is a 4L-minimal congruence. The analogous results hold for 4R-coextensions and therefore we omit them.

Let $\varphi: T \to S$ be a 4L-coextension of (S,J) such that T and S are Green semigroups, J is a $\mathcal{J}$-class of S. Denote by $\varphi^{-1}(J) = J'$, then J' is a $\mathcal{J}$-class of T. Choose an initialization of coordinates y, $\{c_R;\ R \in \mathcal{R}(J')\}$, $\{d_L;\ L \in \mathcal{L}(J')\}$ of J' in T such that for $R,R' \in \mathcal{R}(J)$, $(R,R') \in Ker(\varphi_R^{J'})$ we have $\varphi(c_R) = \varphi(c_{R'})$. Then $x = \varphi(y)$, $\{a_R = \varphi(c_{R'});\ R \in \mathcal{R}(S),\ R' \in \mathcal{R}(T),\ R \subseteq J,\ \varphi(R') = R\}$, $\{b_L = \varphi(d_{\varphi^{-1}(L)});\ L \in \mathcal{L}(J)\}$ is an initialization of coordinates of J in

S. Since Ker $\varphi\cap R = \blacktriangle$ we can assume that the Schutzenberger group $\Psi(H(x)) = (H(x),\Psi)$ of $H(x)$ in S and the Schutzenberger group of $H(y)$ in T are isomorphic and that φ induces a bijection from $\{L; L\in L(J')\}$ onto $\{L; L\in L(J)\}$. Thus we can assume that elements of J are triples of $\Psi\times R(J)\times L(J)$ and elements of J' are triples of $\Psi\times R(J')\times L(J)$ where $\varphi(g,R,L) = (g,\varphi(R),L)$. By Proposition 3.3 there exist mappings $\mu:\{(b,R);\ b\in S-J,\ R\in R(J)$ with $bR\in R(J)\}\to\Psi$ such that the domain-range restriction of f_b, $b\in S-J$, onto R and bR is $f_{\mu(b,R),R,bR}$ and $\nu:\{(b,L);\ b\in S-J,\ L\in L(J)$ with $Lb\in L(J)\}\to\Psi$ such that the domain-range restriction of g_b, $b\in S-J$, onto L and Lb is $g_{\nu(b,L),L,Lb}$. In Chapter III, $R(S,J)$ and $L(S,J)$ are defined. Set $X = R(J)\cup\{0\}$ and for every $b\in S$ such that $bJ\cap J \neq 0$ define a mapping $\xi(b)$ of X into itself, for $R\in R(J)$, $\xi(b)(R) = bR$ if $bR\in R(J)$, $\xi(b) = 0$ otherwise and $\xi(b)(0) = 0$. Analogously, if $R(J')\cup\{0\} = Z$ then for $b\in T$ with $bJ'\cap J' \neq 0$ define $\beta(b)$ such that for $R\in R(J')$, $\beta(b)(R) = bR$ if $bR\in R(J')$, $\beta(b)(R) = 0$ otherwise, and $\beta(b)(0) = 0$. Finally, set $Y = L(J)\cup\{0\}$ and define for $b\in S$ with $Jb\cup J \neq 0$ a mapping $\omega(b)$ such that for $L\in L(J)$ we have $\omega(b)(L) = Lb$ if $Lb\in L(J)$, $\omega(b)(L) = 0$ otherwise, and $\omega(b)(0) = 0$. By Propositions 3.1 and 3.3 we have for $(a,R,L)\in J$, $b\in S-J$ either $b(a,R,L)\notin J$ or $b(a,R,L) =$ $= (\mu(b,R)a,\xi(b)(R),L)$, and either $(a,R,L)b\notin J$ or $(a,R,L)b =$ $= (a\nu(b,L),R,\omega(b)(L))$ and for $(a,R,L)\in J'$, $b\in T-J'$ either $b(a,R,L)\notin J'$ or $b(a,R,L) = (\mu(b,R)a,\beta(b)(R),L)$ and either $(a,R,L)b\notin J$ or $(a,R,L)b = (a\nu(b,L),R,\omega(b)(L))$ (we use that φ is a 4L-coextension), moreover for $b\in T-J$ we have $\varphi(\beta(b)(R)) =$ $= \xi(\varphi(b))(\varphi(R))$ for every $R\in R(J')$.

Thus if Λ is a special coextension family of (S,J) such that $\mathrm{Coex}(\Lambda)$ and φ are isomorphic we can assume that there exist a 0-transitive semigroup (X,Φ), a congruence τ on (X,Φ), and an isomorphism $F:(X,\Phi)/\tau\to R(S,J)$ such that $T' = \Psi\times X\times L(J)$, $k'(a,x,L) = (a,F(\tau x),L)$. Further there exist a family $\{h'_a:V_{J,a}\to X;\ a\in J\}$ of mappings and a mapping $\theta:\{b;\ b\in S-J$ with $bJ\cap J \neq 0\}\to\Phi$ such that $h_a(b,c) = (\Psi(a),h'_a(b,c),L(a))$ for every $a\in J$, $(b,c)\in V_{J,a}$ and for $b\in S-J$, $R\in R(J)$ with $bR\in R(J)$ we have $f_b^R(a,x,L) =$ $= (\mu(b,R)a,\theta(b)(x),L)$ where $F(\tau x) = R$. The other translations fulfil: if $F(\tau x) = R$ and $(a,R,L)R'\subseteq J$ for $R'\in R(J)$ then $f_{(a,x,L)}^{R'}(b,y,L') = (ab,x,L')$ for every $b\in\Psi$ and every $y\in X$ with $F(\tau y) = R'$; if $b\in S-J$, $L\in L(J)$ with $Lb\in L(J)$ then for every $a\in\Psi$,

$x \in X$ define $g^L_b(a,x,L) = (a\nu(b,L),x,\omega(b)(L))$, if $(b,R',L') \in J$ with $L(b,R',L') \subsetneq J$ then for every $y \in X$ with $F(\tau y) = R'$ and every $a \in \Psi$, $x \in X$ define $g^L_{(b,y,L')}(a,x,L) = (ba,y,L)$.

Thus $\mathrm{Coex}(\Lambda)$ is a 4L-coextension if and only if the following hold:

1) (X,Φ) is a 0-transitive semigroup, τ is a minimal congruence on (X,Φ) such that $\{0\}$ is a class of τ, and $F:(X,\Phi)/\tau \to \mathcal{R}(S,J)$ is an isomorphism;
2) Φ-{constant mapping to $0\} = \mathrm{Im}\ \theta \cup \{f:X \to X;\ \mathrm{Im}\ f = \{x,0\}$ for some $x \in X-\{0\}$, $f(0) = 0$ and there exist $a \in \Psi$, $L \in \mathcal{L}(J)$ such that for every $y \in X-\{0\}$ if $(a,F(\tau x),L)F(\tau y) = F(\tau x)$ then $f(y) = x$ else $f(y) = 0\}$;
3) for every $a \in J$, $\sigma_{J,a}$ is finer than h'_a and $F(\tau(\mathrm{Im}\ h'_a)) = \{R(a)\}$;
4) for $a,b,c \in S-J$ with $abc, bc \in J$ we have
   if $ab \in J$ then $\theta(a)(h'_{bc}(b,c)) = h'_{ab}(a,b)$,
   if $ab \notin J$ then $\theta(a)(h'_{bc}(b,c)) = h'_{abc}(ab,c)$;
5) for every $b,c \in S-J$ with $bJ \cap J \neq 0$ we have
   if $bc \in S-J$ then either $\theta(bc) = \theta(b) \circ \theta(c)$, or $bcJ \cap J = 0$ and $\theta(b) \circ \theta(c)$ is a constant to 0,
   if $bc \in J$ then either $\mathrm{Im}\ (\theta(b) \circ \theta(c)) = \{0, h'_{bc}(b,c)\}$ and for $y \in X-\{0\}$, $\theta(b) \circ \theta(c)(y) = h'_{bc}(b,c)$ if and only if $bc(F(\tau y)) \subsetneq J$ or $\theta(b) \circ \theta(c)$ is constant to 0 and $bcJ \cap J = 0$.

Indeed, Proposition 8.1 implies 1), Condition 2) follows from the fact that $\mathcal{R}(T,J')$ is isomorphic to (X,Φ). Condition 3) follows from g') and c'), 4) is implied by h'), and 5) is equivalent to i').

On the other hand, if we define Λ from $F,\{h'_a;\ a \in J\}$ and θ fulfilling 1)-5) as above then a'),b'),d'), and e') are evidently fulfilled. Since the supplement condition of c') follows from 3) we obtain c'). From the properties of $M(\Psi,X,\mathcal{L}(J))$ we clearly obtain f") and j') and moreover from the definition of f^R_a and g^R_a we have f"') because J is a $\mathcal{J}$-class of S. Further g') follows from 3). Finally, if we use the concrete definition of Λ we obtain by a direct inspection that h') is a consequence of 4). Thus we say that a transformation semigroup (X,Φ), a congruence τ, mappings F,θ, and a family $\{h'_a:V_{J,a} \to X;\ a \in J\}$ fulfilling 1)-5) is a 4L-*coextension family*. Then we summarize:

<u>Theorem 8.6: Let</u> S <u>be a Green semigroup,</u> J <u>be a</u> *J*<u>-class of</u> S<u>. Then every 4L-coextension</u> $\varphi:T\to S$ <u>of</u> (S,J) <u>such that</u> T <u>is a Green semigroup is determined, up to isomorphism, by a 4L-coextension family. Further, every 4L-coextension family determines a 4L-coextension</u> $\varphi:T\to S$ <u>such that</u> T <u>is a Green semigroup.</u>∎

If we apply this result we obtain:

<u>Theorem 8.7: Let</u> S <u>be a Green semigroup with a regular</u> *J*<u>-class</u> J <u>of</u> S<u>. Then</u> (S,J) <u>is 4L-coextendable.</u>

Proof: Let $\mathcal{R}(S,J) = (Y,\Psi)$. Choose an element $u\notin Y$ and set $X = Y\cup\{u\}$. Choose $x\in Y$ and define $\Phi = \{f:X\to X;$ either there exists $g\in\Psi$ such that $f(y) = g(y)$ for every $y\in Y$, $f(u) = g(x)$ or for a $g\in\Psi$ such that $\mathrm{Im}\ g = \{x,0\}$ we have $f(y) = u$ if $g(y) = x$, $f(y) = 0$ if $g(y) = 0$, $f(u) = f(x)\}$. Let τ be the smallest equivalence of X with $(u,x)\in\tau$ and define $F(\tau z) = z$ if $\{z\}$ is a singleton class of τ, $F(\tau x) = x$. For $a\in J$, $(b,c)\in V_{J,a}$ define $h_a'(b,c) = R(bc)$. Finally, define a mapping $\theta(b)\in\Phi$ for $b\in S-J$ with $bJ\cap J \neq 0$ such that for $R\in\mathcal{R}(J)$, $\theta(b)(R) = bR$ if $bR\subseteq J$, $\theta(b)(R) = 0$ if $bR\cap J = 0$, $\theta(b)(u) = \theta(b)(x)$. Since (X,Φ) is a 0-transitive transformation semigroup, since τ is a minimal congruence on (X,Φ), and since $F:(X,\Phi)/\tau\to\mathcal{R}(S,J)$ is an isomorphism we have 1). 2) is a consequence of the definition of $\mathcal{R}(S,J)$, 3),4), and 5) follow from the definition of h_a' and θ, and from the fact that S is a semigroup. Thus Theorem 8.6 completes the proof.∎

Analogously as for 3-coextensions, the situation for non-regular *J*-classes is more complicated - see the following examples: Consider the semigroups given by the multiplication tables:

S	a	b	c	0
a	a	a	c	0
b	b	b	c	0
c	0	0	0	0
0	0	0	0	0

T	a	b	d	e	0
a	a	a	d	d	0
b	b	b	e	e	0
d	0	0	0	0	0
e	0	0	0	0	0
0	0	0	0	0	0

Define $k:T \to S$, $k(x) = x$ for $x \in \{a,b,0\}$, $k(d) = k(e) = c$. By a direct inspection we obtain that k is a 4L-coextension of (S,J) for $J = \{c\}$. J is a non-regular $\mathcal{J}$-class of S. Set $J' = \{d,e\}$, then J' is a non-regular $\mathcal{J}$-class of T. We show that (T,J') is not 4L-coextendable. Assume contrary, let $h:U \to T$ be a 4L-coextension of (T,J'). Then $|h^{-1}(J')| > |J'| = 2$ thus by 2), $|\{b;\ b \in T-J',\ bJ' \cap J' \neq 0\}| \geq |h^{-1}(J')|$ but $|T-TJ'T| = 2$ - a contradiction. Thus (T,J') not 4L-coextendable.

IX. Congruence of type 5

We describe 5-minimal congruences and 5-coextensions in the class of Green semigroups.

Proposition 9.1: Let S be a Green semigroup. Then a congruence τ is a 5-minimal congruence if and only if the following conditions hold:

(i) $\tau \neq \Delta$, $\tau \cap \mathcal{R} = \tau \cap \mathcal{L} = \Delta$ and there exists a $\mathcal{J}$-class J of S with $\mathrm{Car}(\tau) \subseteq J$;

(ii) there exist a non-singleton class A of τ and an element $a \in A$ such that for every $b \in A-\{a\}$, A is a weak component of the relation $\{(cad,cbd);\ c,d \in S^1\}$.

If τ is a 5-minimal congruence then the following hold:

A) for every non-singleton class A of τ and for every $a \in S^1$ either aA (or Aa) is a singleton class or $(ax,ay) \notin \mathcal{R}$ (or $(xa,ya) \notin \mathcal{L}$, respectively) for every distinct elements $x,y \in A$;

B) for every non-singleton class A of τ and for every $a \in S^1$, aA and Aa are classes of τ;

C) if $\mathrm{Car}(\tau) \subseteq J \in \mathcal{J}(S)$ then J is non-regular;

D) all non-singleton classes of τ have the same size;

E) if $a,b,c \in S$ such that $(a,c) \in \tau$, $(a,b) \in \mathcal{L}$, $(b,c) \in \mathcal{R}$ then $(\tau a, \tau b) \in \mathcal{H}$ in S/τ;

F) congruences τ_R^J on $\mathcal{R}(S,J)$ and τ_L^J on $\mathcal{L}(S,J)$ have exactly one singleton class consisting of 0, the other classes of τ_R^J and τ_L^J have the same cardinality as the non-singleton class of τ.

Proof: Assume that τ is a 5-minimal congruence. By Corollary 2.5 we immediately obtain (i). If A is a non-singleton class of τ, $a,b\in A$ are different elements of A then by Proposition 1.1, $\tau = \varepsilon(a,b)$. But $\varepsilon(a,b)$ is the reflexive, symmetric, and transitive closure of the relation $\{(cad,cbd);\ c,d\in S^1\}$. Hence we immediately obtain (ii).

Assume that τ fulfils (i). If x,y are different elements with $(x,y)\in\tau$ then for every $a\in S^1$ either $ax = ay$ or $(ax,ay)\notin R$. Moreover, if $(x,y),(y,z)\in\tau$ for $x \neq y \neq z$ and $ax \neq ay = az$ then $(y,z)\notin L$ implies by Proposition 2.4 that $(ay,y)\notin J$ thus $(ax,ay)\in\tau$ is a contradiction with the fact that $Car(\tau)\subseteq J$ ($J = J(y)$). Hence A) follows. B) is a consequence of A) and (i). If $(x,y)\in\tau$, $x \neq y$ then either $(x^2,x)\notin J$ or $(xy,y)\notin J$ or $(x^2,xy)\in R$, thus A) implies C). E) is obvious. D) follows from B) and (i). F) is a consequence of D).

Finally, assume that τ fulfils (i) and (ii). Let $\nu \neq \Delta$ be a congruence finer than τ. Then ν also fulfils (i). Let $(x,y)\in\nu$, $x \neq y$ and let A be a non-singleton class of τ, $a\in A$ fulfilling (ii). Since A is non-singleton we have by (i) that $(x,a)\in J$ thus there exist $c,d\in S^1$ with $cxd = a$. Then by A) and B), $(cx,cy)\notin R$ and hence $cxd \neq cyd$. Since ν is finer than τ we have $cyd\in A$ and by (ii), A is a class of ν. By B) we easily obtain that $\nu = \tau$. Thus τ is minimal. By (i), τ is a 5-minimal congruence.∎

<u>Theorem 9.2: Let S be a Green semigroup and let x,y be different elements of S. Then $\varepsilon(x,y)$ is a 5-minimal congruence if and only if the following hold:</u>

<u>(i)</u> $(x,y)\in J$ <u>but</u> $(x,y)\notin R$, $(x,y)\notin L$;

<u>(ii)</u> <u>for every</u> $a\in S^1$ <u>if</u> $(ax,x)\notin J$ <u>or</u> $(ay,y)\notin J$ <u>then</u> $ax = ay$, <u>if</u> $(xa,x)\notin J$ <u>or</u> $(ya,y)\notin J$ <u>then</u> $xa = ya$;

<u>(iii)</u> <u>for every</u> $a\in S^1$ <u>if</u> $(ax,x)\in J$ <u>or</u> $(ay,y)\in J$ <u>then</u> $(ax,ay)\notin R$ <u>and there exists</u> $b\in S^1$ <u>with</u> $bax = x$, $bay = y$, <u>if</u> $(xa,x)\in J$ <u>or</u> $(ya,y)\in J$ <u>then</u> $(xa,ya)\notin L$ <u>and there exists</u> $b\in S^1$ <u>with</u> $xab = x$, $yab = y$;

<u>(iv)</u> <u>if</u> a,b <u>are elements of the same weak component of the relation</u> $\{(cxd,cyd);\ c,d\in S^1\}$ <u>then</u> $a = b$ <u>whenever</u> $(a,b)\in L$ <u>or</u> $(a,b)\in R$;

(v) *if* $a \neq x$ *is an element of a weak component containing* x *of the relation* $\{(cxd,cyd);\ c,d\in S^1\}$ *then* x *and* y *belong to the same component of the relation* $\{(cxd,cad);\ c,d\in S^1\}$.

Proof: Assume that $\varepsilon(x,y)$ is a 5-minimal congruence. Then by (i) of Proposition 9.1 and by Proposition 2.4 we obtain (i) and (ii). (iii) follows from B) and (i) of Proposition 9.1 because $(bay,y)\in J$ implies $(bay,y)\in L$. Since $\varepsilon(x,y)$ is the reflexive, symmetric, and transitive closure of the relation $\{(cxd,cyd);\ c,d\in S^1\}$ and since $\varepsilon(x,y)\cap R = \varepsilon(x,y)\cap L = \Delta$ we get (iv), and $\varepsilon(x,a) = \varepsilon(x,y)$ concludes (v).

Conversely, if (i), (ii), (iii), (iv), and (v) hold then $\varepsilon(x,y) \neq \Delta$, by (ii) and Proposition 2.4, $\mathrm{Car}(\varepsilon(x,y))\subseteq J(x)$, by (iv), $\varepsilon(x,y)\cap L = \varepsilon(x,y)\cap R = \Delta$. Thus $\varepsilon(x,y)$ fulfils (i) of Proposition 9.1, (v) implies (ii) of Proposition 9.1 and Proposition 9.1 completes the proof that $\varepsilon(x,y)$ is a 5-minimal congruence.∎

Proposition 9.3: *Let* τ *be a congruence on a Green semigroup* S *such that* $\tau \neq \Delta$, $\tau\cap R = \tau\cap L = \Delta$ *and* $\mathrm{Car}(\tau)\subseteq J$ *for some* $J\in J(S)$. *Let* J' *be the* J*-class of* S/τ *corresponding to* J. *Denote by* $\Psi(H) = (H,\Psi)$ *(or* $\Psi(H') = (H',\Phi)$*) the Schutzenberger group of* $H\in H(S)$ *(or* $H'\in H(S/\tau)$*) with* $H\subseteq J$ *(or* $H'\subseteq J'$*, respectively). Then the group* Ψ *(with the operation of composition) is a subgroup of* Φ *(with the operation composition).*
Moreover, τ *is a 5-minimal congruence if and only if* Ψ *is a maximal subgroup of* Φ *(i.e. if* Λ *is a subgroup of* Φ *with* $\Psi\subset\Lambda$ *then* $\Lambda = \Phi$*).*

Proof: Let A be a non-singleton class of τ. Choose an R-class $R\subseteq J$ with $A\cap R \neq 0$. Define $\{a\} = A\cap R$ and $R' = \{x;\ x\in R,\ L(x)\cap A \neq 0\}$. Clearly, $R'\supseteq H(a)$. We show that $\tau R'$ is an H-class of S/τ and that $\tau x \neq \tau y$ for $x \neq y$, $x,y\in R$. The second statement immediately follows from $\tau\cap R = \Delta$. For every $x\in R'$ there exists $u\in A$ ($\{u\} = L(x)\cap A$) with $(x,u)\in L$ - thus $(\tau x,\tau a)\in L$ - in S/τ. Since $R'\subseteq R$ we conclude that $\tau R'$ is a subset of an H-class of S/τ. Assume that $(\tau x,\tau a)\in H$ in S/τ. Then there exist $c,d\in S^1$,

$y,z\in A$ with $xc = y$, $zd = x$. Hence $(y,z)\in R$ and $R\cap\tau = \Delta$ implies $y = z$. Then $(ad,x)\in\tau$ and $ad\in R$. Further there exist $c',d'\in S^1$, $u,v\in A$ with $c'ad = u$, $d'v = ad$. Hence $(u,v)\in L$ and therefore $u = v$. Since $(ad,u)\in L$ we conclude that $ad\in R'$.

Set $G = \{$the domain-range restriction of g_y on R'; $y\in S$, $R'y\subseteq R'\}$, $G' = \{g;\ g\in G,\ g(H(a))\subseteq H(a)\}$. Then G (with the operation composition) is isomorphic to Φ, $G' = \Psi$. Clearly, G' is a subgroup of G. Obviously, each mapping $g\in G$ maps H-classes of S onto H-classes of S. Assume that G'' is a subgroup of G such that $G'\subset G''\subset G$. Set $R'' = \{x;\ x\in R',\ g(a) = x$ for some $g\in G''\}$. Since $H(a)\subset R''\subset R'$ we conclude that R' is saturated by H-classes. Set $B = \{x;\ x\in A,\ L(x)\cap R'' \neq 0\}$, then $H(a)\subset R''$ implies that B is not a singleton and $R''\subset R'$ implies that $B\subset A$. For every $x\in S^1$ either the restriction of g_x on R' belongs to G'' then $Bx\subseteq\bigcup\{L(b);\ b\in B\}$ or the restriction of g_x on R' does not belong to G'' then $Bx\cap(\bigcup\{L(b);\ b\in B\}) = 0$. Since $|A\cap L|\leq 1$ for every $L\in L(S)$, we have that for every $x,y\in S^1$, $yBx\cap B \neq 0$ implies $yBx\subseteq B$ - indeed, it suffices to use that $B\subseteq A$ and that $yAx\cap A \neq 0$ implies $yAx\subseteq A$ for every $y,x\in S^1$. Thus, if ν is the smallest congruence on S such that $(y,x)\in\nu$ for every $y,x\in B$ we have that B is a class of ν. Thus $\nu \neq \tau$ but clearly ν is finer than τ. Thus τ is not minimal. If G' is a maximal subgroup of G and ν is a finer congruence than τ then $\nu\cap R = \nu\cap L = \Delta$ and $Car(\nu)\subseteq J$. If $\nu \neq \Delta$ then define $R'' = \{x;\ x\in R,\ L(x)\cap\nu a \neq 0\}$. We have $H(a)\subset R''\subseteq R'$ and thus $G'' = \{g;\ g\in G,\ g(R'')\subseteq R''\}$ is a subgroup of G containing G' - moreover, $H(a)\subset R''$ implies $G'\subset G''$. By the maximality of G' we have $G = G''$, thus $R'' = R'$ whence $\nu = \tau$. We proved that τ is a 5-minimal congruence.∎

By Theorem 2.2 and Proposition 9.1 we have:

<u>Corollary 9.4: Let</u> S <u>be a Green semigroup with a 5-minimal congruence</u> τ<u>. Then</u> S/τ <u>is also a Green semigroup.</u>∎

We shall now investigate 5-coextensions of Green semigroups.

Let $\varphi:T\to S$ be a 5-coextension of (S,J) such that T and S are Green semigroups and J is a non-regular J-class of S. Denote $\varphi^{-1}(J) = J'$, then J' is a non-regular J-class of T. Choose an initialization of coordinates y, $\{c_R;\ R\in R(J')\}$, $\{d_L;\ L\in L(J')\}$ of

J' in T such that $(z,c_Rd_L)\in H$, and $\varphi(y) = \varphi(z)$ imply $z = c_Rd_L$. Set $x = \varphi(y)$. For every $R\in R(J)$ choose $P_R\in R(J')$ with $\varphi(P_R)\subseteq R$; for every $L\in L(J)$ choose $P_L\in L(J')$ with $\varphi(P_L)\subseteq L$ such that $P_{R(x)} = R(y)$, $P_{L(x)} = L(y)$. Then x, $\{a_R = \varphi(cp_R);\ R\in R(J)\}$, $\{b_L = \varphi(dp_L);\ L\in L(J)\}$ is an initialization of coordinates of J in S. Denote by $\Psi(H(y)) = (H(y),F)$, $\Psi(H(x)) = (H(x),G)$ the Schutzenberger groups. Set $R^0 = \{z; z\in R(y),\ x\in\varphi(L(z))\}$, $L^0 = \{z;\ z\in L(y),\ x\in\varphi(R(z))\}$. Define $\Psi^1 = (R^0,G^1)$, $\Psi^2 = (L^0,G^2)$ where $G^1 = \{g:R^0\to R^0;\ \exists a\in T,\ g$ is the domain-range restriction of g_a to $R^0\}$, $G_2 = \{f:R^0\to R^0;\ \exists a\in T$, f is the domain-range restriction of f_a to $R^0\}$. Then the restriction of φ on R^0 induces an isomorphism from Ψ^1 onto $\Psi(H(x))$ and the restriction of φ on L^0 induces an isomorphism from Ψ^2 onto $\Psi^d(H(x))$. Denote $F' = \{g\in G;\ \exists z\in H(y),\ g(x) = \varphi(z)\}$, then F' is the subgroup of G isomorphic to F (F,G,F' are groups where the operation is the composition). Every orbit of F' is an image of an H-class in R^0. Analogously, if we set $F'' = \{f\in G^d;\ \exists z\in H(y),\ f(x) = \varphi(z)\}$ then F'' is the subgroup of G^d and the orbits of F'' correspond to the H-classes in L^0. Moreover, F' and F'' are dually isomorphic. Further by Proposition 9.1, every inner right translation g_a is either one-to-one on R^0 or $g_a(R^0)\cap J' = 0$ and every inner left translation f_a is either one-to-one on L^0 or $f_a(L^0)\cap J' = 0$. Thus every R-class R of T with $R\subseteq J'$ corresponds to the order pair $(Q(R),A)$ where $Q(R)$ is an R-class of S with $\varphi(R)\subseteq Q(R)$ and A is a left coset of G^2 by F^d such that $G(\varphi(c_Ry))\in A$. Analogously, L-classes L of T with $L\subseteq J'$ correspond to the order pairs $(Q(L),A)$ where $Q(L)$ is an L-class of S with $\varphi(L)\subseteq Q(L)$ and A is a rigth coset of G^1 by F such that $G(\varphi(yd_L))\in A$. The both correspondences are bijections. Moreover, for every left coset A of G^2 by F^d there exists a right coset $\theta(A)$ of G^1 by f such that $x\in\varphi((R(x),A)\cap(L(x),\theta(A))$. By Proposition 9.1, θ is a bijection. Then $\varphi(a,(R,A),(L,B)) = (bab',R,L)$ where $b = \varphi(c_{(R(y),A)}y)$, $b' = \varphi(yd_{(L(y),B)})$. Hence if Λ is a special coextension family of (S,J) (the cross decomposition of J is induced by R- and L-classes) such that $\mathrm{Coex}(\Lambda)$ and φ are isomorphic then we can assume there exists a maximal subgroup F of G (denote by $\mathrm{Cos}(F/G)$ the set of all left cosets of F in G, $\mathrm{Cos}(G\backslash F)$ the set of all right cosets of F in G) such that $\Lambda_0 = T' = F\times(R(J)\times\mathrm{Cos}(F/G))\times(L(J)\times\mathrm{Cos}(G\backslash F))$. Furthermore, for every $A\in F/G$ there exists $c_A\in A$ ($c_A = G(\varphi(c_{(R(y),A)}y))$) and for every

$B \in G \backslash F$ there exists $d_B \in B$ ($d_B = G(\varphi(yd_{(L(y),B)})))$ such that $c_F = d_F$ is the unity of G and $k'(a,(R,A),(L,B)) = (c_A a d_B, R, L)$ for every $(a,(R,A),(L,B)) \in T'$ where $\Lambda_1 = \{k'\}$. Further there exist a bijection θ from $Cos(F/G)$ onto $Cos(G \backslash F)$ and a family of mappings $\{h'_a : V_{J,a} \to Cos(F/G);\ a \in J\}$ such that $\Lambda_2 = \{h_a : V_{J,a} \to T';\ a \in J\}$, where $h_a(b,c) = (c^{-1}_{h'_a(b,c)} G(a) d^{-1}_{\theta(h'_a(b,c))}, (R(a), h'(b_a c)), (L(a), \theta(h'(b_a c))))$ for every $a \in J$, $(b,c) \in V_{J,a}$. For $b \in S-J$ with $bJ \cap J \neq 0$ (or $Jb \cap J \neq 0$) denote by $\mu(b,R)$ (or $\nu(b,L)$) an element of G such that $b(a,R,L) = (\mu(b,R)a, bR, L)$ (or $(a,R,L)b = (a\nu(b,L), R, Lb)$, respectively) where $R \in \mathcal{R}(J)$, $L \in \mathcal{L}(J)$, $a \in G$. Then

$$f^R_b(a,(R,A),(L,B)) = (c^{-1}_{\mu(b,R)A}\mu(b,R)c_A G(a), (bR, \mu(b,R)A), (L,B))$$

$$g^L_b(a,(R,A),(L,B)) = (G(a)d_B \nu(b,L) d^{-1}_{B\nu(b,L)}, (R,A), (Lb, B\nu(b,L)))$$

for every $(a,(R,A),(L,B)) \in T'$. Then Λ is a special coextension family if the following hold:

1) F is a maximal subgroup of G;
2) for every $a \in J$, $\sigma_{J,a}$ is finer than $\mathrm{Ker}\ h_a$;
3) for every $A \in Cos(F/G)$, we have $c_A \in A$, for every we have $B \in Cos(G \backslash F)$, $d_B \in B$, and $c_F = d_F$ is the unity of G;
4) for $a,b,c \in S-J$ such that $abc \in J$ we have

   if $ab \in J$, $bc \notin J$ then

   $$c^{-1}_{h'_{ab}(a,b)} G(ab)\nu(c,L(ab)) d^{-1}_{\theta(h'_{ab}(a,b))\nu(c,L(ab))} = c^{-1}_{h'_{abc}(a,bc)} G(abc) d^{-1}_{\theta(h'_{abc}(a,bc))},$$

   if $ab \notin J$, $bc \in J$ then

   $$c^{-1}_{h'_{abc}(a,bc)} G(abc) d^{-1}_{\theta(h'_{abc}(a,bc))} = c^{-1}_{\mu(a,R(bc))h'_{bc}(b,c)} \mu(a,R(bc)) G(bc) d^{-1}{}_{\theta(h'_{bc}(b,c))},$$

   if $ab, bc \in J$ then

   $$c^{-1}_{h'_{ab}(a,b)} G(ab)\nu(c,L(ab)) d^{-1}_{\theta(h'_{ab}(a,b))\nu(c,L(ab))} = c^{-1}_{\mu(a,R(bc))h'_{bc}(b,c)} \mu(a,R(bc)) G(bc) d^{-1}{}_{\theta(h'_{bc}(b,c))}.$$

Indeed, by the properties of Λ we immediately obtain that a'), b'), c'), and d') hold. e') and f') evidently hold because J is non-regular; i') follows from the definition of Λ_3 and Λ_4 because S is a semigroup and G is a group, j') is a consequence of Chapter III. Further g') is equivalent with 2), and by a direct inspection h') is equivalent with 3). Since $Car(\mathrm{Ker}(Coex(\Lambda))) \subseteq T' = J'$ which is a $\mathcal{J}$-class and since $\Delta \neq \mathrm{Ker}(Coex(\Lambda))$, $\mathrm{Ker}(Coex(\Lambda)) \cap \mathcal{L} = \mathrm{Ker}(Coex(\Lambda)) \cap \mathcal{R} = \Delta$ we obtain by Proposition 9.3 that 1) is equivalent with the fact that $Coex(\Lambda)$ is a 5-coextension. A subgroup

F of G, families $\{c_A;\ A\in Cos(F/G)\}$, $\{d_B;\ B\in Cos(G\backslash F)\}$, and a family $\{h'_a: V_{J,a}\rightarrow Cos(F/G);\ a\in J\}$ of mappings fulfilling 1)-4) is called a 5-*coextension family*. Thus we can summarize:

Theorem 9.5: Let S be a Green semigroup, J be a non-regular *J*-class of S. Then every 5-coextension $\varphi:T\rightarrow S$ of (S,J) such that T is a Green semigroup is determined, up to isomorphism, by a 5-coextension family. Furthermore, every 5-coextension family determines a 5-coextension $\varphi:T\rightarrow S$ such that T is a Green semigroup.■

It is very complicated to determine when (S,J) is 5-coextendable, because condition 3) is very difficult to verify. We end this chapter by an example of a 5-minimal congruence. Consider the semigroup

S	a	b	c	d	e	f	0
a	a	b	c	d	e	f	0
b	b	a	e	f	c	d	0
c	c	d	0	0	0	0	0
d	d	c	0	0	0	0	0
e	e	f	0	0	0	0	0
f	f	e	0	0	0	0	0
0	0	0	0	0	0	0	0

Let τ be the smallest congruence with $(c,f),(d,e)\in\tau$. By a direct inspection τ is a 5-minimal congruence on S.

X. Conclusions

We summarize our results:

Corollary 10.1: If S is a Green semigroup and τ is a minimal congruence then S/τ is also a Green semigroup.■

On the other hand, this statement cannot be strengthened for an arbitrary congruence as was shown in Chapter II. In the foregoing chapters we describe minimal coextensions $\varphi:T\rightarrow S$ such that both T and S are Green semigroups. The following example shows a 5-

coextension $\varphi:T\to S$ such that T is not a Green semigroup but S is: Let T be a bicyclic semigroup, S be a semigroup of all integers with the operation addition. If a,b are generators of T with $ab = 1$ then define $h(b^m a^n) = m-n$. Then h is a surjective homomorphism from T onto S - in fact it is a minimal coextension - see [3]. By a direct inspection we show that h is a 5-coextension.

For the study of congruence lattices of semigroups, it is important to know the relation "covering" - see [10]. Clearly it holds:

Proposition 10.2: Let S be a semigroup, τ,σ be congruences of S. If τ is finer than σ then σ covers τ if and only if the corresponding congruence of σ in S/τ is minimal.∎

This fact can be used for a description of the relation "covering" in quasiperiodical (or finite, commutative, periodical etc.) semigroups because any quotient of a quasiperiodical semigroup is again quasiperiodical (and thus a Green semigroup). Since by Proposition 2.10, a quotient of a Green semigroup need not be a Green semigroup we cannot apply this statement for a description of the relation "covering" in Green semigroups. But it is an open question whether the relation "covering" can be described in an analogous way as the minimal congruences. The second question which can be solved for quasiperiodical semigroups by Proposition 10.2 is the description of join irreducible elements in a congruence lattice. We conjecture that the relation "covering" and the join irreducible congruences in Green semigroups can be described by the method used in this paper.

Another important application of the minimal congruences is the characterization of subdirectly irreducible, or simple semigroups. An analogous way for the solution of this question was used in [4]. The results proved in this paper can be used to obtain an effective algorithm for the construction of all minimal congruences. In [2] a general algorithm for the construction of all minimal congruences in a given algebra was described and its time complexity was estimate (it requires $O(n^3)$ time where n is the size of the input algebra). In [9] it was shown that for a finite semigroup such that the Green relations coincide there exists a linear time algorithm (i.e. it uses $O(n^2)$ time) constructing all minimal congruences (the input is given by the multiplication table of a given semigroup). This algorithm is based on the characterization theorems given in Chapters V, VI, and VII (if $\mathcal{J} = \mathcal{H}$ then there exists no minimal congruences of types 4) and 5)). It is an open question whether for a general finite semigroup there exists a linear time algorithm constructing the minimal congruences.

Finally, we can apply our results to special classes of semigroups. Consider, e.g. the finite semigroups, then analogously as Rhodes [15] we obtain:

Theorem 10.3: If S is a finite semigroup with $|S|>1$ then there exists a minimal congruence τ of S which is either of type 2, or 3, or 4.

Proof: It is well-known that S has a kernel $K(S)$ = the smallest non-empty two-sided ideal. If the Schutzenberger group of an $\mathcal{H}$-class $H \subseteq K(S)$ is non-trivial then it has a minimal normal subgroup and by Proposition 7.1 there exists a 3-minimal congruence on S. If the Schutzenberger group is trivial and $K(S)$ is not trivial then by Propositions 8.1 and 8.4 there exists a 4-minimal congruence on S. If $K(S)$ is trivial choose a $\mathcal{J}$-class J of S such that $J \neq K(S)$ and $S^1JS^1 = J \cup K(S)$. If the Schutzenberger group of an $\mathcal{H}$-class $H \subseteq J$ is non-trivial then it has a minimal normal subgroup and by Proposition 7.1 there exists a 3-minimal congruence τ on S with $\mathrm{Car}(\tau) \subseteq J$. If the Schutzenberger group of H is trivial then either $\eta(J) = \Delta$ and by Proposition 6.1 there exists a 2-minimal congruence τ on S with $\mathrm{Car}(\tau) \subseteq J \cup K(S)$ or $\eta(J) \neq \Delta$ and by Propositions 8.1 and 8.4 there exists a 4-minimal congruence τ on S with $\mathrm{Car}(\tau) \subseteq J$.■

Thus we obtain:

Corollary 10.4: Every finite semigroup results from a finite sequence of 2-coextensions, 3-coextensions, and 4-coextensions beginning at a singleton semigroup.■

The second special case considered in this paper is the class of finite semigroups in which the Green relations coincide. We have:

Corollary 10.5: If S is a finite semigroup such that the Green relations coincide then there exists a minimal congruences on S either of type 2 or 3. Every finite semigroup such that the Green relations coincide results from a finite sequence of 2-coextensions and 3-coextensions beginning at a singleton semigroup.

Proof follows from Theorem 10.3 and Corollary 10.4 because if $H = J$ in S then S has no minimal congruence of type 4 and if τ is a minimal congruence of S then S/τ fulfils $H = J$, too.∎

Moreover, for this class the description of 2-minimal congruences is simpler and it is more similar to that for 1-minimal congruences. If we apply Proposition 6.3 we get:

Corollary 10.6[9]: If S is a semigroup such that the Green relations coincide then τ is a 2-minimal congruence on S if and only if the following hold:

(i) $\tau \cap J = \Delta$;

(ii) there exist two different J-classes J_0, J_1 with $\mathrm{Car}(\tau) \subseteq J_0 \cup J_1$ and $S^1 J_0 S^1$ and $S^1 J_1 S^1$ are comparable.

Moreover, if τ is a 2-minimal congruence then every non-singleton class of τ has exactly two elements.∎

Analogously as Corollaries 5.3 and 5.4 we obtain:

Corollary 10.7[9]: Let S be a semigroup such that the Green relations coincide. Then $\varepsilon(x,y)$ is a 2-minimal congruence if and only if

(i) either $S^1xS^1 = S^1yS^1 \cup J(x)$ or $S^1yS^1 = S^1xS^1 \cup J(y)$ and simultaneously $(x,y) \notin J$;

(ii) for every $a,b \in S^1$, $ax = bx$ if and only if $ay = by$, $xa = xb$ if and only if $ya = yb$, $xa = bx$ if and only if $ya = by$;

(iii) for every $a \in S^1$ if $(ax,x) \notin \mathcal{J}$ or $(ay,y) \notin \mathcal{J}$ then $ax = ay$ if $(xa,x) \notin \mathcal{J}$ or $(ya,y) \notin \mathcal{J}$ then $xa = ya$.∎

Corollary 10.8[9]: Let S be a semigroup such that the Green relations coincide. Let J_0, J_1 be $\mathcal{J}$-classes of S such that $S^1 J_0 S^1 = S^1 J_1 S^1 \cup J_0$. Let ξ be the greatest equivalence on $J_0 \cup J_1$ such that $(a,b) \in \xi$ if and only if the following conditions hold for every $x,y \in S^1$:

if $(ax,a) \notin \mathcal{J}$ or $(bx,b) \notin \mathcal{J}$ then $ax = bx$,

if $(xa,a) \notin \mathcal{J}$ or $(xb,b) \notin \mathcal{J}$ then $xa = xb$,

$xa = ya$ if and only if $xb = yb$,

$xa = ay$ if and only if $xb = by$,

$ax = ay$ if and only if $bx = by$.

Choose $u \in J_0$. Then $\{\varepsilon(u,v);\ (u,v) \in \xi,\ v \in J_1\}$ is the set of all 2-minimal congruences τ with $Car(\tau) \subseteq J_0 \cup J_1$.∎

References

[1] A.H. Clifford, G.B. Preston: *The algebraic theory of semigroups*, Amer. Math. Soc. Providence R.I. 1961.

[2] J. Demel, M. Demlová, V. Koubek: Fast algorithms constructing minimal subalgebras, congruences and ideals in a finite algebra, Theor. Comp. Sci. 36(1985), 203-216.

[3] M. Demlová: On factorizations onto the bicyclic semigroup, Semigroup Forum 15(1977), 103-118.

[4] M. Demlová, V. Koubek: Subdirectly irreducible semigroups with minimal left and right ideals, Algebraic Theory of Semigroups (Proc. Conf. Szeged 1976), Colloq. Math. Soc. J. Bolyai vol.20 North Holland, Amsterdam (1979),73-111.

[5] M. Demlová, P. Goralčík, V. Koubek: Inner injective transextensions of semigroups, Acta Sci. Math.(Szeged) 44(1982), 215-237.

[6] P. Goralčík: On translations of semigroups III. Transformations with increasing elements and transformations with an irregular surjective part, Mat. časop., 4(1968), 273-282 (in Russian).

[7] P. Goralčík, Z. Hedrlín: On translations of semigroups II. Surjective transformations, Mat. časop.,4(11968), 263-272 (in Russian).

[8] P. Goralčík, V. Koubek: Translational extensions of semigroups, Algebraic Theory of Semigroups (Proc. Conf. Szeged 1976), Colloq. Math. Soc. J. Bolyai vol.20, North Holland, Amsterdam (1979), 173-218.

[9] P. Goralčík, V. Koubek, J. Ryšlinková: Linear time consuming algorithm for syntactivity in a special class of semigroups, manuscript 1982.

[10] G. Gratzer:*Universal algebra*, Van Nostrand, Toronto 1968.

[11] O.A. Grillet: Building semigroups from groups (and reduced semigroups), Semigroup Forum 4(1972), 327-334.

[12] Z. Hedrlín, P. Goralčík: On translations of semigroups. Periodical and quasiperiodical transformations, Mat. časop., 3(1968), 161-176 (in Russian).

[13] G. Lallement: *Semigroups and combinatorial applications*, Wiley, New York 1979.

[14] M. Petrich: *Introduction to semigroups*, E.Merrill Publ.Comp. Columbus 1973.

[15] J. Rhodes: A homomorphism theorem for finite semigroups, Math. System Theory 4(1969), 289-304.

ENDOMORPHISM MONOIDS OF ACTS ARE WREATH PRODUCTS OF MONOIDS WITH SMALL CATEGORIES

Vladimir Fleischer and Ulrich Knauer

Fakultät für Mathematik
Staatsuniversität Tartu
202400 Tartu, Estland
USSR

Fachbereich Mathematik
Universität Oldenburg
Carl-von-Ossietzky-Straße
D-2900 Oldenburg

The endomorphism monoid of any act can be represented as the wreath product of a monoid with a small category. The same result is also proved for acts with zero. As a consequence we get a definability result for projective acts by their endomorphism monoids.

Introduction

The aim of this paper is to generalize the well known result that endomorphism monoids of free acts can be represented as wreath products of monoids by an act (see for example [5,6,7,8]). As has been noted before, (see [5]), even for endomorphism monoids of projective acts a representation as wreath product of monoids is not possible. Here we present wreath products of a monoid R with a small category K whose object set is considered as left R-act, thereby generalizing usual wreath products of two monoids R and S by a left R-act A. We prove that the endomorphism monoid of any act with or without zero can be represented as wreath product of a monoid with a small category. As a consequence we can determine to which extent the endomorphism monoid of a projective act defines this projective act.

First we give some definitions and notations. Let K be a small category, i.e. the objects X of K form a set. By $M(x,y)$, $x,y \in X$ denote the set of morphisms from x to y in K; in this case x is the domain and y the codomain of any $\alpha \in M(x,y)$, $x = \mathrm{dom}\,\alpha$, $y = \mathrm{codom}\,\alpha$. If $x,y \in X$ are sets, then $\alpha(x) \subset y$ the image of α is denoted by $\mathrm{Im}\,\alpha$. Set

$M(x) = \bigcup_{y \in X} M(x,y)$, $M^{-1}(x) = \bigcup_{y \in X} M(y,x)$, $M = \bigcup_{x \in X} (M(x) \cup M^{-1}(x))$.

By $|X|$ we denote the cardinality of the set X, by $F(X,Y)$ all mappings from the set X to the set Y.

Let R be a monoid. A set $_RA$ or just A is called left R-act if $ra \in A$, $(pr)a = p(ra)$, and $1a = a$ for all $p, r \in R$, $a \in A$. Right R-acts are defined analogously. Subacts are defined in the natural way. If A and B are left R-acts and $f: A \to B$ a mapping, then f is called R-homomorphism if $f(ra) = rf(a)$ for all $r \in R$, $a \in A$. The set of R-homomorphisms is denoted by $Hom_R(A,B)$ and $End_R(A) = Hom_R(A,A)$. The categories of left R-acts or right R-acts are denoted by R-Act or Act-R. If all R-acts have a zero element, and all R-homomorphisms preserve the zero element the categories are denoted by R-Act_0 or Act_0-R.

A free object F in R-Act is of the form $F \cong \coprod_I {}_RR$ where the coproduct $\coprod$ is the disjoint union. A free object F in R-Act_0 is of the form $F \cong \coprod_I {}_RR$ where the coproduct $\coprod$ is the disjoint union of $|I|$ copies of $_RR$, where all zero elements are identified.

1. Wreath product of monoids with categories

Let R be a monoid with identity 1 and K a small category, whose set X of objects forms a left R-act.

Consider

$$A = \{(r,f) \mid r \in R, \quad f \in F(X,M), \quad f(x) \in M(x,rx)\}.$$

Then for $(r,f),(p,g) \in A$ we can define

$$(r,f)(p,g) = (rp, f_p g)$$

where $(f_p g)(x) = f(px)g(x)$ for any $x \in X$. Here $f(x)g(x)$ is the composition of morphisms in K.

With this multiplication A becomes a monoid with identity $(1,e)$ where $e \in F(X,M)$ is such that $e(x)$ is the identity morphism id_x of x in K.

The monoid A with the above multiplication is called the wreath product of the monoid R with the category K and is denoted by $(R \text{ wr } K)$ or $(R \text{ wr } K |_R X)$.

This construction was introduced in [1].

First we show that this construction generalizes the usual wreath product. Let R,S be monoids and ${}_RA$ a left R-act. Then (see for example [7,8])

$$T(R,S,A) = R \times F(A,S)$$

with the multiplication

$$(r,f)(p,g) = (rp, f_p g)$$

where $(f_p g)(a) = f(pa)g(a)$ for $a \in A$ is the <u>wreath product of the monoids</u> R <u>and</u> S <u>by the left</u> R-<u>act</u> A. Here the identity is $(1,c_1)$, where $c_1(a) = 1 \in S$ for all $a \in A$.

1.1. LEMMA. <u>Let</u> R,S <u>be monoids and</u> A <u>a left</u> R-<u>act</u>. <u>Set</u>

$$X = \{aS \mid aS \cong S_S \text{ a free right } S\text{-act}, \ a \in A\}$$

<u>and for</u> $x,y \in X$ <u>set</u> $M(x,y) = \mathrm{Hom}_S(x,y)$ <u>in</u> Act-S <u>the category of right</u> S-<u>acts</u>. <u>Define</u> $r(aS) = (ra)S$ <u>for</u> $r \in R$, $aS \in X$. <u>Then</u> $(R \text{ wr } K)$ <u>is defined if</u> K <u>has</u> X <u>as set of objects and</u> $M(x,y)$ <u>as sets of morphisms for</u> $x,y \in X$.

PROOF is obvious.

1.2. PROPOSITION. <u>Let</u> K <u>be defined as in</u> 1.1. <u>Then the monoid</u> $(R \text{ wr } K)$ <u>is isomorphic to</u> $T(R,S,A)$.

PROOF. Define

$$\varphi: T(R,S,A) \longrightarrow (R \text{ wr } K) \quad \text{by}$$
$$(r,f) \longmapsto (r,f^*)$$

where $$f^*(aS) : aS \longrightarrow (ra)S \quad \text{with}$$
$$as \longmapsto (ra)f(a)s.$$

Then φ is a mapping and

$$f^*(aS) \in \mathrm{Hom}_S(aS, raS) .$$

Conversely, define

$$\psi: (R \text{ wr } K) \longrightarrow T(R,S,A) \quad \text{by}$$
$$(r,f^*) \longmapsto (r,f)$$

where

$$f(a) = s_a \in S, \quad \text{for every} \quad a \in A, \quad \text{if}$$

$$f^*(aS)(a) = (ra)s_a.$$

Then φ and ψ are mutually inverse mappings, i.e. φ is bijective. To show that φ is a homomorphism of monoids take

$$(r,f),(p,g) \in T(R,S,A)$$

and consider

$$\varphi(r,f)\varphi(p,g) = (rp, f^*_p g^*)$$

and

$$\varphi(rp, f_p g) = (rp, h^*)$$

and show that

$$f^*_p g^* = h^*.$$

Let $a \in A$ then $h^*(aS)\colon aS \longrightarrow (rpa)S$ with

$$as \longmapsto (rpa)(f_p g)(a)s = (rpa)f(pa)g(a)s \ , \quad \text{for} \quad s \in S$$

and

$$(f^*_p g^*)(aS)\colon aS \longrightarrow (rpa)S$$

where

$$(f^*_p g^*)(aS) = f^*(paS)g^*(aS)$$

is the composition and morphism in K. Now

$$f^*(paS)g^*(aS)(as) = f^*(paS)((pa)g(a)s) = (rpa)f(pa)g(a)s \ ,$$

which was to be proved. Moreover

$$\varphi(1,c_1) = (1,c_1^*) \ ,$$

where

$$c_1^*(aS)\colon aS \longrightarrow aS \quad \text{with} \quad as \longmapsto a1s = as \ ,$$

that is, $c_1^* = e$.

1.3. REMARK. Using the fact that every endomorphism monoid of a free right S-act can be represented as a wreath product of monoids, we can represent it also as a wreath product of a monoid with a small category by 1.2.

Now we can show the main result of this section. Recall that every right S-act M is uniquely decomposable into a coproduct of indecomposable S-act$_S$ M_i (see [3]) $i \in I$, I a suitable index set.

1.4. THEOREM. Let M be a right S-act. Then the monoid $\mathrm{End}_S M$ is isomorphic to the monoid $(R \,\mathrm{wr}\, K)$ for some small category K and some monoid R. In particular $\varphi \in \mathrm{End}_S M$ corresponds to (r,f) if $\varphi(M_i) \subset M_{ri}$ with $f(i) = \varphi \mid M_i$ where $M \cong \coprod_{i\in I} M_i$ is the decomposition of M into indecomposable S-acts.

PROOF. Consider the decomposition of M into indecomposable right S-acts $M \cong \coprod_{i\in I} M_i$. First construct the small category K. Take I as the set of objects of the category K and for $i,j \in I$ take the morphism set $K(i,j) = \mathrm{Hom}_S(M_i,M_j)$.

Next we construct the monoid R. Apparently any $\varphi \in \mathrm{End}_S M$ induces a transformation r_φ of I, i.e. $\varphi(M_i) \subset M_{ri}$, where the index φ is omitted. Note that the above inclusion is granted since M_i is indecomposable. Thus by

$$\Theta\colon \mathrm{End}_S M \longrightarrow F(I,I)$$
$$\varphi \longmapsto r_\varphi$$

there is defined a mapping, which is a homomorphism of monoids. Now take $R = \Theta(\mathrm{End}_S M)$, which becomes a monoid using the composition of the induced mappings as multiplication and, moreover, I becomes a left R-act in the obvious way. Now we prove $\mathrm{End}_S M \cong (R \,\mathrm{wr}\, K)$.

Define

$$w\colon \mathrm{End}_S M \longrightarrow R \,\mathrm{wr}\, K \quad \text{by}$$
$$\varphi \longmapsto (r,f)$$

where

$$r = r_\varphi \quad \text{and} \quad f(i) = \varphi|M_i .$$

Injectivity of w is obvious. To show surjectivity take $(r,f) \in (R \,\mathrm{wr}\, K)$. If now $\varphi \in \mathrm{End}_S M$ is such that $\varphi|M_i = f(i)$ for all $i \in I$ then $w(\varphi) = (r,f)$, since $f(i) \in K(i,ri)$, and thus, in particular, $r = r_\varphi$. It is obvious that w preserves the identity. Finally we prove that w preserves multiplication. Let $w(\varphi) = (r,f)$, $w(\psi) = (p,g)$, $\varphi,\psi \in \mathrm{End}_S M$, and let $w(\varphi\psi) = (q,h)$. To show $(q,h) = (rp, f_p g)$ note that

$$\varphi\psi(M_i) \subset \varphi(M_{pi}) \subset M_{rpi}$$

which implies $q = rp$. Moreover, for $i \in I$ we get

$$(f_p g)(i) = f(pi)g(i) = \varphi|_{M_{pi}} \psi|_{M_i} = \varphi\psi|_{M_i} = h(i).$$

This completes the proof.

2. Endomorphism monoids of projective acts

In this section we give an example showing directly that the endomorphism monoid of a projective act may not have a representation as a wreath product of monoids, although this is true for free acts. Moreover, we investigate to which extent a projective act is determined by its endomorphism monoid.

We shall say that a monoid H <u>has a non trivial representation as a wreath product of monoids</u> if there exist monoids R,S and a left R-act A with $|R|, |S|, |A| > 1$ such that $H \cong T(R,S,A)$.

2.1. EXAMPLE. Let $H = \{1,r,p\}$ a right zero monoid, i.e. $rp = p^2 = p$, $pr = r^2 = r$. Consider the right H-act $P \cong rH \coprod H$, which is projective by [3], as it is the coproduct of idempotently generated cyclic acts. Now there are only two H-homomorphisms with domain rH, i.e. one with codomain H and one with codomain rH as $\varphi(r) = \varphi(r^2) = \varphi(r)r = r$ for any φ with $\mathrm{dom}\,\varphi = rH$. There exist five H-homomorphisms with domain H_H, three with codomain H_H and two with codomain rH. Consequently $|\mathrm{End}_H P| = 10$. But as $T(R,S,A) = R \times F(A,S)$ and

$$|R \times F(A,S)| = |R| \cdot |S|^{|A|}$$

10 cannot be reached if $|R|, |S|, |A| > 1$. Consequently $\mathrm{End}_H P$ has no non trivial representation as wreath product of monoids by an act. However, according to Theorem 1.4, we have the representation $\mathrm{End}_H P \cong (R \times K)$ where the set of objects of K is $X = \{1,2\}$ and $|M(1,1)| = |M(1,2)| = 1$, $|M(2,1)| = 2$, $|M(2,2)| = 3$, the morphisms being those mentioned before with the respective domains and codomains. The monoid R is $F(\{1,2\},\{1,2\})$ the full monoid of transformations.

Putting together the result of Theorem 1.4 and a definability result of [1] we can analyze to which extent projective right S-acts are defined by their endomorphism monoids. Free right S-acts are defined completely by their endomorphism monoids, see for example [2,3]. For projective right S-acts the situation is different.

2.2. RESULT [1]. *Let* R *and* R' *be monoids*, K *and* K' *small categories with object sets* X *and* X' *which are left* R- *or* R'-*acts respectively. Suppose that the following conditions for* $r \in R$ *or* R' *and all* $x,y,u,v \in X$ *or* X' *respectively are fulfilled*

(i) $|X| \geq 2$, $|X'| \geq 2$;

(ii) $M(x,y) \neq \emptyset$ *for all* x,y;

(iii) *for every* x *there exists exactly one* $r \in R$ *with* $ry = x$ *for all* y;

(iv) *for all* u,v,x,y, $x \neq y$, *there exists* $r \in R$ *such that* $rx = u$, $ry = v$.

If the wreath products $(R \operatorname{wr} K)$ *and* $(R' \operatorname{wr} K')$ *are isomorphic monoids, then* K *and* K' *are isomorphic categories.*

2.3. THEOREM. *Let* P *be a projective right* S-*act, i.e.*, $P \cong \coprod_{i\in I} e_iS$, $e_i^2 = e_i \in S$, $|I| \geq 2$, *and let* P' *be a projective right* S'*act, i.e.* $P' = \coprod_{i\in I'} e_i'S'$, $e_i'^2 = e_i' \in S'$, $|I'| \geq 2$. *If* $\operatorname{End}_S P$ *and* $\operatorname{End}_{S'} P'$ *are isomorphic, then* $|I| = |I'|$ *and-up to the corresponding bijection of indices - the monoids* e_iSe_i *and* $e_i'Se_i'$ *are isomorphic.*

PROOF. First we use Theorem 1.4 to represent $\operatorname{End}_S P$ and $\operatorname{End}_S P'$ as wreath products of monoids R and R' with categories K and K'. Then I will be the object set of K. Moreover R will be the full transformation monoid of I, i.e. $R \cong F(I,I)$ since $\operatorname{Hom}_S(e_iS,e_jS) \neq \emptyset$ for any $i,j \in I$. In fact, by $g(e_i) = e_je_i$ there is defined one such S-homomorphism, i.e. for any $i,j \in I$ there exists $r \in R$ with $ri = j$. This observation and $|I| \geq 2$ grant already (i) and (ii) of 2.2. Moreover, the above r corresponds to the constant mapping $c_j \in F(I,I)$ with $c_j(i) = j$ for all $i \in I$. Therefore r is the unique element required in (iii) of 2.2. From $R \cong F(I,I)$ it is also clear that (iv) of 2.2 is fulfilled. Consequently we can apply 2.2 to $(R \operatorname{wr} K) \cong (R' \operatorname{wr} K')$. Thus K and K' are isomorphic categories. This implies $|I| = |I'|$ and - omitting the corresponding bijection - $\operatorname{Hom}_S(e_iS,e_jS) \cong \operatorname{Hom}_{S'}(e_i'S',e_j'S')$ which for $i = j$ is an isomorphism of monoids. Using that $\operatorname{Hom}_S(e_iS,e_iS) \cong e_iSe_i$ (see for example [3]), we get that $e_iSe_i \cong e_i'S'e_i'$, again an isomorphism of monoids.

Finally we show by an example that in general it will not be possible to get a better result than Theorem 2.3 about the definability of projective acts by their endomorphism monoids.

2.4. EXAMPLE. Let S and S' be disjoint left zero monoids, i.e. $st = s$ for all $s,t \in S$ (or $\in S'$) with $|S| \neq |S'|$. Then all non free cyclic projectives are one element acts and $\mathrm{End}_S(\coprod_{i\in I} e_iS) \cong \mathrm{End}_{S'}(\coprod_{i\in I} e_i'S)$ for any $1 \neq e_i \in S$, $1' \neq e_i' \in S'$, where $1,1'$ are the respective identiies. But obviously S and S' are not isomorphic.

3. 0-wreath products of monoids with categories

Now we consider the corresponding generalization for 0-wreath products (cf. [5,6]). Let now ${}_RA$ be a left R-act with 0, S a monoid with zero 0 and consider $F_o(A,S) = \{f\colon A \to S \mid f(0) = 0\}$. Then $R \times F_o(A,S) \subset R \times F(A,S)$ is a submonoid. On $R \times F_o(A,S)$ we consider the relation σ such that $(r,f)\sigma(p,g)$ if and only if $dr \cap df = dp \cap dg$ and

$$ra = pa, \quad f(a) = g(a) \quad \text{for all} \quad a \in dr \cap df ,$$

where

$$dr = \{a \in A \mid ra \neq 0\}, \quad df = \{a \in A \mid f(a) \neq a\} .$$

Then $T_\sigma = (R \times F_o(A,S))/_\sigma$ by [6] is a monoid with zero, where the class $(1,c_1)_\sigma$ is the identity and the class $(1,c_o)_\sigma$ is the zero element, c_o being the mapping such that $c_o(a) = 0$ for all $a \in A$. T_σ is called 0-<u>wreath product of</u> R <u>and</u> S <u>by</u> A.

A subset $L \subset M$ is called <u>ideal in the category</u> K if $\lambda \in L$ implies $\mu\lambda, \lambda\nu \in L$ for any appropriate $\mu,\nu \in M$.

Now let L be an ideal of K and $Y \subset X$ an R-subact of X (either of them may be empty). In the wreath product $(R \,\mathrm{wr}\, K)$ of the monoid R with the category K define

$$d(r,f) = \{x \in X \mid rx \notin Y, \quad f(x) \notin L\}$$

for $(r,f) \in (R \,\mathrm{wr}\, K)$, and define a relation

$$((r,f),(p,g)) \in \sigma(Y,L)$$

if and only if $d(r,f) = d(p,g)$ and $f(x) = g(x)$ for every $x \in d(p,g)$.

3.1. LEMMA. <u>The relation</u> σ <u>is a congruence on the monoid</u> $(R\,\mathrm{wr}\,K)$.

PROOF. Obviously σ is an equivalence relation on $(R\,\mathrm{wr}\,K)$. Let $(r,f),(p,g),(q,h) \in (R\,\mathrm{wr}\,K)$ and assume $((r,f),(p,g)) \in \sigma(Y,L)$. Consider $(r,f)(q,h) = (rq,f_q h)$ and $(p,g)(q,h) = (pq,g_q h)$. Now $x \in d(rq,f_q h)$ means $rqx \notin Y$ and $f(qx)h(x) \notin L$ and consequently $f(qx) \notin L$, i.e., $qx \in d(r,f) = d(p,g)$ and moreover, $f(qx) = g(qx)$ by assumption. Thus $pqx \notin Y$ and $g(qx)h(x) = f(qx)h(x) \notin L$, i.e., $x \in d(pq,g_q h)$ and $(f_q h)(x) = (g_q h)(x)$. Analogously, $x \in d(pq,g_q h)$ implies $x \in d(rq,f_q h)$ and $(g_q h)(x) = (f_q h)(x)$. This shows that $((rq,f_q h),(pq,g_q h)) \in \sigma(Y,L)$. Now consider $(q,h)(r,f) = (qr,h_r f)$ and $(q,h)(p,g) = (qp,h_p g)$. Then $x \in d(qr,h_r f)$ means $qrx \notin Y$ and $h(rx)f(x) \notin L$ and consequently $rx \notin Y$, $f(x) \notin L$, i.e. $x \in d(r,f) = d(p,g)$ and moreover $f(x) = g(x)$ by assumption. This, by definition of $A = (R\,\mathrm{wr}\,K)$, implies $rx = px$ and consequently $qpx \notin Y$, $h(px)g(x) \notin L$, i.e. $x \in d(qp,h_p g)$ and $(h_r f)(x) = (h_p g)(x)$. Analogously, $x \in d(qp,h_p g)$ implies $x \in d(qr,h_r f)$ and $(h_r f)(x) = (h_p g)(x)$. This shows that $((qr,h_r f),(qp,h_p g)) \in \sigma(Y,L)$. This proves the lemma.

The factor monoid $A|_{\sigma(Y,L)} = (R\,\mathrm{wr}\,K)|_{\sigma(Y,L)}$ will be called <u>a zero-wreath product of the monoid with the category</u> K <u>through</u> Y <u>and</u> L. It will be denoted by $(R\,\mathrm{wr}\,K|Y,L)$. Thus $(R\,\mathrm{wr}\,K) = (R\,\mathrm{wr}\,K|\emptyset,\emptyset)$

3.2. LEMMA. <u>Let</u> $(r,f)_\sigma$ <u>denote the image of</u> $(r,f) \in (R\,\mathrm{wr}\,K)$ <u>under the natural epimorphism</u> $(R\,\mathrm{wr}\,K) \to (R\,\mathrm{wr}\,K)|_\sigma$. <u>Then</u> $(1,e)_\sigma$ <u>is the identity of</u> $(R\,\mathrm{wr}\,K)|_\sigma$, <u>where</u> $e(x) = id_x$ <u>for all</u> $x \in X$. <u>If</u> $\lambda \in L$ <u>then</u>

$$(1,c_\lambda)_\sigma \supset \{(r,n) \mid r \in R,\ n(x) \in L \text{ for all } x \in X\}$$

<u>is the zero of</u> $(R\,\mathrm{wr}\,K)|_\sigma$, <u>where</u> $c_\lambda(x) = \lambda$ <u>for all</u> $x \in X$.

PROOF. The statement about the identity if obvious. A zero element (r,f) is characterized by $d(r,f) = \emptyset$, all these elements are in one $\sigma(Y,L)$-class, that is the zero element in $(R\,\mathrm{wr}\,K)|_\sigma$ is unique. It can be represented as $(1,c_\lambda)_\sigma$ if $L \neq \emptyset$.

Now we show that this construction generalizes the 0-wreath product of two monoids R and S, S with 0 by the left R-act A with 0_A.

3.3. LEMMA. Let R,S be monoids, S with zero 0, A a left R-act with zero 0_A. Set

$$X = \{aS \mid aS \cong S_S \in \mathrm{Act}_0\text{-}S \text{ with zero element } 0_a = a0,\ 0_A \neq a \in A\} \cup \{0_A\}$$

and for $x,y \in X$ set $M(x,y) = \mathrm{Hom}_S(x,y)$ in Act_0-S. Define $r(aS) = (ra)S$ and $r0_A = 0_A$. Then $(R \,\mathrm{wr}\, K_0)$ is defined, if K_0 has X as set of objects and $M(x,y)$ as sets of morphisms for $x,y \in S$. Moreover,

$$M_0 = \{\alpha \in M \mid \mathrm{Im}\,\alpha = \{0_a\} \text{ for some } 0_A \neq a \in A, \text{ or } \mathrm{dom}\,\alpha = 0_A \text{ or } \mathrm{codom}\,\alpha = 0_A\}$$

is an ideal in K_0.

PROOF. Obviously by the above definition X becomes a left R-act with zero 0_A, which as one element set can also be considered as a right S-act in the trivial way. It is obvious $M_0 \subset M$ is an ideal in the category K_0.

3.4. LEMMA. There exists an isomorphism of monoids $R \times F_0(A,S) \cong (R \,\mathrm{wr}\, K_0 | \emptyset, \emptyset)$ where K_0 is defined as in Lemma 3.3.

PROOF. As in the proof of Proposition 1.2 φ and ψ are constructed where now $f \in F_0(A,S)$ implies that f^* preserves zero elements and vice versa. The rest goes as in the proof of 1.2.

3.5. PROPOSITION. Let K_0 and M_0 be defined as in Lemma 3.3. Then the monoids $(R \,\mathrm{wr}\, K_0 | 0_A, M_0)$ and $T_\sigma = (R \times F_0(A,S))|\sigma$ are isomorphic.

PROOF. The isomorphism constructed in the proof of 1.2 induces an isomorphism between the monoids considered here, since $(\varphi(r,f),\varphi(r',f')) \in \sigma(0_A,M_0)$ if and only if $(r,f)\sigma(r',f')$ in $R \times F_0(A,S)$.

Now we can prove the analogue of Theorem 1.4 in the category of right S-acts with zero. Recall that every right S-act M with 0 is uniquely decomposable into a coproduct of indecomposable right S-acts M_i with 0, that is, M is the disjoint union of the M_i, where all zeros are identified, $i \in I$, I a suitable index set.

3.6. THEOREM. <u>Let</u> M <u>be a right</u> S-<u>act with zero. Then the monoid</u> $\mathrm{End}_S M$ <u>of endomorphisms in</u> Act_0-S <u>is isomorphic to the monoid</u> $(R \,\mathrm{wr}\, K_0 | 0_X, M_0)$ <u>for some small category</u> K_0, <u>some monoid</u> R, 0_X <u>the zero subact of the object set</u> X <u>of</u> K_0, X <u>being a left</u> R-<u>act, and</u> M_0 <u>an ideal of</u> K_0. <u>In particular</u> $(r,f)_\sigma \in (R \,\mathrm{wr}\, K_0 | 0_X, M_0)$ <u>corresponds to</u> $\varphi \in \mathrm{End}_S M$ <u>with</u> $\varphi(M_i) = f(M_i)$ <u>if</u> $M_i \in d(r,f)$ <u>and</u> $\varphi(M_i) = 0_M$ <u>otherwise</u>.

PROOF. First consider the decomposition of $M \cong \coprod_{i\in I} M_i$ where M_i are indecomposable right S-acts with 0. By 0_M and 0_i denote the respective zero elements. First construct the small category K_0. Take $X = \{M_i \mid i \in I\} \cup \{0_X\}$ as the set of objects of K_0, where 0_X is an element not in $\{M_i \mid i \in I\}$ which will be considered as a one element right S-act. For $x, y \in X$ take the morphism set

$$M(x,y) = \mathrm{Hom}_S(x,y) ,$$

in the category of right S-acts with zero. Next we construct the monoid R. In analogy to the proof of Theorem 1.4. any $\varphi \in \mathrm{End}_S M$ induces a transformation r_φ of X where now

$$r_\varphi(M_i) = \begin{cases} 0_X & \text{if } \varphi(M_i) = 0_M \\ M_j & \text{if } \varphi(M_i) \subset M_j \end{cases} , \quad i \in I$$

and

$$r_\varphi(0_X) = 0_X .$$

Thus by

$$\Theta\colon \mathrm{End}_S M \longrightarrow F(X,X)$$
$$\varphi \longmapsto r_\varphi$$

there is defined a mapping which is homomorphism of monoids. Now take $R = \Theta(\mathrm{End}_S M)$. Then R becomes a monoid with zero and X a left R-act with zero 0_X.

Now define the ideal M_0 of the category K_0 in analogy to Lemma 3.3

$$M_0 = \{\alpha \in M \mid \mathrm{Im}\,\alpha = \{0_i\},\ i \in I,\ \text{ or } \mathrm{dom}\,\alpha = 0_X \text{ or } \mathrm{codom}\,\alpha = 0_X\}$$

Then M_0 obviously is an ideal in K_0.

We prove that the monoids $\mathrm{End}_S M$ and $(R \,\mathrm{wr}\, K_0 | 0_X, M_0)$ are isomorphic.

Define

$$\nu: (R \text{ wr } K_0 \mid 0_X, M_0) \longrightarrow \text{End}_S M \quad \text{by}$$

$$(r,f)_\sigma \longrightarrow \nu((r,f)_\sigma) = \varphi$$

with

$$\varphi(M_i) = \begin{cases} f(M_i), & \text{if } M_i \in d(r,f) \\ 0_M, & \text{if } M_i \notin d(r,f) \end{cases} .$$

Obviously, ν is a mapping. To prove injectivity take $(r,f)\sigma \neq (p,g)\sigma \in (R \text{ wr } K_0 \mid 0_X, M_0)$. If now $d(r,f) \neq d(p,g)$, then there exists M_i, $i \in I$ such that $M_i \in d(r,f)$, $M_i \notin d(p,g)$ and then $\nu((r,f)_\sigma)(M_i) \neq 0_M$ but $\nu((p,g)_\sigma)(M_i) = 0_M$. But if $d(r,f) = d(p,g)$ then in any case $f(M_i) \neq g(M_i)$ for some $M_i \in d(r,f)$ and thus $\nu((r,f)_\sigma)(M_i) \neq \nu((p,g)_\sigma)(M_i)$. So from both cases it follows that $\nu((r,f)_\sigma) \neq \nu((p,g)_\sigma)$. To prove surjectivity take $\varphi \in \text{End}_S M$. Set $d\varphi = \{P_i \mid \varphi(P_i) \neq \{0_M\},\ i \in I\}$. Consider $(r,f) \in R \times F(X,M)$ with $r = O(\varphi)$ and $f(M_i) = \varphi(M_i)$ for $M_i \in d\varphi$, and $f(M_i) = 0_X$ if $M_i \notin d\varphi$. It is clear that $(r,f) \in (R \text{ wr } K_0)$ and, in addition, $\nu((r,f)_\sigma) = \varphi$. This shows surjectivity of ν. Thus ν is a bijection.

It remains to show that ν is a homomorphism of monoids. It is clear that ν preserves identity and zero. Let $\nu((r,f)_\sigma) = \varphi$, $\nu((p,g)_\sigma) = \psi$ and $\nu((rp, f_p g)_\sigma) = \chi$, and show that $\varphi\psi = \chi$. In the case where $M_i \in d(rp, f_p g)$ we get $\chi(M_i) = (f_p g)(M_i) = f(p(M_i))g(M_i)$. Then $M_i \in d(rp, f_p g)$ implies $M_i \in d(p,g)$ and $p(M_i) \in d(r,f)$, i.e. $g(M_i) = \psi(M_i)$ and $f(p(M_i)) = \varphi(p(M_i))$. Thus $\chi(M_i) = \varphi(p(M_i))\psi(M_i) = \varphi\psi(M_i)$. In the case where $M_i \notin d(rp, f_p g)$, i.e. $\chi(M_i) = 0_M$, it follows that $rp(M_i) = 0_X$ or $(f_p g)(M_i) \in M_0$. From $rp(M_i) = 0_X$ we get either $p(M_i) = 0_X$ and thus $M_i \notin d(p,g)$ which implies $\psi(M_i) = 0_M$, or $p(M_i) \notin d(r,f)$, i.e. $\varphi(p(M_i)) = 0_M$. So, in both cases, we have $\varphi\psi(M_i) = 0_M$. From $(f_p g)(M_i) \in M_0$ it follows that $\varphi\psi(M_i) = 0_M$, too. This proves that $\chi = \varphi\psi$, thereby completing the proof.

REFERENCES

1. Fleischer, V.G., On the wreath product of monoids with categories, Izv. AN ESSR, to appear (in Russian).

2. Fleischer, V.G., Definability of free acts by their endomorphism semigroups, Uch. Zap. Tartusk. Univ., 366(1975), 27-41 (in Russian).

3. Knauer, U., Projectivity of acts and Morita equivalence of monoids, Semigroup Forum, 3(1972), 359-370.

4. Knauer, U., Column Monomic Matrix Monoids, Math. Nachr., 74(1976), 135-141.

5. Knauer, U., Mikhalev, A., Endomorphism monoids of free acts and 0-wreath products of monoids. I. Annihilator Properties, Semigroup Forum, 19(1980), 177-187.

6. Knauer, U., Mikhalev, A., Endomorphism monoids of free acts and 0-wreath products of monoids. II. Regularity, Semigroup Forum 19(1980), 189-198.

7. Skornjakov, L.A., Regularity of the wreath product of monoids, Semigroup Forum, 18(1979), 83-86.

8. Skornjakov, L.A., On the wreath product of monoids, Universal algebra and applications, Banach Center Publ., 9(1982), 181-185.

FREE RIGHT h-ADEQUATE SEMIGROUPS

John Fountain
Dept. of Mathematics
University of York
Heslington
York YO1 5DD
England

ABSTRACT. Right adequate semigroups can be regarded as semigroups with a unary operation. Taking this view, we give a description of the free objects in a class of right adequate semigroups. By making use of the normal form representation of the elements we derive a number of properties enjoyed by these free objects.

Introduction

On a semigroup S the relation $\mathcal{L}^*$ is defined by the rule that a $\mathcal{L}^*$ b if and only if the elements a,b of S are related by Green's relation $\mathcal{L}$ in some oversemigroup of S. It is well-known that in a monoid S, every principal right ideal is projective if and only if every $\mathcal{L}^*$- class of S contains an idempotent. Following the terminology of [8] we say that any semigroup (with or without an identity) which satisfies the latter condition and has commuting idempotents is <u>right-adequate</u>.

Inverse semigroups are right adequate and so are left cancellative monoids. Further, if S is a subsemigroup of an inverse semigroup T and S, T have the same idempotents, then S is right (and left) adequate. Another example is provided by taking the semigroup of those endomorphisms of a semilattice with identity whose images are principal ideals.

It is noted in [8] that in a right adequate semigroup each $\mathcal{L}^*$-class contains just one idempotent. The idempotent in the $\mathcal{L}^*$-class containing the element a of a right adequate semigroup S will be denoted by a*. The set E of all idempotents in a right adequate semigroup S forms a subsemilattice of S and for each element a of S, the mapping $\alpha_a : E^1 \to E^1$ defined by $x\alpha_a = (xa)^*$ is isotone ([8] , Lemma 2.1). We say that S is a <u>right</u> h-<u>adequate</u> semigroup when S is right adequate and α_a is a semigroup homomorphism for each element a of S. Although not all right adequate semigroups are right h-adequate as is illustrated by Example 2.2 of [8] , the class of right h-adequate semigroups is quite extensive. For example, this class contains every right adequate semigroup in which the semilattice of idempotents is a chain. Right type A semigroups are also members of this class. A right adequate semigroup S is right type A when $ea = a(ea)^*$ for all elements a of S and all idempotents e in S. Such semigroups may also be described as those right adequate semigroups S in which $eS \wedge aS = eaS$ for any element a and any idempotent e of S. These semigroups have been studied in [7] and [8] .

We may regard a right adequate semigroup as an algebra with two operations: the binary operation multiplication and the unary operation *. By a *-semigroup we shall mean an algebra (S,.,*) with an associative binary operation · and a unary operation *, no restriction being placed on *. In [2] , Clifford calls such algebras, unary algebras. Our interest centres on those *-semigroups in which a^* is an idempotent for each element a. Such semigroups are called γ-semigroups in [1] which is devoted to their study. We point out in Section 1 that the classes of right adequate, right h-adequate and right type A are all quasi-varieties of *-semigroups. Consequently free objects exist in these classes. The present paper is concerned with the study of free right h-adequate semigroups.

The free *-semigroup F_X^* on a set X has been described in [2] and our free objects on X could be described as quotients of F_X^*. We adopt a different approach, however, by taking a quotient of the free product of F_X and E_X where F_X is the free semigroup on X and E_X is a semilattice formed from certain finite subsets of F_X. This enables us to find a normal form for the elements of P_X, the free right h-adequate semigroup on the set X.

The existence of a normal form for the elements of P_X allows us to obtain a number of properties of P_X reminiscent of those enjoyed by free inverse semigroups as given in [17] , [20] and [22] . This we do in Section 3 where among other things we show that Green's relations on P_X are all trivial, that P_X satisfies various maximal conditions and is residually finite.

Section 4 is concerned with results about free generators in right h-adequate semigroups and is inspired by corresponding results in the inverse case due to Reilly [19] . The free semigroup F_X is a subsemigroup of P_X and one of our results relates our work to the theory of codes. Namely, a subset of F_X freely generates a *-subsemigroup of P_X if and only if it is a suffix code over X.

In a subsequent paper devoted to right type A semigroups our description of P_X is used to obtain the free right type A semigroup on X as a certain quotient of P_X.

I would like to thank Victoria Gould and Mario Petrich for reading earlier versions of this paper and their helpful comments. I would also like to record my thanks to an anonymous referee for noticing some errors and obscurities and providing suggestions for corrections and clarifications.

1. Preliminaries

For basic facts about semigroups we refer to [3] , [13] or [14] and for universal algebra we refer to [4] , [10] or [15] .

We begin by listing some elementary results concerning right adequate semigroups. In the introduction we have defined the relation $\mathcal{L}^*$ on a semigroup S by the rule that $a\mathcal{L}^*b$ if and only if the elements a,b of S are $\mathcal{L}$- related in some oversemigroup of S. The relation $\mathcal{R}^*$ is defined dually. Alternative and more useful characterisations of $\mathcal{L}^*$ are provided by the following lemma from [16] and [18] .

LEMMA 1.1 *Let* S *be a semigroup and let* a,b *be elements of* S. *Then the following conditions are equivalent:*

(1) $a\mathcal{L}^*b$,

(2) *for all* $x,y\in S^1$, $ax = ay$ *if an only if* $bx = by$,

(3) *there is an* S^1-isomorphism $\phi : aS^1 \to bS^1$ *with* $a\phi = b$.

As an easy consequence we have

COROLLARY 1.2 *If* e *is an idempotent of a semigroup* S, *then the following are equivalent for an element* a *of* S:

(1) $e\mathcal{L}^*a$,

(2) $ae = a$ *and for all* $x,y\in S^1$, $ax = ay$ *implies* $ex = ey$.

From the definition and Lemma 1.1 it follows that $\mathcal{L}^*$ is a right congruence and that $\mathcal{L}\subseteq\mathcal{L}^*$. It is well-known and easy to see that for regular elements a,b of S we have $a\mathcal{L}^*b$ if and only if $a\mathcal{L}b$. In particular, if S is a regular semigroup, then $\mathcal{L} = \mathcal{L}^*$.

In a right adequate semigroup, the idempotents commute and therefore each $\mathcal{L}^*$-class contains a unique idempotent. Denoting the idempotent in the $\mathcal{L}^*$-class of S which contains the element a by a^* we thus have a unary operation * on S. Hence a right adequate semigroup is a *-semigroup.

Using Lemma 1.1, Corollary 1.2 and the above remarks it is not difficult to show that a *-semigroup S is a right adequate semigroup if and only if the following identities and quasi-identities hold:

(I) $(xy)z = x(yz)$,

(II) $x^2 = x \wedge y^2 = y \Rightarrow xy = yx$,

(III) $xx^* = x$,

(IV) $xy = xz \Rightarrow x^*y = x^*z$,

(V) $xy = x \Rightarrow x^*y = x^*$.

We note, in particular, that (V) applied to (III) gives that x^* is an idempotent; that (III), (V) and (II) give $x = x^*$ whenever x is an idempotent. Thus $(x^*)^* = x^*$.

We mention now the following elementary facts from [8] which will be used throughout the paper without further mention.

PROPOSITION 1.3 If S is a right adequate semigroup with semilattice of idempotents E, then

(1) for all $a,b \in S$, $a \mathcal{L}^* b$ if an only if $a^* = b^*$,

(2) for all $a,b \in S$, $(ab)^* = (a^*b)^*$,

(3) for all $a,b \in S$, $(ab)^* \leqslant b^*$ where $\leqslant$ is the usual ordering on E.

As a consequence of (2) we note that $(xy^*)^* = (x^*y^*)^* = x^*y^*$ since x^*y^* is an idempotent. We observe next that the class of right h-adequate semigroups is described by (I)-(V) together with

(VI) $(xz)^*(yz)^* = (x^*y^*z)^*$.

The fact that this is equivalent to the definition given in the introduction is a consequence of the following observations. In view of (2) of Proposition 1.3, (VI) is equivalent to the identity

$(x^*z)^*(y^*z)^* = (x^*y^*z)^*$.

Since $\{x^*:x \in S\}$ is the set of idempotents of S, this identity is simply asserting that $(ez)^*(fz)^* = (efz)^*$ for all idempotents e,f of S and all elements z in S.

The class of right type A semigroups is described by (I) - (V) together with

(VII) $x^*y = y(xy)^*$.

Again variations are possible and what we use most often is the property: $xy = y(xy)^*$ when x is an idempotent.

From Lemma 2.1 of [8] we have in our present terminology

LEMMA 1.4 A right type A semigroup is right h-adequate.

By a *-subsemigroup U of a *-semigroup S we mean a subsemigroup which satisfies $u \in U$ implies $u^* \in U$. The *-subsemigroup of S generated by a subset Y of S is the intersection of those *-subsemigroups which contain Y. As noted in [5] it is clear that if S is right adequate, then so is any *-subsemigroup. Similar remarks apply when S is right h-adequate or right type A. If S, T are *-semigroups, a function $\theta:S \to T$ is called a *-homomorphism if it is a semigroup homomorphism which satisfies $s^*\theta = (s\theta)^*$ for all s in S. Clearly $S\theta$ is a *-subsemigroup of T. A congruence ρ on S is a *-congruence if the natural map from S onto S/ρ is a *-homomorphism.

We denote the free *-semigroup on a set X by F_X^*. We refer the reader to [2] for a construction of F_X^* contenting ourselves with remarking that, in the terminology of [4], it consists of all $\{\cdot,*\}$-words (called polynominals in [2]). The binary operation is simply concatenation of words.

We have noted above that each of the classes of right adequate, right h-adequate and right type A semigroups can be described by means of identities and quasi-identities. Thus each of these classes is a quasi-variety of *-semigroups and so has free objects (see [4] or [15]). If we let η denote the intersection of all *-congruences ρ on F_X^* for which F_X^*/ρ is right adequate, then it is readily verified that F_X^*/η is right adequate. Furthermore, if S is any right adequate semigroup and α is a function from X into S, then there is a unique *-homomorphism θ from F_X^* into S with $\theta|X = \alpha$. Now $F_X^*\theta$ is right adequate and so $\eta \subseteq \ker\theta$. Hence there is a unique *-homomorphism ψ from F_X^*/η into S such that θ factorizes as $\theta = \eta^{\natural}\psi$. Thus F_X^*/η is the free right adequate semigroup on X provided that $\eta^{\natural}|X$ is injective. But this is easily seen to be the case by choosing an S for which α is injective. Similar descriptions can be given of the free right h-adequate and the free right type A semigroups on X. These descriptions, however, do not give an explicit form for the elements of the semigroups and do not allow us to discover their properties.

Following [9] we define a left [right] *-ideal of a semigroup S to be a left [right] ideal of S which contains the $\mathcal{L}^*$-class [$\mathcal{R}^*$-class] of each of its elements. That is, a left [right] ideal is a left [right] *-ideal if and only if it is a union of $\mathcal{L}^*$-classes [$\mathcal{R}^*$-classes]. In the case of a right adequate semigroup S, a left ideal I is a left *-ideal if and only if $a \in I$ implies $a^* \in I$.

PROPOSITION 1.5 *Let S be a right adequate semigroup and I be an ideal which is a left *-ideal. Then the Rees quotient semigroup S/I is right adequate and the natural mapping $\nu: S \to S/I$ is a *-homomorphism. Furthermore, if S is right h-adequate or right type A, then so is S/I.*

Proof. Clearly the idempotents of S/I form a subsemilattice of S/I. Also, if a is an element of S/I other than I, then $a^* \notin I$ so that $a^* \in S/I$ and $aa^* = a$. Since $\mathcal{L}^*$ is a right congruence and a, a^* are $\mathcal{L}^*$-related in S, we have ax, a^*x are $\mathcal{L}^*$-related in S. Hence $ax \in I$ if and only if $a^*x \in I$. From this observation, it is easy to see, using Corollary 1.2, that a and a^* and $\mathcal{L}^*$-related in S/I. That ν is a *-homomorphism is now clear.

The final part of the Proposition now follows because a *-homomorphism will preserve the equations (VI) and (VII).

2. The semigroup P_X

Let X be a non-empty set and let F_X be the free semigroup on X. Partially order F_X by putting $u \leqslant v$ if and only if u is a final segment of v. For any subset A of F_X, we write

$$\max A = \{a \in A : a \text{ is maximal in } A \text{ under } \leqslant \}.$$

Now let

$$E_X = \{A : A \subseteq F_X, A \text{ is finite and non-empty}, A = \max A\}.$$

Thus E_X is the set of all finite suffix codes over X. For $A, B \in E_X$ define $AB = \max(A \cup B)$. Then E_X is a semilattice, in fact, if we consider F_X as a partially ordered set under the dual of the above ordering, then E_X is the free semilattice on this partially ordered set [11]. We note that the following statements are equivalent for members A,B of E_X where we use $\leqslant$ for the order relation in E_X as well as that in F_X:

$A \leqslant B$; $AB = A$; $\max(A \cup B) = A$; for each $b \in B$, there is an $a \in A$ such that $b \leqslant a$; each element in B is a final segment of some element in A.

For $w \in F_X$, $A \in E_X$, we put $A.w = \{aw : a \in A\}$. Clearly $A.w \in E_X$ and we have an action of F_X on E_X. Furthermore, if $w \in F_X$, $A,B \in E_X$, then it is routine to verify that

$(AB).w = (A.w)(B.w)$,

and consequently the action is order-preserving.

For each element w of F_X we define w^* to be the singleton $\{w\} \in E_X$. We note that if $A = \{w_1,\dots,w_k\} \in E_X$, then $A = w_1^* \dots w_k^*$ so that E_X is generated by the set $\{w^* : w \in F_X\}$. We also observe that for any $A \in E_X$, $w \in F_X$ we have $A.w \leqslant \{w\} = w^*$.

Consider the free product $F_X * E_X$. Its elements can be written uniquely as words $a = s_1 \dots s_n$ where $s_i \in F_X \cup E_X$ and for $i = 1,\dots,n-1$ the elements s_i, s_{i+1} are not both in the same factor F_X or E_X. The number n is the length of a.

We extend * from F_X to $F_X * E_X$ as follows: for $A \in E_X$, we put $A^* = A$ and if $a = s_1 \dots s_n$ as above, $b = s_1 \dots s_{n-1}$, b^* has been defined and $b^* \in E_X$, then

$$a^* = \begin{cases} b^* s_n & \text{if } s_n \in E_X \\ b^*.s_n & \text{if } s_n \in F_X. \end{cases}$$

Let ρ be the relation $\{(aa^*,a) : a \in F_X * E_X\}$ and let $\sim$ be the congruence on $F_X * E_X$ generated by ρ.

We put $P_X = (F_X * E_X)/\sim$ and denote by ν the natural homomorphism from $F_X * E_X$ onto P_X. We will find it convenient to work with the monoid P_X^1. In order to derive properties of P_X^1 we first consider normal forms for its elements. Define a <u>normal form</u> to be a sequence $(n \geqslant 0)$

$$(w_o, A_1, w_1, A_2, w_2, \dots, A_n, w_n)$$

where

(i) $w_o, w_n \in F_X \cup \{1\}$,

(ii) $w_1,\dots,w_{n-1} \in F_X$,

(iii) $A_1,\dots,A_n \in E_X$,

(iv) $A_i < (w_o A_1 \dots w_{i-1})^*$ for $i = 1,\dots,n$.

For (iv), $A_1 < (1)^*$ will simply be taken to mean that $A_1 \in E_X$.

Given a normal form $\alpha = (w_o, A_1, \dots, A_n, w_n)$, then the product $\bar{\alpha} = w_o A_1 \dots A_n w_n$ is either 1 or an element of $F_X * E_X$ and so is a representative of an element of P_X^1. Observe that if $\bar{\alpha} \neq 1$, then either $w_n = 1$ and $\bar{\alpha}^* = A_n$ or $w_n \neq 1$ and $\bar{\alpha}^* = A_n.w_n$. We now show that any element in P_X^1 has a representative $\bar{\alpha}$ where α is a normal form, that is, any element in $(F_X * E_X)^1$ is related by $\sim$ to an $\bar{\alpha}$ for some normal form α. Certainly this holds for 1 and we assume inductively that it holds for any word in $F_X * E_X$ of length at most m. Let $a = s_1 \dots s_{m+1}$ have length m + 1. We have $s_1 \dots s_m \sim \bar{\alpha}$ for some normal form $\alpha = (w_o, A_1, \dots, A_n, w_n)$. Since $\sim$ is a congruence, $a \sim \bar{\alpha} s_{m+1}$. If $s_{m+1} \in F_X$, then $\beta = (w_o, A_1, \dots, A_n, w_n s_{m+1})$ is a normal form and $a \sim \bar{\beta}$. Otherwise, we have $s_{m+1} \in E_X$ and there are several possibilities: (a) when $w_n = 1$, $\beta_1 = (w_o,A_1,\ldots,A_n s_{m+1},w_n)$ is a normal form and $a \sim \bar{\beta}_1$; (b) when $w_n \neq 1$ and $\bar{\alpha}^* \leqslant s_{m+1}$, we have $a \sim \bar{\alpha}s_{m+1} \sim \bar{\alpha}\bar{\alpha}^*s_{m+1} \sim \bar{\alpha}\bar{\alpha}^* \sim \bar{\alpha}$; (c) when $w_n \neq 1$ and $\bar{\alpha}^* \nleqslant s_{m+1}$, we have $\bar{\alpha}^*s_{m+1} < \bar{\alpha}^*$ so that $a \sim \bar{\alpha}s_{m+1} \sim \bar{\alpha}\bar{\alpha}^*s_{m+1} \sim \bar{\beta}_2$ where $\beta_2 = (w_o,A_1,\ldots,A_n,w_n,\bar{\alpha}^*s_{m+1},1)$ is a normal form.

We have thus proved the existence part of the following theorem.

THEOREM 2.1 <u>Every element of P_X^1 can be represented uniquely by an element $\bar{\alpha}$ where α is a normal form.</u>

<u>Proof</u>. To prove the uniqueness, we construct a homomorphism from P_X^1 into $\mathcal{T}(\mathcal{N})$, the semigroup of all transformations of the set $\mathcal{N}$ of normal forms. For $w \in F_X$, we define $\psi(w)$ by

$$(w_o,A_1,\ldots,A_n,w_n)\psi(w) = (w_o,A_1,\ldots,A_n,w_nw).$$

Clearly $\psi(ww') = \psi(w)\psi(w')$ so that ψ is a homomorphism from F_X into $\mathcal{T}(\mathcal{N})$

For $B \in E_X$, we define $\psi(B)$ as follows:

(1) $\psi(B) = (1,B,1)$

and for $\alpha = (w_o,A_1,\ldots,A_n,w_n) \neq (1)$,

$$\alpha\psi(B) = \begin{cases} (w_o,A_1,\ldots,A_nB,w_n) & \text{if } w_n = 1 \\ (w_o,A_1,\ldots,A_n,w_n) & \text{if } w_n \neq 1 \text{ and } \bar{\alpha}^* \leqslant B \\ (w_o,A_1,\ldots,A_n,w_n,\bar{\alpha}^*B,1) & \text{otherwise.} \end{cases}$$

Let $B_1,B_2 \in E_X$. It is clear that for a normal form $\alpha = (w_o,A_1,\ldots,A_n,w_n)$ we have

$$\alpha\psi(B_1)\psi(B_2) = \alpha\psi(B_1B_2) \tag{2.2}$$

if $w_n = 1$. Assume that $w_n \neq 1$. If $\bar{\alpha}^* \leqslant B_1$ and $\bar{\alpha}^* \leqslant B_2$, it is again clear that (2.2) holds. If $\bar{\alpha}^* \leqslant B_1$ but $\bar{\alpha}^* \nleqslant B_2$, then $\bar{\alpha}^* \nleqslant B_1B_2$ and we have

$$\begin{aligned}\alpha\psi(B_1)\psi(B_2) &= \alpha\psi(B_2)\\ &= (w_o,A_1,\ldots,A_n,w_n,\bar{\alpha}^*B_2,1)\\ &= (w_o,A_1,\ldots,A_n,w_n,\bar{\alpha}^*B_1B_2,1)\\ &= \alpha\psi(B_1B_2).\end{aligned}$$

Finally, if $\bar{\alpha}^* \nleqslant B_1$, then $\bar{\alpha}^* \nleqslant B_1B_2$ and

$$\begin{aligned}\alpha\psi(B_1B_2) &= (w_o,A_1,\ldots,A_n,w_n,\ \bar{\alpha}^*B_1B_2,1)\\ &= (w_o,A_1,\ldots,A_n,w_n,\ \bar{\alpha}^*B_1,1)\psi(B_2)\\ &= \alpha\psi(B_1)\psi(B_2)\end{aligned}$$

so that (2.2) holds for all $\alpha \in \mathcal{N}$ and hence ψ is a homomorphism from E_X into $\mathcal{T}(\mathcal{N})$.

The universal property of free products now ensures that we have a homomorphism $\psi: F_X * E_X \to \mathcal{T}(\mathcal{N})$ defined by $\psi(s_1\ldots s_n) = \psi(s_1)\ldots\psi(s_n)$.

We next show that $\sim \subseteq \ker\psi$ so that there is an induced homomorphism $\psi^*: P_X \to \mathcal{T}(\mathcal{N})$. Since $\ker\psi$ is a congruence on $F_X * E_X$, it suffices to show that $\rho \subseteq \ker\psi$, that is we want $\psi(aa^*) = \psi(a)$ for all $a \in F_X * E_X$.

From the definitions, we see that for $\alpha \in \mathcal{N}$ and $B \in E_X$, we have $\alpha\psi(B) = \alpha$ if and only if $\bar{\alpha}^* \leqslant B$. Since we want $\alpha\psi(a)\psi(a^*) = \alpha\psi(a)$ for all $\alpha \in \mathcal{N}$ and all $a \in F_X * E_X$, it thus suffices to show that

$$\overline{\alpha\psi(a)}^* < a^* \tag{2.3}$$

for all $\alpha \in \mathcal{N}$ and all $a \in F_X{*}E_X$. First we note that from the definitions of ψ and $*$, we have $\overline{\alpha\psi(s)}^* = \bar{\alpha}^*s$ for $s \in E_X$ and $\overline{\alpha\psi(s)}^* = \bar{\alpha}^*.s$ for $s \in F_X$ and any normal form α. Now let $a = bs$ where $b \in (F_X{*}E_X) \cup \{1\}$ and $s \in F_X \cup E_X$. Let $\gamma = \alpha\psi(b)$ ($= \alpha$ if $b = 1$) and assume that $\bar{\gamma}^* \leqslant b^*$. If $s \in E_X$, we have

$$\overline{\alpha\psi(a)}^* = \overline{\gamma\psi(s)}^* = \bar{\gamma}^*s \leqslant b^*s = a^*$$

and if $s \in F_X$, we have

$$\overline{\alpha\psi(a)}^* = \overline{\gamma\psi(s)}^* = \bar{\gamma}^*.s < b^*.s = a^*.$$

Hence (2.3) follows by induction.

We now have a homomorphism $\psi^*: P_X \to \mathcal{T}(\mathcal{N})$ such that $\psi^*\nu = \psi$. Extend ψ^* to P_X^1 by mapping 1 to the identity map of $\mathcal{N}$. For a normal form α, it is clear from the definitions that

$$(1)\ \psi^*\nu(\bar{\alpha}) = (1)\psi(\bar{\alpha}) = \alpha$$

from which it follows that ψ^* is injective. Hence each element of P_X^1 is represented by a unique normal form and the proof is complete.

Following the customary practice in such matters we shall, from now on, be less careful than hitherto in distinguishing elements of P_X, elements of $F_X{*}E_X$ and normal forms. In fact, we shall regard words $s_1 \ldots s_n$ ($s_i \in F_X \cup E_X$) as elements of P_X, use $=$ to mean equality as elements of P_X and say that an element $w_oA_1 \ldots A_nw_n$ of P_X is in normal form when $(w_o, A_1, \ldots, A_n, w_n)$ is a normal form.

We use the symbol $\equiv$ to denote identity of words in $F_X{*}E_X$. Given a word a in $F_X{*}E_X$ we write $n(a)$ for the unique normal form associated with a so that in this notation, $a = n(a)$ and if b is in $F_X{*}E_X$, then $a = b$ if and only if $n(a) \equiv n(b)$. Whenever we write a word $v_oB_1 \ldots B_nv_n$ we understand that either (or when $n \geqslant 1$, both) of v_o, v_n may be 1 and that otherwise the v_i's belong to F_X and the B_i's belong to E_X.

Notice that with these conventions E_X and F_X are subsemigroups of P_X. In fact, E_X is the set of idempotents of P_X as we now show. For a word $a \equiv w_oA_1 \ldots A_kw_k$ we define the <u>content</u> of a to be the element $c(a) = w_o \ldots w_k$ of F_X^1. It is clear that c is actually a homomorphism from $F_X{*}E_X$ onto F_X^1 and also that $c(a) = c(aa^*)$. It follows that we may regard c as defined on P_X and that it is a homomorphism from P_X onto F_X^1. Since 1 is the only idempotent in F_X^1, we have $c(a) = 1$ for any idempotent a in P_X. It follows that a is in E_X and that E_X is the set of idempotents of P_X so that, in particular, the idempotents of P_X form a subsemilattice of P_X.

As a first application of normal forms we show that P_X can be made into a $*$-semigroup by defining $\nu(a)^*$ to be $\nu(a^*)$ for any word a in $F_X{*}E_X$. First, we prove that $a^* \equiv n(a^*)$. Let $a \equiv s_1 \ldots s_m$ where each s_i is in $F_X \cup E_X$ and consecutive s_i are not both in the same factor F_X or E_X. When $m = 1$ it is clear that $a^* \equiv s_1 \equiv n(a)^*$ if s_1 is in E_X and that $a^* \equiv \{s_1\} \equiv n(a)^*$ if s_1 is in F_X. Suppose inductively that $1 < m$ and $c^* \equiv n(c)^*$ where $c \equiv s_1 \ldots s_{m-1}$. Let $n(c) \equiv w_oA_1 \ldots A_nw_n$. If s_m is in F_X, then

$n(a) \equiv w_oA_1 \dots A_nw_ns_m$ and $n(a)^* \equiv A_n.w_ns_m \equiv (A_n.w_n).s_m \equiv n(c)^*.s_m \equiv c^*.s_m \equiv a^*$. Suppose that s_m is in E_X. Then $a^* \equiv c^*s_m \equiv n(c)^*s_m$ and consideration of the three possibilities for $n(a)$ shows that $a^* \equiv n(a)^*$.

Now if a,b are words in $F_X{*}E_X$ and if $\nu(a) = \nu(b)$, then the uniqueness of normal forms gives $n(a) \equiv n(b)$. Thus $a^* \equiv n(a)^* \equiv n(b)^* \equiv b^*$ and so certainly $\nu(a^*) = \nu(b^*)$. Hence putting $\nu(a)^* = \nu(a^*)$ gives a well-defined unary operation on P_X and with this operation ν is a *-homomorphism, that is, $\sim$ is a *-congruence.

Our next objective is to show that each $\mathcal{L}^*$-class of P_X contains an idempotent. Since $aa^* = a$ for every element a of P_X all we have to do is prove that $ax = ay$ implies $a^*x = a^*y$ for all x,y in P_X^1. Let $n(a) \equiv w_oA_1 \dots A_kw_k$ and suppose that $x = v_oB_1 \dots B_mv_m$ is in P_X. Then

$$ax = w_oA_1 \dots A_kw_kv_oB_1 \dots B_mv_m,$$
$$a^*x = 1(A_k.w_k)v_oB_1 \dots B_mv_m.$$

Following the procedure for reducing elements to normal form we see that if $w_k \neq 1$, then

$$n(ax) \equiv w_oA_1 \dots A_kw_ku_oC_1 \dots C_nu_n$$

for some element $u_oC_1 \dots C_nu_n$ in normal form and that

$$n(a^*x) \equiv \begin{cases} 1(A_k.w_k)u_o & \text{if } n = 0 \\ 1(A_k.w_k)u_oC_1 \dots C_nu_n & \text{if } n \neq 0,\ u_o \neq 1 \\ 1((A_k.w_k)C_1)u_1 \dots C_nu_n & \text{if } n \neq 0,\ u_o = 1 \end{cases}$$

Of course, if $k = 0$, then $A_k.w_k$ is replaced by w_o^* in the above. Thus $n(a^*x)$ is determined by $n(ax)$ and we see that if $ax = ay$ for some y in P_X, then $n(ax) \equiv n(ay)$ so that $n(a^*x) \equiv n(a^*y)$ and consequently $a^*x = a^*y$. Similar considerations show that this is also the case if $w_k = 1$ in the normal form of a. Finally, if $ax = a$, then we also have $ax = aa^*$ so that by the above, $a^*x = a^*a^* = a^*$. Hence for all x,y in P_X^1, $ax = ay$ implies $a^*x = a^*y$ and each $\mathcal{L}^*$-class of P_X contains an idempotent. We have now proved that P_X is right adequate.

Let $A,B \in E_X$, $a \in P_X$ and suppose that $a = bs$ where $s \in F_X \cup E_X$. If we assume that $(ABb)^* = (Ab)^*(Bb)^*$, then we have

$$(ABa)^* = (ABbs)^* = ((ABb)^*s)^* = ((Ab)^*(Bb)^*s)^*$$

and

$$(Aa)^*(Ba)^* = (Abs)^*(Bbs)^* = ((Ab)^*s)^*((Bb)^*s)^*.$$

If $s \in E_X$, it is clear that we obtain $(ABa)^* = (Aa)^*(Ba)^*$ so that the fact that P_X is right h-adequate will follow by induction on word length if we have $(ABw)^* = (Aw)^*(Bw)^*$ for all $A,B \in E_X$ and all $w \in F_X$. But this requirement is simply that $(AB).w = (A.w)(B.w)$ and it was noted that this holds when we defined the action of F_X on E_X. Thus we have established the first part of the following theorem.

THEOREM 2.6 Let X be a non-empty set. Then

(1) the semigroup P_X is right h-adequate and its semilattice of idempotents is E_X;

(2) P_X is free on the set X in the class of right h-adequate semigroups.

Proof. To prove (2) we have to show that any function from X into a right h-adequate semigroup S can be extended uniquely to a $*$-homomorphism from P_X into S. Since such a function can be extended uniquely to a homomorphism from F_X into S, we shall show that any homomorphism $\theta:F_X \to S$ can be uniquely extended to a $*$-homomorphism $P_X \to S$.

The first step is to define by means of θ, a homomorphism $\psi:E_X \to S$. If $A \in E_X$, then $A = \{w_1,\ldots,w_k\}$ is a finite non-empty subset of F_X such that no member of A is a final segment of any other member of A. We may write A as $w_1^* \ldots w_k^*$. On the other hand, if $v_1,\ldots,v_t \in F_X$, then $v_1^* \ldots v_t^* = \max\{v_1,\ldots,v_t\}$ so that $A = v_1^* \ldots v_t^*$ if and only if each v_i is a final segment of some w_j and each w_j occurs as some v_i. Now if v is a final segment of w, say $w = uv$, then $\theta(w) = \theta(u)\theta(v)$ so that $\theta(w)\theta(v)^* = \theta(w)$ giving $\theta(w)^*\theta(v)^* = \theta(w)^*$. It follows from these remarks that if we put

$$\psi(A) = \theta(v_1)^* \ldots \theta(v_t)^*$$

for any $v_1,\ldots,v_t \in F_X$ such that $v_1^* \ldots v_t^* = A$, then ψ is well-defined. It is clear that ψ is a homomorphism.

We immediately obtain a homomorphism $\phi:F_X{*}E_X \to S$ defined by

$$\phi(a) = \phi(w_oA_1\ldots A_nw_n) = \theta(w_o)\psi(A_1)\ldots\psi(A_n)\theta(w_n) \qquad (2.7)$$

where if $w_i = 1$ for $i = 0$ or $i = n$, we take $\theta(w_i) = 1$. We show by induction on the length of words that $\phi(aa^*) = \phi(a)$ for all words a. First, for $A \in E_X$, we have $\phi(AA^*) = \phi(A)$ and for $w \in F_X$, we have $\phi(ww^*) = \theta(w)\psi(w^*) = \theta(w)\theta(w)^* = \theta(w) = \phi(w)$ so that the claim holds for words of length one. Assume that the claim is true for all words of length n. If a is a word of length $n + 1$, then $a = bs$ where b has length n and $s \in F_X \cup E_X$. If $s \in E_X$, we have

$$\begin{aligned}\phi(aa^*) &= \phi(bsb^*s) = \phi(bb^*s) = \phi(bb^*)\phi(s)\\ &= \phi(b)\phi(s) = \phi(bs) = \phi(a).\end{aligned}$$

If $s \in F_X$, we have

$$\begin{aligned}\phi(aa^*) &= \phi(bs(b^*.s)) = \phi(b)\phi(s(b^*.s))\\ &= \phi(b)\theta(s)\psi(b^*.s)\\ &= \phi(bb^*)\theta(s)\psi(b^*.s)\\ &= \phi(b)\psi(b^*)\theta(s)\psi(b^*.s)\end{aligned}$$

Let $b^* = \{v_1,\ldots,v_k\} = v_1^* \ldots v_k^*$ so that

$$b^*.s = \{v_1s,\ldots,v_ks\} = (v_1s)^*\ldots(v_ks)^*$$

and

$$\psi(b^*)\theta(s)\psi(b^*.s) = \theta(v_1)^*\ldots\theta(v_k)^*\theta(s)\theta(v_1s)^*\ldots\theta(v_ks)^*.$$

However, $\theta(v_is)^* = (\theta(v_i)\theta(s))^* = (\theta(v_i)^*\theta(s))^*$ so that $\theta(v_i)^*\theta(s)\theta(v_is)^* = \theta(v_i)^*\theta(s)$ for $i = 1,\ldots,k$ and we conclude that $\psi(b^*)\theta(s)\psi(b^*.s) = \psi(b^*)\theta(s)$.

Hence

$$\phi(aa^*) = \phi(b)\psi(b^*)\theta(s) = \phi(b)\phi(b^*)\phi(s)$$
$$= \phi(bs) = \phi(a).$$

Thus the relation ρ is contained in kerϕ and consequently we may regard ϕ, defined by (2.7), as a homomorphism with domain P_X.

Finally, we wish to show that $\phi: P_X \to S$ is actually a *-homomorphism. For $w \in F_X$, $\phi(w^*) = \phi(w)^*$ by definition, and for $A \in E_X$, $\phi(A^*) = \phi(A) = \phi(A)^*$. Now, for $A = \{v_1,\ldots,v_k\} \in E_X$ and $w \in F_X$ we have

$$\phi((Aw)^*) = \phi(A.w)$$
$$= \phi((v_1w)^*\ldots(v_kw)^*)$$
$$= \phi(v_1w)^*\ldots\phi(v_kw)^*$$
$$= (\phi(v_1)\phi(w))^*\ldots(\phi(v_k)\phi(w))^*$$
$$= (\phi(v_1)^*\phi(w))^*\ldots(\phi(v_k)^*\phi(w))^*$$

and since S is right h-adequate, this gives

$$\phi((Aw)^*) = (\phi(v_1)^*\ldots\phi(v_k)^*\phi(w))^*$$
$$= (\phi(A)\phi(w))^*$$
$$= \phi(Aw)^*. \tag{2.8}$$

Now let $a = bs \in P_X$ where $b \in P_X$, $s \in E_X \cup F_X$. Assuming inductively that $\phi(b)^* = \phi(b^*)$, we have

$$\phi(a)^* = \phi(bs)^* = (\phi(b)\phi(s))^* = (\phi(b)^*\phi(s))^*$$
$$= (\phi(b^*)\phi(s))^* = \phi(b^*s)^*.$$

If $s \in E_X$, then $a^* = b^*s$ is idempotent so that we have $\phi(a)^* = \phi(b^*s) = \phi(a^*)$. If $s \in F_X$, then using (2.8) we have $\phi(a)^* = \phi((b^*s)^*) = \phi(b^*.s) = \phi(a^*)$. Thus ϕ is a *-homomorphism.

It is straightforward to verify that $\phi : P_X \to S$ is the only *-homomorphism which extends θ and hence we have shown that P_X is free on X in the class of right h-adequate semigroups.

As we have pointed out already, the elements of E_X may be written in the form $w_1^*\ldots w_k^*$ where the w_i are members of F_X. Thus the elements of P_X may be regarded as elements of F_X^*. On the other hand, every element of F_X^* represents a member of P_X and the word problem for P_X is the problem of deciding when two elements of F_X^* represent the same element of P_X.

Given an element a of F_X^*, by repeatedly applying the basic properties $(uv)^* = (u^*v)^*$, $(u^*)^* = u^*$, $(uv^*)^* = u^*v^*$, $(uw)^*(vw)^* = (u^*v^*w)^*$ of right h-adequate semigroups, we may reduce a to a member of $F_X{}^*E_X$ in a finite number of steps. That is, we may reduce a into a word $s_1\ldots s_n$ where each s_i is in $F_X \cup E_X$ and s_i is in F_X if and only if s_{i+1} is in E_X. Here we are thinking of E_X as the set of all finite products $w_1^*\ldots w_k^*$ where w_i is in F_X. To see this note that it is enough to be able to transform words of the form $(s_1\ldots s_n)^*$ where $s_1\ldots s_n \in F_X{}^*E_X$ into words of the form $w_1^*\ldots w_k^*$ where $w_i \in F_X$. Clearly repetition of such a process will reduce an arbitrary element of F_X^* to a member of $F_X{*}E_X$. Easy induction arguments show that $(u_1^*\ldots u_m^*)^* = u_1^*\ldots u_m^*$, $(u_1^*\ldots u_m^* w)^* = (u_1 w)^* = (u_1 w)^*\ldots(u_m w)^*$ and $(wu^*_1\ldots u_m^*)^* = w^* u_1^*\ldots u_m^*$ where $u_1,\ldots,u_m,w \in F_X$. It is not difficult to see that using these rules the desired transformation may be carried out in a finite number of steps. For example,

$$\begin{aligned}(v_1^*\, v_2\, v_3^*\, v_4^*\, v_5\, v_6^*\, v_7)^* &= ((v_1^*\, v_2)^*\, v_3^*\, v_4^*\, v_5\, v_6^*\, v_7)^* \\ &= ((v_1 v_2)^*\, v_3^*\, v_4^*\, v_5\, v_6^*\, v_7)^* \\ &= (((v_1 v_2)^*\, v_3^*\, v_4^*\, v_5)^*\, v_6^*\, v_7)^* \\ &= ((v_1 v_2 v_5)^*\, (v_3 v_5)^*\, (v_4 v_5)^*\, v_6^*\, v_7)^* \\ &= (v_1 v_2 v_5 v_7)^*\, (v_3 v_5 v_7)^*\, (v_4 v_5 v_7)^*\, (v_6 v_7)^*.\end{aligned}$$

Now suppose that we have an element $a = w_o A_1 w_1 \ldots A_n w_n$ of $F_X{*}E_X$ where $w_o, w_n \in F_X^1$, $w_1,\ldots,w_{n-1} \in F_X$ and $A_1,\ldots,A_n \in E_X$, each A_i being a word of the form $w_{i1}^*\ldots w_{ik}^*$ with $w_{ij} \in F$. We show that we can effectively reduce such a word to a word in normal form. First, we observe that for A, B in E_X we have a finite procedure for testing $A \leqslant B$. This follows from the fact that $A \leqslant B$ if and only if each element in B is a final segment of some element in A. Now w_o is certainly in normal form. Suppose that $b = w_o A_1 \ldots A_t w_t$ is in normal form and let $c = w_o A_1 \ldots A_{t+1} w_{t+1}$. Note that as above, we may effectively calculate b^*. If $b^* \leqslant A_{t+1}$, then

$$c = w_o A_1 \ldots A_t w_t w_{t+1}$$

and this expression is in normal form. If $b^* \nleqslant A_{t+1}$, then putting $B_{t+1} = b^* A_{t+1}$ and noting that $bb^* = b$, we have

$$c = w_o A_1 \ldots A_t w_t B_{t+1} w_{t+1}$$

and this expression is in normal form. A finite number of such steps produces a normal form for a.

Thus there is an effective procedure for reducing a word in F_X^* into the normal form of an element of P_X. Two words of F_X^* represent the same member of P_X if and only if they reduce to the same normal form. Clearly there is a finite procedure for checking whether two normal forms are identical. Thus we have

PROPOSITION 2.10 <u>The word problem for</u> P_X <u>is solvable</u>.

3. Properties of P_X

We start by considering the relations $\mathcal{L}^*$ and $\mathcal{R}^*$ on P_X.

LEMMA 3.1 <u>Let a be an element of</u> P_X <u>with normal form</u> $w_o A_1 \ldots A_n w_n$. <u>Then a is right cancellable in</u> P_X^1 <u>if and only if</u> $w_o \neq 1$.

<u>Proof</u>. If $w_o = 1$, then for $B \in E_X$ with A_1, B incomparable, we have $A_1 B \neq B$ but $A_1 B a = Ba$. Suppose that $w_o \neq 1$ and that $ba = ca$ where b, c have normal forms $v_o B_1 \ldots B_m v_m$, $u_o C_1 \ldots C_k u_k$ respectively. By considering the procedure for obtaining normal forms we see that the normal forms for ba and ca start with $v_o B_1 \ldots B_m v_m w_o$, $u_o C_1 \ldots C_k u_k w_o$ respectively. If $k < m$, then since normal forms are

unique, it follows that

$$v_oB_1\ldots B_kv_k = u_oC_1\ldots C_ku_kw_o$$

so that $v_i = u_i$ ($i = 0,1,\ldots,k-1$) and $v_k = u_kw_o$. On the other hand, by comparing contents, we see that $u_k = v_k\ldots v_m$ and we conclude that $v_k = v_k\ldots v_mw_o$, a contradiction since $w_o \neq 1$. It follows that $k = m$ and then consideration of normal forms leads to $b = c$.

LEMMA 3.2 <u>Let a be an element of</u> P_X <u>with normal form</u> $w_oA_1\ldots A_nw_n$ <u>and let</u> $A \in E_X$. <u>Then</u> $a\,\mathcal{R}^*A$ <u>if and only if</u> $w_o = 1$ and $A_1 = A$.

<u>Proof</u>. If $a\,\mathcal{R}^*A$, then $Aa = a$ so that clearly $w_o = 1$ and $AA_1 = A_1$. Since $w_o = 1$, we have $A_1a = a$ so that $A_1A = A$ giving $A = A_1$.

Suppose now that $w_o=1$ and $A_1 = A$. Certainly $Aa = a$. Let $ba = ca$ where $b,c \in P_X^1$ with $b = v_oB_1\ldots B_mv_m$ and $c = u_oC_1\ldots C_ku_k$ in normal form.

Since $c(b)c(a) = c(ba) = c(ca) = c(c)c(a)$, we have $c(b) = c(c)$, that is

$$u_o\ldots u_k = v_o\ldots v_m \tag{3.2.1}$$

If $c(b) = c(c) = 1$, then when $b = 1$, $c \neq 1$ we see that $n(ba)$ begins with $1Aw_1$ and that $n(ca)$ begins with $1C_1Aw_1$ so that $A = C_1A$, that is $bA = cA$. The same is true when $b \neq 1$, $c = 1$ and similar reasoning shows that if $b = B_1$, $c = C_1$, then we get $bA = B_1A = C_1A = cA$.

Now suppose that $c(b) = c(c) \neq 1$. We consider first the case when $m < k$.

If $v_m = 1$, then $n(ba)$ begins with $v_oB_1\ldots v_{m-1}$ and as $n(ca)$ begins with $u_oC_1\ldots C_{k-1}u_{k-1}$ we see that

$$v_o\ldots v_m = v_o\ldots v_{m-1} = u_o\ldots u_{m-1}.$$

Hence by (3.2.1), $u_o\ldots u_{m-1} = u_o\ldots u_k$, a contradiction since $m - 1 < k - 1$ but $u_{k-1} \neq 1$.

Hence $v_m \neq 1$. Now $n(ba)$ begins with $v_oB_1\ldots B_mv_m$ so that comparing with $n(ca)$ and using (3.2.1) we obtain

$$u_o\ldots u_k = v_o\ldots v_m = u_o\ldots u_m.$$

Thus $u_k = 1$ and $m = k - 1$. So $v_oB_1\ldots B_mv_m \equiv u_oC_1\ldots C_{k-1}u_{k-1}$ and the next factor in $n(ba)$ is $(B_m\cdot v_m)A$ and that in $n(ca)$ is C_kA. Hence $(B_m\cdot v_m)A = C_kA$ and so

$$bA = v_oB_1\ldots B_mv_mA = c_oC_1\ldots C_kA = cA.$$

The case when $k < m$ is treated similarly.

We now suppose that $k = m$. Comparison of $n(ba)$ and $n(ca)$ leads to $v_i = u_i$ for $i = 0,1,\ldots,k - 1$ and so in view of (3.2.1) we also have $v_m = u_k$. If $v_m \neq 1$, then comparing $n(ba)$ and $n(ca)$ yields $b = c$ so that certainly, $bA= cA$. When $v_m = 1$, $n(ba)$ begins with $v_oB_1\ldots B_mAw_1$ and $n(ca)$ begins with $u_oC_1\ldots C_mAw_1$ so that again we have $bA = cA$.

We have now shown that for any b,c in P_X^1, $ba = ca$ implies $bA = cA$. This demonstrates that $a\,\mathcal{R}^*A$.

PROPOSITION 3.3 Let a,b be elements of P_X with normal forms $w_oA_1\ldots A_nw_n, v_oB_1\ldots B_mv_m$ respectively. Then

(1) a $\mathcal{L}^*$ b if and only if $A_n.w_n = B_m.w_m$;

(2) a $\mathcal{R}^*$ b if and only if either w_o, v_o are both different from 1 or $w_o = v_o = 1$ and $A_1 = B_1$.

Proof. It is clear that (1) holds since we have seen that P_X is right adequate with $a^* = A_n.w_n$, $b^* = B_m.w_m$. That (2) holds follows readily from Lemma 3.1 and 3.2

COROLLARY 3.4 The monoid P_X^1 is left h-adequate.

The relation $\mathcal{D}^*$ on a semigroup S is defined to be the join $\mathcal{L}^* \vee \mathcal{R}^*$ of the relations $\mathcal{L}^*$ and $\mathcal{R}^*$ in the lattice of equivalence relations on S.

COROLLARY 3.5 On P_X the relation $\mathcal{D}^*$ is the universal relation, that is P_X is $\mathcal{D}^*$-simple.

Proof. Let $A \in E_X$ and $w \in A$. Then either $w^* = A$ and $w \mathcal{L}^* A$ or $A < w^*$ and wA is in normal form. In the latter case, $w \mathcal{R}^* wA \mathcal{L}^* A$. Hence always we have $w \mathcal{D}^* A$. Thus every idempotent is $\mathcal{D}^*$-related to a right cancellable element. Since two such elements are $\mathcal{R}^*$-related, it follows that all idempotents are in a single $\mathcal{D}^*$-class. The result follows.

Next we point out that Green's relations are all trivial on P_X.

PROPOSITION 3.6 On P_X, $\mathcal{J} = \iota$

Proof. Suppose that $a \mathcal{J} b$. Then a = cbd and b = sat for some members $c,d,s,t \in P_X^1$. Comparison of contents shows that c,d,s,t are all idempotent. Hence a = cad so that b = sat = scadt = csatd = cbd = a.

Since on any semigroup S the intersection of a set of left cancellative congruences is left cancellative, it follows that there is a minimum left cancellative congruence on S. When S is right adequate and $\phi:S \to T$ is a homomorphism of S onto a left cancellative semigroup T, then T is right adequate, that is, T is a monoid and ϕ is a *-homomorphism. For, if e, f are idempotents in S, then from $e\phi f\phi = e\phi(ef)\phi$ we obtain $f\phi = (ef)\phi$ and similarly we have $e\phi = (ef)\phi$. Hence all idempotents in S are mapped onto a single element, say k, of T. Now if $t \in T$, then $t = a\phi$ for some $a \in S$ so that $tk = a\phi a^*\phi = aa^*\phi = a\phi = t$ and k is a right identity for T. Since T is left cancellative, k is an identity. It is now clear that ϕ is a *-homomorphism. We shall denote the minimum left cancellative congruence on a right adequate semigroup by σ. Our next result describes σ on P_X.

PROPOSITION 3.7 <u>On P_X, we have</u>

(1) $(a,b) \in \sigma$ if and only if $c(a) = c(b)$;

(2) $P_X/\sigma \cong F_X^1$.

<u>Proof</u>. Obviously (2) follows from (1). Certainly ker c is a (left) cancellative congruence so that to establish (1) we have only to prove the claim that if ρ is a left cancellative congruence on P_X and $c(a) = c(b)$, then $(a,b) \in \rho$. From the remarks preceding the Proposition, it is clear that this is the case when $c(a) = c(b) = 1$. Assume inductively that the claim is true whenever $c(a)$ has length less than t (as a word in F_X^1). Let a,b have normal forms $w_oA_1\ldots A_nw_n, v_oB_1\ldots B_mv_m$ respectively and $c(a) = c(b)$ have length t. Now if w_o, v_o are both non-empty, let $w_o = xw_o'$, $v_o = xv_o'$ where $x \in X$. Then $a = xc$, $b = xd$ and $c(c) = c(d)$ so that by the induction hypothesis, $(c,d) \in \rho$ and hence $(a,b) \in \rho$. If $w_o = 1 = v_o$, then the same argument shows that $(w_1A_2\ldots A_nw_n, v_1B_2\ldots B_mv_m) \in \rho$ from which $(a,b) \in \rho$ follows. In the case where $w_o = xw_o'$ $(x \in X)$ and $v_o = 1$ we have $v_1 = xv_1'$ and $a = xc$, $b = B_1xd$ for some $c,d \in P_X^1$ with $c(c) = c(d)$ having length less that t. By the induction hypothesis $(c,d) \in \rho$ so that $(xc,xd) \in \rho$. From the remarks preceding the Proposition, $B_1\rho$ is the identity of P_X/ρ so that $(xd,B_1xd) \in \rho$ and hence $(a,b) \in \rho$. The case where $w_o = 1$ and $v_o \neq 1$ is, of course, similar.

We next make a few comments on the semilattice of P_X. Not surprisingly, E_X enjoys properties similar to those of the semilattice of the free inverse semigroup on X. For this latter semilattice is constructed from the free group on X in essentially the same way as E_X is constructed from F_X. (See e.g. [17], [20], [21]).

We point out first that E_X^1 is a distributive lattice. This follows easily from the fact that $E_X^1 \cong (I, \cup)$ where

$$I = (A : A \subseteq F_X,\ A \text{ is finite and } w \in A,\ v \leqslant w \Rightarrow v \in A)$$

and the isomorphism $I \to E_X^1$ is $A \to \max A$. The least upper bound of elements A,B of E_X^1 is just $\max(A \cap B)$.

It is easy to see that E_X satisfies the ascending chain condition and that the maximal elements of E_X are the elements $\{x\}$ where $x \in X$. Thus E_X has $|X|$ maximal elements and hence $P_X \cong P_Y$ implies $|X| = |Y|$.

As in [12], the maximal condition on principal right (left, two-sided) ideals is denoted by $M^R(M^L, M^J)$. It is easy to see that any semigroup in which $\mathcal{J} = \iota$ and which satisfies M^J also satisfies M^L and M^R. In the next proposition we use the fact that E_X satisfies the ascending chain condition to prove that P_X satisfies M^J. In view of [9], the principal left *-ideals of P_X are just the idempotent generated principal left ideals. Also, there is just one principal right *-ideal other than the idempotent generated principal right ideals, namely the principal right *-ideal generated by any right cancellable element. It will follow from the next proposition, then, that P_X satisfies the maximal condition for principal left (right) *-ideals.

PROPOSITION 3.8 P_X satisfies M^L, M^R and M^J.

Proof. Since $\mathcal{J} = \iota$ on P_X, it suffices to prove that M^J holds.

Let $I_1 \subseteq I_2 \subseteq \ldots$ be an increasing sequence of principal ideals, say $I_j = P_X^1 a_j P_X^1$. Then for $j \geq 2$, there are elements s_j, t_j such that $a_{j-1} = s_j a_j t_j$. Since $c(a_{j-1}) = c(s_j)c(a_j)c(t_j)$ and F_X^1 satisfies M^J, we see that for some positive integer k, $c(a_k) = c(a_{k+1}) = \ldots$

Thus for $k + 1 \leq j$, the elements s_j, t_j are idempotent so that $a_{j-1}a_j^* = s_j a_j t_j = a_{j-1}$ from which we obtain $a_{j-1}^* \leq a_j^*$. Now E_X satisfies the ascending chain condition so that $a_n^* = a_{n+1}^* = \ldots$ for some positive integer n. Since we also have $a_{j-1}t_j = a_{j-1}$, we get $a_{j-1}^* \leq t_j$ so that for $n + 1 \leq j$ we have $a_j^* \leq t_j$ and $a_{j-1} = s_j a_j$.

As P_X^1 is left adequate it is clear that a dual argument will show that $a_m = a_{m+1} = \ldots$ for some positive integer m and thus the sequence of principal ideals must terminate.

We complete this section by considering the question of residual finiteness. Our aim is to show that P_X is residually finite as a right adequate semigroup, that is, for any two distinct elements a,b of P_X we can find a finite right adequate semigroup S and a *-homomorphism ϕ from P_X onto S such that $a\phi = b\phi$.

We start by defining, for an element $w_oA_1\ldots A_nw_n$ of $F_X{*}E_X$, the set $\mathcal{C}(w_oA_1\ldots A_nw_n)$ to consist of all letters which occur in a word which is a member of $\{w_o,\ldots,w_n\} \cup \left(\bigcup_{i=1}^{n} A_i\right)$ Then we have the following lemmas.

LEMMA 3.9 If $w_oA_1\ldots A_nw_n$ is in normal form and if $x \in \mathcal{C}(w_oA_1\ldots w_{i-1})$, then $x \in \mathcal{C}(A_i)$.

Proof. If $i = 1$, then since $A_1 \leq w_o^* = \{w_o\}$, the result is clearly true. Assume inductively that the lemma holds for $i = k - 1$ and suppose $x \in \mathcal{C}(w_oA_1\ldots w_{k-1})$. Then $x \in \mathcal{C}(w_oA_1\ldots A_{k-2}w_{k-2}) \cup \mathcal{C}(A_{k-1}w_{k-1})$ and so by the induction hypothesis, $x \in \mathcal{C}(A_{k-1}w_{k-1})$. Clearly $\mathcal{C}(A_{k-1}w_{k-1}) = \mathcal{C}(A_{k-1}\cdot w_{k-1})$ and since $A_k \leq A_{k-1}\cdot w_{k-1}$ it follows that $x \in \mathcal{C}(A_k)$.

LEMMA 3.10 Let $a \in F_X{*}E_X$ and let $n(a)$ denote the normal form of a in P_X. If $x \in \mathcal{C}(a)$, then $x \in \mathcal{C}(n(a))$

Proof. Let $a \equiv s_1\ldots s_m s_{m+1}, s_i \in F_X \cup E_X$, s_i, s_{i+1} not both in F_X or both in E_X. If $m = 0$, then as $a \equiv n(a)$, the result is obvious. Assuming the result for $s_1\ldots s_m$, consider $a = n(s_1\ldots s_m)s_{m+1}$.

Let $n(s_1\ldots s_m) = w_oA_1\ldots A_nw_n$. If $s_{m+1} \in F_X$, then $n(a) = w_oA_1\ldots A_nw_ns_{m+1}$, so that if $x \in \mathcal{C}(a)$, then either $x \in \mathcal{C}(s_1\ldots s_m)$ or $x \in s_{m+1}$ and certainly therefore, $x \in \mathcal{C}(n(a))$. Otherwise $s_{m+1} \in E_X$ and $n(a)$ ends in $B1$ for some $B \in E_X$ with $B \leq s_{m+1}$ and $B \leq A_n\cdot w_n$. Now if $x \in \mathcal{C}(s_1\ldots s_m)$, then $x \in \mathcal{C}(w_oA_1\ldots A_nw_n)$ so that by Lemma 3.9, $x \in \mathcal{C}(A_nw_n)$ whence $x \in \mathcal{C}(A_n\cdot w_n)$. It follows that $x \in \mathcal{C}(s_1\ldots S_{m+1})$ implies that $x \in \mathcal{C}(B)$, so that certainly $x \in \mathcal{C}(n(a))$.

COROLLARY 3.11 <u>If</u> $x \in \mathcal{b}(a)$, <u>then</u> $x \in \mathcal{b}(a^*)$.

For a subset Y of X, define I(Y) by

$a \in I(Y)$ if and only if $\mathcal{b}(n(a))$ contains a member of Y.

LEMMA 3.12 I(Y) <u>is an ideal of</u> P_X <u>and is a left</u> *-<u>ideal</u>.

<u>Proof</u>. Let $a \in I(Y)$, $b \in P_X$. Certainly, both $\mathcal{b}(n(a)b)$ and $\mathcal{b}(bn(a))$ contains members of Y, so that by Lemma 3.10 so do $\mathcal{b}(n(n(a)b))$ and $\mathcal{b}(n(bn(a)))$. Hence $n(a)b$, $bn(a) \in I(Y)$ and I(Y) is an ideal. Let $a \in I(Y)$ and let $b \in P_X$ be such that $b\,\mathcal{L}^*\,a$. Let $n(a) = w_oA_1\ldots A_nw_n$, $n(b) = v_oB_1\ldots B_mv_m$. Then $A_n.w_n = n(a)^* = n(b)^* = B_mv_m$. Let $y \in \mathcal{b}(n(a))$. From Lemma 3.9, it follows that $y \in \mathcal{b}(A_n.w_n) = \mathcal{b}(B_m.v_m)$. Clearly, then, $y \in \mathcal{b}(n(b))$ and so $b \in I(Y)$.

LEMMA 3.13 <u>For a positive integer</u> k <u>let</u> J_k <u>be the ideal of</u> E_X <u>consisting of all</u> A's <u>which contain a word of length at least</u> k. <u>Now let</u>

$T_k = \{a \in P_X : a\,\mathcal{L}^*A \text{ for some } A \in J_k\}$.

<u>Then</u> T_k <u>is an ideal of</u> P_X <u>which is a left</u> *-<u>ideal</u>.

<u>Proof</u>. Certainly T_k is a union of $\mathcal{L}^*$-classes. Let $a \in T_k$, say $a\,\mathcal{L}^*A$ with $A \in J_k$ and let $b \in P_X$. Then $(ba)^* < a^* = A$ so that $(ba)^* \in J_k$ and $ba \in T_k$. On the other hand, $ab\,\mathcal{L}^*\,Ab$ and it is not difficult to see that $(Ab)^*$ must contain a word of length at least k, that is $(Ab)^* \in J_k$ and so $ab \in T_k$.

PROPOSITION 3.14 P_X <u>is residually finite in the class of right adequate semigroups</u>.

<u>Proof</u>. Let a,b be distinct elements of P_X with normal forms $w_oA_1\ldots A_nw_n$, $v_oB_1\ldots B_mv_m$ respectively. Choose an integer k greater than the length of any member of $A_n.w_n$ and $B_m.v_m$. Put $Y = X \setminus (\mathcal{b}(a) \cup \mathcal{b}(b))$. Let $I = I(Y) \cup T_k$. It is clear that P_X/I is finite and since I is a left *-ideal, we have by Proposition 1.5, that the natural homomorphism $\nu: P_X \to P_X/I$ is a *-homomorphism. On the other hand, neither a nor b is a member of I so that $a\nu \neq b\nu$.

We recall that an algebra is <u>hopfian</u> if all its surjective endomorphisms are automorphisms. In [8] Evans proves that any finitely generated resudually finite algebra in a variety of algebras is hopfian. Of course an algebra which is finitely generated and residually finite in a quasi-variety of algebras retains these properties in the variety generated by the quasi-variety. Thus every finitely generated residually finite right h-adequate semigroup must be hopfian. Thus we obtain the following corollary to Proposition 3.14.

COROLLARY 3.15 For a finite set X, P_X is hopfian.

4. Sets of free generators.

For a subset Y of a right adequate semigroup S we denote by $\langle Y\rangle^*$ the *-subsemigroup of S generated by Y. As we observed in Section 1, if S is right h-adequate then so is $\langle Y\rangle^*$. If $\langle Y\rangle^* = P_Y$, we say that Y is a set of free generators for $\langle Y\rangle^*$.

Inspired by the corresponding results of Reilly [21] in the inverse case we consider under what conditions Y is a set of free generators for $\langle Y\rangle^*$. This allows us to characterise those subsets Y of F_X which are sets of free gnerators for $\langle Y\rangle^*$ as the suffix codes over X. Consequently, if X has at least two elements, then P_X contains the free right h-adequate semigroup on a countably infinite set of generators as a *-subsemigroup. We also show that any non-idempotent a of P_X is a free generator of $\langle a\rangle^*$. We begin by observing that P_X determines X.

PROPOSITION 4.1 X is the only set of free generators for P_X.
Proof. For $A \subseteq P_X$ let $A^* = \{a^* : a \in A\}$. We can characterise X as the set of elements in P_X for which $X \cap X^* = \emptyset$ and X^* is the set of maximal members of E_X.

PROPOSITION 4.2 A subset Y of a right h-adeuate semigroup S is a set of free generators for Y * if and only if the following conditions hold where c_i, e_i are elements of $\langle Y\rangle^* \cup \{1\}$ with $e_i^2 = e_i$ and $e_i c_i = c_i$ for i = 1, 2:

(1) If $b_1, b_2 \in \langle Y\rangle \cup \{1\}$ are such that $e_i < b_i^*$ for i = 1,2 and if $b_1c_1 = b_2c_2$, then $b_1 = b_2$.

(2) If $b_1, b_2 \in \langle Y\rangle$ and h_1, h_2 are idempotents in $\langle Y\rangle^*$ such that $h_1b_1c_1 = h_2b_2c_2$, then $h_1 = h_2$; further if for each i either $c_i = 1$ or $e_i < (h_ib_i)^*$, then $b_1c_1 = b_2c_2$.

(3) If
$$(y_t \cdots y_1)^* > \prod_{j=1}^{m} (y_{jp(j)} \cdots y_{j1})^*$$
where the y_i and y_{jk} are elements of Y, then there is a j such that $y_i = y_{ji}$ for i = 1,...,t.

Proof. Let X be a set in one-one correspondence with Y and let $\theta: X \to Y$ be a bijection. Then θ extends uniquely to a *-homomorphism, which we shall also denote by θ, from P_X^1 onto $\langle Y\rangle^* \cup \{1\}$. Thus Y is a set of free generators for $\langle Y\rangle^*$ if and only if θ is injective.

Suppose first that θ is injective. We identify X with Y and P_X with $\langle Y\rangle^*$. Condition (1) follows from the uniqueness of normal forms because for i = 1, 2 we have $n(b_ic_i) = b_in(c_i)$ and $n(c_i)$ must begin with an idempotent since $e_ic_i = c_i$ and $e_i \neq 1$. Sinimilarly, condition (2) is a consequence of the uniqueness of normal forms because the normal form of $h_ib_ic_i$ must begin with h_i. If the additional condition holds, then we see that $n(h_i b_i c_i) = h_i b_i n(c_i) = h_i n(b_i c_i)$. If the hypothesis of condition (3) holds, then from the definition of P_X we see that $y_t \ldots y_1$ is a final segment of $y_{jp(j)} \ldots y_{j1}$ for some j. Hence the condition (3) holds.

Now suppose that conditions (1), (2) and (3) hold. Let A,B be idempotents of P_X. Then there are elements $v_1, \ldots, v_m$, $w_1, \ldots, w_n$ of F_X such that

$$A = v_1^* \ldots v_m^*, \; B = w_1^* \ldots w_n^*$$

and we may assume that

(α) for $i,j \in \{1,\ldots,m\}$, $v_j^* \leqslant v_i^* \Rightarrow i = j$,

for $k,\ell \in \{1,\ldots,n\}$, $w_k^* \leqslant w_\ell^* \Rightarrow k = \ell$.

Let $v_j = x_{j,p(j)} \ldots x_{j1}, w_k = x'_{k,r(k)} \ldots x'_{k1}$ where $j = 1,\ldots,m$; $k = 1,\ldots,n$; $p(j)$, $r(k)$ are positive integers and $x_{ji}, x'_{k\ell} \in X$ for $1 \leqslant i \leqslant p(j)$, $1 \leqslant \ell \leqslant r(k)$. If $A\theta = B\theta$, then

$$(v_1\theta)^* \ldots (v_m\theta)^* = A\theta = B\theta = (w_1\theta)^* \ldots (w_n\theta)^*.$$

Hence $B\theta \leqslant (v_j\theta)^*$ so that by condition (3), there is a number $j\sigma$ in $\{1,\ldots,n\}$ such that $x_{ji}\theta = x'_{j\sigma,i}\theta$ for $i = 1,\ldots,p(j)$. Since the restriction of θ to X is injective, we see that v_j is a final segment of $w_{j\sigma}$. Similarly $w_{j\sigma}$ is a final segment of some v_i. Condition (α) ensures that $i = j$. As each w_k must be a final segment of some v_j it follows from (α) that σ is a bijection so that $m = n$ and $A = B$. Thus θ restricted to E_X is injective.

Now if v,w are element of F_X and $v\theta = w\theta$, then $v^*\theta = (v\theta)^* = (w\theta)^* = w^*\theta$ so that $v^* = w^*$ and hence $v = w$.

Now let $w \in F_X$, $A \in E_X$ and suppose that

(β) $w\theta = A\theta$.

Let $w = x_1 \ldots x_k$ where $x_i \in X$. Put $y_i = x_i\theta \in Y$ for $i = 1,\ldots,k$. Then $y_1 \ldots y_k$ is idempotent and hence

$$(y_1 \ldots y_k y_1 \ldots y_k)^* = (y_1 \ldots y_k)^*$$

but this is impossible by condition (3) so that (β) cannot hold.

Now let w,A be as above and let b be an element of P_X with normal form $v_o B_1 \ldots B_m v_m$. We assume that b is not a member of $E_X \cup F_X$. Suppose that

(γ) $A\theta = b\theta$.

Taking $b_1 = 1$, $b_2 = v_o\theta$ and using the fact that $B_1 < v_o^*$ gives $B_1\theta < (v_o\theta)^*$, we see that condition (1) gives $1 = v_o\theta$ so that $v_o = 1$. It now follows from (γ) that $(B_1\theta)(A\theta) = (A\theta)$ and hence $B_1 A = A$. On the other hand,

$$(AB_1)\theta(v_1 \ldots B_m v_m)\theta = (A\theta)^2 = (B_1\theta)(v_1 \ldots B_m v_m)\theta$$

and since $v_1\theta \in \langle Y\rangle$, condition (2) applies giving $(AB_1)\theta = B_1\theta$.

Thus $A = AB_1 = B_1$. Now

$$A\theta = A^*\theta = b^*\theta = (B_m \cdot v_m)\theta$$

so that $A = B_m \cdot v_m$. However, by the definition of normal form we have $B_{i+1} < B_i \cdot v_i$ for $i = 1,\ldots,m-1$ and hence

$$A = B_m \cdot v_m \leqslant B_1.v_1 \ldots v_m = A.v_1 \ldots v_m.$$

This is clearly impossible since it requires each element in $A.v_1 \ldots v_m$ to be a final segment of some element in A. Thus an equation in the form (γ) cannot hold.

Suppose that

(δ) $w\theta = b\theta$.

As above we have $B_1\theta < (v_o\theta)^*$; choosing an element u in F_X which has w as a proper final segment we have $u^* < w^*$ so that $(u\theta)^* < (w\theta)^*$. In view of

$$(w\theta)(u\theta)^* = (v_o\theta)((B_1v_1\ldots B_mv_mu^*)\theta),$$

condition (1) gives $w\theta = v_o\theta$ so that $w = v_o$. But (δ) gives $w^* = b^* = B_m.v_m < \{v_o\ldots v_m\}$, a contradiction and thus equations of the form (δ) cannot hold.

Finally, let a be an element of P_X with normal form $w_oA_1\ldots A_nw_n$ and suppose that a is not a member of $E_X \cup F_X$. If $a\theta = b\theta$, then since $A_1\theta < (w_o\theta)^*$, $B_1\ \theta < (v_o\theta)^*$, condition (1) gives $w_o\theta = v_o\theta$ so that $w_o = v_o$. Now S is right adequate and

$$(w_o\theta)(A_1\theta)(w_1\ldots A_nw_n)\theta = (w_o\theta)(B_1\theta)(v_1\ldots B_mv_m)\theta$$

so that

$$(w_o\theta)^*(A_1\theta)(w_1\ldots A_nw_n)\theta = (w_o\theta)^*(B_1\theta)(v_1\ldots B_mv_m)\theta$$

and hence

$$(A_1\theta)(w_1\ldots A_nw_n)\theta = (B_1\theta)(v_1\ldots B_mv_m)\theta.$$

If $a = w_oA_1$ and $b = v_oB_1$ we now have $a = b$. Otherwise since equations of the form (γ) cannot hold we must have $w_1 \neq 1$ and $v_1 \neq 1$. Hence condition (2) gives $A_1\theta = B_1\theta$ so that $A_1 = B_1$.

If $a \neq w_oA_1w_1$, then $A_2 < (A_1w_1)^*$ so that $A_2\theta < ((A_1w_1)\theta)^*$. Similarly, either $b = v_oB_1v_1$ or $B_2\theta < ((B_1v_1)\theta)^*$. Hence from condition (2) we have

$$(w_1\ldots A_nw_n)\theta = (v_1\ldots B_mv_m)\theta$$

and it is now clear that repeated use of conditions (1) and (2) will yield $a = b$. This concludes the proof.

We now consider which subsets of F_X are sets of free generators for the *-subsemigroups of P_X which they generate. We first observe that conditions (1) and (2) are necessarily true for any subset Y of F_X. This follows by arguments about normal forms which are essentially the same as those used to prove the necessity of these conditions in Proposition 4.2.

Thus condition (3) is sufficient for Y to be a set of free generators for $\langle Y\rangle^*$. We use this observation in the following corollary. First, we recall that a subset C of F_X is a <u>suffix code over</u> X if it satisfies $F_XC \cap C = \emptyset$. We refer the reader to Chapter 5 of [16] for the essential facts about suffix codes and, in particular, for a proof that if C is a suffix code, then $\langle C\rangle$ is the free semigroup on C.

COROLLARY 4.3. <u>Let</u> X <u>be a set and</u> Y <u>be a subset of</u> F_X. <u>The *-subsemigroup</u> $\langle Y\rangle^*$ <u>of</u> P_X <u>is freely generated by</u> Y <u>if and only if</u> Y <u>is a suffix code over</u> X.

<u>Proof</u>. Suppose that $\langle Y\rangle^*$ is freely generated by Y. If $w \in F_XY \cap Y$, then w has a proper final segment, say u, in Y. Now $u^* = \{u\} > \{w\} = w^*$ but $u \neq w$ contradicting condition (3). Hence $F_XY \cap Y = \emptyset$ and Y is a suffix code over X.

Now suppose that Y is a suffix code over X. As remarked above we need only prove that condition (3) holds. Suppose then that the hypotheses of this condition are satisfied. Then $y_t \cdots y_1$ is a final segment of an element in $\prod_{j=1}^{m} (y_{jp(j)} \cdots y_{j1})^*$, that is, $y_t \cdots y_1$ is a final segment of $y_{jp(j)} \cdots y_{j1}$ for some j. Since Y is a suffix code, neither y_{j1} nor y_1 can be a proper final segment of the other and hence $y_1 = y_{j1}$ so that $y_t \cdots y_2$ is a final segment of $y_{jp(j)} \cdots y_{j2}$. Similarly, we obtain $y_i = y_{ji}$ for $i = 1, \dots, t$. Thus condition (3) holds.

Over sets with at least two elements there are infinite suffix codes and hence we have

COROLLARY 4.4 <u>If</u> $2 \leq |X|$, <u>then there is a countably infinite subset</u> Y <u>of</u> F_X <u>such that</u> Y <u>is a set of free generators for the *-subsemigroup</u> $\langle Y \rangle^*$ <u>of</u> P_X.

We now turn our attention to right h-adequate semigroups which are generated as *-semigroups by a single element. We shall show that any non-idempotent element a of P_X is a free generator for $\langle a \rangle^*$. Along the way we shall observe that if X has only one element, then no pair of distinct elements of P_X are free generators for the *-subsemigroup which they generate.

LEMMA 4.5 <u>Let</u> S <u>be a right h-adequate semigroup and let</u> $S = \langle a \rangle^*$. <u>Then</u>

$$E(S) = \{(a^k)^* : k \in N\}.$$

<u>Proof</u>. Consider P_X with $X = \{x\}$. We know that $E(P_X) = E_X$ is generated by the set $\{w^* : w \in F_X\}$. In our case $F_X = \{x^k : k \in N\}$. If $m \leq n$, then x^m is a final segment of x^n so that $(x^n)^* \leq (x^m)^*$. Hence

$E_X = \{(x^k)^* : k \in N\}$ and E_X is an ω-chain.

There is a surjective *-homomorphism $\theta : P_X \to S$ with $x\theta = a$. If e is an idempotent of S, then $e = e^*$ so that if $e = z\theta$ where $z \in P_X$, then

$$e = (z\theta)^* = (z^*)\theta = (x^k)^*\theta = (a^k)^*$$

for some $k \in N$.

From the proof of this lemma we see that if X has one element and $a, b \in P_X$, then one of $a^* \leq b^*$, $b^* \leq a^*$ holds. Hence, if $a \neq b$, then condition (3) of Proposition 4.2 is not satisfied by $\{a,b\}$ and we have shown

COROLLARY 4.6 <u>If</u> $|X| = 1$ <u>and</u> $a, b \in P_X$ <u>with</u> $a \neq b$, <u>then</u> a,b <u>are not free generators for</u> $\langle a,b \rangle^*$.

LEMMA 4.7 <u>Let</u> X <u>be a set and</u> a <u>be an element of</u> P_X <u>with normal form</u> $w_o A_1 \dots A_n w_n$. <u>Suppose that</u> a <u>is not idempotent. Then</u>

(1) $(a^{k+1})^* = (A_n . w_n)(A_n . w_n c(a)) \dots (A_n . w_n c(a)^k)$ <u>where</u> c(a) <u>is the content of</u> a

and $k \in N$,

(2) $h \neq k$ implies $(a^h)^* \neq (a^k)^*$,

(3) $w_o = 1$ and $h \neq k$ imply $A_1(a^h)^* \neq A_1(a^k)^*$.

Proof. By definition and the fact that F_X acts on E_X by semilattice homomorphisms, we have that for an idempotent B,

$$\begin{aligned}(Ba)^* &= (Bw_oA_1w_1\ldots A_nw_n)^*\\ &= (\ldots(((B.w_o)A_1).w_1)\ldots)A_n).w_n\\ &= (B.w_ow_1\ldots w_n)(A_1.w_1\ldots w_n)\ldots(A_n.w_n).\end{aligned}$$

Since $w_oA_1\ldots A_nw_n$ is in normal form, $A_{i+1} < A_i.w_i$ for $i = 1,\ldots,n-1$ and so

$$(Ba)^* = (B.c(a))(A_n.w_n).$$

Taking $B = (a^k)^*$ and assuming that (1) holds for $k - 1$, we have

$$\begin{aligned}(a^{k+1})^* &= ((a^k)^*a)^*\\ &= ((a^k)^*.c(a))(A_n.w_n)\\ &= (A_n.w_n)(A_n.w_nc(a))\ldots(A_n.w_nc(a)^k).\end{aligned}$$

Hence (1) holds by induction.

Notice that for any positive integer t the element $(a^t)^*$ is a set of elements of F_X and that any word of greatest length in $(a^t)^*$ is a member of $A_n.w_nc(a)^{t-1}$. Since $c(a) \neq 1$, it is thus clear that if $h < k$, then $(a^k)^*$ contains a word of greater length than any word in $(a^h)^*$. Hence $(a^h)^* \neq (a^k)^*$ and so (2) holds.

Now $a^* = A_n.w_n \leqslant A_1.c(a)$ so that each element in $A_1.c(a)$ is a final segment of some element in a^*. Hence, for any t, $(a^t)^*$ contains words of greater length than any word in A_1. Consideration of the lengths of the longest words in $A_1(a^h)^*$ and $A_1(a^k)^*$ shows that if $h \neq k$, then $A_1(a^h)^* \neq A_1(a^k)^*$. Hence (3) holds.

PROPOSITION 4.8 Let a be a non-idempotent of P_X. Then $\langle a\rangle^*$ is freely generated by a.

Proof. Using Lemma 4.5, it is easy to show that a typical element of $\langle a\rangle^*$ has the form

$$b = a^{k_o}(a^{h_1})^*a^{k_1}\ldots a^{k_p}$$

where k_o, k_p are non-negative integers and $k_1,\ldots,k_{p-1},h_1,\ldots,h_p$ are positive integers. To see that condition (1) of Proposition 4.2 holds, suppose that $0 \leqslant k_o < h_1$ and that

$$b = a^{t_o}(a^{s_1})^*a^{t_1}\ldots a^{t_m}$$

where $0 \leqslant t_o < s_1$. We have to show that $k_o = t_o$. If $k_o < t_o$, then we have

$$(a^{k_o})^*(a^{h_1})^*a^{k_1}\ldots a^{k_p} = (a^{k_o})^*a^{t_o-k_o}(a^{s_1})^*\ldots a^{t_m}.$$

Let a have normal form $w_oA_1\ldots A_nw_n$. If $w_o \neq 1$, then by uniqueness of normal forms we have

$$(a^{h_1})^* = (a^{k_o})^*(a^{h_1})^* = (a^{k_o})^*$$

so that by (2) of Lemma 4.7, $h_1 = k_o$, a contradiction. If $w_o = 1$, then since a is not idempotent, $w_1 \neq 1$ and now comparing normal forms gives

$$(a^{h_1})^*A_1 = (a^{k_o})^*(a^{h_1})^*A_1 = (a^{k_o})^*A_1 .$$

Now (3) of Lemma 4.7 gives $h_1 = k_o$, a contradiction. Similarly $t_o < k_o$ is impossible and we conclude that $k_o = t_o$.

For condition (2) of Proposition 4.2 we have to consider the equation

$(a^{h_1})^*c = (a^{s_1})^*d$

where $c = a^{k_1}$ or $c = a^{k_1}(a^{h_2})^* \ldots a^{k_p}$ with $(a^{h_2})^* < ((a^{h_1})^*a^{k_1})^* = (a^{h_1+k_1})^*$, and $d = a^{t_1}$ or $d = a^{t_1}(a^{s_2})^* \ldots a^{t_m}$ with $(a^{s_2})^* < (a^{s_1+t_1})^*$. If $w_o \neq 1$, then each of c,d has a left factor in F_X so that applying condition (2) to P_X gives $(a^{h_1})^* = (a^{s_1})^*$.

Similarly, if $w_o = 1$, then since $w_1 \neq 1$ we obtain $(a^{h_1})^*A_1 = (a^{s_1})^*A_1$ so that by (3) of Lemma 4.7 we have $(a^{h_1})^* = (a^{s_1})^*$.

From $(a^{h_1})^*c = (a^{h_1})^*d$ we obtain $a^{h_1}c = a^{h_1}d$ so that if $c = a^{k_1}$ and $d = a^{t_1}$ we have $a^{h_1+k_1} = a^{h_1+t_1}$ so that $k_1 = t_1$ and $c = d$. We next show that it is impossible to have $c = a^{k_1}$ and $d = a^{t_1}(a^{s_2})^* \ldots a^{t_m}$ with $s_2 > h_1+t_1$. Notice first that we may assume that $(a^{s_3})^* < ((a^{s_2})^*a^{t_2})^*$ so that $s_3 > s_2 + t_2$ and in general, $s_i > s_{i-1}+t_{i-1}$ for $i = 3,\ldots, m-1$. Hence

$s_{m-1} + t_m > h_1 + t_1 + \ldots + t_m.$

Now $(a^{h_1}d)^* = ((a^{s_{m-1}})^*a^{t_m})^*$ and $(a^{h_1}c)^* = (a^{h_1+k_1})^*$ so that $h_1 + k_1 = s_{m-1} + t_m$ and hence $k_1 > t_1 + \ldots + t_m$. But comparing contents gives $c(a^{k_1}) = c(a^{t_1}) \ldots c(a^{t_m})$ so that $k_1 = t_1 + \ldots + t_m$, a contradiction.

Similarly, it is impossible to have $d = a^{t_1}$ and $c = a^{k_1}(a^{h_2})^* \ldots a^{k_p}$ with $h_2 > h_1 + k_1$.

Now suppose that $c = a^{k_1}(a^{h_2})^* \ldots a^{k_p}$ with $h_2 > h_1 + k_1$ and that $d = a^{t_1}(a^{s_2})^* \ldots a^{t_m}$ with $s_2 > h_1 + t_1$. Then from $a^{h_1}c = a^{h_1}d$, by arguing as above, we obtain $k_1 = t_1$ and we thus deduce

$(a^{h_1+k_1})^*(a^{h_2})^*c_1 = (a^{h_1+k_1})^*(a^{s_2})^*d_1$

where $c_1 = 1$ or $c_1 = a^{k_2}$ or $c_1 = a^{k_2}(a^{h_3})^* \ldots a^{k_p}$ with $h_3 > h_2 + k_2$ and $d_1 = 1$ or $d_1 = a^{t_2}$ or $d_1 = a^{t_2}(a^{s_3})^* \ldots a^{t_m}$ with $s_3 > s_2 + t_2$. Thus

$(a^{h_2})^*c_1 = (a^{s_2})^*d_1$

and since a cannot be a factor of an idempotent, either $c_1 = d_1 = 1$ and $c = d$ or both $c_1 \neq 1$ and $d_1 \neq 1$. In the latter case repetition of the above arguments eventually leads to $c = d$.

Finally, if $(a^t)^* > \prod_{j=1}^{m} (a^{p(j)})^*$, then since the product is $(a^k)^*$ where $k = \max\{p(1),\ldots,p(m)\}$, we see that $t \leqslant k$ and so condition (3) of Proposition 4.2 holds.

The result now follows.

REFERENCES

1. A. Batbedat, 'Les demi-groups idunaires ou gamma-demi-groups' Cahiers Mathématiques 20, Montpellier, 1981.
2. A.H. Clifford, 'The free completely regular semigroup on a set', J. Algebra 59 (1979), 434-451.
3. A.H. Clifford and G.B. Preston, The algebraic theory of semigroups. Math. Surveys of the Amer. Math. Soc., 7 (Providence, R.I., 1961 (vol.1) and 1967 (vol.2)).
4. P.M.Cohn, Universal algebra. Harper and Row (New York, 1965).
5. A. El-Qallali, Structure theory for abundant and related semigroups (D.Phil. Thesis, University of York, 1980).
6. T. Evans, 'Finitely presented loops, lattices, et cetera are hopfian', J. London Math. Soc. 44 (1969), 551-552.
7. J.B. Fountain, 'A class of right PP monoids', Quart. J. Math. Oxford (2) 28 (1977), 285-300.
8. J.B. Fountain, 'Adequate semigroups', Proc. Edinburgh Math. Soc. 22 (1979), 113-125.
9. J.B. Fountain, 'Abundant semigroups', Proc. London Math. Soc. 44 (1982), 103-129.
10. G. Grätzer, Universal algebra. Van Nostrand (Princeton, N.J., 1968).
11. A. Horn and N. Kimura 'The category of semilattices', Algebra Universalis 1 (1971), 26-38.
12. E. Hotzel, 'On semigroups with maximal conditions', Semigroup Forum, 11 (1975/76), 337-362
13. J.M. Howie, An introduction to semigroup theory. Academic Press (London 1976).
14. G. Lallement, Semigroups and combinatorial applications. Wiley (New York 1979).
15. A.I. Mal'cev, Algebraic systems. Springer-Verlag (Berlin, 1973).
16. D. B. McAlister, 'One-to-one partial right translations of a right cancellative semigroup', J. Algebra 43 (1976), 231-251.
17. W.D. Munn, 'Free inverse semigroups', Proc. London Math. Soc. 29 (1974),385-404.
18. F. Pastijn, 'A representation of a semigroup by a semigroup of matrices over a group with zero'. Semigroup Forum 10 (1975), 238-249.
19. N.R. Reilly, 'Free generators in free inverse semigroups', Bull. Austral. Math. Soc. 7 (1972), 407-424.
20. N.R. Reilly, 'Free inverse semigroups', Algebraic Theory of Semigroups (G. Pollak, editor, Colloquia Mathematica Societatis Janos Bolyai, 20 North-Holland, 1979, pp.247-275).
21. H.E. Scheiblich, 'Free inverse semigroups', Proc. Amer. Math. Soc. 38 (1973),1-7.
22. B.M. Schein, 'Free inverse semigroups are not finitely presentable'. Acta Math. Acad. Sci. Hungar. 26 (1975), 41-52.

GROUP AND SEMIGROUP THEORETIC CONSIDERATIONS INSPIRED BY INVERSE PROBLEMS OF THE ADDITIVE NUMBER THEORY

Gregory A. Freiman and Boris M. Schein

School of Mathematical Sciences, Tel-Aviv University,
Ramat-Aviv, 69978 Tel-Aviv, Israel
Department of Mathematical Sciences, University of Arkansas,
Fayetteville, Arkansas 72701, USA

This paper contains an expanded version of a talk on algebraic systems with small squaring given by the first author during the traditional Mathematical Colloquium at the University of St. Andrews, Scotland, in the summer of 1984. New results for groups and semigroups are added. The collaboration of the authors on these topics started in Kääriku, Estonia, in 1976, continued in Moscow, Russia, in 1977-1979, in St. Andrews, Scotland, in 1984, Fayetteville, Arkansas, in 1985, and Tel-Aviv, Israel, in 1986. The authors are grateful to Graham Higman, Joseph Rotman and Elliot Weinberg for fruitful discussions.

1. Let K be a finite set of integers, $K \subset \mathbf{Z}$, $k = |K|$, $T = |K + K|$. It is easy to see that

$$2k - 1 \leq T \leq k(k+1)/2 \ . \tag{1}$$

Suppose that K has small squaring (which means that T is small). What is the structure of K?

Why may such problems be interesting? In the additive number theory one usually studies the possibility of representation of integers as sums of summands of a certain form. (These are so-called direct problems of the additive number theory; for example, the Waring problem.) Thus, it is desirable for T to be large. If we know the structure of K for small T, that is, if the inverse problem is solved, we may see when a given set of summands does not possess a desired property, and hence we may prove that T is large.

Certain inverse additive problems of this kind are considered in [4]. For example, if $T = 2k - 1$, then K is an arithmetic progression. If $T < 3k - 3$, then K is a subset of an arithmetic progression of length $T - k + 1$. A description of the structure of K is known [4] for $T < Ck$, where C is a positive constant not depending on k and k is sufficiently large.

Similar problems can be naturally raised for subsets K of groups or semigroups. In this case we replace $K + K$ by $K^2 = K \cdot K$ (i.e., we use multiplicative rather than the additive notation). For example, the following result holds:

THEOREM 1 (see [5]). The inequality $|K^2| < \frac{3}{2}|K|$ holds for every finite subset K of a group G if and only if either

(1) the subgroup of G generated by K has order $|K^2|$ (that is, K^2 is the subgroup);

or

(2) K is contained in a coset of G modulo a normal subgroup of order $|K^2|$.

In [5] an analogous result is obtained for $C = 1.6$. This problem for larger values of C, or for arbitrary C, is quite natural, albeit considerably more difficult.

Let G be a torsion-free group.

Let K be a "progression" with k elements

$$K = \{a, aq, \ldots, aq^{k-1}\} . \tag{2}$$

If

$$aq = qa , \tag{3}$$

then $K^2 = \{a^2, a^2q, \ldots, a^2q^{2k-2}\}$. If

$$aq = q^{-1}a , \tag{4}$$

then (2) implies $K^2 = \{a^2q^s : -(k-1) \le s \le k-1\}$. Thus $|K^2| = 2|K| - 1$ in both cases.

CONJECTURE 1. If $|K^2| = 2|K| - 1$, then K has the form (2) and either (3) or (4) holds.

REMARK. One can consider "progressions" of the form

$$M = \{b, qb, \ldots, q^{m-1}b\} . \tag{5}$$

However, each one of (3) or (4) ensures that progressions of the form (2) are also of the form (5) and vice versa.

Suppose that K and M are finite nonempty subsets of the torsion-free group G, that (2) holds for K, and (5) holds for M for some $a, b, q \in G$. Then $KM = \{aq^sb : 0 \le s \le k+m-2\}$ and $|KM| = |K| + |M| - 1$.

CONJECTURE 2. If $|K| \ge 2$, $|M| \ge 2$, and

$$|KM| \le |K| + |M| - 1,$$

then K and M are "progressions" of the form (2) and (5).

PROPOSITION 1. Conjecture 2 for $k = m > 1$ implies Conjecture 1.

Proof. Suppose that $K = M$ and $|K^2| = 2|K| - 1 > 1$. By Conjecture 3, $K = \{a, aq, \ldots, aq^{k-1}\}$ $= \{b, qb, \ldots, q^{k-1}b\}$ for some $a, b, q \in G$. Thus, for every $i = 0, 1, \ldots, k-1$, there exists j_i such that $aq^i = q^{j_i}b$ and $0 \le j_i \le k-1$. It follows that $q^{j_{i+1}}b = aq^{i+1} = aq^iq = q^{j_i}bq$, hence $q^{j_{i+1}-j_i}b = bq$. Therefore, $j_{i+1} - j_i$ does not depend on i, i.e., $j_0, j_1, \ldots, j_{k-1}$ is an arithmetic progression. This is possible only if either $j_i = i$ for all i or $j_i = k-(i+1)$ for all i. In the former case $a = aq^0 = q^{j_0}b = b$, $aq = qb = qa$ and so (3) holds. In the latter case $a = aq^0 = q^{k-1}b$ and $aq = q^{k-2}b = q^{-1}(q^{k-1}b) = q^{-1}a$, so that (4) holds.

PROPOSITION 2. Conjecture 2 holds in a torsion-free group if either $|K| = 2$ or $|M| = 2$.

Proof. As explained in the very beginning of our proof of Theorem 2, without loss of generality we may assume that $|K| \geq |M|$. Therefore we may assume that $|M| = 2$. Let $K = \{a_1, \ldots, a_k\}$ and let $M = \{b, c\}$ with $b \neq c$. Then $|KM| \leq k + 2 - 1 = k + 1$. Since Kb contains k different elements, Kc contains at most one element, say $a_k c$, which does not belong to Kb. Therefore, $\{a_1, \ldots, a_{k-1}\}c \subset Kb$, that is, $a_i c = a_{f(i)} b$ for every $i = 1, \ldots, k-1$, where $f(i) \in \{1, \ldots, k\}$. Obviously, f is a one-to-one mapping, for if $f(i) = f(j)$, then $a_i c = a_{f(i)} b = a_{f(j)} b = a_j c$ which implies $a_i = a_j$ and $i = j$. Let $q = cb^{-1}$. If $f^r(i) = i$ for some i and some positive r, then $a_i c = a_{f(i)} b$ implies $a_i q = a_{f(i)}$. Therefore $a_i q^2 = a_{f(i)} q = a_{f^2(i)}$, $a_i q^3 = a_{f^2(i)} q = a_{f^3(i)}, \ldots, a_i q^r = a_{f^r(i)} = a_i$, and hence $q^r = 1$, so that $q = 1$ and $b = c$ contrary to our assumption. Thus $f^r(i) \neq i$ for all i and all positive r. If f maps $\{1, \ldots, k-1\}$ onto itself, then f is a permutation and f^r is the identity mapping for a suitable positive r. As we have just seen, this is impossible. Thus k belongs to the range of f. We can always renumerate the elements $a_1, \ldots, a_{k-1}$ so that $f^{-1}(k) = k-1$, $f^{-1}(k-1) = k-2, \ldots, f^{-1}(2) = 1$. Thus $a_i q = a_{i+1}$ for all $i = 1, \ldots, k-1$, and K has form (2), while, $M = \{b, qb\}$. If $|KM| = k$ then $Kb = Kc$, and hence $K = Kq$. It follows that $K = Kq^n$ for every n, so that $a_1 q^n \in K$ for all n. This contradicts the finiteness of K. The case when $|KM| < k$ is impossible. This completes the proof.

PROPOSITION 3. Conjecture 2 holds in linearly orderable groups.

Proof. Let G be a linearly orderable group. Choose an order relation $<$ on G. Of course, G is torsion-free. Let $K = \{x_1, \ldots, x_k\}$ and $M = \{y_1, \ldots y_m\}$, where $x_1 < \ldots < x_k$ and $y_1 < \ldots < y_m$. Then, for every $i = 1, \ldots, m-1$,

$$x_1 y_1 < x_1 y_2 < \ldots < x_1 y_i < x_2 y_i < x_2 y_{i+1} < \ldots < x_2 y_m < x_3 y_m < \ldots < x_k y_m \,. \tag{6}$$

The number of different terms of the form $x_i y_j$ in (6) is $k + m - 1$. Therefore the inequalities $x_1 y_i < x_1 y_{i+1} < x_2 y_{i+1}$ imply $x_2 y_i = x_1 y_{i+1}$, so that $y_{i+1} y_i^{-1} = x_1^{-1} x_2$. The right-hand side of this equality does not depend on i. Let $p = x_1^{-1} x_2$. Then $y_2 = p y_i$, $y_3 = p y_2, \ldots, y_m = p y_{m-1}$, whence $y_i = p^{i-1} y_1$ for $i = 1, 2, \ldots, m$. Analogously, $x_i^{-1} x_{i+1} = p$ for $i = 1, 2, \ldots, k-1$, whence $x_i = x_1 p^{i-1}$ for $i = 1, 2, \ldots, k$.

REMARK. Conjecture 2 holds in torsion-free abelian groups because such groups are linearly orderable. It also holds in any cancellative linearly orderable groupoid [= an algebra with a binary operation such that any one of $ab = ac$ or $ba = ca$ implies $b = c$]. In such groupoids $a < b$ implies both $ac < bc$ and $ca < cb$, which are the only properties we used in our proof of Proposition 3. Thus, Conjecture 2 holds in free groups, free loops, and free semigroups.

We prove Conjecture 2 in the case when k is large with respect to m.

THEOREM 2. Suppose K and M are finite subsets of a torsion-free group G, $|K| = k \geq 2$, $|M| = m \geq 2$, and

$$|KM| \leq k + m - 1 \,. \tag{7}$$

If

$$k \geq 3(m-1)^5 \,, \tag{8}$$

then there exist $a, b, p \in G$ such that

$$K = \{a, ap, \ldots, ap^{k-1}\} \tag{9}$$

and

$$M = \{b, pb, \ldots, p^{m-1}b\}\ . \tag{10}$$

Proof. We may assume that $k \geq m$ in Conjecture 3. Indeed, let Theorem 2 hold for $k \geq m$. Let $A^{-1} = \{a^{-1} : a \in A\}$ for any subset A of G. If $k < m$, replace K by M^{-1} and M by K^{-1}. Then $|M^{-1}| = |M| = m > k = |K| = |K^{-1}|$ and $|M^{-1}K^{-1}| = |(KM)^{-1}| = |KM| \leq k+m-1$. Thus, M^{-1} has the form (9) and K^{-1} form (10). Then K and M have forms (9) and (10), respectively, where a is replaced by b^{-1}, b is replaced by a^{-1}, and p is replaced by p^{-1}.

If we replace K and M by cK and Md, respectively, for some $c, d \in G$, the conditions of Theorem 2 still hold. Therefore we may assume that $1 \in K$ and $1 \in M$. Let $M = \{y_1, \ldots, y_m\}$ with $y_1 = 1$. Assume that $q = y_j$, the element y_j to be chosen later. Let $Q = \langle q \rangle$ denote the cyclic subgroup of G generated by q. For every $a \in G$ we order the set aQ linearly as follows: $aq^u \leq aq^v \leftrightarrow u \leq v$. Let $a_1Q, \ldots, a_nQ$ be those left cosets of Q which intersect K. In the sequel of the proof we use a finite number of left cosets of Q which are disjoint from K. All of them can be numbered, and so we assume that if $j > n$ for a left coset a_jQ, then $a_jQ \cap K = K_j = \emptyset$. Clearly, $n > 0$. Let $c_i = a_iq^v$ be the maximal element of $K \cap a_iQ$ for $i = 1, \ldots, n$. Then $c_i \in K, q \in M, c_iq \in KM$, and $c_iq \notin K$. Thus, the elements $c_iq, \ldots, c_nq$ belong to $KM \backslash K$. Now $1 \in M$ implies $K \subset KM$. Therefore $KM| \geq |K| + |KM\backslash K| \geq k + n$. By (7)

$$n \leq m-1. \tag{11}$$

Our choice of $q = y_j$ determines n. We choose $q = y_j$ in such a way as to make the value of n *minimal*.

Suppose that $n = 1$. Then $1 \in K$ implies $K \subset Q$. If there exists $y_i \in M\backslash Q$, then $K \cap Ky_i = \emptyset$, whence $|KM| \geq 2k > k + m - 1$ which contradicts (7). Thus $M \subset Q$. Now, Q is a linearly orderable group. Applying Proposition 3 we prove that (9) and (10) hold for some $a, b, p \in Q$.

Now we assume that $n > 1$. By (11)

$$m \geq 3. \tag{12}$$

Let $V_i = \{v \in \mathbf{Z} : a_iq^v \in K_i\}$, where $K_i = K \cap a_iQ$ and $1 \leq i \leq n$. A nonempty segment of $\mathbf{Z}$

$$\{v : e \leq v \leq f\} \subset V_i \tag{13}$$

is called ***maximal*** if $e - 1 \notin V_i$ and $f + 1 \notin V_i$. Let g be the number of maximal segments whose union is V_i. For each of the maximal segments (13) we have $a_iq^{f+1} = a_iq^f \cdot q \in KM$. Therefore $|KM\backslash K| \geq g$, and, by (7),

$$g \leq m-1. \tag{14}$$

Let p be a natural number such that $p \leq m$. Deleting at most $p-1$ elements from each of the g segments (13) we can turn them into (possibly, empty) segments of lengths divisible by p. It follows from (14) that the number of elements deleted from V_i does not exceed $g(p-1) \leq (m-1)(m-1) = (m-1)^2$. After that each V_i turns into a set W_i which is a disjoint union of segments (not necessarily maximal ones) of length p. If $p = m$ then, as follows from (8) and (11), there exists i such that

$$|K_i| \geq \frac{3(m-1)^5}{m-1} = 3(m-1)^4 > (m-1)^2, \tag{15}$$

so that W_i contains at least one segment of length m. Suppose that such a segment consists of the following elements of K_i:

$$a_iq^u, a_iq^{u+1}, \ldots, a_iq^{u+m-1}. \tag{16}$$

By (12) there exists y_i such that $y_i \neq 1$. We shall prove that

$$y_i^s = q^r \tag{17}$$

for some $s > 0$ and $0 < |r| < m$.

For each $t, u \leq t \leq u+m-1$, let v be the maximal nonnegative integer depending on t such that $a_iq^ty_i^v \in K$. Clearly, such v exists. Then $a_iq^ty_i^{v+1} \in (KM\backslash K) \cap a_iq^t\langle y_i\rangle$. If left cosets $a_iq^t\langle y_i\rangle$ are different for all t, then $KM\backslash K$ has at least m elements. This contradicts (7). Therefore there exist t and x such that $u \leq t < t+x \leq u+m-1$ and $a_iq^{t+x}\langle y_i\rangle = a_iq^t\langle y_i\rangle$, i.e. $q^x \in \langle y_i\rangle$. Thus (17) holds. Clearly, s in (17) determines r in a unique way.

Let s_i be the minimal positive s in (17) such that

$$y_i^{s_i} = q^{r_i}. \tag{18}$$

Then $s_i|s$ for every s which satisfies (17) for some r. Thus, s and r in (17) are of the form $s = s_it$ and $r = r_it$ for $t \in \mathbf{Z}$.

We say that a_iQ is *connected* with a_jQ for some i and j whenever $a_iQ \cap a_jQ = \emptyset$ and $a_iQM \cap a_jQ \neq \emptyset$, that is,

$$a_iq^sy \in a_jQ \tag{19}$$

for some $s \in \mathbf{Z}$ and $y \in M$. Clearly, $y \neq 1$. We say that every such element a_iq^s *connects* a_iQ with a_jQ. This concept makes sense for any two cosets a_iQ and a_jQ, not necessarily for those with non-empty K_i and K_j.

Our immediate goal is to prove that if a_iQ is connected with a_jQ then

$$|K_j| \geq |K_i| - 3(m-1)^2. \tag{20}$$

If $i > n$, that is, if K_i is empty, then (20) holds. So we may suppose that $K_i \neq \emptyset$. Assume that (19) holds for some a_i. Consider the set $K_i \cap a_iq^s\langle q^{r_i}\rangle$, where r_i satisfies (18). Each element of this set connects a_iQ with a_jQ. Indeed, each element of this set has the form $a_iq^{s'} = a_iq^sq^{tr_i}$, where $a_iq^{s'} \in K_i$, so that $s' \in V_i$. By (19) $a_iq^sy \in a_jQ$, so that $a_iq^{s'}y = a_iq^sq^{tr_i}y = a_iq^sy^{ts_i}y = a_iq^syy^{ts_i} = a_iq^syq^{tr_i} \in a_jQq^{tr_i} = a_jQ$. Therefore the element $a_iq^{s'}$ connects a_iQ with a_jQ.

To estimate $|K_i \cap a_iq^s\langle q^{r_i}\rangle|$ we recall that no more than $(m-1)^2$ elements have been deleted from V_i to obtain W_i. Each segment of length r_i in W_i contains one element s' such that $s' \equiv s \pmod{|r_i|}$. Thus, W_i contains at least $\frac{|K_i|-(m-1)^2}{|r_i|}$ of such elements s'. It follows that

$$|K_i \cap a_iq^s\langle q^{r_i}\rangle| \geq \frac{|K_i| - (m-1)^2}{|r_i|}. \tag{21}$$

Suppose that $a_iq^{s'} \in K_i$ and $a_iq^{s'}y \notin K_j$. Since $a_iq^{s'}y \in a_jQ$, the condition $a_iq^{s'}y \notin K_j$ means that $a_iq^sy \notin K$. However, $a_iq^sy \in KM$. It follows that $a_iq^sy \in KM\backslash K$. But by (7), $|KM\backslash K| \leq m-1$. Therefore the number of elements of the form $a_iq^{s'}$ which belong to K_i and such that $a_iq^{s'}y \notin K_j$ does not exceed $m-1$. As we have seen, $a_iq^{s'}y \in a_iq^sy\langle q^{r_i}\rangle$. Let $P = |K_j \cap a_iq^sy\langle q^{r_i}\rangle|$. Then

$$|P| \geq \frac{|K_i| - (m-1)^2}{|r_i|} - (m-1) \geq \frac{|K_i| - 2(m-1)^2}{|r_i|}. \tag{22}$$

Every element of P has the form a_jq^v. Suppose that w is the least integer, $0 < w < |r_i|$, such that $a_jq^{v+w} \notin K$. Since $q \in M$, we have $a_jq^{v+w-1}q \in KM\backslash K$. By (7), $|KM\backslash K| \leq m-1$, therefore there are at most $m-1$ integers w satisfying the above requirement. Thus there exist at least $|P|-(m-1)$ elements $a_jq^v \in P$ for which all $|r_i|$ terms of the sequence $a_jq^{v+t}, t = 0, 1, \ldots, |r_i|-1$, belong to K. All terms of these sequences belong to a_jQ, hence they belong to $K_j = K \cap a_jQ$. Moreover, by the definition of P, all terms of all such sequences are different. Therefore K_j contains at least $[|P|-(m-1)]\cdot|r_i|$ elements. Applying (22) and $|r_i| < m$ we obtain $|K_j| \geq |K_i| - 2(m-1)^2 - (m-1)|r_i| \geq |K-i| - 3(m-1)^2$. This proves (20).

Let $H = H_0$ denote the union of all a_iQ, $1 \leq i \leq n$. We say that a coset a_iQ is connected with a subset $D \subset G$ whenever it is connected with some coset $a_jQ \subset D$. Let H_1 denote a subset of H obtained by deleting from H all cosets connected with $G\backslash H$. We define the sequence

$$H_0 \supset H_1 \supset \cdots \supset H_i \supset \cdots \tag{23}$$

by induction. Suppose that H_i has been defined. Then H_{i+1} is the subset of H_i obtained by deleting from H_i all cosets connected with $G\backslash H_i$. Since H contains a finite number n of cosets, the sequence (23) is finite. The last nonempty set H_e either

(1) consists of cosets each of which is connected with $G\backslash H_e$, so that all subsequent terms of the sequence (23) are empty sets,

or

(2) H_e consists of the classes each of which is not connected with $G\backslash H_e$.

In the former case, for every coset $b_eQ \subset H_e$ there exists a sequence

$$b_eQ, b_{e-1}Q, \ldots, b_0Q, bQ$$

of cosets such that $b_iQ \subset H_i\backslash H_{i+1}$ for $0 \leq i \leq e, bQ \not\subset H$, and each of these cosets is connected with the subsequent coset. Let $K^i = K \cap b_iQ$. Applying (20) successively we obtain $|K^i| \leq |K^{i-1}| + 3(m-1)^2 \leq |K^{i-2}| + 3\cdot 2(m-1)^2 \leq \cdots \leq |K^1| + 3(i-1)(m-1)^2 \leq |K^0| + 3i(m-1)^2 \leq 3(i+1)(m-1)^2 \leq 3(e+1)(m-1)^2$. Clearly, $e+1 \leq n$. By (11),

$$|K^i| \leq 3(m-1)^3. \tag{24}$$

It follows from (11) and (24) that

$$k = |K| \leq n \cdot max_{1 \leq i \leq n}|K_i| \leq n \cdot 3(m-1)^3 \leq 3(m-1)^4,$$

which contradicts (8). This shows that case (1) is impossible, so that case (2) takes place. This means that $H_eM \subset H_e$.

We can assume that, in this case, $H_e = H_0$, that is, H_0 does not contain cosets connected with $G\backslash H_0$. To prove that consider $K' = K \cap H_e$. This set is obtained from K by deleting subsets of the form $K \cap b_iQ$. By (24) each of these subsets contains $3(m-1)^3$ elements at most. Since K intersets n cosets and k' is not empty, we obtain, by (11),

$$|K'| \geq k - (n-1)\cdot 3(m-1)^3 > k - 3(m-1)^4. \tag{25}$$

Now, $K' \subset K'M$, since $1 \in M$. Obviously, $K'M \subset KM$. It follows from $K' \subset H_e$ and $H_eM \subset H_e$ that $K'M\backslash K' \subset KM\backslash K$. Suppose that the set $H_0\backslash H_e$ is not empty. Then it contains a coset a_iQ for which $K_i \neq \emptyset$. In the proof of formula (11) we saw that $c_iq \in KM\backslash K$. However, $c_iq \notin H_e$, and so $c_iq \notin K'M$. It follows that the inclusion $K'M\backslash K' \subset KM\backslash K$ is proper. By (7) $|KM\backslash K| \leq m-1$, so that $|K'M\backslash K'| \leq m-2$. Therefore

$$|K'M| \leq |K'| + m - 2. \tag{26}$$

Now we proceed by induction on n. For $n = 1$ the theorem holds, as we have noticed just before formula (12) was proved. Suppose that the theorem holds for any number of cosets less than n.

Let y be any element of m different from 1 and q. Then (26) and (25) show that K' and $M\backslash\{y\}$ satisfy conditions (7) and (8) in which k and m are replaced by $|K'|$ and $m-1$, respectively. Since K' intersects fewer than n cosets the inductive assumption works. Therefore, K' satisfies (9). Since $1 \in K'$, we see that $K' \subset Q$, so that K' is contained in a single coset. Analogously, $M\backslash\{y\} \subset Q$. It follows that if $y \notin Q$, then, by (25),

$$|KM| \geq |K'M| \geq |K'| + |K'y| = 2|K'| > 2[k - 3(m-1)^4] > k + m - 1.$$

This contradicts (7), and so $M \subset Q$. Every K_i is contained in a single left coset of Q, and, as we have seen just after formula (11) was introduced, in this case $|K_iM| \geq k + m - 1$. Therefore, if $n > 1$, then

$$|KM| = |\bigcup_{i=1}^{n} K_iM| = \sum_{i=1}^{n}(|K_i| + m - 1) = \sum_{i=1}^{n}|K_i| + n(m-1) = k + n(m-1) > k + m - 1,$$

which contradicts (7). So $n = 1$. In this case, as we have seen, the theorem is true. Therefore we can assume that $H_0\backslash H_e = \emptyset$, that is, $H_e = H_0$. Therefore, $H_y \subset H$ for every $y \in M$. By (18), $y^{-s} \in Q$, so that $Hy^{-1} = Hy^{-s}y^{s-1} \subset Hy^{s-1} \subset H$. It follows that

$$H\langle M\rangle = H. \tag{27}$$

(27) implies that $\langle M\rangle \subset H$ because $1 \in Q \subset H$. Also $Q \subset \langle M\rangle$ because $q \in M$.

If $b\langle M\rangle \cap H \neq \emptyset$ for some $b \in G$, that is, if $by = h$ for some $y \in \langle M\rangle$ and $h \in H$, then $b = hy^{-1}$ and, by (27), $b \in H$ and $b\langle M\rangle \subset H$. Thus, H is a union of left cosets of $\langle M\rangle$. Thus the number n of left cosets of Q in H is divisible by the number u of left cosets of Q in $\langle M\rangle$. Assume that

$$\langle M\rangle \neq H. \tag{28}$$

Then $\frac{n}{u} \geq 2$ and so

$$u \leq \frac{n}{2}. \tag{29}$$

The number of right cosets of Q in M is u. by (29) there exists a right coset Qb which contains at least $\frac{m}{\frac{n}{2}} = \frac{2m}{n}$ elements of M.

Let $M_1 = Mb^{-1}$ and let $M_0 = M_1 \cap Q$. Then $|M_0| \geq \frac{2m}{n}$ and $|KM| = |KM_1| \geq |KM_0|$. Since $M_0 \subset Q$, it is easily seen that $|KM_0| = \sum_{i=1}^{n}|K_iM_0|$. Now, $a_i^{-1}K_i \subset Q$ and $M_0 \subset Q$. By Proposition 3, $|K_iM_0| = |a_i^{-1}K_iM_0| \geq |a_i^{-1}K_i| + |M_0| - 1 = |K_i| + |M_0| - 1$. It follows that

$$|KM| \geq \sum_{i=1}^{n}|K_iM_0| \geq \sum_{i=1}^{n}(|K_i| + |M_0| - 1) = k + n(|M_0| - 1) \geq$$
$$\geq k + n\left(\frac{2m}{n} - 1\right) = k + 2m - n > k + m,$$

which contradicts (7). Thus (29) is impossible, and hence $\langle M\rangle = H$.

Consider the group $\langle M\rangle$ and its subgroups $Q = \langle q\rangle$ and $\langle y_i\rangle$. It follows from $\langle M\rangle = H$ that $[\langle M\rangle : Q] = n$. By (18), $[Q : (\langle q\rangle \cap \langle y_i\rangle)] = |r_i|$ and $[\langle y_i\rangle : (\langle q\rangle \cap \langle y_i\rangle)] = s_i$. Therefore $[\langle M\rangle : (\langle q\rangle \cap \langle y_i\rangle)] = n \cdot |r_i|$ and

$$n_i = [\langle M\rangle : \langle y_i\rangle] = \frac{n \cdot |r_i|}{s_i}.$$

Now we define a sequence

$$R_1 \subset R_2 \subset \cdots \subset R_j \subset \cdots$$

of subsets of H. We define $R_1 = a_i Q$, where $|K_i| > \frac{3(m-1)^5}{n} \geq 3(m-1)^4$. For $j \geq 1$ we define $R_{j+1} = R_j MQ$. Thus R_{j+1} contains R_j and all left cosets of Q connected with those in R_j. Obviously, $R_j \subset \langle M\rangle$ for all j. Since $[\langle M\rangle : Q]$ is finite and each R consists of cosets mod Q, there exists j such that $R_{j+1} = R_j$, where $j \leq n$.

Let $h \in H$. Then $h \in K_i\langle M\rangle$, and hence $h = a_i y_{i_1}^{\varepsilon_1} \cdots y_{i_u}^{\varepsilon_u}$, for some $y_{i_1}, \ldots, y_{i_u} \in M$ and $\varepsilon_i = \pm 1$. If $\varepsilon_v = +1$, then $y_v^{\varepsilon_v} \in Qy_v$. If $\varepsilon_v = -1$, then, by (18), $y_v^{\varepsilon_v} = y_v^{-s_v} y_v^{s_v - 1} = q^{-r_v} y_v^{s_v - 1} \in Qy_v^{s_v - 1} \subset (Qy_v)^{s_v - 1} = Qy_v Qy_v \cdots Qy_v$. Therefore, $h \in a_i Qy_{j_1} Qy_{j_2} \cdots Qy_{j_w}$ for some $y_{j_1}, \ldots, y_{j_w} \in M$. Since $H = \langle M\rangle$, we see that there exist $a_{t_1}, \cdots a_{t_w}, a_t \in K$ such that $a_i Qy_{j_1} \cap a_{t_1} Q \neq \emptyset$, $a_{t_1} Qy_{j_2} \cap a_{t_2} Q \neq \emptyset, \ldots, a_{t_w} Qy_{j_w} \cap a_t Q \neq \emptyset$ and $h \in a_t Q$. Thus, in the sequence $a_i Q, a_t, Q, a_{t_1} Q, \ldots, a_t Q$ every term is connected with the previous one. Of all such sequences beginning with $a_i Q$ and ending with $a_t Q$ choose one of the minimal length. Clearly, such minimal sequence has no repetitions and hence $w + 2 \leq r$. By (11) and (20) $w + 1 \leq m - 2$ and

$$|K_t| \geq |K_{t_w}| - 3(m-1)^2 \geq |K_{t_{w-1}}| - 3 \cdot 2(m - 1^2 \geq \cdots \geq |K_{t_1}| - 3w(m-1)^2 \geq$$
$$|K_i| - 3(w+1)(m-1)^2 \geq |K_i| - 3(m-2)(m-1)^2 > 3(m-1)^4 - 3(m-1)^3 > 3(m-1)^4.$$

Thus, for every $K_t \subset H$,

$$|K_t| > 3(m-1)^4 \tag{30}$$

Let

$$m = nt + d, \qquad 0 \leq d < r. \tag{31}$$

The set M consists of m elements and, since $M \subset \langle M\rangle = H$, it is contained in the union of r right cosets of Q. Let $Qb_1, Qb_2, \ldots, Qb_n$ be these right cosets. Furthermore, let $M_j = M \cap Qb_j$. Choose M_j which has the maximal number of elements. Let $|M_j| = v$. Then $v \geq \frac{m}{n}$. Suppose that $M_j = \{q^{\alpha_1}, \ldots, q^{\alpha_v}\} \cdot b_j$, where $\alpha_1 < \cdots < \alpha_v$. Let $\overline{M} = Mb_j^{-1}q^{-\alpha_1}$ and let $\overline{M}_j = M_j b_j^{-1} q^{-\alpha_1}$. Then

$$|KM| = |K\overline{M}| \geq |K\overline{M}_j| = \sum_{i=1}^{n} |K_i \overline{M}_j|. \tag{32}$$

Suppose that $|K_i\overline{M}_j| \geq |K_i| + v$ for every i. Then (32) implies $|KM| \geq \sum_{i=1}^{n}(|K_i| + v) = k + vn$. Since $v \geq \frac{m}{n}$, we obtain $vn \geq m$ and $|KM| \geq k + m$ contrary to (7). Thus $|K_i\overline{M}_j| \leq |K - i| + v - 1$ for some i. In this case $|a_i^{-1}K\overline{M}_j| \leq |a_i^{-1}K| + v - 1$ and, since $a_i^{-1}K$ and $\overline{M}_j$ are subsets of Q, we obtain by Proposition 3

$$\overline{M}_j = \{1, q, q^2, \ldots, q^{v-1}\}. \tag{33}$$

Indeed, we have proved above that for every set K_i which satisfies (15) the set W_i contains a segment of length m (see the line right after formula (15)). Thus, by (30), the set $a_i^{-1}K_i$, contains two successive terms from Q. Therefore, the difference of the progression which, by Proposition 3, is formed by the numbers $\alpha_1, \alpha_2, \ldots, \alpha_v$ is 1. The sets K and $\overline{M}$ satisfy all conditions of Theorem 2, so it suffices to prove Theorem 2 for K and $\overline{M}$ rather than for K and M. Notice that $\overline{M}$ contains both 1 and q. We chose $q \in M$ which minimized n. If we replace M by $\overline{M}$, this choice of q may not minimize n. In such a case we replace q by another element of $\overline{M}$ which minimizes n. Repeating this process finitely many times we find q which produces minimal r both for M and $\overline{M}$. Of course, this can be proved by an inductive argument. In the sequel we write M rather than $\overline{M}$. Thus, $1, q, \ldots, q^{v-1} \in M$.

Now we prove that, for every $y_i \in M, y_i \neq 1$, the number r_i which appears in (18) is positive. Choose some i. To simplify notation we omit the index i in y_i, r_i, and s_i. Clearly, $r \neq 0$. Suppose that $r < 0$. Consider a left coset $b\langle y \rangle$ in the group $H = \langle M \rangle$ modulo the subgroup $\langle y \rangle$. Choose a coset $a_i Q$ for which $a_i Q \cap b\langle y \rangle \neq \emptyset$. This intersection has the form $a_i q^w \langle q^r \rangle$ for some $w < |r|$.

By (21), (30), and $|r| < m$, W_i contains no less than $\frac{|K_i| - (m-1)^2}{|r|} \geq \frac{3(m-1)^4 - (m-1)^2}{m-1} > m$ segments of length $|r|$. Each of these segments contains a number of the form $w + tr$ for some t. Therefore, there exists more than m elements of the form $a_i q^{w+tr}$ belonging to K_i. For each of these elements consider the sequence

$$a_i q^{w+tr}, a_i q^{w+tr} y, \ldots, a_i q^{w+tr} y^{s-1}. \tag{34}$$

Suppose that each of these sequences contains an element which does not belong to K. Since $a_i q^{w+tr} \in K$, and $y \in M$, each of these sequences contains an element from $KM \setminus K$. All our sequences are disjoint, and there are more than m of them. Therefore $|KM \setminus K| > m$ which contradicts (7). Thus, there exists a sequence of the form (34) which is contained in K. Fix one such sequence.

We know that $a_i q^{w+tr} y^{s-1} \in K$. Find minimal $a \geq s$ such that $a_i q^{w+tr} y^{a-1} \in K$ and $a_i q^{w+tr} y^a \notin K$. Then $a_i q^{w+tr} y^a \in KM \setminus K$. Let $a = ps + u$ for $0 \leq u < s$ and $p \geq 1$. Then $a_i q^{w+tr} y^a = a_i q^{w+tr} (y^s)^p y^a = a_i q^{w+tr} q^{rp} y^a = a_i q^{w+(t+p)r} y^a$. Now, $a_i q^{w+tr} y^u$ and $a_i q^{w+(t+p)r} y^u \in a_i q^w y^u Q$, since $q^r = y^s$ commutes with y^u. Since $r < 0$, we see that $a_i q^{2+tr} y^a < a_i q^{w+tr} y^u$ in $a_i q^w y^u Q$. Now, $a_i q^{w+tr} y^u$ occurs in (34). Thus, every coset $b\langle y \rangle$ contains an element $a_i q^{w+tr} y^a$ which belongs to $KM \setminus K$ and which is less than one of the elements of K. Since the index of $\langle y \rangle$ in H is n_i which is not less than n, we find at least n different elements of $KM \setminus K$ of the above type.

Recall that c_i is the maximal element of $K_i = K \cap a_i Q$. We know that $q, q^2, \ldots, q^{v-1} \in M$, and hence $c_i q, c_i q^2, \ldots, c_i q^{v-1} \in KM \setminus K$. There are n different sets K_i, for each one of them we found $v - 1$ elements in $KM \setminus K$. Each of these elements belongs to $a_i Q$ for some i and it is greater than all elements of K_i. Therefore, these elements are different from n elements of $KM \setminus K$ found earlier. Thus, $|KM \setminus K| \geq n + n(v-1) = nv \geq m$ which contradicts (7). Thus $r > 0$.

Suppose that there exists $i \neq j$ such that $|M_i| = |M_j| = v$. An argument analogous to that used in the proof of (33) shows that M_i has the form

$$\{1, q, \ldots, q^{v-1}\} y \tag{35}$$

for some $y \in M, y \neq 1$.

Consider the set $C = \bigcup_{i=1}^{n} \{c_i q, c_2 q^2, \ldots, c_i q^{v-1}\}$. This set has $n(v-1)$ elements and is contained in $MK \setminus K$. Consider a left coset $b\langle y \rangle$ in $\langle M \rangle$, where y is the element which appears in (35). Let by^w be an element, where w is a minimal number with respect to the following property: if $by^\alpha \in a_\ell Q$ for some ℓ, then $c_\ell q^{v-1} < by^\alpha$ in $a_\ell Q$. Thus, if $z < \alpha$, then $by^z \leq c_\ell q^{v-1}$, where $a_\ell Q$ is the coset which contains by^z.

If $by^{\alpha-1} \in C$, then $by^{\alpha-1} = c_i q^p$ for some $p, 1 \le p \le v-1$, and some i. Thus

$$by^\alpha = c_i q^p y. \tag{36}$$

However, $c_i \in K$ and, by (35), $q^p y \in M$, so that $by^\alpha \in KM$, where $by^\alpha \notin C$ and so $by^\alpha \notin K$.

If $by^{\alpha-1} \notin C$, then $by^{\alpha-1} \le c_i$ for some i. If $by^{\alpha-1} \in K$ then $by^\alpha = by^{\alpha-1}y \in KM\backslash K$. If $by^{\alpha-1} \notin K$ then there exists minimal $u < \alpha$ such that $by^u \notin K$ and $by^{u-1} \in K$. Then $by^u \in KM\backslash K$. Since $by^{w-1} \notin C$, we have $by^u \notin C$.

In either case $b\langle y\rangle$ has an element belonging to $KM\backslash K$ and not belonging to C. The index of $\langle y\rangle$ in $\langle M\rangle$ is not less than n, so that

$$|KM| \ge |K| + |C| + n \ge K + n(v-1) + n = k + nv \ge k + m,$$

which contradicts (7).

Thus we may assume that if $i \ne j$, then $|M_i| < v$. If $(v-1)n \ge m$ then

$$|KM| \ge |KM_j| = |\bigcup_{i=1}^{n} K_i M_j| = \sum_{i=1}^{n} |K_i M_j| \ge \sum_{i=1}^{n} (|K_i| + v - 1) = k + n(v-1) \ge k + m$$

which contradicts (7). Here we used formula (33) by which $K_i M_j \subset a_i Q$, so that the sets $K_i M_j$ are disjoint for different i.

Thus $(v-1)n < m$ or, by (31), $v < t + \frac{d}{n} + 1$. On the other hand, $nt + d = m \le vn$, so that $t + \frac{d}{n} \le v$. Therefore,

$$v = \begin{cases} t, & \text{if } d = 0; \\ t+1, & \text{if } d > 0. \end{cases}$$

If $d \ne 1$ in (31), then, since $|M_i| < v$ for all $i \ne j$, we have $|M| = |M_j| + \sum_{i \ne j} |M_i| \ge v + (v-1)(n-1) = (v-1)n + 1$. If $d = 0$, then $v = t$ and $(v-1)n + 1 = (t-1)n + 1 = tn - n + 1 = m - n + 1 < m$. If $d > 1$ then $v = t+1$ and $(v-1)n + 1 = tn + 1 < m$. thus, $m = |M| < m$ which shows that $d = 1$.

By (34), $v = t + 1$. M contains a subset of the form (33). Formula (31) can be written as $m - 1 = n(v-1)$, and C contains $m - 1$ elements. The exponents u of the elements $a_i q^u$ in K_i form a segment of the natural series. Indeed, were it otherwise, there would have existed $a_i q^u \in K_i$ such that $a_i q^{u+1} \in KM\backslash K$ and $a_i q^u \le c_i$. Then $|KM| \ge |K| + |C| + 1 = k + m - 1 + 1 = k + m$ which contradicts (7). Thus, $KM = K \cup C$.

Next we show that a nonempty intersection of any right coset $Qy \ne Q$ and M is of the form

$$Qy \cap M = \{1, q, q^2, \ldots, q^{v-2}\}y. \tag{37}$$

If y denotes the minimal element of $Qy \cap M$, with respect to the natural order in Qy, then the maximal element of this set has the form $q^\beta y$. If $\beta \ge v$ then

$$|KM| \ge \sum_{i=1}^{n} |K_i(Qy \cap M)| \ge \sum_{i=1}^{n} (k_i + v) = k + vn > k + m$$

which contradicts (7).

If $\beta = v - 1$ we can arrive at a contradiction repeating our argument for $d \ne 1$ in which (36) is replaced by $by^\alpha = c_i q^p \cdot y = c_i q^{p-(v-1)} \cdot q^{v-1} y \in KM\backslash K$, the remaining part of the argument is unchanged.

Thus $\beta \le v-2$ and $|Qy \cap M| \le v-1$. If the latter inequality is strict in at least one case then $m = |M| \le v + (n-2)(v-1) + v - 2 = nv - n = nt = m - 1 < m$. Therefore $|Qy \cap M| = v - 1$ for $Qy \ne Q$ and $Qy \cap M \ne \emptyset$, which proves (37). Now we prove that, for some i,

$$by^{\alpha-1} = c_i q^{v-1}. \tag{38}$$

As we have seen above, $a_i Q \cap b\langle y\rangle = a_i q^w \langle q^r\rangle$. Suppose that $by^{\alpha-1}$ is less than the least element of K_i. For any β, $by^\beta \in a_i Q$ only if $\beta - (\alpha - 1)$ is a multiple of r. If $\beta \le \alpha - 1$, then $by^\beta \notin K_i$ because $by^\beta \le by^{\alpha-1}$. If $\beta > \alpha - 1$, then $\beta \ge \alpha + r$ and $by^\beta \notin K_i$ by the definition of α. Thus $b\langle y\rangle \cap K_i = \emptyset$. However, as we have seen above, W_i contains more than m segments of length r which shows that $b\langle y\rangle \cap K_i \ne \emptyset$. Thus $by^{\alpha-1}$ is not less than the least element of K. It follows from the definition of α that $by^{\alpha-1} \in K \cup C$.

If $by^{\alpha-1} \in K$ then $by^\alpha \in KM$. If $by^{\alpha-1} \in C$ and $by^{\alpha-1} = c_i q^a$ for $1 \le a < v-1$ then, by (37), $by^\alpha = c_i q^a y \in KM$. However, $KM = K \cup C$, and hence $by^\alpha \in K \cup C$ contrary to the definition of α. Thus, (38) holds.

Now, $[H : \langle y\rangle] = n_i \ge n$ by the minimality of n. Formula (38) holds for every $b \in H$. If $b_1\langle y\rangle \ne b_2\langle y\rangle$ then $b_1 y^{\alpha_1 - 1}$ and $b_2 y^{\alpha_2 - 1}$ belong to different cosets $a_i Q$. Here α_1 and α_2 are exponents of y corresponding to b_1 and b_2 in (38). It follows that $n_i \le n$. Thus, $n_i = n$.

For every $b \in H$ consider the set of numbers β such that there exist i and p for which $c_i q^{v-1} < by^\beta$ and $by^{\beta-1} \le c_p q^{v-1}$. Here the order relations $<$ and $\le$ are considered in $a_i Q$ and $a_p Q$, respectively. A proof analogous to that of formula (38) for β shows that $by^{\beta-1} = c_i q^{v-1}$. Indeed, $b\langle y\rangle \cap a_i Q \ne \emptyset$, and hence $b\langle y\rangle \cap K_i \ne \emptyset$ because W_i contains segments of length r. If $by^{\beta-1}$ is less than the least element of K_i, then $by^{\beta-1}y^{ra} \in K_i$ for a suitable $a > 0$. Then $by^{\beta-1}y^{ra}y \in KM = K \cup C$. However, $by^{\beta-1}y^{ra}y = by^\beta y^{ra} = by^\beta q^{sa} > c_i q^{v-1}$, and hence $by^{\beta-1}y^{ra}y \notin K \cup C$. This contradiction shows that $by^{\beta-1}$ is not less than the least element of K_i. The remaining part of the proof of (38) is repeated without changes.

Take n elements $b_1, \ldots, b_n$ belonging to different left cosets of H mod $\langle y\rangle$. Choose $\beta_1, \ldots, \beta_n$ for each of these b_i. Then $b_i y^{\beta_i - 1} = c_{p_i} q^{v-1}$. Since there are precisely n elements of the form $c_j q^{v-1}$, there are n elements of the form $b_i y^{\beta_i - 1}$. It follows that for every b there exists precisely one element β satisfying the above properties.

Let $a \ge \beta$. If $by^a \le c_j q^{v-1}$ for some j, then there exists $a' > a$ such that $c_\ell q^{v-1} < by^{a'}$ and $by^{a'-1} \le c_p q^{v-1}$ for some ℓ and p. Then $a' = \beta$ and $\beta \le a < a' = \beta$ which is a contradiction. Thus $by^a > c_j q^{v-1}$ for $a \ge \beta$. Analogously, $by^a \le c_j q^{v-1}$ for $a < \beta$ and j such that $by^a \in a_j Q$.

Now consider the set

$$C_1 = \{c_i q^j : 1 \le i \le n; v - 1 \le j < v - 1 + r\},$$

where $r = r_i$ is determined by formula (18), and the sequence

$$c_1 q^{v-1}, c_1 q^{v-1}(yq^{-1}), c_1 q^{v-1}(yq^{-1})^2, c_1 q^{v-1}(yq^{-1})^3, \cdots \tag{39}$$

We prove that all terms of (39) belong to C_1. Indeed, $c_1 q^{v-1} \in C_1$. Let $b = c_1 q^{v-1}(yq^{-1})^a \in C_1$, i.e., let $c_1 q^{v-1}(yq^{-1})^a = c_i q^j$ for $1 \le i \le n$ and $v - 1 \le j < v - 1 + r$. Then $b \in H$, so that for b there exists the number β defined above. Suppose that $by \le c_\ell q^{v-1}$ for some ℓ. Then $1 < \beta$, because 1 is the exponent of y in by. In this case $0 < \beta$, and $b = by^0 < c_i y^{v-1}$ contrary to our assumption. Thus, $by > c_\ell q^{v-1}$. Therefore $byq^{-1} \ge c_\ell q^{v-1}$. Now, $b = c_i q^j < c_i q^{v-1+r}$, and hence $bq^{-r} < c_i q^{v-1}$ or, equivalently, $by^{-s} < c_i q^{v-1}$. Thus $-s < \beta$. Since $by^{-s} \neq c_i q^{v-1}$, we see that $-s \neq \beta - 1$. Therefore $-s < \beta - 1$. Thus, $1 - s < \beta$ and hence $by^{1-s} = byy^{-s} = byq^{-r} \leq c_\ell q^{v-1}$ or, equivalently, $by \leq c_\ell q^{v-1} q^r$. Therefore $byq^{-1} < c_\ell q^{v-1} q^r = c_\ell q^{v-1+r}$. It follows that $byq^{-1} = c_1 q^{v-1}(yq^{-1})^{a+1} \in C_1$. Thus, all terms of (39) are in C_1.

The sequence (39) is infinite and the set C_1 finite. Therefore, there exist a and b, $a \neq b$, such that $c_1 q^{v-1}(yq^{-1})^a = c_1 q^{v-1}(yq^{-1})^b$ or, equivalently, $(yq^{-1})^{a-b} = 1$. Since G is a torsion free group, $yq^{-1} = 1$ or $y = q$ contrary to our assumption about y. This contradiction completes the proof of Theorem 2.

2. Suppose that in a group the small squaring condition holds for all subsets which have a certain cardinality. What is the structure of the group?

Here are two results of this kind.

THEOREM 3. (see [6]). Let $|K^2| \leq 3$ for all $K \subset G$ with $|K| = 2$. Then either G is abelian or $G = Q \times E$, where Q is the quaternion group and E is an elementary abelian 2-group.

THEOREM 4. (see [1]). Let $|K^2| \leq 6$ for all $K \subset G$ with $|K| = 3$. Then either G is an abelian group or $G = S_3$.

In [3] all finite groups G are described in which, for every K such that $|K| = 2$, the inequality $|K^3| \leq a$ holds (for any $a \leq 7$). The most difficult case when $a = 7$ and G is a 2-group remains unsolved.

GENERAL PROBLEM. Describe the structure of finite groups G such that $|K^a| \leq b < k^a$ for all $K \subset G$ with $|K| = k$.

3. This sort of problems can be considered for algebraic structures more general than groups. Further in this paper, we consider the case when G is a semigroup in which, for every $K \subset G$ with $|K| = 2$, we have $|K^2| \leq 2$ or $|K^2| \leq 3$. Here we give a few comments on a similar problem for a quasigroup G. In this case, because $K^2 = \{a^2, ab, ba, b^2\}$ and $\{a^2, b^2\} \cap \{ab, ba\} = \emptyset$, we have $a^2 = b^2$ and $ab = ba$ for any $a, b \in G$. Thus G is a commutative quasigroup with a single element e such that $a^2 = e$ for every $a \in G$. If $|G| = 2$, then G is a group. The case $|G| = 3$ is impossible. If $|G| = 4$, the Cayley multiplication table of G is of the form

	e	x	y	z
e	e	a	b	c
x	a	e	c	b
y	b	c	e	a
z	c	b	a	e

where $G = \{e, a, b, c\}$ and the sequence (x, y, z) coincides with one of the sequences (a, b, c), (a, c, b), or (b, c, a). The case $|G| = 5$ is impossible. There are five different quasigroups G of order 6. However, a description of all commutative quasigroups in which all elements have the same square is unknown.

The next step would be to describe semigroups and quasigroups - or at least some special classes of them - in which $|K^2| \leq 3$ for every two-element subset K.

4. It is possible to introduce a concept of isomorphism of subsets K of G (cf. [4], p.8, and [6], p.14). If $|K| = 2$, the following types of multiplication tables in groups and quasigroups are possible (here equal products are denoted by a same letter, while distinct letters stand for unequal products):

	a	b		a	b		a	b		a	b
a	A	B	a	A	B	a	A	B	a	A	B
b	B	A	b	B	C	b	C	A	b	C	D
	1			2			3			4	

Let $P_i(G)$, where $i = 1, 2, 3, 4$, denote the number of multiplication tables of the corresponding type. We have already stated results for the case $P_4(G) = 0$.

Which are the results for groups G satisfying the condition $P_3(G) = 0$?

THEOREM 5 (Ya.G. Berkovich, published in [2]). Let H be a Sylow 2-subgroup of a finite group G. Then $P_3(G) = 0$ if and only if H is normal and $P_3(H) = 0$.

This reduces the problem to the case when G is a 2-group. If $|G| = 8$ and G is not abelian, then $P_3(G) > 0$ as is easily seen.

We can find an example when $P_3(H) = 0$ and H is a nonabelian 2-group in the following way. Every nonabelian group contains a nonabelian subgroup whose all proper subgroups are abelian. Rédei [10] listed all such groups. They are:

(1) The quaternion group H of order 8. In this case $P_3(H) > 0$.

(2) $H = < x, y : x^{2^m} = 1 = y^{2^n}, x^{-1}yx = y^{1+2^{n-1}}, m > 0, n > 1 >$. If $n = 2$, then $(yx)^2 = x^2$ but $yx \cdot x \neq x \cdot yx$. If $m = 1$ and $n \geq 3$, then $(y^{2^n-1}x)^2 = (y^{2^n-1})^2$ but $y^{2^n-1}x \cdot y^{2^n-1} \neq y^{2^n-1} \cdot y^{2^n-1}x$. In these cases $P_3(H) > 0$. If $m > 1$ and $n > 2$ then $P_3(H) = 0$.

(3) $H = < x, y, z : z = [x, y], 1 = x^{2^m} = y^{2^n} = z^2, 1 = [x, z] = [y, z], m > 0, n > 0 >$. In this case $P_3(H) = 0$ always except when $m = n = 1$.

Theorem 5 makes it plausible that the condition $H < G$ will play an important role in the description of groups G such that $P_3(G) = 0$.

5. We say that a semigroup S is *tight* if $|A^2| \leq 2$ for every subset A of S such that $|A| \leq 2$.

Clearly, $1 \leq |A^2| \leq 4$ for every A such that $|A| \leq 2$. Thus, a tight semigroup S can be characterized by the condition: for all $x, y \in S$

$$\begin{aligned} x^2 = xy = yx \text{ or } y^2 = xy = yx \text{ or } x^2 = xy = y^2 \text{or } x^2 = yx = y^2 \text{ or } (x^2 = xy \ \& \ y^2 = yx) \\ \text{or } (x^2 = yx \ \& \ y^2 = xy) \text{ or } (x^2 = y^2 \ \& \ xy = yx). \end{aligned} \tag{40}$$

Obviously, this class of semigroups is closed under subsemigroups and homomorphic images. We use this fact in the sequel without special references.

Suppose $\langle a \rangle$ is a finite monogenic semigroup. Then there exist positive integers m and n such that $a^{m+n} = a^m$. The smallest possible m and n always exist, they form a pair (m, n) which is called the *characteristic* of $\langle a \rangle$. If $n = 1$, then a is called a *nil* element in this case a^m is a zero of $\langle a \rangle$ (however, a^m need not be a zero of S). If $m = 1$, then a is called a *group* element (in this case $\langle a \rangle$ is a cyclic group). Usually, "nil" takes precedence over "group": if $m = n = 1$, we call the element nil rather than group.

LEMMA 1. Each element a of S is of finite order and has one of the following characteristics: $(1, 1)$, $(1, 2)$, $(2, 1)$, $(2, 2)$, $(3, 1)$.

Proof. Let $A = \{a, a^2\}$. Then $A^2 = \{a^2, a^3, a^4\}$. Since $|A^2| \leq 2$, we have $a^2 = a^3$ or $a^2 = a^4$ or $a^3 = a^4$. The statement of Lemma 1 follows.

It follows that $a^4 \in E$, where E denotes the set of all idempotents of S. As explained earlier, we call an element $s \in S$ a group element if its type is (1,2).

LEMMA 2. For all $x, y \in S, (xy)^2$ is an idempotent, i.e. $(xy)^4 = (xy)^2$.

Proof. We examine all terms of the disjunction (40). If $xy = x^2$, then $(xy)^2 = x^4 \in E$. If $xy = y^2$, then $(xy)^2 = y^4 \in E$. If $xy = yx$ and $x^2 = y^2$, then $(xy)^2 = x^2y^2 = x^2x^2 = x^4 \in E$. There remains the case when $x^2 = yx = y^2$. In this case $(xy)^2 = x(yx)y = xx^2y = x^5y = x^3x^2y = x^3y^2y = xx^2y^2y = x(yx)(yx)y = (xy)^3 \in E$. Here we used two facts: $s^3 = s^5$ for every $s \in S$, and $s^3 = s^2 \rightarrow s^2 \in E$. This completes our proof of Lemma 2.

LEMMA 3. S satisfies the identity

$$(xy)^2 = x^4y^4. \tag{41}$$

Proof. We use Lemma 2 without further references. Let $x, y \in S$. Then the subset $\{x, y\}^2 = \{x^2, xy, yx, y^2\}$ contains at most two distinct elements. Four cases are possible.

Case 1. Let $xy = yx$. In this case (41) is obvious.

Case 2. Let $x^2 = xy$, $y^2 = yx$. Then $(xy)^2 = x(yx)y = xy^2y = xyy^2 = x^2y^2 = x^4y^4$.

Case 3. Let $x^2 = yx$, $y^2 = xy$. Then $(xy)^2 = x(yx)y = xx^2y = x^2xy = x^2y^2 = x^4y^4$.

Case 4. Let $x^2 = y^2 = yx$. Then $(xy)^2 = x(yx)y = xy^2y = xyy^2 = xyyx = xy^2x = xx^2x = x^2x^2 = x^2y^2 = x^4y^4$.

An endomorphism f of S is called *a retraction* if $f^2 = f$ (i.e. if $(sf)f = sf$ for all $s \in S$).

A semigroup is called *orthodox* if its subset E of idempotents forms a subsemigroup. (In [12], where the orthodox semigroups were introduced, they were called "semigroups with weak involution". The term "orthodox" belongs to Hall (see [7] and [8]).

LEMMA 4. The mapping $q : S \rightarrow S$ defined by $sq = s^4$ for every $s \in S$ is a retraction. S is orthodox, and E coincides with the range of q.

Proof. It follows from Lemma 1 that $s^4 \in E$ for every $s \in S$. Also, if $e \in E$, then $eq = e^4 = e$, hence $sqq = (sq)^4 = sq$ for all $s \in S$. Thus, E is the range of q. By Lemmas 2 and 3, q is an endomorphism. This completes the proof.

It is well known that every semigroup S is a semilattice Y of its subsemigroups $\{S_i\}_{i \in Y}$ which are semilattice indecomposable. Thus, $S_iS_j \subset S_{ij}$ for all $i, j \in Y$, and ij is the product of i and j in Y. As usual, ij may be interpreted as the greatest lower bound of $\{i, j\}$ in Y under the natural (partial) order in Y. If the natural order of Y is linear, i.e. if $ij = i$ or $ij = j$ for all $i, j \in Y$, then Y is called *a chain semilattice* (or merely *a chain*). Y is called *the structure semilattice* of S, it is defined uniquely up to isomorphism. The subsemigroups $S_i, i \in Y$, are called *the semilattice components* of S. If $|Y| = 1$, S is called *semilattice indecomposable*.

If a semigroup satisfies the identity $xy = x$ $[xy = y]$, it is called *a left [right] zero semigroup*. A semigroup is called *singular* if it is a right zero semigroup or a left zero semigroup. If Y is a chain semilattice and $\{S_i\}_{i\in Y}$ is a family of disjoint semigroups, we can define the following multiplication on $S = \sum\{S_i : i \in Y\}$. For all $x \in S_i$ and $y \in S_j, i, j \in Y$, we define

$$xy = yx = x, \quad \text{if} \quad i < j \quad \text{in} \quad Y;$$
$$xy \text{ is the product of } x \text{ and } y \text{ in } S_i \quad \text{if} \quad i = j.$$

Then S is a semigroup which is called *the ordinal sum* of the family $\{S_i\}_{i\in Y}$. Lyapin [9] calls it *a successively annihilating band* of semigroups S_i. Here "*band*" stands for "idempotent semigroup".

LEMMA 5. A band is tight if and only if it is an ordinal sum of singular semigroups.

Proof. In a band E $e^2 = e$ for every $e \in E$. It follows from condition (40) that $xy = x$ or $xy = y$ for all $x, y \in E$, where E is a tight band. Conversely, if $xy = x$ or $xy = y$ for all $x, y \in S$ for a semigroup S, then $xx = x$, i.e. S is a band, and $\{x, y\}^2 = \{x, y, xy, yx\} = \{x, y\}$, i.e. S is tight.

Thus, tight bands are precisely the semigroups S which satisfy the condition

$$xy = x \quad \text{or} \quad xy = y \tag{42}$$

for all $x, y \in S$.

It is easy to see that these are precisely the semigroups in which each nonempty subset is a subsemigroup. It is known (see Rédei [11], Theorem 50) that these are precisely ordinal sums of singular semigroups.

Let S be a semilattice Y of semilattice indecomposable subsemigroups $S_i, i \in Y$. Clearly, Y is a homomorphic image of S. Suppose that S is a tight semigroup. Then Y is a semilattice with small squares. Therefore, Y is a chain. Let $i, j \in Y, i < j$, $s \in S_i$, $t \in S_j$. Then $t^2 \in S_j$ and $s^2, st, ts \in S_i$. Since $\{s, t\}^2$ cannot contain more than two elements, $s^2 = st = ts$. Replacing t by t^2 we obtain $s^2 = st^2 = stt = s^2t = sst = ss^2 = s^3$. It follows that if i is not the greatest element of Y, then $s^2 = s^3$ for every $s \in S_i$. Therefore, $sq = s^2$ for such S.

LEMMA 6. Let S be a tight semigroup, E be its subsemigroup of idempotents. S is semilattice indecomposable if and only if E is semilattice indecomposable.

Proof. If $f : E \to Y$ is a homomorphism of E onto a semilattice Y, then $qf : S \to Y$ is a homomorphism of S onto Y. Thus, if E is not semilattice indecomposable, then S is not. Conversely, suppose that E is semilattice indecomposable. As a subsemigroup of S, E is a tight semigroup. By Lemma 5, E is a singular semigroup. Suppose that $S \to Y$ is a homomorphism of S onto a semilattice Y. Then Ef is a singleton, because Ef is both a singular semigroup and a subsemilattice of Y. If $s \in S$, then $(sf) = (sf)^2$ because sf is an indempotent in Y. On the other hand, $(sf)^2 = s^2f \in Ef$ because $s^2 \in E$. Therefore, Sf is a subset of Ef, i.e. Y is a singleton. It follows that S is semilattice indemposable. Lemma 6 is proved.

The argument preceding Lemma 6 settles the problem of multiplication of elements belonging to different semilattice components of S. We see that S is a chain of such components, and $st = ts = s^2 = sq$ for any s and t belonging to different components of S with s belonging to a component below the component containing t. It remains to find the structure of tight semilattice indecomposable semigroups. Let S be such a semigroup. Then, by Lemma 6, E is a singular semigroup.

Suppose that $|E| > 1$. Assume that E is a left zero semigroup. Let $s,t \in S$, $sq = e$, $tq = f$ for different $e,f \in E$. Then $\{s,f\}^2 = \{s^2, sf, fs, f\}$. This subset has at most two different elements. If $s^2 = f$, then $e = (s^2)^2 = f^2 = f$ contrary to our assumption. If $sf = fs$, then $e = ef = s^4 f^4 = (sf)^2 = (fs)^2 = f^4 s^4 = fe = e$, contrary to our assumption. If $fs = s^2$ then $s^2 = fs = ffs = fs^2 = fss = s^2 s = s^3$. The only remaining case is $fs = f$. In this case $sf = s^2$ (because sf and fs are different elements). It follows that $s^3 = s^2 s = sfs = sf = s^2$. Thus, in all cases $s^2 = s^3$ and $s^2 = e$. Analogously, $t^2 = t^3 = f$. It follows that $\{s,t\}^2 = \{e, st, ts, f\}$. If $st = ts$, then $e = ef = s^2 t^2 = (st)^2 = (ts)^2 = t^2 s^2 = fe = f$, contrary to our assumption. Therefore, s and t do not commute. If $st = f$ then $e = ef = s^2 t^2 = s(st)t = sft = stf = ff = f$, contrary to our assumption. Therefore, $st = e$. It follows that $ts = f$. We have proved the following result:

LEMMA 7. Let S be a tight semilattice indecomposable semigroup containing more than one idempotent. Let E be the subsemigroup of idempotents of S. Then $s^3 = s^2 \in E$ for every $s \in S$ and one of the following two cases holds:

(i) E is a left zero semigroup and $st = s^2$ for any $s,t \in S$ such that s^2 differs from t^2;
(ii) E is a right zero semigroup and $st = t^2$ for any $s,t \in S$ such that s^2 differs from t^2.

Case (ii) is proved analogously to case (i) if we assume that E is a nontrivial right zero semigroup.

Let $S_e = \{s \in S : s^2 = e\}$ for every $e \in E$. It follows from Lemma 4 that S_e is a subsemigroup of S. Lemma 7 describes products of elements belonging to different subsemigroups S_e of S. It remains to establish the structure of S_e. Clearly, S_e is a tight semigroup containing only one idempotent. Thus, we must describe the structure of tight semigroups with single idempotents.

Suppose that S is a tight semigroup with a single idempotent e. If $s^2 = e$ for all $s \in S$, then e is the zero of S. Indeed, $es = se = s^3 = s^2 = e$. Therefore, we write 0 instead of e. We have $s^2 = 0$ for every $s \in S$. Thus $\{s,t\}^2 = \{0, st, ts\}$ for all $s,t \in S$. Therefore S is characterized by the following conditions:

(i) S is a semigroup with zero 0 and $s^2 = 0$ for every $s \in S$;
(ii) for every $s,t \in S$ we have one of the following: $st = 0$, or $ts = 0$, or $st = ts$.

We call such semigroups S *semigroups of the first type*.

EXAMPLE. Clearly, every commutative semigroup with zero which satisfies the identity $s^2 = 0$ is of the first type. Let A be a nonempty set and S be the set consisting of A and all nonempty subsets of A. Define the following multiplication in S: for any $s,t \in S$

$$st = \begin{cases} s \cup t & \text{if } s \cap t = \emptyset, \\ A & \text{if } s \cap t \neq \emptyset. \end{cases}$$

We will not go further in our study of the structure of S. Note that S satisfies the identity $sts = 0$ for all $s,t \in S$. Indeed, if either $st = 0$ or $ts = 0$, then $sts = 0$. If $st = ts$, then $sts = s^2 t = 0$. Another interpretation for the semigroups of the first type follows.

A (partial) groupoid is any nonempty set endowed with a (partial) operation. Suppose that v and w are groupoid words and $v = w$ is an identity. We say that a partial groupoid T *strongly satisfies* $v = w$ if, for any values of variables from v and w, both v and w are defined or not defined in T simultaneously and, when v and w are defined, then $v = w$ in S. We say that S *weakly satisfies* $v = w$ if it satisfies it for such values of the variables occurring in v and w for which both v and w are defined simultaneously. If we add a new element 0 to T and define $st = 0$ for all s and t belonging to T or equal to 0 for which st has not been defined in G, we obtain a groupoid. This groupoid is called an *0-extension* of T. It is easy to see that T strongly satisfies an identity $v = w$ if and only if the 0-extension of T satisfies this identity. For example, a groupoid T strongly satisfies the identity of associativity $(xy)z = x(yz)$ if and only if its 0-extension satisfies this identity, that is, its 0-extension is a semigroup. In this case we say that T is *strongly associative*. Strongly associative partial groupoids are also called *semigroupoids* [14]. On the other hand, T weakly satisfies the identity of commutativity $xy = yx$, if, for all $s, t \in T$ such that both st and ts are defined in T, we have $st = ts$. In this case we say that T is *weakly commutative*.

We say that a partial groupoid T is *without squares* if s^2 is never defined for all $s \in T$. The following Proposition becomes obvious:

PROPOSITION 4. A semigroup S is a semigroup of the first type if and only if it is a 0-extension of a weakly commutative semigroupoid without squares.

There remains a case of a tight semigroup S with a single idempotent e which may not satisfy the condition $s^2 = e$ for all $s \in S$. We call it *a semigroup of the second type*. Let S be a semigroup of the second type. Let G be the set of all group elements of S (including e), Z be the set of all $s \in S$ of type (2,2), i.e. such that $s^2 = s^4$, and all the elements in $\{s, s^2, s^3\}$ different. Clearly, if $z \in Z$, then z^3 is a group element.

Suppose that S has at least one group element g. If $s^3 = s^4$ for some $s \in S$, then $s^3 = e$ and $se = e$, so that $gs = (ge)s = g(es) = ge = 0$. Analogously, $sg = g$, hence $\{g, s\}^2 = \{g^2, gs, sg, s^2\} = \{e, g, s^2\}$. If $s^2 = g$, then $g = gs$ implies $g = gs = gs^2 = g^2 = e$, contrary to our choice of g. Therefore, $s^2 = s$. Thus, if S has group elements, it has no elements of type (3,1). We call an element $s \in S$ *nil* if $s^3 = e$. Let N be the set of all nil elements of S.

LEMMA 8. *G is a subgroup of exponent 2 of the semigroup S. N is a subsemigroup of S and $gn = ng = g$ for all $g \in G$ and $n \in N$. If G is nontrivial, then N is a semigroup of the first type.*

Proof. Clearly, G is a group of exponent 2 if it is a singleton. Suppose that it is not, ie. S has some group elements. Then, as we have seen above, $gn = ng = g$ for all group elements g. Also, $en = ne = e$ and $n^2 = e$. Let $s, t \in G$. Then $es = s$, and hence $e(st) = st$. It follows that st does not belong to Z. If $st \in N$, then $e = e(st) = st$, hence $st \in G$. Thus, G is a subsemigroup of S. Since $g^2 = e$ for all $g \in G$, G is a group of exponent 2. It remains to prove that N is a subsemigroup of S. Let $m, n \in N$. Then $(mn)^3 = (mn)^2 m_n = emn = en = e$, hence mn does not belong to Z. If $mn \in G$, then $mn = emn = e \in N$. Therefore $mn \in N$. Lemma 8 is proved.

We do not consider further properties of semigroups of the second type.

6. Let us consider certain semigroups S for which a set $\{s, t\}^2$ always contains less than four elements. This means that at least two of the elements of the set $\{s^2, st, ts, t^2\}$ must be equal.

Suppose that S is a cancellative semigroup. If $s \neq t$ then s^2 and t^2 cannot coincide with st or ts. Therefore, either $st = ts$ or $s^2 = t^2$. In other words, noncommuting elements have equal squares.

Let Z denote the center of S. If s is not central, then s does not commute with at least one element t of S. Therefore, $s^2 = t^2$. If s^3 and t commute then $t^2 st = s^2 st = s^3 t = ts^3 = tss^2 = tst^2$. Using the cancellativity we obtain $ts = st$ which contradicts our assumption. Thus s^3 and t do not commute. It follows that $s^6 = (s^3)^2 = t^2 = s^2$. Again using cancellativity we see that S contains an identity element 1 and $s^4 = 1$. Of course, $t^4 = 1$ as well. Thus both s and t belong to the group of units of S and $s^{-1} = s^3, t^{-1} = t^3$. If s commutes with st then $sst = sts$. Cancelling s on the right we obtain $st = ts$. Therefore, s and st do not commute. It follows that $ss = stst$ and, by the cancellativity, $s = tst$. Analogously, $t = sts$. This shows that the subsemigroup $\langle s,t\rangle$ of S generated by s and t consists of the elements $\{s, s^2, s^3, 1, t, t^3, st, ts\}$. If all these elements are different, we see that $\langle s,t\rangle$ is a group isomorphic to Q, the quaternion group. If some elements of the above set coincide, $\langle s,t\rangle$ must be isomorphic to a proper homomorphic image of Q. Since all proper homomorphic images of Q are commutative and $\langle s,t\rangle$ is not, we see that $\langle s,t\rangle$ is isomorphc to a quaternion group. We have proved the following result:

LEMMA 9. If S is a cancellative semigroup and $\{s,t\}^2$ always has less than four elements, then any two elements s and t of S either commute or generate a subsemigroup $\langle s,t\rangle$ of S which is isomorphic to the quaternion group Q.

PROPOSITION 5. S is a cancellative semigroup in which $\{s,t\}^2$ contains less than four elements for any s and t in S if and only if S is either a commutative semigroup or it is isomorphic to a direct product of Q, the quaternion group, and an elementary abelian 2-group.

Proof. The "if" part. Clearly, every commutative and cancellative semigroup satisfies our condition. If S is isomorphic to a direct product of Q and B, an elementary abelian 2-group, then it is easy to check that S again satisfies our condition. In this case S is, of course, a special case of a Hamiltonian group: a Hamiltonian group without elements of odd order.

The "only if" part. Suppose that S is a cancellative semigroup satisfying our condition. If S is commutative, we are done. Therefore, assume that S is not commutative. Then, as we have seen in the proof of Lemma 9, S contains a subgroup isomorphic to Q. Suppose that s is a non-central and z central elements of S. Then $s^4 = 1$. If s and t do not commute, then sz and t do not commute. Therefore, $s^2z^2 = (sz)^2 = t^2 = s^2$, and hence $z^2 = 1$. Thus, $x^4 = 1$ for every element x of S. It follows that S is a group without elements of odd order. Let H be a subgroup of S and s an element of S. If h is an element of H then either s and h commute or not. If they commute, then $s^{-1}hs = h$. If they do not commute, then they generate a subgroup of S isomorphic to Q. Therefore, $s^{-1}hs = h^{-1}$. In both cases $s^{-1}hs$ is an element of H. Therefore, H is a normal subgroup of S. Since all subgroups of S are normal, S is a Hamiltonian group without elements of odd order. Thus S is isomorphic to a direct product of Q and an elementary abelian 2-group [see, for example, Theorem 9.7.4 of [13]]. This completes the proof.

Theorem 3 follows from Proposition 5 as an obvious corollary.

Next we consider left cancellative semigroups satisfying the condition: $\{s,t\}^2$ always has less than four elements. Suppose that s and t are two non-commuting (and hence different) elements of S. Since $\{s,t\}^2 = \{s^2, st, ts, t^2\}$ and $s^2 \neq st$, $ts \neq t^2$ and $st \neq ts$, we see that

$$s^2 = ts, \qquad \text{or} \qquad st = t^2, \qquad \text{or} \qquad s^2 = t^2. \tag{43}$$

If s and st commute, then $sst = sts$. Cancelling on the left we obtain $st = ts$ which contradicts our assumption. Therefore, s and st do not commute. Substituting st for t in (43) we obtain $s^2 = sts$, or $sst = (st)^2$, or $s^2 = (st)^2$. Applying left cancellation, we obtain $s = ts$, or $st = tst$, or $s = tst$. If $s = ts$, then $st = tst$, i.e. the first of these equalities implies the second one. Thus

$$st = tst \qquad \text{or} \qquad s = tst. \tag{44}$$

Next we consider the possibilities implied by (43) and (44).

Case of $s^2 = ts$ and $st = tst$. In this case $s^3 = ss^2 = sts = tsts = s^2s^2 = s^4$, and hence $s = s^2$, by left cancellativity. Thus, s is idempotent. Idempotents of S are precisely its left identity elements. Indeed, if e is a left identity of S, then $e^2 = e$. Conversely, if e is an idempotent, then $e^2u = eu$ for every u in S. Cancelling on the left we obtain $es = s$, i.e. e is a left identity. Since s is a left identity, $t = st = tst = t^2$, i.e. t is idempotent as well.

Case of $s^2 = ts$ and $s = tst$. In this case $(ts)^2 = (tst)s = ss = ts$, i.e. ts is idempotent. Thus ts is a left identity, and $s = (ts)t = t$ which is impossible because $st \neq ts$.

Case of $st = t^2$ and $st = tst$. In this case $t^2 = st = tst$. By left cancellativity, $t = st = t^2$, i.e. t is idempotent and hence a left identity of S. Therefore $ts = s$ and $s^2 = sts = t^2s = ts = s$ and s is idempotent as well.

Case of $st = t^2$ and $s = tst$. In this case $s = t(st) = t^3$, and hence $st = ts$ which contradicts our assumption.

Case of $s^2 = t^2$ and $st = tst$. In this case $t^2s = s^3 = st^2 = (st)t = (tst)t = tst^2 = tss^2 = ts^2s = tt^2s = t^3s$. By left cancellativity $s = ts$. It follows that $t^2 = s^2 = ts^2 = tt^2$. By left cancellativity, $t = t^2$. Thus $t = s^2$ and $st = ts$ contradicting our assumption.

Case of $s^2 = t^2$ and $s = tst$. In this case the subsemigroup $\langle s,t\rangle$ of S generated by s and t is isomorphic to Q, the quaternion group. Indeed, $t^2 = s^2 = tsttst = tss^2st = ts^4t = tt^4t = t^6$. It follows that $t^4 = e$ is an idempotent, and hence a right identity of S. Since $e = s^4$, this idempotent e commutes both with s and t, i.e. it is the identity of the subsemigroup $\langle s,t\rangle$. It is easily seen that $\langle s,t\rangle =$

$$\{t, t^2, t^3, e, s, s^3, ts, t^3s\}. \tag{45}$$

Indeed, all powers of s and t are in (45) as well as all elements of the form t^ks. If $m > 1$, then $t^ks^m = t^ks^2s^{k-2} = t^kt^2s^{k-2} = t^{k+2}s^{k-2}$ which shows that each element of the form t^ks^m equals t^n or t^ns for suitable n. Finally, $st = est = t^4st = t^3(tst) = t^3s$ which shows that every element of the form t^ms^n is in (45). It is easy to see that if all the elements listed in (45) are different, then $\langle s,t\rangle$ is isomorphic to Q. If some of the elements in (45) are equal, then $\langle s,t\rangle$ is a proper homomorphic image of Q, i.e. $\langle s,t\rangle$ is an abelian group. It follows that $st = ts$ contrary to our assumption. We have proved

LEMMA 10. Let S be a left cancellative semigroup in which $\{s,t\}^2$ has less that four elements for any s and t. For any s and t in S one of the following possibilities holds:

(i) $st = ts$;

(ii) s and t are left identities of S;

(iii) the subsemigroup $\langle s,t\rangle$ of S generated by s and t is isomorphic to the quaternion group Q.

Suppose that S does contain an isomorphic copy of Q. Each idempotent of S is a left identity of S. Thus, if S has a central idempotent, it is a two-sided identity of S and there are no other idempotents in S. If e is a non-central idempotent, then e does not commute with an element t. Now, $\langle e,t\rangle$ cannot be isomorphic to Q. By Lemma 10, t is an idempotent. It follows that e commutes with every nonidempotent element of S. Thus, e commutes with all elements of Q except, maybe, its identity element 1. But then e commutes with 1 (because $1 = s^4$ for every s in Q). Since 1 is an idempotent, it is a left identity of S, hence $s = 1s = s1 = 1$. Thus, in all cases S contains a single idempotent. This idempotent 1 is the identity of S. Indeed, applying Lemma 10 to 1 and an arbitrary element s of S we see that $\langle 1,s\rangle$ cannot be a quaternion group. If s is an idempotent, then $s = 1$. Therefore $1s = s1$, and hence 1 is a two-sided identity of S. Suppose that $xs = ys$ for some x, y and

s in S. If s is a central element, then $sx = sy$, and we obtain $x = y$ by left cancellativity. If s is not central, then, by Lemma 10, it belongs to a quaternion subgroup of S, in which case $s^4 = 1$. Therefore $x = x1 = (xs)s^3 = (ys)s^3 = y1 = y$. Thus S is right cancellative. By Proposition 5, S is isomorphic to a direct product of Q and an elementary abelian 2-group.

Now suppose that S does not contain a subgroup isomorphic to Q. By Lemma 10, each element of S is either central or idempotent. The idempotents of S form a left zero subsemigroup of S. Denote it by L. Of course, L may be empty. Let Z denote the center of S (which may be empty). If Z and L overlap, then S has a central idempotent and, as we have seen, S has an identity element. In this case $S = Z$. Thus, if S is not commutative, Z and L do not overlap and $L \neq \emptyset$. In this case S is an ordinal sum of Z and L (in this order). Indeed, for every z in Z and e in L $ze = ez = z$. We have proved the "if" part of the following theorem, the "only if" part of which is trivial:

PROPOSITION 6. S is a left cancellable but not right cancellable semigroup in which $\{s,t\}^2$ has less than four elements for any s and t, if and only if S is an ordinal sum of a commutative semigroup and a left zero semigroup (each of which may be empty).

Of course, one readily obtains a full description of right but not left cancellative semigroups with the above property.

REFERENCES

[1] Ya.G. Berkovich and G.A. Freiman, On the connection between some numeric characteristics of a finite group and the structure of the group (manuscript).

[2] V. Brailovsky and G.A. Freiman, On two-element subsets in groups, Ann. New York Acad. Sci. 373(1981), 183-190.

[3] V. Brailovsky and G.A. Freiman, Groups with small cardinality of the cubes of their two-elements subsets, Ann. New York Acad. Sci. 410(1983), 75-82.

[4] G.A. Freiman, Foundations of a structural theory of set addition, Kazan', 1966 [Russian]; English translation: Translations of Mathematical Monographs, vol. 37, Amer. Math. Soc., Providence, R.I. 1973.

[5] G.A. Freiman, Groups and the inverse problems of the additive set theory. In the book: Number-theoretic investigations on the Markov spectrum and the structure theory of set addition, Kalinin University, Moscow, 1973, pp. 175-183, [Russian].

[6] G.A. Freiman, On two- and three-element subsets of groups, Aequationes Math. 22(1981), 140-152.

[7] T.E. Hall, On regular semigroups whose idempotents form a subsemigroup, Bull, Austral. Math. Soc. 1(1969), 195-208.

[8] T.E. Hall, Addenda to [7], ibid. 3(1970), 287-288.

[9] J.H.B. Kemperman, On complexes in a semigroup, Indagat. Math. 18(1956), 247-254.

[10] E.S. Lyapin, Semigroups, 3rd edition, Translations of Mathematical Monographs, Vol. 3, Amer. Math. Soc., Providence, R.I., 1974.

[11] L. Rédei, Das "Schiefe Produkt" in der Gruppentheorie, Comment. Math. Helvet. 20(1947), 225-264.

[12] L. Rédei, Algebra, 1. Teil, Geest & Portig, Leipzig, 1959.

[13] B.M. Schein, On the theory of inverse semigroups and generalized grouds, Theory of Semigroups and Its Appl., Saratov State University, Saratov 1(1965), 286-324 [Russian. English translation in Amer. Math. Society Translations (2) 113(1979), 89-122].

[14] W.R. Scott, Group Theory. Prentice Hall, 1964.

[15] V.V. Wagner [Vagner], Algebra of binary relations and its applications in differential geometry, Differential Geometry 4(1979), Saratov University Press, Saratov, 15-131 [Russian].

CORRESPONDENCES OF SEMIGROUPS

Simon M. Goberstein
Department of Mathematics
California State University, Chico
Chico, CA 95929

With any universal algebra (or, more generally, with any mathematical structure) one can associate various derived algebraic structures, such as the group of all automorphisms, the semigroup of all endomorphisms, the inverse semigroup of all partial automorphisms, the lattice of all subalgebras of the given algebra, etc. A natural problem is to investigate how well universal algebras from a certain class are characterized by their derived algebraic structures. These derived structures may be quite different from one another in nature and in the amount of information that they carry about the original universal algebra. Therefore, it seems expedient to look for some general derived structure (containing more information about the given algebraic system than other derived structures) as a unifying framework for research. An interesting candidate for such a general derived structure was suggested by A. G. Kurosh.

Let (A, Ω) be a universal algebra. A subalgebra (perhaps empty) of the direct product $A \times A$ is called a <u>correspondence</u> (or a <u>stable relation</u>) of the algebra A. A correspondence of A can be considered as a "partial multivalued endomorphism" of the algebra A. Let $C(A)$ denote the set of all correspondences of A. According to the Birkhoff-Frink theorem, $(C(A), \subseteq)$ is an algebraic lattice. With respect to composition $(\circ)$ of binary relations, $C(A)$ is a semigroup with an identity Δ_A (= the equality relation on A) and a natural involution $({}^{-1})$ (i.e., the operation of taking the inverse, ρ^{-1}, of any correspondence ρ of A). Following A. G. Kurosh [13, 14] we will say that the system $(C(A), \circ, {}^{-1}, \subseteq)$ is the <u>bundle of correspondences</u> of A. (If there is no danger of confusion, it will be denoted simply by $C(A)$.) The idea of studying bundles of correspondences (and related derived structures) of algebras of various types was promoted by A. G. Kurosh at a number of algebra conferences in the USSR, as well as in his books [13, 14] and his lectures and seminars at Moscow University. In this paper we will give a brief survey of results about bundles of correspondences of universal algebras, groups, and semigroups. In the last section, devoted to semigroups, some new results are announced.

1. UNIVERSAL ALGEBRAS

Recall that a lattice L is said to be algebraic if it is complete and compactly generated (i.e., every element of L is a sup of compact elements). G. Birkhoff and O. Frink proved in [1] (see also [9]) that the subalgebra lattice of a universal algebra is algebraic, and, conversely, any algebraic lattice is isomorphic to the lattice of all subalgebras of some universal algebra. The first part of this result implies that for any algebra A, $(C(A), \subseteq)$ is an algebraic lattice. The converse part of the Birkhoff-Frink theorem was strengthened by A. A. Iskander as follows:

THEOREM 1 [10]. *Any algebraic lattice is isomorphic to the lattice of all correspondences of some universal algebra.*

In fact, A. A. Iskander proved in [10] a somewhat stronger theorem and established a number of other interesting facts. In [11] he also studied lattices of correspondences of partial algebras.

We will say that algebras (A, Ω) and (B, Ω) are *C-isomorphic* if their bundles of correspondences are isomorphic, and any isomorphism of $C(A)$ onto $C(B)$ will be called a *C-isomorphism* of A onto B. Let Φ be a C-isomorphism of A onto B. It is obvious that all properties of correspondences that can be expressed through $\circ$, $^{-1}$ and $\subseteq$ are preserved under any C-isomorphism. In particular, $(A \times A)\Phi = B \times B$, $\Delta_A\Phi = \Delta_B$, $\emptyset\Phi = \emptyset$. Furthermore, if ρ is a congruence, an endomorphism, or a partial automorphism (i.e., an isomorphism between subalgebras) of A, then $\rho\Phi$ is a congruence, an endomorphism, or a partial automorphism of B, respectively. Thus Φ induces isomorphisms between the congruence lattices, the endomorphism semigroups, the partial automorphism semigroups (and thus between the subalgebra lattices and the automorphism groups) of algebras A and B.

Let K be a class of algebras of the same type and $L \subseteq K$. We say that L is *C-closed in* K (or simply *C-closed* if K is the class of all algebras of the given type) if for all $A \in L$ and $B \in K$, $C(A) \cong C(B)$ implies $B \in L$.

Let ω be an n-ary operation on a set A. An n-ary operation $\bar{\omega}$ on A is called an *isomer* of ω if there exists a permutation $(i_1, \ldots, i_n)$ of $\{1, \ldots, n\}$ such that for all $a_1, \ldots, a_n \in A$, $a_1 \ldots a_n\omega = a_{i_1} \ldots a_{i_n}\bar{\omega}$. (In particular, the only isomers of a binary operation are the operation itself and its dual.) Let (A, Ω) be an algebra. Let us choose an isomer $\bar{\omega}$ for each operation $\omega \in \Omega$ and let $\bar{\Omega}$ consist of all such operations $\bar{\omega}$. Then we say that algebras (A, Ω) and $(A, \bar{\Omega})$ are *isomeric*. It is obvious that isomeric algebras are C-isomorphic. Thus a C-closed class M of algebras must contain all algebras isomeric to each algebra A in M. On the other hand, D. A. Bredihin [2] gave an example of a 3-element commutative groupoid (A, ω) which is C-isomorphic to a noncommutative groupoid $(A, \overline{\omega})$. Thus (A, ω) and $(A, \overline{\omega})$ are C-isomorphic but not isomeric. It follows also that the class of commutative groupoids is not C-closed in the class of all groupoids.

The first work devoted to the study of bundles of correspondences and C-isomorphisms of universal algebras and semigroups was the paper [18] by G. I. Žitomirskiǐ. After making obvious general remarks, including some of those contained in the last three paragraphs, he established a number of results concerning C-closed classes of algebras and semigroups. Some of them will be formulated below.

Let K be a class of algebras. Its inverse homomorphic image consists of all inverse homomorphic images of algebras from K. The class of all overalgebras for K comprises all algebras having at least one subalgebra in K.

THEOREM 2 [18]. *If K is a C-closed class of algebras, then the inverse homomorphic image for K and the class of all overalgebras for K are also C-closed.*

Let ε be a congruence on a universal algebra A. Let $C_\varepsilon(A) = \{\rho \in C(A) : \rho = \rho \circ \varepsilon = \varepsilon \circ \rho\}$. Then [18] $(C_\varepsilon(A), \circ, ^{-1}, \subseteq) \cong (C(A/\varepsilon), \circ, ^{-1}, \subseteq)$. (Žitomirskiǐ used this to prove the first part of Theorem 2.) It follows [2] that if Φ is a C-isomorphism of an algebra A onto an algebra B and ε is a congruence on A, then A/ε and $B/(\varepsilon\Phi)$ are C-isomorphic. Using this fact, D. A. Bredihin proved

THEOREM 3 [2]. *Let A and B be C-isomorphic algebras.*

(1) *If A is a subdirect product of the family of algebras $(A_i)_{i\in I}$, then B is a subdirect product of the family of algebras $(B_i)_{i\in I}$ where, for each $i \in I$, the algebras A_i and B_i are C-isomorphic.*

(2) *If $A = A_1 \times \ldots \times A_n$ for some natural number n, then $B = B_1 \times \ldots \times B_n$ and, for every $i = 1, \ldots, n$, the algebras A_i and B_i are C-isomorphic.*

For any algebra A, let $\Delta(A)$ denote the set of all partially identical correspondences of A (i.e., all correspondences of the form Δ_H where H is a subalgebra of A).

By definition a C-isomorphism of an algebra A onto an algebra B is an isomorphism of $(C(A), \circ, ^{-1}, \subseteq)$ onto $(C(B), \circ, ^{-1}, \subseteq)$. A natural problem is to find the situations in which an isomorphism of $(C(A), \circ)$ onto $(C(B), \circ)$, or an isomorphism of $(C(A), \circ, \subseteq)$ onto

$(C(B), \circ, \subseteq)$, etc., is already a C-isomorphism of A onto B. An interesting result in this direction was obtained by D. A. Bredihin:

THEOREM 4 [8]. Let M be a congruence-permutable variety of algebras and $A, B \in M$. Let Φ be a mapping from $C(A)$ to $C(B)$. The following conditions are equivalent:

(1) Φ is a C-isomorphism of A onto B;

(2) Φ is an isomorphism of $(C(A), \circ, \subseteq)$ onto $(C(B), \circ, \subseteq)$;

(3) Φ is an isomorphism of $(C(A), \circ)$ onto $(C(B), \circ)$ such that $(\Delta(A))\Phi = \Delta(B)$;

(4) Φ is an isomorphism of $(C(A), \circ)$ onto $(C(B), \circ)$ such that $(A \times A)\Phi = B \times B$.

2. GROUPS

Throughout this section a group is considered as an algebra $(G, \cdot, ^{-1}, 1)$. In particular, here a correspondence of a group G is a subgroup of $G \times G$.

The class of groups is one of the most important congruence-permutable classes of algebras. Thus Theorem 4 can be applied to this class. However a somewhat stronger result holds for groups. D. A Bredihin showed in [8] that Theorem 4 remains true for groups A,B and a mapping Φ from $C(A)$ to $C(B)$ if condition (3) is replaced by the following:

(3') Φ is an isomorphism of $(C(A), \circ)$ onto $(C(B), \circ)$ such that $\{(1_A, 1_A)\}\Phi = \{(1_B, 1_B)\}$.

A correspondence ρ of a group G is called a multi-automorphism of G if ρ is a subdirect square of G (i.e., if $pr_1\rho = pr_2\rho = G$). Let $M(G)$ denote the set of all multi-automorphisms of G. Since the composition of two multi-automorphisms is again a multi-automorphism, $(M(G), \circ)$ is a subsemigroup of $(C(G), \circ)$. In fact, $(M(G), \circ)$ is an inverse semigroup [17]. An M-isomorphism of a group G onto a group H is any isomorphism of $(M(G), \circ)$ onto $(M(H), \circ)$. It is easy to see that any C-isomorphism of groups induces their M-isomorphism.

THEOREM 5 [7]. If a group H is M-isomorphic to an abelian group G, then G and H are isomorphic.

Let K be a class of groups and $L \subseteq K$. A group $G \in L$ is said to be C-determined in K if for any group $H \in K$, $C(G) \cong C(H)$ implies $G \cong H$. A similar definition can be given for every derived structure of a group. Thus Theorem 5 states that any abelian group is

M-determined in the class of all groups.

An RL-lattice is defined as an algebra $(S, \wedge, \vee, R, L)$ where $(S, \wedge, \vee)$ is a lattice and R, L are endomorphisms of S such that $L \circ R = L$ and $R \circ L = R$ (see [4]). Let G be a group. Then the lattice of correspondences of G can be considered as an RL-lattice if we define R and L as follows: $\rho R = \Delta_{pr_1 \rho}$ and $\rho L = \Delta_{pr_2 \rho}$ for any $\rho \in C(G)$.

THEOREM 6 [4]. Let G and H be groups with isomorphic RL-lattices of correspondences. If G is solvable (nilpotent) and has an abelian (central) series $1 = G_0 \subset \ldots \subset G_n = G$, then H is also solvable (nilpotent) and has an abelian (central) series $1 = H_0 \subset \ldots \subset H_n = H$ such that G_{i+1}/G_i and H_{i+1}/H_i are isomorphic for each $i = 0, \ldots, n-1$. In particular, any abelian group is determined by its RL-lattice of correspondences in the class of all groups.

A symmetric lattice [12] is an algebra $(S, \wedge, \vee, ')$ where $(S, \wedge, \vee)$ is a lattice and $'$ is a unary operation on S which is an involutory automorphism of $(S, \wedge, \vee)$. Let G be a group. Then the lattice of correspondences of G becomes a symmetric lattice if we define $\rho' = \rho^{-1}$ for any $\rho \in C(G)$. D. A. Bredihin announced in [6] that Theorem 6 remains true if the phrase "RL-lattices of correspondences" is replaced by "symmetric lattices of correspondences."

3. SEMIGROUPS

In this section a correspondence of any semigroup S is a subsemigroup of $S \times S$ (including the empty one).

The bundle of correspondences contains a great amount of information about the given semigroup. Therefore it is natural to expect that many important classes of semigroups are C-closed. Indeed, a number of such classes were described in [18].

THEOREM 7 [18]. The following classes of semigroups are C-closed in the class of all semigroups: (a) all groups; (b) all completely regular semigroups; (c) all simple semigroups; (d) all completely [0-]simple semigroups.

Several other C-closed classes of semigroups were described in [18] and additional such classes can be obtained if Theorem 2 is applied. However the following fact appears to be new: the class of all (completely) semisimple semigroups is C-closed in the class of all semigroups.

Let K be a class of semigroups and $L \subseteq K$. A semigroup $S \in L$ is said to be C-determined in K if for any semigroup $T \in K$ such that $C(S) \cong C(T)$, semigroups S and T are isomorphic or antiisomorphic (of course, the latter possibility becomes obsolete if S is antiisomorphic to itself).

THEOREM 8 [2]. The following semigroups are C-determined in the class of all semigroups: (a) monogenic semigroups; (b) bands; (c) commutative cancellative semigroups; (d) nonperiodic cancellative semigroups; (e) nonperiodic commutative inverse semigroups.

Let ϕ be a bijection of a semigroup S onto a semigroup T and let Φ be a C-isomorphism of S onto T. We say that Φ is induced by ϕ if for any $\rho \in C(S)$, $\rho\Phi = \phi^{-1} \circ \rho \circ \phi$, i.e., $\rho\Phi = \{(x\phi, y\phi) : (x,y) \in \rho\}$.

Let S be a semigroup. Recall that if an element a of S has finite order, then the index of a (to be denoted by ind a) is a least positive integer m for which there exists an integer $n > m$ such that $a^m = a^n$. If a has infinite order, set ind a = ∞. Finally, let ind S = $\sup\{\text{ind } a : a \in S\}$. The following result proved to be very useful in the study of C-isomorphisms of semigroups.

THEOREM 9 [3]. Let S and T be semigroups and ind S > 1. Then any C-isomorphism Φ of S onto T is induced by a unique bijection ϕ of S onto T.

Using Theorems 3, 8, and 9 and a description of finite subdirectly irreducible commutative semigroups [16], D. A. Bredihin proved the following

THEOREM 10 [3]. The class of all commutative semigroups is C-closed.

This answered one of the two open questions raised in [18]. The second question (whether the class of inverse semigroups is C-closed) was also answered in [3]. Corollaries 2.1 and 2.2 of that paper assert that the classes of all regular and all inverse semigroups, respectively, are C-closed. The latter and Theorem 7(a) imply that the class of all inverse semigroups, which do not contain subgroups of order 2, is also C-closed. At the same time a much stronger result can be proved for this class of semigroups.

The set of all partial automorphisms (i.e., isomorphisms between subsemigroups) of a semigroup S is an inverse semigroup (with respect to composition). Let us denote it by PA(S). Any isomorphism of PA(S) onto PA(T) will be called a PA-isomorphism of a semigroup S onto a semigroup T. It is clear that a C-isomorphism of semigroups S and T induces their PA-isomorphism. Thus the affirmative answer to the second question raised in [18] is a corollary to Theorem 7(a) and the following

THEOREM 11. The class of inverse semigroups, which do not contain subgroups of order 2, is PA-closed in the class of all semigroups.

Since a two-element group is PA-isomorphic to a two-element nilsemigroup, the class of all inverse semigroups is not PA-closed in the class of all semigroups. The proof of Theorem 11 uses, in particular, some results from [15] corrected and complemented by the present author.

Returning to C-determined semigroups, it is worthwhile to mention

THEOREM 12 [5]. Any nilsemigroup is C-determined in the class of all semigroups.

Let S and T be semigroups and ϕ a bijection of S onto T. Recall that ϕ is said to be a half-isomorphism of S onto T if for any $a, b \in S$, either $(ab)\phi = a\phi \cdot b\phi$ or $(ab)\phi = b\phi \cdot a\phi$. A half-isomorphism ϕ of S onto T is called strong if ϕ^{-1} is a half-isomorphism of T onto S.

THEOREM 13 [3]. Let S and T be semigroups, ϕ a C-isomorphism of S onto T and $\operatorname{ind} S > 2$. Then ϕ is induced by a unique strong half-isomorphism of S onto T. If S is commutative, then ϕ is induced by a unique isomorphism of S onto T.

At the same time the question whether any commutative semigroup is C-determined is still open.

An interesting series of examples of finite nonisomorphic groups which are C-isomorphic was constructed by D. A. Bredihin in his dissertation (private communication):

EXAMPLE [Bredihin(*)]. Let p, q, r be primes such that $p > q > r \geq 3$ and $p \equiv q \equiv 1 (\mathrm{mod}\ r)$. Then there exist natural numbers $1 < k < p$ and $1 < \ell < q$ such that $k^r \equiv 1 (\mathrm{mod}\ p)$, $\ell^r \equiv 1 (\mathrm{mod}\ q)$, and $k^n \equiv 1 (\mathrm{mod}\ p)$ $[\ell^n \equiv 1 (\mathrm{mod}\ q)]$ implies $n \equiv 0 (\mathrm{mod}\ r)$ (see, for example, G. H. Hardy and E. M. Wright, An Introduction to the Theory of Numbers, Oxford Univ. Press, 4th ed., 1960, Section 7.5). Choose a pair of such numbers, k and ℓ, and denote by $Q(p,q,r;k,\ell)$ a group having the following presentation:

$$Q(p,q,r;k,\ell) = \langle a,b,c \mid a^p = b^q = c^r = 1,\ ab = ba,\ cac^{-1} = a^k,\ cbc^{-1} = b^\ell \rangle.$$

(*) D. A. Bredihin, Multiplicative algebras of correspondences, Candidate (=Ph.D.) dissertation, Saratov State University, 1977 (in Russian).

Let $Q(p,q,r)$ be the set of all groups of the form $Q(p,q,r;k,\ell)$ for various choices of k and ℓ. Then all groups from $Q(p,q,r)$ have isomorphic bundles of correspondences. However not all of these groups are isomorphic. For example, $Q(13, 7, 3; 3, 2) \not\cong Q(13, 7, 3; 3, 4)$.

With the exception of nonperiodic commutative inverse semigroups and some classes of groups, the problem of C-determinability of inverse semigroups was not considered before. Since (finite) groups, in general, are not C-determined, it is natural to concentrate on the study of C-determinability of fundamental inverse semigroups and to hope that certain nice conditions would distinguish C-determined semigroups from that class. Surprisingly enough no such conditions are needed. Our main new result is the following

THEOREM 14. *Let* S *be any fundamental inverse semigroup and* T *any semigroup. If* $C(S) \cong C(T)$, *then* $S \cong T$. *In other words, any fundamental inverse semigroup is* C-*determined in the class of all semigroups. Moreover, any* C-*isomorphism of* S *onto* T *is induced by a (unique) isomorphism or antiisomorphism.*

The proofs of new results announced in this paper will appear elsewhere.

ACKNOWLEDGEMENT

I would like to thank the organizers of the conference, Professors H. Jürgensen, G. Lallement, and H. J. Weinert, for inviting me to take part, and the administration of the Mathematical Research Institute, Oberwolfach, for creating excellent conditions for the meeting.

REFERENCES

1. G. Birkhoff and O. Frink, Representations of lattices by sets, Trans. Amer. Math. Soc., 64(1948), 299-316.

2. D. A. Bredihin, Involuted semigroups of stable binary relations, Studies in Algebra, Saratov Univ. Press, No. 4, 1974, 3-12 (in Russian).

3. D. A. Bredihin, Bundles of correspondences of semigroups, Contemporary Algebra, Leningrad, No. 4, 1976, 31-47 (in Russian).

4. D. A. Bredihin, RL-lattices of correspondences of groups, Ordered Sets and Lattices, Saratov Univ. Press, No. 5, 1978, 7-11 (in Russian).

5. D. A. Bredihin, On determinability of nilsemigroups by their bundles of correspondences, Studies in Contemporary Algebra, Ural Univ. Press, Sverdlovsk, 1978, 3-9 (in Russian).

6. D. A. Bredihin, Symmetric lattices of correspondences of groups, XV All-Union Conference on Algebra. Summaries of Talks, Part I, Krasnoyarsk, 1979, 26 (in Russian).

7. D. A. Bredihin, On determinability of abelian groups by their inverse semigroups of multi-automorphisms, Theory of Semigroups and Its Applications, Saratov Univ. Press, 1983, 58-63 (in Russian).

8. D. A. Bredihin, Bundles of correspondences and $\mathcal{R}$-isomorphisms of congruence-permutable algebras, Theory of Semigroups and Its Applications, Saratov Univ. Press, 1984, 4-9 (in Russian).

9. G. Grätzer, Universal Algebra, 2nd edition, Springer-Verlag, New York, 1979.

10. A. A. Iskander, The lattice of correspondences of a universal algebra, Izv. Akad. Nauk SSSR, Ser. Mat., 29 (1965), 1357-1372 (in Russian).

11. A. A. Iskander, Partial universal algebras with preassigned lattices of subalgebras and correspondences, Mat. Sb., 70(112):3 (1966), 438-456 (in Russian).

12. P. G. Kontorovich and K. M. Kutyev, Symmetric lattices, Sibirsk. Mat. Ž., 10(1969), 537-548 (in Russian).

13. A. G. Kurosh, The Theory of Groups, 3rd edition, Nauka, Moscow, 1967.

14. A. G. Kurosh, General Algebra (Lectures of the 1969-70 academic year), Nauka, Moscow, 1974.

15. A. L. Libih, Local automorphisms of monogenic inverse semigroups, Theory of Semigroups and Its Applications, Saratov Univ. Press, No. 4, 1978, 54-59 (in Russian).

16. P. E. McNeil, Finite commutative subdirectly irreducible semigroups, Trans. Amer. Math. Soc., 172(1972), 57-67.

17. B. M. Schein, On some classes of semigroups of binary relations, Sibirsk, Mat. Ž., 6(1965), 616-635 (in Russian).

18. G. I. Žitomirskiĭ, Stable binary relations on universal algebras, Mat. Sb., 82(124):2 (1970), 163-174 (in Russian).

ON UNIVERSALITY OF EXTENSIONS

by P. Goralčík and V. Koubek

Computing Center of Charles University
Faculty of Mathematics and Physics
118 00 Praha 1, Czechoslovakia

A category $\underset{\sim}{C}$ is called universal if it contains a full subcategory isomorphic to the category $\underset{\sim}{G}$ of graphs, that is to say, if there is a full embedding functor $F : \underset{\sim}{G} \to \underset{\sim}{C}$, injective on both objects and morphisms, such that for an arbitrary $\underset{\sim}{C}$-morphism $f : F(A) \to F(B)$, $A, B \in \underset{\sim}{G}$, there is an $\underset{\sim}{G}$-morphism $g : A \to B$ with $F(g) = f$.

The interest of universal categories comes from their representing power. In a universal category $\underset{\sim}{C}$, an arbitrary monoid can be represented as $\mathrm{End}(A)$ by an object $A \in \underset{\sim}{C}$. The cost to pay is that $\mathrm{End}(A)$ then practically carries no information on the representing object A since there are, generally, arbitrarily many non-isomorphic such.

For both the positive and the negative aspect of universality, it is worthwhile to ask about the category you are just dealing with, whether it is universal or not. To prove that a given category is universal, it is no more necessary to try to fully embed the category $\underset{\sim}{G}$ of graphs into it; any one on the long list of universal categories (cf. [3]) can replace $\underset{\sim}{G}$ in the definition of a universal category.

The category of all semigroups and all semigroup homomorphisms enjoys a honorable position in this list since the pioneering paper [1] by Hedrlín and Lambek. Later Koubek and Sichler [2] gave us a nice criterion for a semigroup variety $\underset{\sim}{V}$ to be universal: $\underset{\sim}{V}$ is universal if and only if it contains all commutative semigroups and fails the identity $x^n y^n = (xy)^n$ for every $n > 1$. Implicit in their proof is a somewhat stronger result we get in the bargain just by observing that the full embedding $F : \underset{\sim}{G} \to \underset{\sim}{V}$ they construct has the following additional property: for every graph $G \in \underset{\sim}{G}$, in the corresponding semigroup $F(G) \in \underset{\sim}{V}$ no element has a left or right zero, i. e. we have $x \neq xy \neq y$ for all $x, y \in F(G)$. Thus in every universal semigroup variety $\underset{\sim}{V}$, the semigroups in $\underset{\sim}{V}$ in which no element has a left zero form a universal full subcategory of $\underset{\sim}{V}$.

The categories we want here to try for universality arise very naturally as all possible semigroup extensions of a given semigroup S. More exactly, we examine the category $\mathrm{Ext}(S)$ of all semigroups X containing an isomorphic copy of S as a subsemigroup, and of all semigroup homomorphisms between pairs of objects of $\mathrm{Ext}(S)$. Our aim is to establish the following result:

THEOREM. $\mathrm{Ext}(S)$ *is universal if and only if* S *has no idempotents.*

Proof. If S contains an idempotent then End(S) has the constant endomorphisms for one-sided zeroes and Ext(S) can no more be universal.

Let there be given a semigroup S with no idempotents (and thus infinite). Let $(\text{card}S)^+$ denote the cardinal successor of card$S$ and choose a cardinal A, $A > (\text{card}S)^+$. (We work under the usual convention that cardinals are the initial ordinals, so we have in A the natural numbers $0, 1, 2, \ldots$ and A is naturally well ordered.)

Let W_A be the semigroup of reduced forms for the presentation $\langle A;\ \{ba = b;\ a, b \in A,\ a < b\}\rangle$, i. e. of the non-decreasing words $w = a_1 a_2 \ldots a_n, a_1 \leq a_2 \leq \ldots \leq a_n$, over alphabet A. The defining relations imposed to the set A of generators of W_A enforce that for every $a \in A$, the set $\{b \in A;\ a < b\}$ is formed by left zeroes of a and has cardinality greater than card$S$.

We now fully embed into Ext(S) the universal category $\underset{\sim}{U}$ of all semigroups in which no element has a left zero. To this end, for every semigroup $T \in \underset{\sim}{U}$ construct the coproduct $W_A \coprod S \coprod T$ formed by the non-void finite sequences $(x_1, \ldots, x_m)$ such that $x_i \in W_A \cup S \cup T$ for all $i = 1, \ldots, m$ and x_i, x_{i+1} do not both belong to the same one of the three semigoups W_A, S, T. The product $(x_1, \ldots, x_m)(y_1, \ldots, y_n)$ of two elements of the coproduct equals $(x_1, \ldots, x_{m-1},\ x_m y_1,\ y_2, \ldots, y_n)$ if x_m, y_1 both belong to the same one of the semigroups W_A, S, T, and simply their concatenation $(x_1, \ldots, x_m,\ y_1, \ldots, y_n)$ if this is not the case.

Let $G = (V(G), E(G))$ be a connected rigid graph (i. e. one with End(G) reduced to the identity) with the set of vertices $V(G) = A$ ($E(G)$ denotes the set of edges of G), let s_0 be an arbitrary element of S, and let $S \to A : s \to a_s$ be an injective assignment with $a_s \neq 0$ for all $s \in S$. Let R_T be the least congruence on $W_A \coprod S \coprod T$ such that

(1) $(0, s_0, ab, s_0, a, s_0, 0) R_T (0, s_0, ab, s_0, b, s_0, 0)$ for all $a, b \in A$ such that $a < b$ and $\{a, b\} \in E(G)$,

(2) $(0, s_0^2, 0, s, a_s, s_0^2, 0) R_T (0, s_0^2, 0, s^2, a_s, s_0^2, 0)$ for all $s \in S$,

(3) $(0, s_0^3, 0, t, 0, s_0^3, 0) R_T (0, s_0^3, 0, t', 0, s_0^3, 0)$ for all $t, t' \in T$.

Observe that two R_T-congruent elements $(x_1, \ldots, x_m), (y_1, \ldots, y_n)$ of $W_A \coprod S \coprod T$ must have the same length $m = n$. If $x_i \neq y_i$ and $x_i \in W_A$ then $x_{i-2} = ab$ for some $a, b \in A$ with $a < b$ and $\{x_i, y_i\} = \{a, b\} \in E(G)$. If $x_i \neq y_i$ and $x_i \in S$ then $x_{i+1} = a_s$ for a unique $s \in S$ and $\{x_i, y_i\} = \{s, s^2\}$.

We are now ready to define a functor $F : \underset{\sim}{U} \to Ext(S)$: for $T \in \underset{\sim}{U}$, set $F(T) = (W_A \coprod S \coprod T)/R_T$; for a homomorphism $h : T \to U$, $T, U \in \underset{\sim}{U}$, let $F(h) : F(T) \to F(U)$ be the homomorphism taking $(x_1, \ldots, x_m)/R_T$ to $(y_1, \ldots, y_m)/R_U$, where $y_i = x_i$ if $x_i \in W_A \cup S$, and $y_i = h(x_i)$ if $x_i \in T$.

Let us see that the definition of F is correct. First note that the respective isomorphic copies of W_A, S, and T, canonically embedded into the coproduct $W_A \coprod S \coprod T$, are not affected by R_T, so we still have them in $F(T)$, and, in particular, $F(T)$ is in Ext(S). For this reason also, the functor F is injective on objects. Further, $F(h)$ is obtained from $\text{id}_{W_A} \coprod \text{id}_S \coprod h$ by factorization, which is correct, since R_T-classes are taken to R_U-classes, as can be seen by applying $\mathrm{id}_{W_A} \coprod \mathrm{id}_S \coprod h$ to the generating pairs for R_T. Again, $F(h)$ is an extension of h, hence F is injective on morphisms.

The rest of the proof is devoted to proving that $F : \underset{\sim}{U} \to \mathrm{Ext}(S)$ is a full embedding. To this end, take an arbitrary homomorphism $f : F(T) \to F(U)$, $T, U \in \underset{\sim}{U}$. To be done, we have to show that $f = F(h)$ for some homomorphism $h : T \to U$.

We start by showing that $f(a) = a$ for every $a \in A$, and that $f(s_0) = s_0$. First, f must be injective on A : if $f(a) = f(b)$ for $a, b \in A$, $a < b$, then $f(b) = f(ba) = f(b)f(a) = f(b)f(b)$, but there is no idempotent in $F(U)$. Next, for every $a \in A$, $f(a)$ must have a set of left zeroes of cardinality greater than card$S$, since a has such. Now, the only elements in $F(U)$ having left zeroes must be of length 1 and not in T, thus only can be in $W_A \cup S$. However, for every $s \in S$, all left zeroes of s are in S, thus s has at most card$S$ left zeroes. It follows that $f(a) \in W_A$ for every $a \in A$.

For every $w \in W_A$, let w^0 denote the last letter of w. Then for $a, b \in A$, $a < b$, we have $f(b)f(a) = f(ba) = f(b)$, hence $f(a)^0 < f(b)^0$, and, since A is well ordered, $a \leq f(a)^0$.

Assume that $f(s_0) = (x_1, \ldots, x_m)/R_U$. Choose $\{a, b\} \in E(G)$ with $a < b$, and such that $x_m^0 < b$ if $x_m \in W_A$. Then by (1), $(0, s_0, ab, s_0, a, s_0, 0) R_T (0, s_0, ab, s_0, b, s_0, 0)$, whence $f(0)(x_1, \ldots, x_m) f(ab)(x_1, \ldots, x_m) f(a)(x_1, \ldots, x_m) f(0)$ is R_U-congruent to $f(0)(x_1, \ldots, x_m) f(ab)(x_1, \ldots, x_m) f(b)(x_1, \ldots, x_m) f(0)$. By the choice of a and b, we have either $x_m \in W_A$ and $\left(x_m f(a)\right)^0 = x_m^0$ or $f(a)^0$, both strictly less than $f(b)^0 = \left(x_m f(b)\right)^0$, or $x_m \notin W_A$ and $f(a)^0 < f(b)^0$. In both cases $x_m f(a) \neq x_m f(b)$. Since $f(a), f(b) \in W_A$ the above R_U-congruent pair must be so according to (1) (with T replaced by U). Consequently, $f(a), f(b) \in A$, and $x_m = s_0$. Let us now go through the same reasoning with an arbitrary edge $\{a, b\} \in E(G)$, $a < b$. As above we establish that $f(a), f(b) \in A$, $\{f(a), f(b)\} \in E(G)$, $f(a) < f(b)$, hence by rigidity of G, $f(a) = a$, $f(b) = b$. We show that $m = 1$. Indeed, if $m \neq 1$ then by force $x_{m-1} = ab$. Taking another edge $\{a', b'\} \in E(G)$ with $a' < b'$, $\{a, b\} \neq \{a', b'\}$, replacing a by a' and b by b' in (1), and taking images by f, we arrive at an R_U-congruent pair formed by

$$(0, x_1, \ldots, ab, s_0, a'b', x_1, \ldots, ab, s_0, a', x_1, \ldots, ab, s_0, 0)$$

$$(0, x_1, \ldots, ab, s_0, a'b', x_1, \ldots, ab, s_0, b', x_1, \ldots, ab, s_0, 0)$$

whence $ab = a'b'$, a contradiction.

We next show that $f(s) = s$ for all $s \in S$. Assume that $f(s) = (y_1, \ldots, y_n)/R_U$. Applying f to the pair in (2), we obtain that $(0, s_0^2, 0)(y_1, \ldots, y_n)(a_s, s_0^2, 0) R_U (0, s_0^2, 0)(y_1, \ldots, y_n)^2 (a_s, s_0^2, 0)$. By the length argument we get $n = 1$, then the difference $y_1 \neq y_1^2$ is situated between two subsequent occurences of s_0^2, hence by (2), $f(s) \in S$, and the entry a_s to the right of it identifies it as $f(s) = s$ (recall that the assignment $s \to a_s \in A$ is injective).

Finally, we show that $f(T) \subseteq U$. By (3), we have for every $t \in T$, $(0, s_0^3, 0, t, 0, s_0^3, 0) R_T (0, s_0^3, 0, t^2, 0, s_0^3, 0)$. Then again by the length argument we conclude that $f(t) \in W_A \cup S \cup U$. Since $f(t) \neq f(t^2)$, only $f(t) \in U$ is possible.

It is now clear that $f = F(h)$, where $h : T \to U$ is a restriction of f to T and U. This completes the proof of the theorem. ∎

Reference

[1] Z. Hedrlín and J. Lambek: *How comprehensive is the category of semigroups?*, J. Algebra 11 (1969), 195 – 212.

[2] V. Koubek and J. Sichler: *Universal varieties of semigroups,* J. Austral. Math. Soc. (Series A) 36 (1984), 143 – 152.

[3] A. Pultr and V. Trnková: *Combinatorial, algebraic and topological representations of groups, semigroups and categories,* North Holland, Amsterdam, 1980.

ON ADDITIVELY OR MULTIPLICATIVELY IDEMPOTENT SEMIRINGS AND PARTIAL ORDERS

U. Hebisch
Technische Universität Clausthal
Erzstraße 1
D-3392 Clausthal-Zellerfeld

L.C.A. van Leeuwen
University of Groningen
Postbus 800
NL-9700 AV Groningen

§ 1 Introduction

A *semiring* $(S,+,\cdot)$ is an algebra such that $(S,+)$ and $(S,\cdot)$ are semigroups and $a(b+c) = ab+ac$ and $(b+c)a = ba+ca$ hold for all $a,b,c \in S$. According to Def.3.1 in [3], $(S,+,\cdot,\leq)$ is called a *weak partially ordered (weak p.o.) semiring* if $(S,+,\cdot)$ is a semiring, $(S,\leq)$ a p.o. set and

$$a < b \quad \text{implies} \quad a+c \leq b+c \quad \text{and} \quad c+a \leq c+b \quad \text{for all } a,b,c \in S.$$

In particular, $(S,+,\cdot,\leq)$ is called a *weak fully ordered (weak f.o.) semiring* if $(S,\leq)$ is a f.o. set.

Now let $(S,\circ,\leq)$ be a semigroup and a p.o. set. We introduce the set $P^\circ(S) = \{ p \in S \mid a \leq p\circ a \text{ and } a \leq a\circ p \text{ for all } a \in S \}$ of all *positive* elements of $(S,\circ,\leq)$ and correspondingly the set $N^\circ(S)$ of all *negative* ones. So for a weak p.o. semiring $(S,+,\cdot,\leq)$ one has the sets $P^+(S)$, $N^+(S)$, $P^\cdot(S)$ and $N^\cdot(S)$. In such a semiring we also define the set of all (multiplicatively) *monotone* elements by

$$M(S) = \{ m \in S \mid a < b \Rightarrow ma \leq mb \text{ and } am \leq bm \text{ for all } a,b \in S \}.$$

A weak p.o. semiring $(S,+,\cdot,\leq)$ satisfying the condition

(A) $\quad P^+(S) \subseteq M(S)$

is called a *partially ordered (p.o.) semiring* (cf. Def.3.3 in [3] and various examples given in that paper). Note that this definition applies to each p.o. ring $(R,+,\cdot,\leq)$ as well as to its positive cone $P = P^+(R)$, a p.o. semiring $(P,+,\cdot,\leq)$. Moreover, whereas $(S,+,\leq)$ is a p.o. semigroup for each weak p.o. semiring $(S,+,\cdot,\leq)$ by definiton, $(S,\cdot,\leq)$ is a p.o. semigroup iff $M(S) = S$ holds. Clearly, the latter implies (A), but not conversely.

In this paper we deal with two questions which turn out to be closely related. For the first, recall that a semiring $(S,+,\cdot)$ may have a neutral o of $(S,+)$ or a neutral e of $(S,\cdot)$, usually called the zero o and the identity e of $(S,+,\cdot)$, respectively. There are various cases such that both elements coincide (cf. § 2), and so we call an element $d \in S$ a *double-neutral* of a semiring $(S,+,\cdot)$ if

$$a+d = d+a = da = ad = a \quad \text{holds for all } a \in S.$$

In § 2 we give necessary and sufficient conditions such that a semiring $(S,+,\cdot)$ can be embedded as a subsemiring into a semiring $(T,+,\cdot)$ which has a double-neutral. If this is the case, T can be obtained from S by adjoining one element d as a double-neutral, and if $(S,+,\cdot,\leq)$ is a weak p.o. semiring, the partial order can be extended such that $(T,+,\cdot,\leq)$ is a weak p.o. semiring, too. Secondly, a semiring $(S,+,\cdot)$ is called a *mono-semiring* if

$$a+b = a\cdot b \quad \text{holds for all } a,b \in S$$

(cf. [4], § 4). Each weak p.o. mono-semiring $(S,+,\cdot,\leq)$ satisfies $M(S) = S$ and is hence a p.o. semiring as remarked above. In § 3 we deal with a certain class of p.o. semirings $(S,+,\cdot,\leq)$. It contains mono-semirings, which are characterized in two different ways. In both cases, one of the corresponding conditions states that $(S,+,\cdot)$ is embeddable into a semiring with a double-neutral. The latter also illustrates statements already given in [2], which are consequences of our results.

§ 2 Weak p.o. semirings with a double-neutral element

For near by hand examples of p.o. semirings $(S,+,\cdot,\leq)$ with a double-neutral d (in fact fully ordered ones) we refer to [3], Expl. 2.1. Others are obtained as follows.

Example 2.1 Let $(H,\cdot,\leq)$ be a semilattice-ordered semigroup in the meaning of [1] (p. 128), i.e. a p.o. semigroup and a, say upper, semilattice such that $\sup(ac,bc) = \sup(a,b)\cdot c$ and $\sup(ca,cb) = c\cdot\sup(a,b)$ hold for all $a,b,c \in H$. For instance, the semigroup $(\mathbb{N},\cdot,|)$ of positive integers with the usual multiplication $\cdot$ and the divisibility relation $|$ is such a semigroup with $\sup(a,b)$ as the least common multiple of a and b, and so are suitable subsemigroups of the latter as e.g. those generated by some primes and 1. Now assume that $(H,\cdot,\leq)$ has an identity e and consider $S = \{ a \in H \mid a \geq e \}$. Then $(S,\cdot,\leq)$ is a p.o. subsemigroup and hence $(S,+,\cdot,\leq)$ a p.o. semiring with respect to $a+b = \sup(a,b)$, for which the identity e is at the same time the zero.

<u>THEOREM 2.2</u> <u>Let</u> $(S,+,\cdot)$ <u>be any semiring. Then it is embeddable into a semiring</u> $(T,+,\cdot)$ <u>with a double-neutral</u> $d \in T$ <u>iff</u>

(B) $(S,+)$ <u>is idempotent and</u>

(C) $a+ab = ab+a = b+ab = ab+b = ab$ <u>holds for all</u> $a,b \in S$.

<u>Proof</u> Let $(S,+,\cdot)$ be a subsemiring of a semiring $(T,+,\cdot)$ with a double-neutral $d \in T$. Then $a(d+b) = ad+ab$ yields $ab = a+ab$ even for all $a,b \in T$, and similarly one obtains the other equations contained in (C). Choosing $b = d$ in the last equation, we get $a+a = a$. Conversely, let $(S,+,\cdot)$ be a semiring satisfying (B) and (C), and $T = S \cup \{d\}$ for an element $d \notin S$. We extend the operations on S to those on T by

$$x+d = d+x = dx = xd = x \quad \text{for all } x \in T.$$

Then one easily checks that $(T,+,\cdot)$ is a semiring with d as a double-neutral containing $(S,+,\cdot)$ as a subsemiring.

In this context we give two statements for semirings with a zero o. The latter may already be a double-neutral, a case clearly without interest for the second statement.

<u>LEMMA 2.3</u> <u>Let</u> $(S,+,\cdot)$ <u>be a semiring satisfying</u> (B) <u>and</u> (C) <u>which has a zero</u> o. <u>Then we have</u>:

a) $(S \setminus \{o\},+,\cdot)$ <u>is a subsemiring of</u> $(S,+,\cdot)$, <u>which also satisfies</u> (B) <u>and</u> (C) <u>as each subsemiring does</u>.

b) <u>Moreover, the zero</u> o <u>of</u> $(S,+,\cdot)$ <u>can be replaced by a double-neutral</u> d <u>to obtain a semiring</u> $((S \setminus \{o\}) \cup \{d\},+,\cdot)$. <u>This means to write</u> d <u>instead of</u> o <u>and to change the products</u> $o \cdot s$ <u>and</u> $s \cdot o$ <u>for all</u> $s \in S \setminus \{o\}$ <u>to</u> $d \cdot s = s \cdot d = s$ <u>and</u> $o \cdot o$ <u>to</u> $d \cdot d = d$.

<u>Proof</u> a) Since $(S,+)$ is idempotent, $a+b = o$ for some $a,b \in S$ implies $a = a+o = a+a+b = a+b = o$. Hence $S \setminus \{o\}$ is closed with respect to the addition. Also $ab = o$ yields $a = o$ directly from (C).

b) Due to a), we can adjoin a double-neutral $d \notin S \setminus \{o\}$ to the semiring $(S \setminus \{o\},+,\cdot)$. In this way one obtains a semiring $((S \setminus \{o\}) \cup \{d\},+,\cdot)$ as described above, where the sums $o+s = s+o = s$ for all $s \in S$ simply change to $d+s = s+d = s$, whereas the products $o \cdot s$, $s \cdot o$ and $o \cdot o$ are in general replaced by other values according to $d \cdot s = s$, $s \cdot d = s$ and $d \cdot d = d$.

Now assume that a weak p.o. semiring $(S,+,\cdot,\leq)$ without a double-neutral is embedded into a semiring $(T,+,\cdot)$ with a double-neutral d as in Thm. 2.2, i.e. $T = S \cup \{d\}$. We want to extend the partial order from S to T such that $(T,+,\cdot,\leq)$ is a weak p.o. semiring. The simplest way to do this is to take the same relation, i.e. leaving d incomparable with any other element. But this implies $P^+(T) = N^+(T) = \{d\}$, which is less satisfactory. On the other hand, assume that $d < s$ holds for some $s \in S$ and an extension $\leq$ from S to T such that $(T,+,\circ,\leq)$ is a weak p.o. semiring. Then it follows $a \leq s+a$ and $a \leq a+s$ for all $a \in S$, hence $s \in P^+(S)$. Similarly, $s < d$ leads to $s \in N^+(S)$. Therefore, provided that $(S,+,\cdot)$ has no zero, the greatest possible extension $\leq$ from S to T is given by $N^+(S) < d < P^+(S)$, i.e.

(2.1) $\quad n < d < p$ for all $n \in N^+(S)$ and all $p \in P^+(S)$.

COROLLARY 2.4 *Let $(S,+,\cdot,\leq)$ be a weak p.o. semiring satisfying (B) and (C) without a zero, and $T = S\cup\{d\}$ the semiring containing $(S,+,\cdot)$ and a double-neutral d. Extending $\leq$ by (2.1), we obtain a weak p.o. semiring $(T,+,\cdot,\leq)$ such that*

$$P^+(T) = P^+(S)\cup\{d\},\quad N^+(T) = N^+(S)\cup\{d\} \ \text{and} \quad M(T) \subseteq M(S)\cup\{d\}$$

hold. But even if $(S,+,\cdot,\leq)$ is a f.o. semiring, commutative with respect to both operations and satisfying $M(S) = S$, the weak p.o. semiring $(T,+,\cdot,\leq)$ needs neither be fully ordered nor satisfy $P^+(T) \subseteq M(T)$.

Proof All statements are checked straightforward except the last one, for which we give the following

Example 2.5. Let $(S,+,\leq)$ be given by $n < n' < a < p$ and the table below, and consider the trivial multiplication for which each product equals n. Then $(S,+,\cdot,\leq)$ is a f.o. semiring without a zero, commutative with respect to both operations, which satisfies (B), (C), $N^+(S) = \{n,n'\}$, $P^+(S) = \{p\}$, and $M(S) = S$.

+	n	n'	a	p
n	n	n	n	n
n'	n	n'	a	a
a	n	a	a	a
p	n	a	a	p

Now let $(T,+,\cdot,\leq)$ be the weak p.o. semiring as considered in Cor. 2.4. According to the right hand diagram, it is not fully ordered. (Recall that (2.1) defines the greatest possible extension $\leq$ from S to T such that $(T,+,\circ,\leq)$ is a weak p.o. semiring, which can be checked directly in this case. In particular, a and d have to be incomparable, since $a < d$ would contradict $a+n' > d+n'$, and $d < a$ would contradict $d+p > a+p$.) Further, $p \in P^+(T)$ is not contained in M(T) because of $d < p$, but $p = dp > pp = n$. This disproves $P^+(T) \subseteq M(T)$ and shows that M(T)

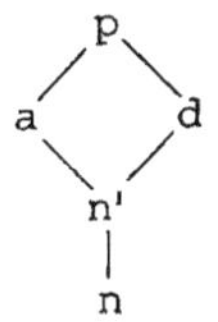

is properly contained in $M(S)\cup\{d\} = S\cup\{d\} = T$.

Continuing the considerations before Cor. 2.4, assume that $(S,+,\cdot)$ has a zero o. Then (2.1) has to be replaced by

(2.2) $N^+(S)\setminus\{o\} < d < P^+(S)$ or $N^+(S) < d < P^+(S)\setminus\{o\}$,

according to the choice of $d < o$ or $o < d$. One checks that both possibilities yield a weak p.o. semiring $(T,+,\cdot,\leq)$, which satisfies similar statements as those given in Cor. 2.4. Another way is described in

COROLLARY 2.6. *Let $(S,+,\cdot,\leq)$ be a weak p.o. semiring satisfying (B) and (C) with a zero o, which clearly may be assumed not to be a double-neutral. Then we change to $T = (S\setminus\{o\})\cup\{d\}$ in replacing o by a double-neutral d according to Lemma 2.3 b), and we also write $s < d$ or $d < s$ in T whenever $s < o$ or $o < s$ hold in S. Then $(T,+,\cdot,\leq)$ is a weak p.o. semiring which satisfies*

$$P^+(T) = (P^+(S)\setminus\{o\})\cup\{d\}, \quad N^+(T) = (N^+(S)\setminus\{o\})\cup\{d\}$$

and $M(T) \subseteq M(S\setminus\{o\})\cup\{d\}$. If S is a weak f.o. semiring, the same holds for T, but even in this case $P^+(S) \subseteq M(S)$ does not imply $P^+(T) \subseteq M(T)$.

Proof By Lemma 2.3 a), $(S\setminus\{o\},+,\circ,\leq)$ is a weak p.o. semiring satisfying (B) and (C). Note that this semiring may have again a zero (cf. Expl. 2.7), hence we cannot apply Cor. 2.4 to $(S\setminus\{o\},+,\cdot,\leq)$. But we can apply Lemma 2.3 b) to $(S,+,\cdot)$ and replace o by a double neutral d. Replacing also o by d with respect to the partial order, clearly $(S,+,\leq)$ and $(T,+,\leq)$ are order-isomorphic semigroups. This yields that $(T,+,\cdot,\leq)$ is a weak p.o. semiring, and fully ordered iff $(S,+,\circ,\leq)$ is, as well as the statements on $P^+(T)$ and $N^+(T)$. Moreover, $M(T)$ is clearly a subset of $M(S\setminus\{o\})\cup\{d\}$. That this inclusion may be a proper one, and the last statement are shown by the following

Example 2.7. Let $(S,+,\leq)$ be given by $n < o < p$ and the table below, again with the trivial multiplication $S^2 = \{n\}$. Then $(S,+,\cdot,\leq)$ is a f.o. semiring with o as zero, commutative with respect to both operations, which satisfies (B), (C), and $M(S) = S$. Note that $S\setminus\{o\}$ has again a zero, namely p, but we

+	n	o	p
n	n	n	n
o	n	o	p
p	n	p	p

may replace o by a double-neutral d adjoined to $S\setminus\{o\}$ as described above. Then we have $n < d < p$ for the weak f.o. semiring $(T,+,\circ,\leq)$. Obviously, we obtain $M(S\setminus\{o\}) = M(S)\setminus\{o\} = \{n,p\}$ and $P^+(T) = \{d,p\}$, but $M(T) = \{n,d\}$, the latter because of $d < p$, but $p = dp > pp = n$. This shows $M(T) \subset M(S\setminus\{o\})\cup\{d\}$

and $P^+(T) \nsubseteq M(T)$. Note in this context that there are also examples such that $M(S) \setminus \{o\}$ is properly contained in $M(S \setminus \{o\})$.

§ 3 Double-neutrals and mono-semirings

Let $(S,+,\cdot)$ be a mono-semiring which satisfies (B) and (C). Then the common operation $a+b = a\cdot b$, say $a\circ b$, is clearly idempotent, and also commutative. The latter follows since (C) implies $a\circ b\circ a = b\circ a\circ b = a\circ b$ for all $a,b \in S$. Conversely, each idempotent and commutative semigroup $(S,\circ)$ defines a mono-semiring $(S,+,\cdot)$ by $a+b = a\cdot b = a\circ b$ which satisfies (B) and (C). Including partial orders in those considerations, we immediately obtain

PROPOSITION 3.1. *Each p.o. mono-semiring $(S,+,\cdot,\leq)$ which is embeddable into one with a double-neutral corresponds to a p.o. semigroup $(S,\circ,\leq)$ which is commutative and idempotent, and conversely.*

One partial order $\leq$ on $(S,\circ)$ in this context is clearly given by $a \leq b \Leftrightarrow a\circ b = b$ and characterized by $P^\circ(S) = S$ ([2], Remark 2.4). This case will be linked with our further investigations, which we start with the following variant of Thm. 2.2.

THEOREM 3.2 *Let $(S,+,\cdot,\leq)$ be a weak p.o. semiring such that $P^+(S) = S$ holds and $(S,+)$ is idempotent. Then $(S,+,\leq)$ is a semilattice according to $a+b = \sup(a,b)$, and one has $M(S) = S$ which implies that $(S,+,\cdot,\leq)$ is a p.o. semiring. Moreover, the following statements are equivalent:*

α) *$(S,+,\cdot)$ is embeddable into a semiring with a double-neutral.*

β) *$P^\bullet(S) = S$ holds.*

γ) *$(S,+,\cdot)$ satisfies (C), which can be reduced to*

(C') $\quad a+ab = ab = ab+b$ *for all* $a,b \in S$

since $(S,+)$ is commutative.

Proof By Thm. 2.3 of [2] we get $a \leq b \Leftrightarrow a+b = b \Leftrightarrow b+a = b$ for all $a,b \in S$. This yields that $a < b$ implies $ac \leq bc$ and $ca \leq cb$ for all $a,b,c \in S$, hence $M(S) = S$. Since $(S,+)$ is idempotent, α) and γ) are equivalent by Thm. 2.2. Finally, β) states that $a \leq ab$ and $b \leq ab$ hold for all $a,b \in S$, which is the same as (C') since $(S,+,\leq)$ is a semilattice.

COROLLARY 3.3 *Let $(S,+,\cdot,\leq)$ be a p.o. semiring as considered in Thm. 3.2 which satisfies the conditions α) - γ). Then we can apply either Cor. 2.4.*

or Cor. 2.6, and in both cases we get a p.o. semiring $(T,+,\cdot,\leq)$ with a double-neutral d, which is the least element of $(T,\leq)$. Moreover, $M(T) = T$ holds and $(T,+,\cdot,\leq)$ is fully ordered iff the same holds for $(S,+,\cdot,\leq)$.

Proof If S has no zero, we apply Cor. 2.4 and consider the weak p.o. semiring $(T,+,\circ,\leq)$ with $T = S\cup\{d\}$. Then we have $d < P^+(S) = S$, which shows the statement on d, $P^+(T) = T$, and that $(T,\leq)$ is fully ordered iff $(S,\leq)$ is. It remains to show $M(T) = T$. For the latter, all cases are trivial except that $d < a$ implies $db \leq ab$ and $bd \leq ba$ for all $a,b \in S$, which is just $P^{\cdot}(S) = S$ according to β). Now assume that S has a zero o and replace the latter by a double-neutral d as in Cor. 2.6. Since o is the least element of S, we get that d is the least element of $T = (S\setminus\{o\})\cup\{d\}$. Again it remains to prove $M(T) = T$. Now $o < a < b$ for some $a,b \in S$ is changed to $d < a < b$, which yields $dc \leq ac \leq bc$ and $cd \leq ca \leq cb$ for each $c \in S$ by $P^{\cdot}(S) = S$ and $M(S) = S$.

Note in this context that Expls. 2.5 and 2.7 are out of order here because of $P^+(S) \neq S$ in both examples. Now we turn to necessary and sufficient conditions such that a semiring as considered in Thm. 3.2 is a mono-semiring.

THEOREM 3.4 Let $(S,+,\cdot,\leq)$ be a (weak) p.o. semiring such that $P^+(S) = S$ holds and $(S,+)$ is idempotent. Then $(S,+,\cdot,\leq)$ is a mono-semiring iff it satisfies one of the conditions α) - γ) and $(S,\cdot)$ is idempotent.

Proof We use statements of Thm. 3.2 without comment, e.g. that α), β) and γ) are equivalent. Now assume that $(S,+,\cdot)$ is a mono-semiring. Then the idempotence of $(S,+)$ yields $a+a+b = a+b = a+b+b$, which is (C') and hence γ) because of $(S,+) = (S,\cdot)$. Also $(S,\cdot)$ is idempotent. Conversely, $M(S) = S$ implies that $(S,\cdot,\leq)$ is a p.o. semigroup, now assumed to be idempotent and to satisfy β), i.e. $P^{\cdot}(S) = S$. So we can apply Thm. 2.3 of [2] and obtain $a\cdot b = \sup(a,b)$ for all $a,b \in S$. This shows $a\cdot b = a+b$ because of $a+b = \sup(a,b)$.

THEOREM 3.5 Let $(S,+,\cdot,\leq)$ be a weak p.o. semiring such that $P^+(S) = S$ holds, but assume that $(S,\cdot)$ is idempotent. Then $(S,+,\cdot,\leq)$ is a mono-semiring iff $P^{\cdot}(S) = S$ and

(C") $a+ab = ab$ for all $a,b \in S$

are satisfied. In the latter case, clearly, $(S,+,\circ,\leq)$ is a p.o. semiring as described in Thm. 3.2 and all statements α) - γ) given there are true.

Proof If $(S,+,\cdot)$ is a mono-semiring the idempotence of $(S,+) = (S,\cdot)$ yields (C') and hence (C") as in the proof above. The idempotence of $(S,+)$ and (C') also imply the conditions $\alpha) - \gamma)$ by Thm. 3.2, in particular $\beta)$, i.e. $P^{\bullet}(S) = S$. Conversely, assume (C") and $P^{\bullet}(S) = S$. Then, for all $a,b \in S$, we have $ab \geqq b$ by $P^{\bullet}(S) = S$, and by $P^{+}(S) = S$ we obtain $ab = a+ab \geqq a+b = (a+b)^2 = a^2+ab+ba+b^2 \geqq ab$. This proves $a\cdot b = a+b$.

Remark 3.6 We note that Thm. 3.5 implies Thm. 4.4 of [2]. The latter gives necessary and sufficient conditions such that a p.o. semiring $(S,+,\circ,\leqq)$ for which $P^{+}(S) = S = P^{\bullet}(S)$ holds and $(S,\circ)$ is idempotent is a mono-semiring. These conditions are, due to the assumptions of Thm. 4.4, all equivalent to the idempotence of $(S,+)$. The latter is a consequence of Thm. 3.5, since (C") is equivalent to the idempotence of $(S,+)$ in this context. Indeed, (C") implies $a+aa = aa$, hence $a+a = a$ by the assumption on $(S,\circ)$. The converse follows from $ab = ab+ab \geqq a+ab \geqq ab$ due to $P^{\circ}(S) = S$ for the first and $P^{+}(S) = S$ for the second inequality.

Remark 3.7 Also Cor. 4.5 of [2] is a consequence of Thm. 3.5 and more understandable in the light of the latter. This corollary states that a p.o. semiring $(S,+,\cdot,\leqq)$ such that $P^{+}(S) = S$ holds and $(S,\cdot)$ is idempotent, which has a double-neutral, is a mono-semiring. This is now clear, since, by Thms. 3.5 and 3.2, it is even enough that such an element can be adjoined to S.

REFERENCES

[1] M.L. Dubreille-Jacotin, L. Lesieur, R. Croisot, Leçons sur la theórie des treillis, des structures algébraiques ordonnées et des treillis géométriques, Gauthier Villars, Paris 1953

[2] Hanumanthachari, J., Venu Raju, K. and Weinert, H.J., Some results on partial ordered semirings and semigroups, in: Algebra and Order, Proc. First Intern. Symp. on Ordered Algebraic Structures, Marseilles, June 1984, Heldermann-Verlag, Berlin 1986, 313 - 322.

[3] Weinert, H.J., Partially ordered semirings and semigroups, same Proceedings as in [2], 265 - 292.

[4] Zeleznekow, J., Regular semirings, Semigroup Forum 23 (1981), 119 - 136

CONGRUENCE SEMIMODULAR VARIETIES OF SEMIGROUPS

Peter R. Jones
Department of Mathematics, Statistics and Computer Science
Marquette University, Milwaukee, WI 53233

Considerable attention has been paid in recent years to congruence modular varieties of algebras. (See, for example, [5, Appendix 3]). Whilst presenting a unified approach to algebras such as groups, rings and lattices, for instance, this theory has little relevance to semigroups: any congruence modular variety of semigroups consists entirely of groups.

There are, however, various scattered results in the literature showing that certain classes of semigroups have *semimodular* congruence lattices. (See §1). Our purpose here is to unify and generalize these results from a varietal point of view and in so doing to demonstrate that significant varieties of semigroups are *congruence semimodular* (CSM).

It is shown in §2 that a *regular* variety V of semigroups (that is, one containing the variety S of semilattices) is CSM if and only if it contains no nontrivial completely simple semigroups, equivalently, if it satisfies identities $x^{n+1} = x^n$, $(x^ny^n)^n = (y^nx^n)^n$ for some positive integer n. In fact a similar theorem is proven which applies to more general classes of semigroups, for instance pseudovarieties of finite semigroups. (The corresponding CSM pseudovarieties have recently been studied from the point of view of language theory [11]).

Every *irregular* variety of semigroups is "completely simple-by-nil". We prove that every "completely simple-by-nilpotent" variety is CSM. Whether *every* irregular variety is CSM is open. However we note that every irregular pseudovariety (of finite semigroups) *is* CSM.

It is shown that a variety V of *inverse* semigroups is CSM if and only if it is a *combinatorial* variety or a *group* variety.

1. *Preliminaries*

A lattice $(L, \wedge, \vee)$ is called *semimodular* if whenever $x \succ x\wedge y$ then $x\vee y \succ y$. (Here $\succ$ denotes the 'covering' relation). This is equivalent to the definition in [4, §IV.2] : the reader is referred there for further details on such lattices. In particular we note that semimodularity is inherited by interval sublattices and by subdirect products. Every *modular* lattice is semimodular (because in such a lattice the intervals $[x\wedge y,x]$ and $[y,x\vee y]$ are isomorphic). The "prototype" semimodular lattice is the lattice $E(X)$ of *equivalence relations* on a set X.

We call an algebra A *congruence semimodular*, abbreviated to CSM, if its congruence lattice $(C(A),\cap,\vee)$ is semimodular. A *class* (in particular a variety) V of algebras is then CSM if each algebra in V is CSM. (Congruence modularity is defined analogously).

Throughout we use the following notation. If C is a class of algebras then $\underset{\sim}{H}C$ consists of all homomorphic images, $\underset{\sim}{S}C$ of all subalgebras and $\underset{\sim}{P}C$ of all direct products of members of C. Further $\underset{\sim}{P}_f C$ consists of all direct products of finitely many members of C.

A useful observation is the following : if C is CSM then so is $\underset{\sim}{H}C$, for if $B \in \underset{\sim}{H}C$ then $B \cong A/\rho$ for some $\rho \in C(A)$ and by an isomorphism theorem $C(B)$ is isomorphic with the interval sublattice $[\rho,\omega]$ of $C(A)$. (Here, and throughout, ι and ω will denote the least and greatest elements of $C(A)$). Thus a variety is CSM if and only if each of its free algebras is. A technically more useful result is the following.

PROPOSITION 1.1. *Let V be a class of algebras closed under* $\underset{\sim}{H}$. *Then V is* CSM *if and only if for each $A \in V$ and each atom ρ of $C(A)$, $\rho\vee\tau \succ \tau$ for every $\tau \in C(A)$, $\rho \nleq \tau$.*

Proof. Necessity is clear, for ρ is an atom if and only if $\rho \succ \iota$. Conversely, suppose V has the given property, let $A \in V$ and ρ,τ be any congruences in $C(A)$ such that $\rho \succ \rho\cap\tau$. Now $\rho/\rho\cap\tau$ is an atom of $C(A/\rho\cap\tau)$, and $A/\rho\cap\tau \in V$, so $(\rho\vee\tau)/\rho\cap\tau = (\rho/\rho\cap\tau) \vee (\tau/\rho\cap\tau) \succ \tau/\rho\cap\tau$. Thus $\rho\vee\tau \succ \tau$ in $C(A)$. Hence A is CSM.

We now concentrate on semigroup varieties, considering first the *atoms* in the lattice L of varieties. These are (see [2]) RZ, the *right zero semigroups*, defined by the identity $xy = y$; LZ, the *left zero semigroups*, defined by $xy = x$; N, the *null* semigroups, defined by $xy = zt$; S, the *semilattices*, defined by $xy = yx$, $x^2 = x$; and, for each prime p, G_p, the *groups of exponent* p, defined by $x^{p+1} = x$, $x^p = y^p$.

THEOREM 1.2. (i) *Each atom of L is* CSM.

(ii) *The only atoms of L whose congruence lattices satisfy a non-trivial lattice identity are the group varieties.*

(iii) *A variety of semigroups is congruence modular if and only if it consists of groups.*

Proof. (i) This is well-known, if not in varietal terms. On any right zero, left zero or null semigroup every equivalence relation is a congruence, so $C(A) = E(A)$, and is therefore semimodular (see above). That semilattices are CSM was proven in [6,3]. Semigroup varieties consisting of groups are congruence modular, whence CSM.

(ii) The class of equivalence lattices $E(X)$, X a set, satisfies no nontrivial lattice identity, eliminating RZ, LZ and N as in (i). For S this was proved in [3].

(iii) Let V be a congruence modular variety of semigroups. Then V contains none of RZ, LZ, N and S, from which it easily follows that V consists entirely of groups.

In the literature there are many results on CSM for special types of semigroups. We mention here some of relevance.

G. Lallement proved that every completely simple semigroup is CSM [10]. A proof appears also in Howie's monograph [7] where further details on completely simple semigroups may be found. The class *CS* of such semigroups is not a variety of semigroups, but contains the varieties *RZ*, *LZ* and the G_p's .

A semigroup S with zero is *nil* if for each x in S there exists $n > 0$ such that $x^n = 0$. In a private communication T. E. Hall showed the author that such semigroups are CSM. A nil *variety* satisfies an identity of the form $x^n = 0$ and is therefore CSM. A semigroup S with zero is *nilpotent* if $S^n = \{0\}$ for some $n > 0$. Nilpotent *varieties* satisfy an identity of the form $x_1 \ldots x_n = 0$ and are thus nil.

D. C. Trueman [13, Theorem 3.4] showed that any finite semigroup which is an ideal extension of a group by a nil semigroup is CSM. The author [8, p.29] showed that any finite combinatorial inverse semigroup is CSM. These two results are encompassed by our general theorems.

To complete this section we review some important facts about the lattice *L* of semigroup varieties. (See [2] for further details.)

A variety *V* is *regular* if it satisfies only identities in which the same variables appear on each side. Equivalently $S \subseteq V$. Otherwise *V* is *irregular* and by a theorem of Chrislock [1], *V* is *completely simple*-by-*nil*. (If *C* and *D* are classes of semigroups, we call a semigroup S *C - by - D* if S is an ideal extension of a member of *C* by a member of *D*; a variety is *C* - by - *D* if each of its members has this property.) Thus each semigroup S in *V* has a completely simple kernel (minimum ideal) K such that S/K is nil.

2. *Regular* CSM *varieties*

The main theorem of this section is the following.

THEOREM 2.1. *A regular variety V of semigroups is* CSM *if and only if it contains no nontrivial completely simple semigroups, that is, $V \cap CS = T$, the trivial variety.*

In fact we prove a rather more general result, that *the theorem holds for any regular class* C *of semigroups which is closed under* $\underset{\sim}{H}, \underset{\sim}{S}$ *and* $\underset{\sim}{P}_f$, where now 'regular' means that *C* contains the *two-element semilattice* $Y_2 = \{0,1 \mid 1 > 0\}$. We will use the term *pseudovariety* for any class of semigroups (not necessarily finite) closed under $\underset{\sim}{H}$, $\underset{\sim}{S}$ and $\underset{\sim}{P}_f$. There is then a *largest* pseudovariety which is CSM. This class, which will be denoted by *ABG*, is characterized below in several ways. This leads to a more illuminating version of the theorem above (see Corollary 2.5 and Proposition 2.4).

LEMMA 2.2. *Let V be a class of semigroups closed under* $\underset{\sim}{H}$ *and* $\underset{\sim}{P}_f$. *If V is*

CSM *and* $Y_2 \in V$ *then* $V \cap CS = T$.

Proof. Let V be such a class and suppose $Y_2 \in V$ and $S \in V$, where S is completely simple and nontrivial. Put $T = S \times Y_2$ and let π_1 and π_2 be the congruences on T induced by the projections onto S and Y_2, respectively. From the simplicity of S it easily follows that π_1 is an atom of $C(T)$. Let ρ denote the Rees congruence modulo the ideal $S \times \{0\}$. Then $\pi_1 \vee \rho = \omega$ but $\rho \subsetneq \pi_2 \subsetneq \omega$, so $\pi_1 \vee \rho$ does not cover ρ and $C(T)$ is not semimodular.

To prove a converse of this lemma we need further properties of pseudovarieties V for which $V \cap CS = T$. Clearly any such V is *periodic*, that is, for any $S \in V$ and any $x \in S$, $x^{n+r} = x^n$ for some $n,r > 0$. (Otherwise V contains an infinite cyclic semigroup and therefore nontrivial finite groups). In any periodic semigroup S some power of each element x is idempotent : denote that idempotent by x^ω, so that $x^\omega = x^n$ for some n, depending on x. (But for any finite subset $\{x_1,x_2,\dots,x_k\}$ of S there is a uniform n such that $x_i^\omega = x_i^n$ for each i, a fact which will be used without comment).

We call a periodic semigroup S *block-group* if every regular $\mathcal{D}$-class of S is inverse (whence the nonzero $\mathcal{D}$-class of a Brandt semigroup). For instance all inverse semigroups and all nil semigroups are block-group. This terminology extends that of [11], where it was applied only to finite semigroups, the corresponding pseudovariety playing an important role. Various equivalent properties were found there, one of which is included in the following. Denote the class of periodic block-group semigroups by BG.

LEMMA 2.3. *For a periodic semigroup* S *the following are equivalent:*

1) $S \in BG$,
2) *for every* $x,y \in S$, $(x^\omega y^\omega)^\omega = (y^\omega x^\omega)^\omega$,
3) *for every* $x,y \in S$ *with* $x \mathcal{J} y$, *if there exist* $s,t \in S^1$ *such that* $sxt = syt \mathcal{J} x$ *then* $x = y$.

Proof. The equivalence of 1) and 2) is contained in [11, Proposition 2.2]. 3)$\Rightarrow$1). Let D be a regular $\mathcal{D}$-class of S, and suppose D contains $\mathcal{R}$-related idempotents e and f. Then $ee = ef = f\mathcal{J}e$, so $e = f$, by 3). Hence each regular $\mathcal{R}$- (and dually $\mathcal{L}$-) class of D contains a unique idempotent, that is, D is inverse. 2)$\Rightarrow$3). Suppose first that $xt = yt \mathcal{J}x\mathcal{J}y$ for some $t \in S^1$. By periodicity, $xt\,\mathcal{R}x$ and $yt\,\mathcal{R}y$, so $x = xta$ and $y = ytb$, for some $a,b \in S^1$, whence $x = x(ta)^\omega$ and $y = y(tb)^\omega$. Thus

$$\begin{aligned} x = x(ta)^\omega &= y(ta)^\omega, \text{ since } xt = yt,\\ &= y(tb)^\omega(ta)^\omega\\ &= x(tb)^\omega(ta)^\omega, \text{ similarly,}\end{aligned}$$

$$\begin{aligned} \text{so that}\quad x &= x((tb)^\omega(ta)^\omega)^\omega\\ &= x((ta)^\omega(tb)^\omega)^\omega, \text{ by 2).}\end{aligned}$$

$$\text{Thus}\quad x = x(tb)^\omega = y(tb)^\omega = y.$$

The dual case is similar and the general case now follows, for if $sxt = syt \mathcal{J}x\mathcal{J}y$ then $(sx)t = (sy)t\ \mathcal{J}sx\ \mathcal{J}sy$, whence $sx = sy\ \mathcal{J}x\ \mathcal{J}y$.

Denote by ABG the class of *aperiodic* periodic block-group semigroups, that is, those with trivial subgroups.

PROPOSITION 2.4. *For a periodic semigroup* S *the following are equivalent:*

1) $S \in ABG$,
2) *every regular* $\mathcal{D}$*-class of* S *is inverse and has trivial subgroups,*
3) S *contains no nontrivial completely simple subsemigroups,*
4) *for every* $x,y \in S$, $x^{\omega+1} = x^{\omega}$ *and* $(x^{\omega}y^{\omega})^{\omega} = (y^{\omega}x^{\omega})^{\omega}$,
5) *for every* $x \in S$, *if there exist* $s,t \in S^1$ *such that* $x \mathcal{J} s^{\omega}xt^{\omega}$ *then* $x = sxt$.

Proof. The equivalence of 1), 2) and 3) is obvious, as is that of 4) upon noting that for any element x of a periodic semigroup S, $x^{\omega+1}$ ($= x^{\omega}x$) belongs to the maximal subgroup of S whose identity element is x^{ω}. 1) => 5). Suppose $x \mathcal{J} s^{\omega}xt^{\omega}$. Then since $s^{\omega+1} = s^{\omega}$ and $t^{\omega+1} = t^{\omega}$, $s^{\omega}(sxt)t^{\omega} = s^{\omega}xt^{\omega} \mathcal{J} x \mathcal{J} sxt$, so applying 3) of the previous lemma, $x = sxt$.

5) => 2). Let e be an idempotent of S, satisfying 5). If $x \in H_e$ then $x = x.x^{\omega}$, so $x = x.x$ and $x = e$. Thus S is aperiodic. If $f^2 = f$ and $f \mathcal{R} e$ then $e \mathcal{J} f = ef$, so $e = ef = f$; similarly if $f \mathcal{L} e$, $f = e$. Thus each regular $\mathcal{D}$-class is inverse.

COROLLARY 2.5. *The class* ABG *is a pseudovariety of semigroups and is the largest pseudovariety whose intersection with* CS *is* T. *Hence a variety* V *of semigroups satisfies* $V \cap CS = T$ *if and only if* V *satisfies identities* $x^{n+1} = x^n$, $(x^ny^n)^n = (y^nx^n)^n$ *for some* $n > 0$.

Proof. That ABG is a pseudovariety is easily verified from 4) of the proposition. That $ABG \cap CS = T$ and that ABG is the largest pseudovariety with this property is clear from 3) of the proposition.

If a variety V satisfies the specified identities, then $x^{\omega} = x^n$ for each $x \in S$, $S \in V$, so $V \subseteq ABG$. Conversely, any variety V such that $V \cap CS = T$ is *uniformly* periodic, that is, there exist $n,r > 0$ such that $x^n = x^{n+r}$ is an identity satisfied in V; see, for instance, [2]. Clearly $r = 1$, whence $x^{\omega} = x^n$ and the identities are immediate from 4).

The proof of Theorem 2.1, in its more general form for pseudovarieties, will be completed by the following, whose proof occupies the remainder of the section.

PROPOSITION 2.6. *The pseudovariety* ABG *is* CSM.

We need first some properties of the *atoms* of the congruence lattice C(S) of an arbitrary semigroup S. The proof of the following result may be found in the article by M. Demlova and V. Koubek elsewhere in this volume. Recall that the set $S/\mathcal{J}$ of $\mathcal{J}$-classes of S may be partially ordered by: $J_a \le J_b$ if $a \in S^1 b S^1$. That elements x and y of a partially ordered set are *incomparable* will be denoted by $x \,||\, y$.

RESULT 2.7. *Let* S *be a semigroup and suppose* ρ *is an atom of* C(S). *Then either*

a) $\rho \subseteq (J \times J) \cup \iota$ *for some* $J \in S/\mathcal{J}$, *or*
b) $\rho \subseteq (J \times K) \cup (K \times J) \cup \iota$ *for some* $J,K \in S/\mathcal{J}$, $J \,||\, K$, *or*

c) $\rho \subseteq (J \times J) \cup (J \times K) \cup (K \times J) \cup \iota$ *for some* $J,K \in S/J$, $K < J$.

In the remainder of the section $S \in ABG$, ρ is an atom in $C(S)$ and $\tau \in C(S)$, $\rho \nsubseteq \tau$. By Proposition 1.1 it is sufficient to show $\rho \vee \tau \succ \tau$. *Suppose this is false.* Then there exist $x,y \in S$ such that $(x,y) \in \rho \vee \tau$, $(x,y) \notin \tau$ and $\rho \nsubseteq (x,y)^* \vee \tau$ (where $(x,y)^*$ denotes the congruence generated by the pair (x,y)). Since $(x,y) \in \rho \vee \tau$ there is a sequence $x = x_0 \to x_1 \to x_2 \to \ldots \to x_n = y$ of elements of S with the property that $(x_{i-1},x_i) \in \rho \cup \tau$, $1 \le i \le n$. Amongst possible choices for x and y, choose them so that *the length of this sequence is minimum.* Thus for any $(x',y') \in \rho \vee \tau$ with $(x',y') \notin \tau$, connected by a *shorter* sequence, $\rho \subseteq (x',y')^* \vee \tau$.

Suppose $x \,\tau\, x_1$. Then $(x_1,y) \in \rho \vee \tau$ and $(x_1,y) \notin \tau$, and the sequence $x_1 \to x_2 \to \ldots \to x_n = y$ has length $n-1$, so $\rho \subseteq (x_1,y)^* \vee \tau \subseteq (x,y)^* \vee \tau$ (since $(x_1,y) \in (x,y)^* \vee (x,x_1)^* \subseteq (x,y)^* \vee \tau$), a contradiction. Hence $x\rho x_1$ and, similarly, $x_{n-1}\rho x$. Moreover, by minimality, all the x_i's are distinct.

Case A. Suppose $J_x \neq J_y$. Since ρ is an atom (and $x\rho$ and $y\rho$ are nontrivial) then from Result 2.7 we may assume that $J_x = K$ and $J_y = J$, with either

(A1) $K \,||\, J$

or (A2) $K < J$.

Note further that since for each i either $x_i \,\rho\, x_{i+1}$ or $x_i \,\rho\, x_{i-1}$ then each $x_i \in J \cup K$ similarly. Moreover since ρ is trivial on K and $x\rho x_1$, $x_1 \in J$ in either case. We now prove that $J_{y\tau} = J_{x_1\tau} \le J_{x\tau}$ in S/τ. (Since $x_1 \in J$, $x_1 \mathcal{J} y$ so clearly $J_{y\tau} = J_{x_1\tau}$ in S/τ).

Consider x_2. Since $x\rho x_1$ and the given sequence is minimal, $x_1 \tau x_2$. If $x_2 \in K$ then $J_{x_1\tau} = J_{x_2\tau} = J_{x\tau}$. Otherwise $x_2 \in J = J_{x_1}$, so $x_2 = px_1q$ for some $p,q \in S^1$. So $x_1 \tau px_1q$ and, inductively, $x_1 \tau p^\omega x_1 q^\omega$. Now since $x_1 \neq x_2$ applying 5) of Proposition 2.4 yields the fact that $p^\omega x_1 q^\omega \notin J_{x_1} = J$. But $p^\omega x_1 q^\omega \rho p^\omega x q^\omega$, so (by Result 2.7) $p^\omega x_1 q^\omega = p^\omega x q^\omega$. Thus $x_1 \tau p^\omega x q^\omega$ and so $J_{x_1\tau} \le J_{x\tau}$ in S/τ.

Now in (A1), by symmetry $J_{x\tau} \le J_{y\tau}$ and in (A2), $J_x = K < J = J_y$, so $J_{x\tau} \le J_{y\tau}$ again. Thus $x\tau \,\mathcal{J}\, y\tau$ in any case.

Next, since $x_1 \mathcal{J} y$ there exist $s,t \in S^1$ such that $x_1 = syt$. We prove, by induction, that $(s^kxt^k, s^kyt^k) \notin \tau$ for all $k \ge 0$ (true by hypothesis for $k = 0$). An important point to be noted is that for any k for which this is true the sequence

$$s^kxt^k = s^kx_0t^k \to s^kx_1t^k \to \ldots \to s^kx_nt^k = s^kyt^k ,$$

where for each i, $(s^kx_{i-1}t^k, s^kx_it^k) \in \rho \cup \tau$, is again 'minimal', (for if there is a shorter such sequence then by the original hypothesis

$\rho \subseteq (s^kxt^k, s^kyt^k)^* \vee \tau \subseteq (x,y)^* \vee \tau$, a contradiction), in which case

$s^kxt^k \in K$ and s^kyt^k and $s^kx_1t^k$ belong to J, as before.

So let $m \geq 1$ and assume $(s^{m-1}xt^{m-1}, s^{m-1}yt^{m-1}) \notin \tau$. Suppose $(s^mxt^m, s^myt^m) \in \tau$. Then, in S/τ,

$$(s^m\tau)(x\tau)(t^m\tau) = (s^m\tau)(y\tau)(t^m\tau) = (s^{m-1}\tau)(x_1\tau)(t^{m-1}\tau),$$

where $s^{m-1}x_1\,t^{m-1} \in J$, so that $(s^{m-1}\tau)(x_1\tau)(t^{m-1}\tau)\ \mathcal{J}\ y\tau\ \mathcal{J}\ x\tau$ in S/τ. But since $S \in \mathit{ABG}$, $S/\tau \in \mathit{ABG}$ and by 3) of Lemma 2.3, $x\tau = y\tau$, a contradiction. This completes the inductive proof that $(s^kxt^k, s^kyt^k) \notin \tau$, and thus $s^kyt^k \in J$, for each $k \geq 0$. In particular $s^\omega yt^\omega \in J = J_y$, so by 5) of Proposition 2.4, $y = syt = x_1$, contradicting minimality of the original sequence.

This completes the proof of Case A.

<u>Case B</u>. Suppose $J_x = J_y$. Here there are four subcases, according to the possibilities inherent in Result 2.7.

(B1) $x,y \in J$ and $\rho \subseteq (J \times J) \cup \iota$,

(B2) $x,y \in J$ and $J \,||\, K$,

(B3) $x,y \in K$ and $J > K$,

(B4) $x,y \in J$ and $J > K$.

In any case $x = syt$ for some $s,t \in S^1$ and the argument is similar to, but slightly more complicated than, the last part of Case A. We again prove by induction that $(s^kxt^k, s^kyt^k) \notin \tau$ for $k \geq 0$. By the original hypothesis it may again be assumed, for any k for which this is true, that in (B1) and (B2) $s^kxt^k \in J$, in (B3) $s^kxt^k \in K$ and in (B4) $s^kxt^k \in J \cup K$. In that final case, should $s^kxt^k \in K$ then we may replace x and y by s^kxt^k and s^kyt^k (for $(s^kxt^k, s^kyt^k) \in (x,y)^*$), falling back to (B3), if $s^kyt^k \in K$ also, or (A2) if $s^kyt^k \in J$. Thus in every case it may be assumed that $s^kxt^k\ \mathcal{J}\ x$.

Now a similar argument to that in Case A completes the induction and yields the contradiction $x\ \tau\ y$, completing the proof.

3. *Irregular varieties*

As remarked in §1 any irregular variety of semigroups is completely simple-by-nil. The nil varieties themselves are in fact contained in ABG, so are CSM. The main result of this section is the following.

THEOREM 3.1. *Every completely simple-by-nilpotent semigroup is* CSM. *Hence every completely simple-by-nilpotent variety is* CSM.

Before proving this theorem we describe such varieties in terms of identities.

PROPOSITION 3.2. *A variety V of semigroups is completely simple-by-nilpotent if and only if V satisfies an identity of the form*

$$x_1\ldots x_n = (x_1\ldots x_n)(x_1\ldots x_n\ y_1\ldots y_n\ x_1\ldots x_n)^r\ , \quad \text{for some } n,r > 0.$$

Proof. If V satisfies such an identity it is clearly irregular, and so is completely simple-by-nil. If $S \in V$ and $x_1,\dots,x_n \in S$ then $x_1\dots x_n = (x_1\dots x_n)^{3r+1}$, so $x_1\dots x_n$ belongs to a subgroup of S, and therefore to the kernel K. Thus $S^n \subseteq K$ and S is completely simple-by-nilpotent.

Conversely if V is completely simple-by-nilpotent it is uniformly periodic, that is, V satisfies an identity $x^{m+r} = x^m$ for some $m,r > 0$. If $S \in V$ then the subgroups of its kernel K must satisfy the identity $x^r = 1$. Since S/K is nilpotent, $S^n \subseteq K$ for some $n > 0$. So if $x_1,\dots,x_n,y_1,\dots,y_n \in S$, then $x_1\dots x_n \in K$ and $y_1\dots y_n \in K$ and, further, by complete simplicity of K, $x_1\dots x_n y_1\dots y_n x_1\dots x_n \; \mathcal{H} \; x_1\dots x_n$. Thus $(x_1\dots x_n y_1\dots y_n x_1\dots x_n)^r$ is the identity element of its maximal subgroup, yielding the specified identity.

Whether *every* irregular variety of semigroups is CSM we do not know. Some slight extensions of the main theorem (omitted here) suggest this may indeed by so. We first reduce the general case to the cases *right zero-by-nil* and *left zero-by-nil* by means of the following lemma.

LEMMA 3.3. *Let* S *be a completely simple-by-nil semigroup. Then* $\mathcal{L}$, $\mathcal{R}$ *and* $\mathcal{H}$ *are congruences on* S *and the map*

$$\alpha \longmapsto (\alpha \cap \mathcal{H},\ \alpha \vee \mathcal{L},\ \alpha \vee \mathcal{R})$$

is an isomorphism of $C(S)$ *onto a subdirect product of its interval sublattices* $[\iota,\mathcal{H}]$, $[\mathcal{L},\omega]$ *and* $[\mathcal{R},\omega]$.

Proof. Denote the kernel of S by K. Let $x,y \in S$, $x \neq y$, and suppose $x\mathcal{L}y$. Then since $\mathcal{J}$ is trivial on $S\backslash K$, $x,y \in K$. For any $s \in S$, $xs \;\mathcal{L}\; ys$ (since $\mathcal{L}$ is always a right congruence) and $sx \;\mathcal{L}\; x \;\mathcal{L}\; y \;\mathcal{L}\; sy$ (since $L_{sx} \leq L_x$, $L_{sy} \leq L_y$ in the completely simple semigroup K : see [7, §III.2]), so $\mathcal{L}$ is a congruence. By duality so is $\mathcal{R}$ and therefore $\mathcal{H} = \mathcal{L} \cap \mathcal{R}$ also.

Let $\alpha \in C(S)$ and $x \in S$. Then $x^n \in K$ for some $n > 0$. Denote by e_x the identity element of the subgroup containing x^n (and thus all powers x^m, $m \geq n$). If $x\alpha a$ for some $a \in K$ then for some $n > 0$ $x^n\alpha a^n$ and $e_x\alpha e_a$, so that $e_x x e_x \alpha e_a a e_a = a$. Thus $x\alpha e_x x e_x \in K$.

Suppose $\alpha,\beta \in C(S)$ and $\alpha \cap \mathcal{H} \subseteq \beta \cap \mathcal{H}$, $\alpha \vee \mathcal{L} \subseteq \beta \vee \mathcal{L}$ and $\alpha \vee \mathcal{R} \subseteq \beta \cup \mathcal{R}$. To prove the specified map is an order isomorphism we must show $\alpha \subseteq \beta$. So let $(x,y) \in \alpha$. Since $\alpha \subseteq \beta \vee \mathcal{L}$ either $x\beta y$ as required, or, since $\mathcal{L}$ is trivial on $S\backslash K$, $x\beta a$ for some $a \in K$ (possibly $x \in K$ already). By the preceding paragraph $x\beta e_x x e_x$; similarly $y\beta e_y y e_y$. But since $x\alpha y$, $e_x x e_x \alpha e_y y e_y$. So *we may assume initially* that $x,y \in K$.

We leave it to the reader to now verify that since $\alpha \subseteq \beta \vee \mathcal{L}$ there exists $s \in K$ such that $x\beta s$ and $s \in R_x \cap L_y$. Similarly since $\alpha \subseteq \beta \vee \mathcal{R}$ there exists $t \in K$ such that $t\beta y$ and $t \in R_x \cap L_y$. Let g be the identity element of $R_x \cap L_y$. Then since $x\alpha y$, $(gxg,gyg) \in \alpha \cap \mathcal{H} \subseteq \beta \cap \mathcal{H}$, that is $gxg\beta gyg$. But $gxg\beta gsg = s\beta x$ and $gyg\beta gtg = t\beta y$, so $x\beta y$, as required.

COROLLARY 3.4. *Let* S *be a completely simple-by-nil semigroup. Then* S *is* CSM *if and only if the right zero-by-nil semigroup* $S/\mathcal{L}$ *and the left zero-by-nil semigroup* S/R *are each* CSM.

Proof. Necessity is clear from the remarks in §1. The converse is evident from: (i) $C(S/L) \cong [L,\omega]$ and $C(S/R) \cong [R,\omega]$; (ii) any lattice of congruences contained in H is modular [9]; (iii) a subdirect product of semimodular lattices is semimodular.

LEMMA 3.5. *Let* S *be a completely simple-by-nil semigroup. If* ρ *is an atom of* C(S) *then either* ρ *is an atom of* E(S) *or* $\rho \subseteq \rho_K$, *the Rees congruence modulo the kernel* K *of* S.

Proof. This is almost immediate from Result 2.7 and the fact that K is the only nontrivial J-class of S. (Atoms in E(S) identify precisely one pair of distinct elements).

This Lemma can be extended in the following case.

LEMMA 3.6. *Let* S *be a right zero-by-nilpotent semigroup. If* ρ *is an atom of* C(S) *then* ρ *is an atom of* E(S).

Proof. By the previous lemma we may assume $\rho \subseteq \rho_K$. Suppose ρ contains a pair (x,y), $x \neq y$, for which $xt = yt$ for all $t \in S\backslash K$. Then since $x,y \in K$ and K is right zero, $xt = yt$ for all $t \in S$ and $(sx,sy) = (x,y)$ for all $s \in S$. Thus $\rho = (x,y)^* = \{(x,y),(y,x)\} \cup \iota$, an atom in E(S).

Hence for any atom ρ of C(S) and any $(a,b) \in \rho$ with $a \neq b$, if ρ is not an atom of E(S) than $at \neq bt$ for some $t \in S\backslash K$. Applying the same argument to the pair $(at,bt) \in \rho$ and repeating it yields a sequence $t = t_1,t_2,\dots$ of elements of $S\backslash K$ such that $at_1t_2\dots t_i \neq bt_1t_2\dots t_i$ for each i. But since S/K is nilpotent, $S^n \subseteq K$ for some $n > 0$. Thus $t_1t_2\dots t_n \in K$ and $at_1\dots t_n = bt_1\dots t_n = t_1\dots t_n$, a contradiction. Hence ρ is an atom of E(S).

Now for any right zero-by-nilpotent semigroup S, atom ρ of C(S) and $\tau \in C(S)$ with $\rho \nsubseteq \tau$, by semimodularity of E(S), $\rho\vee\tau \succ \tau$ in E(S), whence in C(S). Hence the class of such semigroups is CSM, by Proposition 1.1. Theorem 3.1 then follows from Corollary 3.4.

4. *Varieties of inverse semigroups*

In a now standard way (see [12]) inverse semigroups may be regarded as algebras of type $\langle 2,1 \rangle$. As such, a variety of inverse semigroups is therefore a class closed under $\underset{\sim}{H}$, $\underset{\sim}{P}$ and $\underset{\sim}{S}$ (inverse subsemigroups). Such a variety is again *regular* if and only if it contains the variety S of semilattices. Thus Lemma 2.2 still applies and, since an inverse completely simple semigroup is a group, a regular CSM variety V of inverse semigroups must satisfy $V \cap G = T$, where G is the variety of groups. Thus [12, §XII.1] V is a *combinatorial* variety, that is, consists entirely of combinatorial inverse semigroups. Equivalently V satisfies an identity of the form $x^{n+1} = x^n$ for some $n > 0$. Conversely, it is clear from

Proposition 2.4 that any such variety is contained in *ABG* and is therefore CSM. (This also follows from [8, Theorem 3.3]).

THEOREM 4.1. *A variety* V *of inverse semigroups is* CSM *if and only if either 1)* V *is a variety of groups, or 2)* V *is a combinatorial variety.*

5. *Pseudovarieties of finite semigroups*

Denote by *Fin* the pseudovariety consisting of all finite semigroups. In this section we describe all CSM pseudovarieties of finite semigroups. The regular case has already been completed in §2. It is easily verified (c.f. §3) that any irregular pseudovariety of finite semigroups consists of completely simple-by-nilpotent semigroups (any finite nil semigroup is nilpotent). As such, it is CSM by Theorem 3.1.

THEOREM 5.1. *1) A regular pseudovariety of finite semigroups is* CSM *if and only if it is contained in the pseudovariety* $ABG \cap Fin$.

2) Every irregular pseudovariety of finite semigroups is CSM.

Acknowledgement. The author gratefully acknowledges support of a Monash University Visiting Senior Lectureship.

References

1. J. L. Chrislock, A certain class of identities on semigroups, *Proc. Amer. Math. Soc.* 21 (1969), 189-190.
2. T. Evans, The lattice of semigroup varieties, *Semigroup Forum* 2 (1971), 1-43.
3. R. Freese and J. B. Nation, Congruence lattices of semilattices, *Pacific J. Math.* 49 (1973), 51-58.
4. G. Grätzer, *General Lattice Theory*, Academic Press, New York, 1978.
5. G. Grätzer, *Universal Algebra*, Second Edition, Springer-Verlag, New York, 1979.
6. T. E. Hall, On the lattice of congruences on a semilattice, *J. Austral. Math. Soc.* 12 (1971), 456-460.
7. J. M. Howie, *An Introduction to Semigroup Theory*, Academic Press, London, 1976.
8. P. R. Jones, On congruence lattices of regular semigroups, *J. Algebra* 82 (1983), 18-39.
9. G. Lallement, Congruences et équivalences sur un demi-groupe régulier, *C. R. Acad. Sci. Paris*, Sér. A. 262 (1966), 613-616.
10. G. Lallement, Demi-groupes réguliers, *Ann. Mat. pura.ed.appl.* 77 (1967), 47-129.
11. S. W. Margolis and J. E. Pin, Varieties of finite monoids and topology for the free monoid, *Proc. Marquette Conf. on Semigroups*, Marquette University (1984), 113-130.
12. M. Petrich, *Inverse Semigroups*, Wiley, New York, 1914.
13. D. C. Trueman, The lattice of congruences on direct products of cyclic semigroups and certain other semigroups, *Proc. Roy. Soc. Edinburgh* 95A (1983), 203-214.

DECOMPOSITION OF LANGUAGES INTO DISJUNCTIVE OUTFIX CODES ★

Masashi Katsura
Faculty of Science,
Kyoto Sangyo University,
Kyoto 603, Japan

H. J. Shyr
Institute of Applied Mathematics,
National Chung-Hsing University,
Taichung, Taiwan, 400

Let X be a finite alphabet consisting of more than one letter and let X^* be the free monoid generated by X. Let $X^+ = X^* \setminus \{1\}$, where 1 is the empty word. Any subset of X^* is called a *language* over X. A nonempty subset of X^* is said to be a *right (left) ideal* of X^* if $IX^* \subseteq I$ $(X^*I \subseteq I)$. An *ideal* is both a right and left ideal. We call $A \subseteq X^+$ a *prefix (suffix) code* if $A \cap AX^+ = \emptyset$ $(A \cap X^+A = \emptyset)$. $A \subseteq X^+$ is an *outfix code* if $uxv,\ uv \in A$ imply $x = 1$, and an *infix code* if $uxv,\ x \in A$ implies $uv = 1$. It is immediate that every outfix (infix) code is both a prefix and suffix code. Some particular infix codes were recently studied by Guo, Shyr and Thierrin [4].

Let $A \subseteq X^*$. The relation P_A defined on X^* by $x \equiv y\,(P_A)$ if and only if $uxv \in A \Leftrightarrow uyv \in A$ holds for any $u, v \in X^*$ is a congruence on X^*. The language $A \subseteq X^*$ is *regular* if P_A is of finite index and *disjunctive* if P_A is the equality. We call a language A *regular free* if every regular language contained in A is finite.

It is known that for every right ideal I there exists a unique prefix code P such that $I = PX^*$. In [3], Shyr introduced the concept of ordered catenation of two languages and showed that every right ideal I such that $I = PX^*$, with P an infinite prefix code, is a disjoint union of infinitely many regular free disjunctive prefix codes.

The purpose of this paper is to consider a better setting for the above decomposition. We call a language A *DO-splittable* when A is a disjoint union of infinitely many disjunctive outfix code. Since an outfix code is always regular free, our results generalize those of Shyr [3].

We assume that X^* is equipped with the following total order $\leqq$.

If $lg(u) < lg(v)$, then $u < v$.

When $lg(u) = lg(v)$, $u < v$ if and only if u is lexicographically smaller than v.

★ Part of this research has been supported by Grant NSC 73-0204-M005-01 R. O. C.

Following [3] we define the *ordered catenation* of two infinite languages $A = \{a_1 < a_2 < \cdots < a_i < \cdots\}$ and $B = \{b_1 < b_2 < \cdots < b_i < \cdots\}$ by

$$A \triangle B = \{a_1b_1,\ a_2b_2, \cdots, a_ib_i, \cdots\}.$$

Clearly, for three languages A, B and C, we have $(A \triangle B) \triangle C = A \triangle (B \triangle C)$.

We call a language $A \subseteq X^*$ *dense* if for every $x \in X^*$, there exists u and v in X^* such that $uxv \in A$ [1]. Denseness of a language $A \subseteq X^*$ is equivalent to A containing a disjunctive subset, see [2]. No infix code can be dense. Hence there is no disjunctive infix code. However, disjunctive outfix codes do exist. First we show the following.

Lemma 1. *Let $P, A, S \subseteq X^*$ be three infinite languages.*
(1) *If P is a prefix code and S is a suffix code, then $P \triangle A \triangle S$ is an outfix code.*
(2) *If A is dense, then $P \triangle A \triangle S$ is disjunctive.*

Proof. (1) Let $P = \{p_1 < p_2 < \cdots < p_i < \cdots\}$, $A = \{a_1 < a_2 < \cdots < a_i < \cdots\}$ and $S = \{s_1 < s_2 < \cdots < s_i < \cdots\}$. Assume that $P \triangle A \triangle S$ is not an outfix code. Then for some $i < j$ and $w \neq 1$, we have $p_ia_is_i = uv$ and $p_ja_js_j = uwv$. From the first equation, we have either $u = p_ix$ or $v = ys_i$ for some $x, y \in X^*$. If $u = p_ix$, then $p_ja_js_j = p_ixwv$ which contradicts the fact that P is a prefix code. Similarly, $v = ys_i$ contradicts the fact that S is a suffix code. Thus $P \triangle A \triangle S$ is an outfix code.
(2) This follows from Theorem 3 of [3]. □

For an infinite language $A = \{a_1 < a_2 < \cdots\}$ and $i \in \mathbb{N}$, we denote $A_i = \{a_i < a_{i+1} < \cdots\}$. In view of [3], we get

$$AB = (\cup_{i \geq 1} A_i \triangle B_1) \cup (\cup_{j \geq 2} A_1 \triangle B_j).$$

It is easy to see that for the catenation of three infinite languages A, B and C, we have

$$ABC = \cup_{(i,j,k) \in \Omega} A_i \triangle B_j \triangle C_k$$

where

$$\Omega = \{(1, j, k) \mid j, k \geqq 1\} \cup \{(i, 1, k) \mid i \geqq 2, k \geqq 1\} \cup \{(i, j, 1) \mid i, j \geqq 2\}.$$

Note that this is a disjoint union.

Proposition 1. *Let P be an infinite prefix code, A be a dense language and S be an infinite suffix code. Then PAS is DO-splittable.*

Proof. Let $PAS = \cup_{(i,j,k) \in \Omega} P_i \triangle A_j \triangle S_k$, where Ω is the index set given above. Here the union is a disjoint union and each P_i is an infinite prefix code, each A_j is dense and each S_k is an infinite suffix code. By Lemma 1, for each $(i, j, k) \in \Omega$, $P_i \triangle A_j \triangle S_k$ is a disjunctive outfix code. □

Lemma 2. *Let $a \in X$ and let $\{p,q\} \subseteq X^+$ be a prefix code. If for some nonempty language $D \subseteq pX^*a$, D is an outfix code and $D \cup \{q\}$ is not an outfix code, then $q \in X^*a$.*

Proof. From the given conditions, we see that for some $r \in X^*$, $q = uv$, $pra = uwv$ or $pra = uv$, $q = uwv$ where $u, v \in X^*$, $w \in X^+$. Assume $v = 1$. Then $pra = qw$ or $q = praw$. This contradicts the fact that $\{p, q\}$ is a prefix code. Hence $v \neq 1$. It can easily be seen that $q \in X^*a$. □

Proposition 2. *Let P be a prefix code with $|P| > 1$, and let A be a language such that $PX^* \subseteq A \subseteq X^+$. If for every $u \in A$ there exists $p \in P$ such that $\{u,p\}$ is a prefix code, then A is DO-splittable.*

Proof. Let $a, b \in X$ be fixed with $a \neq b$. We can express A as

$$A = (\cup_{p\in P} pa^+bX^*ab^+) \cup (\cup_{p\in P} pa^+bX^*ba^+) \cup B$$

where the union is disjoint and B a language.

Now, for each $p \in P$, since pa^+b is an infinite prefix code and ab^+, ba^+ are infinite suffix codes, by Proposition 1 both $pa^+bX^*ab^+$ and $pa^+bX^*ba^+$ are DO-splittable. Let the decompositions be as follows:

$$pa^+bX^*ab^+ = D_{p,1} \cup D_{p,2} \cup \cdots,$$

$$pa^+bX^*ba^+ = D'_{p,1} \cup D'_{p,2} \cup \cdots.$$

For each $u \in B$, fix $p_u \in P$ such that $\{u, p_u\}$ is a prefix code. We note that $1 \notin A$ and hence $1 \notin B$. Put

$$B_p = \{u \in X^*a \cap B \mid p_u = p\} = \{u_{p,1}, u_{p,2}, \cdots\},$$

$$B'_p = \{u \in X^*(X \setminus a) \cap B \mid p_u = p\} = \{u'_{p,1}, u'_{p,2}, \cdots\}$$

for each $p \in P$. These sets may be infinite, finite or even empty. We then put

$E_{p,i} = D_{p,i} \cup \{u_{p,i}\}$ if $u_{p,i}$ exists, and $E_{p,i} = D_{p,i}$ otherwise,

$E'_{p,i} = D'_{p,i} \cup \{u'_{p,i}\}$ if $u'_{p,i}$ exists, and $E'_{p,i} = D'_{p,i}$ otherwise.

Then we have $A = \cup_{p\in P} \cup_{i\in \mathbb{N}} (E_{p,i} \cup E'_{p,i})$. Here, the union is disjoint. By Lemma 2, $E_{p,i}$ and $E'_{p,i}$ are outfix codes and since the union of a disjunctive set with a finite set is disjunctive, we see that both $E_{p,i}$ and $E'_{p,i}$ are disjunctive. Therefore, A is DO-splittable. □

Corollary 1. *Let I be a right ideal of X^* which is not principal. Then I is DO-splittable.*

Proof. Suppose that I is a right ideal which is not principal. It is known that $I = PX^*$, where P is a prefix code and $|P| > 1$. Now the result follows from the above proposition. □

The following corollaries are immediate.

Corollary 2. *Let I be a right ideal of X^* and let A be a language over X such that $I \subseteq A \subseteq X^+$. If $I \cap aX^* \neq \emptyset$ for every $a \in X$, then A is DO-splittable.* □

Corollary 3. *Let P be a maximal prefix code and let A be a language over X such that $PX^* \subseteq A \subseteq X^+$. Then A is DO-splittable.* □

Corollary 4. *Let I be an ideal of X^* and let A be a language over X such that $I \subseteq A \subseteq X^+$. Then A is DO-splittable.* □

REFERENCES

[1] Lallement, G, Semigroups and combinatorial applications, Wiley, New York, 1979.

[2] Shyr, H. J., Free monoids and languages, Lecture Notes, Soochow University, Taipei, Taiwan, 1979.

[3] Shyr, H. J., Ordered catenation and regular free disjunctive languages, Information and Control 46 (1980), 257-269.

[4] Guo, Y. Q., Shyr, H. J. and Thierrin, G., E-convex infix codes, Order 3 (1986), 55-59.

SOME ALGORITHMS FOR SEMIGROUPS AND MONOIDS PRESENTED BY A SINGLE RELATION

Gerard Lallement
The Pennsylvania State University
University Park, PA 16802, USA

1. General problem and results.

Let $M = \langle A;R \rangle$ be a monoid presented by a set A of generators and a set R of relations. By definition, R is a subset of $A^* \times A^*$ where A^* is the free monoid on A, and M is the quotient of A^* by the congruence γ_R on A^* generated by R. The problem of word equality - or more shortly, the word problem - for the presentation $\langle A;R \rangle$, is said to be decidable if there exists an algorithm allowing to decide for all pairs of words $(w,w') \in A^* \times A^*$ whether $(w,w') \in \gamma_R$ or not. A similar problem can be stated for semigroups.

It is known that, in general, the word problem for semigroups is undecidable (A. Markov, 1947, E. Post, 1947). There is even a semigroup presented by three relations with an undecidable word problem (J. Matijasevitch, 1967). It is generally believed that semigroups and monoids presented by a single relation have a decidable word problem, but this is still an open question, in spite of the fact that the word problem for one-relator groups has been solved positively in 1932 by W. Magnus.

Recently, utilizing the diagram method of J. H. Remmers, (1980), J. Howie and S. J. Pride have shown, among other results, that the problem could be reduced to the following cases:

(1) $\langle A;\ a = wa \rangle$ (2) $\langle A;\ avb = awb \rangle$ (3) $\langle A;\ ava = bwa \rangle$

with $a,b \in A$, $a \neq b$ and v,w arbitrary in A^*. In fact the known results utilizing the standard derivation methods treating a presentation as a Thue system with rewriting rules allow to show that the only case left is (1') $\langle A;v = w \rangle$ with v and w having distinct initial letters and same terminal letters. Thus, taking the results of Howie and Pride into account, it is likely that the word problem for semigroups and monoids presented by a single relation could be reduced to the cases $\langle A;\ a = bwa \rangle$ and $\langle A;\ ava = bwa \rangle$ with $a,b \in A$, $a \neq b$, and $v,w \in A^*$.

The purpose of the present paper is to give an outline of the results of Adjan and Oganesjan reducing the problem to the case (1') above, and also to give some further indications on this case (1').

Throughout we denote by $|w|$ the length of the word w, by $i(w)$ the first letter of w and by $t(w)$ the last letter of w. Equality of words in the free monoid is denoted by $\equiv$.

2. Adjan's classical results.

In 1966, S. I. Adjan proved the following theorem:

Theorem 2.1. For the monoids $\langle A;\ w = 1\rangle$ *the word problem is decidable.*

The proof of this theorem is long but the general idea is relatively simple. Let $M = \langle A;\ w = 1\rangle = A/\gamma_R$ with $R = \{(w,1)\}$. In A^* one verifies easily that the set U of invertible words modulo γ_R is a submonoid satisfying the following property:

For all $u,v,w \in A^*$, $uv \in U$ and $wu \in U$ imply $u,v,w \in U$.

Thus U is a free submonoid of A^* generated by a special type of biprefix code C which is easy to construct algorithmically. One constructs the smallest set C_1 containing w and such that $u_1v \in C_1$, $vu_2 \in C_1$ imply $vu_1 \in C_1$ and $u_2v \in C_1$. Note that for all words $u \in C_1$, $|u| = |w|$; thus C_1 is finite and it is effectively computable from w. C is then the set of all words of minimal length which are left and right factors of words of C_1. For example with $\langle a,b,c;\ abcababc = 1\rangle$ we obtain

$$C_1 = \{\underline{abc}ab\underline{abc},\ \underline{ab}abcabc,\ abcabc\underline{ab},\ \underline{c}abcabab,\ ababcab\underline{c},\ldots\}$$

and $C = \{ab,c\}$. Introducing an alphabet Γ in bijection with C (say through $\varphi : C \longrightarrow \Gamma$) we can rewrite the presentation $\langle A;\ w = 1\rangle$ as $\langle \Gamma;\ \varphi(w) = 1\rangle$ (in the example, with $\varphi(ab) = \alpha$ and $\varphi(c) = \gamma$ one gets $\alpha\gamma\alpha\alpha\gamma = 1$). The presentation $\langle \Gamma;\ \varphi(w) = 1\rangle$ is a presentation of the group G of units of M, and to solve the word problem for M one proceeds as follows:

For each $z \in A^*$ define the (finite) set $\delta(z)$ of direct descendants of z by

$$\delta(z) = \{v \in A^* : z \equiv xuy,\ v \equiv xu'y,$$
$$\text{with } u,u' \in U,\ |u'| \le |u| \text{ and } \varphi(u) = \varphi(u') \text{ in } G\}$$

This set $\delta(z)$ can be constructed algorithmically by Magnus' result. Denote by $\Delta(z)$ the (finite) set of descendants of z, i.e. the transitive closure of $\delta(z)$. Then one proves that $(z_1,z_2) \in \gamma_R$ if and only if $\Delta(z_1) \cap \Delta(z_2) \neq \emptyset$.

The second important result of Adjan is:

Theorem 2.2. For the semigroups $\langle A;\ v = w\rangle$ *where the initial letters of* v *and* w *are distinct, and the terminal letters of* v *and* w *are distinct, the word problem is decidable.*

One shows more precisely that $\langle A;\ v = w\rangle$ is a right [resp. left] cancellative semigroups if and only if the terminal [resp. initial] letters of v and w are distinct. So $\langle A;\ v = w\rangle$ as in Theorem 2.2 is cancellative. This semigroup is even embeddable in the one-relator group having the same presentation, and again Theorem 2.2 can be deduced from Magnus' result on one-relator groups.

3. Some reduction algorithms.

The results contained in this section are due to S. Adjan and G.

Oganesjan, 1978.

A word $s \in A^*$ is called a *primary word* if it is not of the form uvu with $u \neq 1$ (i.e. s is not self-overlapping). The bifactor (factor which is both a prefix and a suffix) of minimal length of any word $w \in A^+$ is a primary word. Also, if s is a primary word then any word $w \in A^+$ can be written uniquely as $w = u_1su_2s\ldots u_ksu_{k+1}$ where k is the number of occurrences of s in w. A relation $v = w$ is called *compressible* if v and w have a common bifactor. For example the relations $\underline{ab}a\underline{ab} = \underline{ab}\,\underline{ab}$, $\underline{a}bab\underline{a} = \underline{a}b\underline{a}$, $\underline{a}b\underline{a} = \underline{a}$ are compressible (the common bifactors for each relation have been underlined).

We shall say that the word problems for two presentations (of different semigroups) are equivalent if they are both decidable or undecidable.

Theorem 3.1. Let $M = \langle A;\ v = w\rangle$, *where* $v = w$ *is a compressible relation. Then there exists* $M' = \langle B;\ \varphi(v) = \varphi(w)\rangle$ *such that* $|\varphi(v)\ \varphi(w)| < |vw|$, *the word problems for* M *and* M' *are equivalent, and* M' *can effectively be constructed from* M.

Proof. Let s be the common primary bifactor of v and w. Let $L = \{r_0, r_1, \ldots, r_i, \ldots\}$ be the (denumerable) list of all words *without* an occurrence of s, and let $B = \{b_0, b_1, \ldots, b_i, \ldots\}$ be an alphabet in bijection with L so that r_i corresponds to b_i for all i. Let $Q_s = sA^*s \cup \{s\}$ and define $\varphi : Q_s \longrightarrow B^*$ by

$$\varphi(s) = 1 \quad \text{and} \quad \varphi(szsr_ns) = \varphi(szs)b_n$$

The mapping φ is a bijection; it is not a homomorphism in general, but it is straightforward to check that φ preserves one-step applications of the rewriting rules from the presentation $\langle A;\ v = w\rangle$ to the presentation $\langle B;\ \varphi(v) = \varphi(w)\rangle$. We can even replace the infinite alphabet B by an alphabet containing only the letters occurring in $\varphi(v)$ and $\varphi(w)$. We call φ a *compression of* the relation $v = w$ in s.

Example. $M = \langle a,b;\ ababa = aba\rangle$. Here $s = a$, and $L = \{b^0, b^1, \ldots b^i, \ldots\} = \{b\}^*$. Take $B = \{b_0, b_1, \ldots, b_i, \ldots\}$. Compressing in a and suppressing the unnecessary letters gives $M' = \langle b_1;\ b_1b_1 = b_1\rangle$. Since M' has a decidable word problem so does M.

Repeating compressions of relations as often as possible gives:

Corollary 3.2. Let $M = \langle A;\ v = w\rangle$ *where* $v = w$ *is a compressible relation. Then there exists* $M' = \langle B;\ \psi(v) = \psi(w)\rangle$ *such that* $\psi(v) = \psi(w)$ *is an incompressible relation, the word problems for* M *and* M' *are equivalent, and* M' *can effectively be constructed from* M.

<u>Corollary 3.3</u>. (G. Lallement, 1974) <u>Let</u> $M = \langle A;\ v = w\rangle$ <u>where</u> v <u>is a bifactor of</u> w. <u>Then the word problem for</u> M <u>is decidable</u>.

<u>Proof</u>. The presentation M' of Corollary 3.2 is $\langle B;\ \psi(w) = 1\rangle$. Applying Corollary 3.2 and Theorem 2.1 gives the result.

Note that the monoids in Corollary 3.3 are exactly those presented by a single relation with idempotents $\neq 1$ (see G. Lallement, 1974) unless $v = 1$ and M is a group.

We do not discuss decidability of the left or right divisibility of a word by another here. However it is worth mentioning that further discussion along the lines above allows to show that in a one-relation monoid the decidability of the divisibility of a word by another on the left and on the right implies the decidability of the word problem.

Suppose now that we have a presentation $M = \langle A;\ v = w\rangle$ with $v, w \neq 1$ and the relation $v = w$ is <u>incompressible</u> (in case v or $w \equiv 1$, see Theorem 2.1). Then it is always possible to find a primary word s_0 (related to the primary bifactor s of v and t of w, see below) having the following special property:

<u>No right</u> [<u>resp. left</u>] <u>factor of</u> s_0 <u>is a left</u> [<u>resp. right</u>] <u>factor of</u> v <u>or</u> w.

The word s_0 is either a left factor of one of s or t and a right factor of the other in case such a word exists, or s_0 is the shortest of s and t. Let us define $\theta : A^* \longrightarrow (A \cup \{x\})^*$, where $x \notin A$, as follows:

$$\theta(z) = z \quad \text{if } s_0 \text{ does not occur in } z,$$

$$\theta(z) = u_1 x u_2 x \dots u_k x u_{k+1} \quad \text{if } z = u_1 s_0 u_2 s_0 \dots u_k s_0 u_{k+1}\,.$$

<u>Theorem 3.4</u>. <u>Let</u> $M = \langle A;\ v=w\rangle$ <u>where</u> $v, w \neq 1$ <u>and</u> $v = w$ <u>is an incompressible relation</u>. <u>Then there exists</u> $M' = \langle A \cup \{x\};\ \theta(v) = \theta(w)\rangle$ <u>with either</u> $i[\theta(v)] \neq i[\theta(w)]$ <u>or</u> $t[\theta(v)] \neq t[\theta(w)]$, <u>the word problems for</u> M <u>and</u> M' <u>are equivalent, and</u> M' <u>can effectively be constructed from</u> M.

The key point in the proof of Theorem 3.4 is again the fact that an elementary transition say $z_1 \equiv \alpha v \beta \longrightarrow \alpha w \beta \equiv z_2$ in M translates into the elementary transition $\theta(z_1) \equiv \theta(\alpha)\theta(v)\theta(\beta) \longrightarrow \theta(\alpha)\theta(w)\theta(\beta) \equiv \theta(z_2)$ in M' and conversely. The special property of s_0 is used in analyzing the occurrences of s_0 in $z_1 \equiv \alpha v \beta$ and in $z_2 \equiv \alpha w \beta$ when computing $\theta(z_1)$ and $\theta(z_2)$. For example, the special property ensures that positions of s_0 as indicated below are impossible.

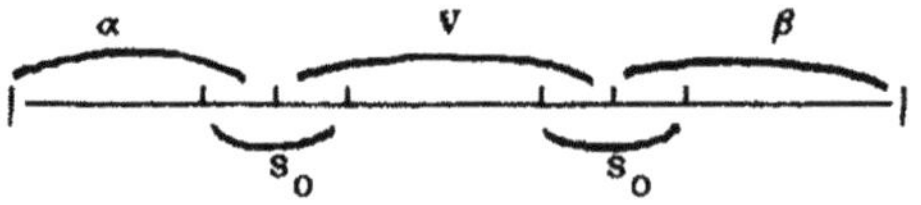

Examples. 1) $M = \langle a,b;\ aababaab = abbabb\rangle$. Here $s = aab$ and $t = abb$. We take $s_0 = ab$ and we obtain $M' = \langle a,b,x;\ axxax = xbxb\rangle$. The word problem for M' is decidable by Theorem 2.2, hence it is also decidable for M.
2) $M = \langle a,b;\ baa = bbaabaabbaa\rangle$. Here $s = baa$ and $t = bbaa$. Taking $s_0 =$ baa gives $M' = \langle a,b,x;\ x = bxxbx\rangle$. The word problem for M' is decidable (see Example 4.3), hence it is also decidable for M.

Corollary 3.5. *The word problem for semigroups and monoids presented by a single relation is decidable if it is decidable for presentations* $\langle A;\ v = w\rangle$ *with* $i(v) \not\equiv i(w)$ *and* $t(v) \equiv t(w)$.

4. *The case of presentations* $\langle A;\ v = w\rangle$ *with* $i(v) \not\equiv i(w)$.

We shall assume that $i(v) \equiv a$ and $i(w) \equiv b$.

Lemma 4.1. *If the divisibility on the left of an arbitrary word* $z \in A^*$ *by* a *and* b *is decidable, then the word problem is decidable*.

Proof. $M = \langle A;\ v = w\rangle$ is left cancellative. In order to decide if $z = z'$ in M or not, we test if z' is divisible on the left by $i(z)$. This is decidable if $i(z)$ is a or b, and in case $i(z)$ is a letter distinct from a and b, z' is left divisible by $i(z)$ if and only if $i(z) \equiv i(z')$. In case z' is left divisible by $i(z)$ we rewrite z' as $i(z)z_1'$. Since $z \equiv i(z)z_1$, then by left cancellativity $z = z'$ in M if and only if $z_1 = z_1'$. But $|z_1| = |z| - 1$, and an induction on $|z|$ completes the proof.

There is a pseudo-algorithm due to Adjan, 1977, to detect if an arbitrary word z is left divisible by a or b. This will be explained on the example $M = \langle a,b;\ aba = baababaa\rangle$.

Suppose we want to detect if the word $z \equiv bbaaababa$ is left divisible by a in M. Since z begins by b, z has a left factor in common with baababaa (right-hand side of the presentation relation). The longest common left factor is b which is the first factor in Adjan's decomposition of z. We write the beginning of the decomposition as b|baaababa and continue to decompose baaababa following the same principle. The second step gives b|baa|ababa and we observe that the third factor of the decomposition is aba which is the complete left-hand side of the presentation relation. We then let dec(z) = b|baa|<u>aba</u>|ba and call aba the *head* of dec(z) in Adjan's decomposition of z.

The pseudo-algorithm consists in doing the following:

1) if a word admits a decomposition without a head then stop;

2) if a word admits a decomposition whose head is on one side of the presentation relation, then replace the head by the word on the other side of the relation.

In the example above, z should be transformed into $z_1 \equiv$ bbaabaababaaba. Adjan's decomposition of z_1 is

$$\text{dec}\,(z_1) = \text{b}|\text{baaba}|\underline{\text{aba}}|\text{baaba} \quad \text{with head aba.}$$

This yields $z_2 \equiv$ bbaababaababaabaaba whose decomposition is:

$$\text{dec}\,(z_2) = \text{b}|\underline{\text{baababaa}}|\text{babaabaaba} \quad \text{with head baababaa.}$$

This, in turn, gives $z_3 \equiv$ bababaabaaba and

$$\text{dec}\,(z_3) = \text{ba}|\text{ba}|\text{ba}|\text{baaba}|\underline{\text{aba}} \quad \text{with head aba.}$$

Then $z_4 \equiv$ babababaababaababaa and

$$\text{dec}\,(z_4) = \text{ba}|\text{ba}|\text{ba}|\underline{\text{baababaa}}|\text{babaa} \quad \text{with head baababaa.}$$

Finally $z_5 \equiv$ bababaabababaa and dec (z_5) = ba|ba|baababa|baa has no head. We stop and applying the next theorem we declare that z is <u>not</u> left divisible by a.

<u>Theorem 4.2</u>. <u>Let</u> M = <A; v = w> <u>with</u> i(v) ≡ a, i(w) ≡ b, <u>and let</u> z ∈ A* <u>with</u> i(z) ≡ b [<u>resp</u>. a]. <u>Then</u> z <u>is left divisible by</u> a [<u>resp</u>. b] <u>in</u> M <u>if and only if Adjan's pseudo-algorithm applied to</u> z <u>yields a word</u> z' <u>beginnning by</u> a [<u>resp</u>. b], <u>after a finite number of steps</u>. <u>In all other cases</u> (i.e. <u>the process stops on a decomposition without a head</u>, <u>or the process goes on indefinitely</u>) z <u>is not left divisible by</u> a [<u>resp</u>. b].

If one admits that decidability of the divisibility of a word by a letter in a left cancellative semigroup presented by one relation is a reasonable conjecture, then an appropriate strategy for solving the word problem would consist in showing that all the cases where Adjan's process goes on indefinitely can be detected in a finite number of steps as in the following examples:

<u>Example 4.3</u>. 1) <a,b; a = baaba>. This case (more precisely the right cancellative analog a = abaab) was mentioned by Howie and Pride as a case escaping their results. It can be shown that here Adjan's pseudo-algorithm is in fact an algorithm since the process to test left divisibility of a word by a never goes on indefinitely. Furthermore, using the algorithm to put the highest power of a to the leftmost in every word z gives a canonical form for z. For the more general case $\langle a,b; a = ba^N ba\rangle$ see D. A. Jackson, 1986.

2) <a,b; a = baaabbabaa>. There are two decompositions where Adjan's process goes on indefinitely: $\text{b}|\text{baaa}|\underline{\text{a}}$ and $\text{b}|\text{ba}|\text{baaa}|\underline{\text{a}}$. When stopping all

processes where one of these two decompositions appears and declaring the corresponding words not left divisible by a we make Adjan's pseudo-algorithm into an algorithm. We do not give a proof here; a proof that this can be done for all presentations of the type studied in this paragraph, provided one side of the relation is a single letter (monadic case) will be given elsewhere.

Utilizing previous results of his (see Oganesjan 1978) and results of Sarkisjan, 1981, Oganesjan showed in 1982 that the left divisibility problem (hence the word problem) for semigroups presented by <a,b; a = bwa> is decidable.

References

S. I. Adjan, 1966, Defining relations and algorithmic problems for groups and semigroups, Tr. Mat. Inst. Steklov, 85, 1-124 (Russian), Am. Math. Soc. Transl., 152, 1967.

S. I. Adjan, 1977, On word transformations in semigroups given by systems of defining relations, Algebra i Logika, 15, no. 6, 611-621 (Russian).

S. I. Adjan, G. U. Oganesjan, 1978, On the problems of equality and divisibility in semigroups with one defining relation, Izv. Akad. Nauk SSSR, Ser. Mat., 42, no. 2, 219-225 (Russian).

James Howie, S. J. Pride, 1986, The word problem for one-relator semigroups, Math. Proc. Cambridge Phil. Soc. 99, 33-44.

D. A. Jackson, 1986, Some one-relator semigroup presentations with solvable word problems, Math. Proc. Cambridge Phil. Soc., 99, 433-434.

G. Lallement, 1974, On monoids presented by a single relation, J. of Algebra, 32, 370-388.

W. Magnus, 1932, Das Identitäts - Problem für Gruppen mit einer definierenden Relation, Math. Ann. 106, 295-307.

A. A. Markov, 1947, On the impossibility of certain algorithms in the theory of associative systems, Dokl. Akad. Nauk, 55, 587-590 and 58, 353-356 (Russian).

J. Matyasevitch, 1967, Simple examples of unsolvable associative calculi, Dokl. Akad. Nauk, 173, 1264-1266; Tr. Mat. Inst. Steklov, 93, 50-88 (Russian).

G. U. Oganesjan, 1978, On the problems of equality and divisibility in semigroups with defining relations of the form a = bA, Izv. Akad. Nauk SSSR, Ser. Mat., 42, no. 3, 602-612 (Russian).

G. U. Oganesjan, 1982, On semigroups with one relation and semigroups without cycles, Izv. Akad. Nauk SSSR, Ser. Mat., 46, no. 1, 88-94 (Russian).

E. L. Post, 1947, Recursive unsolvability of a problem of Thue, J. of Symb. Logic, 12, 1-11.

J. H. Remmers, 1980, On the geometry of semigroup presentations, Adv. in Math., 36, 283-296.

O. A. Sarkisjan, 1981, On the problems of equality and divisibility in semigroups and groups without cycles, Izv. Akad. Nauk SSSR, Ser. Mat. 45, no. 6, 1424-1441 (Russian).

REMARKS ON ACTS AND THE LATTICE OF THEIR TORSION THEORIES

Dedicated with respect and gratitude to
Professor G. Pickert on his 70th birthday

Wilfried Lex
Technische Universität Clausthal
D-3392 Clausthal-Zellerfeld, Germany

A torsion theory for acts or semi-automata was developed in [3] and further investigated in [1], [2] and [4]. The study of lattices of torsion theories begun in [2] will be continued. In order to make this article self-contained we shall recall first the necessary definitions and results, I. Using the fact that the abstract classes of irreducible S-acts form a Boolean sublattice of the lattice t of torsion classes of the category of S-acts under consideration [2], theorem 2, we shall gain some structural properties of t, II, theorem 1 and corollary 1 and 2. As an application of these results we shall completely characterize the non-trivial simple abelian groups - among all groups - by the lattices of the torsion theories of their corresponding group acts in different categories, III, theorem 2.

I

Covering all possible applications - in algebra, e.g. permutation groups, or in automata theory - acts are treated in full generality: an *act* or a *semi-automaton*, more precisely an S-*act* or an S-*semi-automaton*, $A = {}_SA$ is a triple (S,A,δ) where $S\ (\neq \emptyset)$ and A are sets and δ is any mapping from $S \times A$ into A. - Thus an act turns out to be essentially a unary algebra and therefore what is understood by a *subact*, a *congruence*, a *homomorphism*, an *isomorphism* ($\simeq$) etc. should be clear. - We are going to sketch the torsion theory for acts as developed in [3].

For any set $\mathfrak{r} = \{A_\iota \mid \iota \in I\}$ of disjoint subacts $A_\iota = {}_SA_\iota$ $(\iota \in I)$ of an S-act A let $\kappa(\mathfrak{r})$ be the *Rees congruence* of A induced by $\mathfrak{a}$, i.e. the smallest congruence κ of A such that A_ι lies in a class of κ for every $\iota \in I$. We write $A \twoheadrightarrow B$ iff there is a Rees congruence ϱ with $B \simeq A/\varrho$, and $B \rightarrowtail A$ iff $B \simeq C$ for a subact C of A.

We shall work in a category $\mathbf{C}$ where the objects form a *universal class* $\mathbf{U}$ of S-acts, which means that $\mathbf{U}$ is *Rees closed*, i.e. closed under $\twoheadrightarrow$, and *hereditary*, i.e. closed under $\rightarrowtail$. The morphisms of $\mathbf{C}$ are finite compositions of the preorders $\twoheadrightarrow$ and $\rightarrowtail$. - For the following we will assume that $\mathbf{U}$ contains at least one non-empty S-act, thus containing already the class $\mathbf{0}$ of *trivial* S-acts, i.e. acts with at most one element.

A *torsion theory* of $\mathbf{C}$ is a pair $(\mathbf{T}, \mathbf{F})$ of subclasses of $\mathbf{U}$ - the *torsion class* $\mathbf{T}$ and the *torsionfree class* $\mathbf{F}$ - with the following properties:

$\mathbf{T} \cap \mathbf{F} = \mathbf{0}$,
$\mathbf{T}$ is Rees closed,
$\mathbf{F}$ is hereditary,

for every act A of $\mathbf{C}$ there are subacts $A_\iota \in \mathbf{T}$ with $\iota \in I$ and $A/_{\kappa(\{A_\iota \mid \iota \in I\})} \in \mathbf{F}$ for an appropriate index set I.

A subclass $\mathbf{A}$ of $\mathbf{U}$ is said to be *closed under extensions* iff for every $A \in \mathbf{U}$ and subacts $A_\iota \in \mathbf{A}$ with $\iota \in I$ and $A/_{\kappa(\{A_\iota \mid \iota \in I\})} \in \mathbf{A}$ also A lies in $\mathbf{A}$.

Using the operators Γ and Δ defined for any $\mathbf{A} \subseteq \mathbf{U}$ by

$$\Gamma\mathbf{A} = \{A \in \mathbf{U} \mid \forall B \in \mathbf{A}: A \twoheadrightarrow B \Rightarrow B \in \mathbf{0}\}$$

and dually

$$\Delta\mathbf{A} = \{A \in \mathbf{U} \mid \forall B \in \mathbf{A}: B \rightarrowtail A \Rightarrow B \in \mathbf{0}\}$$

we have a characterization of torsion theories:

(A): $(\mathbf{T}, \mathbf{F})$ with $\mathbf{T}, \mathbf{F} \subseteq \mathbf{U}$ is a torsion theory iff the following five properties hold:

a) $\mathbf{T}$ is Rees closed,

b) $\mathbf{T}$ is *inductive*, i.e. for every ascending chain $A_1 \subseteq \ldots \subseteq A_\iota \subseteq \ldots$, where $A_\iota = {}_S A_\iota$ with $\iota \in I$ are subacts of an act $A (\in \mathbf{U})$ and $A_\iota \in \mathbf{T}$ for $\iota \in I$, one has ${}_S(\bigcup_{\iota \in I} A_\iota) \in \mathbf{T}$,

c) $\mathbf{T}$ is closed under extensions,

d) $\mathbf{O} \subseteq \mathbf{T}$

e) $\mathbf{F} = \Delta\mathbf{T}$.

There is also a dual characterization with respect to $\mathbf{F}$, s. [3], theorem 3, p. 272.

Some idea of the multitude and structural dependences of these torsion theories is given by the following result proved as theorem 1 in [2]:

(B): a) The torsion classes and the trosionfree classes of $\mathbf{C}$ form complete lattices $\mathcal{t} = (\mathfrak{t}, \cap, \vee)$ and $\mathcal{f} = (\mathfrak{f}, \cap, \vee)$ with

$$\mathbf{C} \vee \mathbf{D} \doteqdot \Gamma(\Delta\mathbf{C} \cap \Delta\mathbf{D}) \qquad (\mathbf{C}, \mathbf{D} \in \mathfrak{t})$$

and

$$\mathbf{C} \vee \mathbf{D} \doteqdot \Delta(\Gamma\mathbf{C} \cap \Gamma\mathbf{D}) \qquad (\mathbf{C}, \mathbf{D} \in \mathfrak{f})$$

respectively.

b) The torsion theories of $\mathbf{C}$ form a complete sublattice $\mathcal{l} \doteqdot (\mathfrak{l}, \wedge, \vee)$ of $\mathcal{t} \otimes \mathcal{f}'$ - where $\mathcal{f}'$ is the dual lattice to $\mathcal{f}$ -, i. e.

$$\begin{aligned} (\mathbf{C}, \mathbf{D}) \wedge (\mathbf{T}, \mathbf{F}) &= (\mathbf{C} \cap \mathbf{T},\ \mathbf{D} \vee \mathbf{F}) \\ (\mathbf{C}, \mathbf{D}) \vee (\mathbf{T}, \mathbf{F}) &= (\mathbf{C} \vee \mathbf{T},\ \mathbf{D} \cap \mathbf{F}), \end{aligned} \qquad ((\mathbf{C}, \mathbf{D}), (\mathbf{T}, \mathbf{F}) \in \mathfrak{l})$$

c) $\mathcal{t} \simeq \mathcal{f}' \simeq \mathcal{l}$.

II

The lattice $\mathcal{t}$ of (B) is in general rather complicated but $\mathcal{t}$ has some pleasant sublattices; examples require further particulars.

Let $\mathbf{A}$ be a subclass of $\mathbf{U}$, the universal class of all S-acts of $\mathbf{C}$: let us define $\overline{\mathbf{A}} \doteqdot \mathbf{U} \setminus \mathbf{A}$ and $\mathbf{A}' \doteqdot \mathbf{A} \setminus \mathbf{O}$; further, $\mathbf{A}$ is called *abstract* iff $\mathbf{O} \subseteq \mathbf{A}$ and $\mathbf{A}$ is isomorphically closed. - An act $A = {}_{S}A$ is said to be *irreducible* iff ${}_{S}\emptyset$ and A are the only subacts of A, and *reducible* otherwise. - Let $\mathbf{I}$ be the class of irreducible acts of $\mathbf{C}$ and $\mathfrak{a}$ the class of abstract subclasses of $\mathbf{I}$; moreover

$$a \doteqdot (\mathfrak{a}, \cap, \cup, \hat{\ }, \mathbf{I}, \mathbf{O})$$

with $\hat{\mathbf{A}} \doteqdot (\mathbf{I} \setminus \mathbf{A}) \cup \mathbf{O}$ for $\mathbf{A} \in \mathfrak{a}$,

$$b \doteqdot (\mathfrak{b}, \cap, \cup, \check{\ }, \mathbf{U}, \mathbf{I}')$$

with $\mathfrak{b} \doteqdot \Gamma\mathfrak{a}$ and $\check{\mathbf{B}} = \overline{\mathbf{B}'} \cap \mathbf{I}$ for $\mathbf{B} \in \mathfrak{b}$, and

$c \leftrightharpoons (\mathfrak{c}, \cap, \cup, \sim, \mathbf{U}, \mathbf{O})$

with $\mathfrak{c} \leftrightharpoons \mathfrak{a} \cup \mathfrak{b}$ and $\widetilde{\mathbf{C}} \leftrightharpoons \begin{Bmatrix} \Gamma \mathbf{C} \\ \Gamma^{-1} \mathbf{C} \end{Bmatrix}$ for $\mathbf{C} \in \begin{Bmatrix} \mathfrak{a} \\ \mathfrak{b} \end{Bmatrix}$.

Herewith we can formulate:

Proposition: Let t denote the lattice of torsion classes of $\mathbf{C}$ (cf. (B)):

a) If $\mathfrak{a} \neq \mathfrak{b}$ then $\mathfrak{a} \cap \mathfrak{b} = \emptyset$.

b) a, b and c are Boolean algebras.

c) $\gamma \leftrightharpoons \Gamma|\mathfrak{a}$ is an isomorphism of a onto the dual of b and so $a \simeq b$.

d) $(\mathfrak{a}, \cap, \cup) \simeq (\mathfrak{b}, \cap, \cup)$ and $(\mathfrak{c}, \cap, \cup)$ are complete atomistic Boolean sublattices of t.

Proof: a was regocnized as a complete atomistic Boolean algebra of torsion classes in [2], III, proof of th. 2, and $(\mathfrak{a}, \cap, \cup)$ as a sublattice of t, l. c.

As already stated in [3], ex. 8, p. 275, every $\mathbf{A} \in \mathfrak{a}$ is clearly also a trosionfree class and hence $\Gamma \mathbf{A}$ a torsion class; moreover $\Gamma \mathbf{A} = \bar{\mathbf{A}} \cup \mathbf{O}$ which implies the bijectivity of γ (onto $\mathfrak{b}$). - It is easily verified that γ is a homomorphism of a onto the dual of b, which reveals b as an isomorphic image of a and thus as a complete atomistic Boolean algebra too.

Since $\mathbf{A} \vee \mathbf{B}$ is the smallest torsion class containing $\mathbf{A} \cup \mathbf{B}$ we get $\mathbf{A} \vee \mathbf{B} = \mathbf{A} \cup \mathbf{B}$ for $\mathbf{A}, \mathbf{B} \in \mathfrak{b}$, hence $(\mathfrak{b}, \cap, \cup) \leq t$, and so all assertions - except for a) - concerning a and b are proven.

Let $\mathfrak{a} \neq \mathfrak{b}$: if $\mathfrak{b} \subseteq \mathfrak{a}$ then for every $\mathbf{A} \in \mathfrak{a}$ one has $(\mathbf{U} \setminus \mathbf{A}) \cup \mathbf{O} = \gamma \mathbf{A} \subseteq \mathbf{I}$, thus $\gamma \mathbf{A} = \hat{\mathbf{A}}$ and $\gamma(\gamma \mathbf{A}) = \hat{\hat{\mathbf{A}}} = \mathbf{A}$, therefore $\mathfrak{b} = \mathfrak{a}$ which is impossible. Hence $\mathfrak{b} \setminus \mathfrak{a} \neq \emptyset$ and there is a reducible act R in $\mathbf{U}$; for every $\mathbf{B} \in \mathfrak{b}$ we obtain

(1) $\mathbf{B} \supseteq \mathbf{I}' = \gamma \mathbf{I} = (\mathbf{U} \setminus \mathbf{I}) \cup \mathbf{O} \ni R$,

so $R \in \mathbf{B}$, for every $\mathbf{A} \in \mathfrak{a}$ however $R \notin \mathbf{A} \subseteq \mathbf{I}$ and thus $\mathfrak{a} \cap \mathfrak{b} = \emptyset$.

Finally we have to show that c is indeed a Boolean algebra with the asserted properties; let $\mathbf{A}, \mathbf{B} \in \mathfrak{c}$.

$\mathfrak{c}$ is closed under $\cap$: the cases $\mathbf{A}, \mathbf{B} \in \mathfrak{a}$ and $\mathbf{A}, \mathbf{B} \in \mathfrak{b}$ are trivial in view of the first part of b). So without loss of generality let $\mathbf{A} \in \mathfrak{a}$ and $\mathbf{B} \in \mathfrak{b}$; then there is a $\mathbf{D} \in \mathfrak{a}$ with $\gamma \mathbf{D} = \mathbf{B}$ and, taking $\mathbf{O} \subseteq \mathbf{A} \subseteq \mathbf{I}$ into account, this gives

$$\mathbf{A} \cap \mathbf{B} = \mathbf{A} \cap (\overline{\mathbf{D}} \cup \mathbf{O}) = (\mathbf{A} \cap \overline{\mathbf{D}}) \cup \mathbf{O} = (\mathbf{A} \cap (\mathbf{I} \cap \overline{\mathbf{D}})) \cup \mathbf{O} = \mathbf{A} \cap \hat{\mathbf{D}} \in \mathfrak{a} \subseteq \mathfrak{c},$$

and

(2) $\mathbf{A} \cap \mathbf{B} = \mathbf{A} \cap \widehat{\gamma^{-1}\mathbf{B}} \quad (\mathbf{A} \in \mathfrak{a},\ \mathbf{B} \in \mathfrak{b})$.

Similarly one obtains

(3) $\mathbf{A} \cup \mathbf{B} = \check{\gamma\mathbf{A}} \cup \mathbf{B} \quad (\mathbf{A} \in \mathfrak{a},\ \mathbf{B} \in \mathfrak{b})$

and thus $\mathfrak{c}$ is also closed under $\cup$.

(2) and (3) yield $\mathbf{C} \cap \tilde{\mathbf{C}} = \mathbf{O}$ and $\mathbf{C} \cup \tilde{\mathbf{C}} = \mathbf{U}$ for every $\mathbf{C} \in \mathfrak{c}$, resp., and therefore $\mathfrak{c}$ is a Boolean algebra and $(\mathfrak{c}, \cap, \cup)$ a Boolean sublattice of t, which moreover - in view of (2) and (3) - is complete and in addition atomistic.

Using the proposition we obtain

<u>Theorem 1</u>: Let t be the torsion class lattice of $\mathbf{C}$ and $\mathbf{U}$ contain n non-isomorphic non-trivial irreducible acts - where n denotes any cardinality - and

a) no reducible acts. Then $t \simeq (\mathfrak{a}, \cap, \cup)$ and thus t is a complete atomistic Boolean lattice of order 2^n.

b) at least one reducible act. Then t contains a complete atomistic Boolean sublattice of order 2^{n+1}.

<u>Proof</u>:

a) The first part of assertion a) immediately follows from the proposition. Since the atoms of t are exactly the classes containing one non-trivial act, its isomorphic images, and $\mathbf{O}$ - cf. [2], III, proof of th. 2 -, the order of t is 2^n.

b) Let R be a reducible act of $\mathbf{C}$. By (1), proof of the proposition, we have - as there - $\mathfrak{a} \cap \mathfrak{b} = \emptyset$ and hence by the proposition b, c and a) above the assertion.

In order to give some application of theorem 1 we have to recall the following definitions and facts. Let $A = {}_SA$ and $B = {}_SB$ be S-acts.

An S-act $_S D$ is said to be *decomposable* iff it possesses non-empty subacts A and B with $A \cup B = D$ and $A \cap B = \emptyset$, and otherwise *indecomposable*.

Let us call an element a of A a *trap* (of A) iff $Sa (= \{\sigma a | \sigma \in S\}) = \{a\}$. - A is a *reset* - or more precisely an S-*reset* - iff every element of A is a trap. - The class $\mathbf{S}$ of all S-resets of $\mathbf{C}$ is a torsion class, cf. [3], ex. 5 a, p. 274.

Now we can state

<u>Corollary 1:</u> If $\mathbf{U}$ contains n - arbitrary cardinality - non-isomorphic irreducible acts which are not trivial and at least one decomposable act which is not a reset then the order of the torsion class lattice $\mathfrak{t}$ of $\mathbf{C}$ is at least $2^{n+1}+1$; and this bound is in general the best possible.

<u>Proof:</u> Let $\mathbf{U}$ fulfill the above mentioned requirements. - Since a decomposable act is *a fortiori* reducible we can apply th. 1 b which yields $|\mathfrak{t}| \geq 2^{n+1}$.

$\mathbf{U}$ contains a decomposable act D which is not a reset. Since $\mathbf{U}$ is Rees closed there is an act $E = {}_S E$ in the class $\mathbf{S}$ of resets of $\mathbf{C}$ with $D \twoheadrightarrow E$ and $|E| > 1$, hence $\mathbf{S} \neq \mathbf{0}$.

On the other hand we have $D \in \mathbf{I}^{\mathsf{T}}$ as in (1), proof of the proposition, and of course

$$\mathbf{S} \subseteq (\mathbf{U} \setminus \mathbf{I}) \cup \mathbf{0} = \mathbf{I}^{\mathsf{T}},$$

thus $\mathbf{0} \subsetneqq \mathbf{S} \subsetneqq \mathbf{I}^{\mathsf{T}}$ and hence $\mathbf{S} \notin \mathfrak{a} \cup \mathfrak{b} = \mathfrak{c}$ which proves $|\mathfrak{t}| > 2^{n+1}$.

That this bound is the best possible shows for n = 1 already the pentagon of torsion classes for a group of prime order acting as a permutation group within an appropriate category $\mathbf{C}$, s. III, th. 2 b.

Moreover we obtain

<u>Corollary 2:</u> If $\mathbf{C}$ has at least a non-trivial irreducible act and a decomposable one which is no reset, then the lattice $\mathfrak{t}$ of torsion classes of $\mathbf{C}$ is not modular.

Proof: By the proof of Cor. 1 we have the five - distinct! - elements **U**, **O**, **I**, **S**, and **T** of t, which form a sublattice of t, namely a pentagon; hence t cannot be modular.

III

For the application mentioned in the introduction we have to recall some notations and definitions.

Let $\mathbb{P}$ denote the set of primes. - The *order* of an S-act (S,A,δ) is $|A|$. - We will call a G-act $A = {}_GA$ a *group act* iff $G = (G,\cdot,\varepsilon)$ is a group with unit ε such that $\varepsilon a = a$ and

$$\alpha(\beta a) = (\alpha\beta)a \qquad (\alpha,\beta \in G)$$

for all $a \in A$; then A is said to be a *G-act* and we will write also ${}_GA$ for A. As usual we call Ga with $a \in A$ an *orbit*, more precisely a *G-orbit of a on A*; it is *maximal* iff

$$\alpha a = \beta a \Rightarrow \alpha = \beta \qquad (\alpha,\beta \in G).$$

As in I we will use **O** and **U** for the class of trivial and of all S-acts of our category $\mathfrak{C}$, resp., and t for the lattice of torsion classes of $\mathfrak{C}$ (cf. I, (B)). We shall denote the class of reducible and trivial acts of $\mathfrak{C}$ by **R** and that of irreducible ones by **I**, the class of resets of $\mathfrak{C}$ by **S** and the class of those acts of $\mathfrak{C}$ with at least one trap together with ${}_S\emptyset$ by **T**. - Let $\mathbb{B}_n$ stand for the class of Boolean lattices of order n ($\in \mathbb{N}$).

Generalizing - and slightly correcting - theorem 2 of [2] we obtain a full characterization of the non-trivial simple abelian groups, i.e. groups of prime order, - among all groups - via the lattice of torsion theories in various categories.

Theorem 2: Let G be a group of order g - finite or infinite - and $\mathfrak{C}$ a category of G-acts such that **U** contains

a) all acts of order m with $m \leq 2g$. Then $g \in \mathbb{P}$ iff t has the form

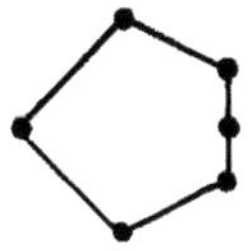

b) no act with two maximal orbits but all (other) reducible acts of order m with $m \leq g+1$. Then $g \in \mathbb{P}$ iff $\mathcal{L}$ is a pentagon.

c) no reducible act with a maximal orbit but all (other) acts of order m with $m \leq g$. Then $g \in \mathbb{P}$ iff $\mathcal{L} \in B_4$.

The different possiblities are shown in the following diagram:

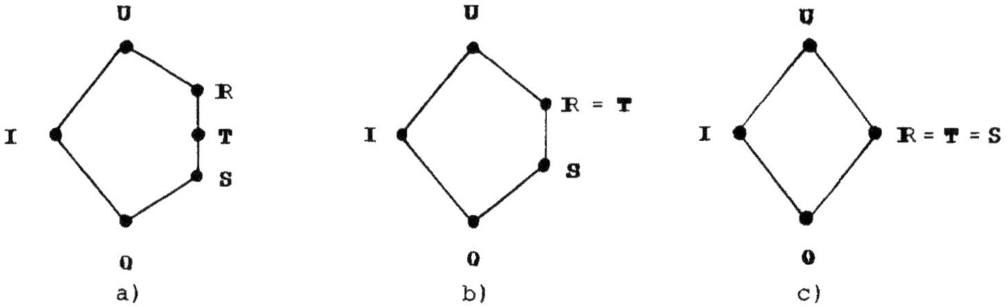

Proof: Let $G = (G,\cdot,e)$ be any group with $g \neq |G|$ and $\mathbf{C}$ a category of G-acts such that the universal class $\mathbf{U}$ of its objects fulfills the above mentioned conditions.

The acts of $\mathbf{C}$ are *fully reducible*, i.e. their supports are disjoint unions of the G-orbits on them. - By definition we have

(4) $\quad \mathbf{O} \subseteq \mathbf{S} \subseteq \mathbf{T} \subseteq \mathbf{R} \subseteq \mathbf{U}$;

moreover, these classes and $\mathbf{I}$ are torsion classes (cf. the proposition, the remark before cor. 1 and [3], exs. 6,7, p. 275).

1. Let $g \in \mathbb{P}$. The irreducible G-acts are either trivial or of order g, and then isomorphic to each other. - Let $\mathbf{U}$ contain

   a) all acts of order m with $m \leq 2g$. Then there is in $\mathbf{U}$ an act $A = {}_G A$ with 2 irreducible subacts $B = {}_G B$ and $C = {}_G C$ of order g, hence $A = B \cup C$ and $B \cap C = \emptyset$.

   Since $\mathbf{U}$ is hereditary we have $B \in \mathbf{U}$. Thus $g > 1$ implies $B \in \mathbf{I} \setminus \mathbf{O}$. - As $\mathbf{U}$ is also Rees closed there are in $\mathbf{U}$ an act $T (\in \mathbf{T})$ of order g+1 with $A \twoheadrightarrow T$ and an act $S (\in \mathbf{S})$ of order 2 with $A \twoheadrightarrow S$. Taking into account $A \in \mathbf{R} \setminus \mathbf{T}$ and (4) we have shown

$$\mathbf{O} \subsetneqq \mathbf{S} \subsetneqq \mathbf{T} \subsetneqq \mathbf{R} \subsetneqq \mathbf{U}$$

   and hence the asserted diagram because one can easily see that there are no more torsion classes in $\mathbf{U}$.

b) no act with 2 maximal orbits but all reducible acts of order m with $m \leq g+1$. Then every reducible act has a trap, i. e. $\mathbf{R} \subseteq \mathbf{T}$, and by (4) hence $\mathbf{R} = \mathbf{T}$.

Analogously to a) one obtains

$$\mathbf{0} \subsetneqq \mathbf{S} \subsetneqq \mathbf{R} = \mathbf{T} \subsetneqq \mathbf{U}$$

and thus the asserted pentagon.

c) no reducible act with a maximal orbit but all acts of order m with $m \leq g$. Then every reducible act is a reset, i. e. $\mathbf{R} \in \mathbf{S}$, and by (4) therefore $\mathbf{R} = \mathbf{T} = \mathbf{S}$.

On the other hand the irreducible act ${}_G G = (G,G,\cdot)$ has order g and thus lies in $\mathbf{U}$. So $g > 1$ implies ${}_G G \in \mathbf{I} \setminus \mathbf{0}$ and one obtains $\mathcal{t} \in \mathbf{B}_4$ as, besides $\mathbf{0}, \mathbf{I}, \mathbf{U}$ and $\mathbf{S}$, no other torsion classes are in $\mathbf{U}$.

2\. We shall prove the other direction by contraposition: thus let $g \notin \mathbb{P}$.

If $g = 1$ one has in all 3 cases $\mathcal{t} \in \mathbf{B}_2$ since there are only resets.

So let us assume $g \notin \mathbb{P} \cup \{1\}$. Then G possesses a proper subgroup $(H,\cdot)$. - Since $\mathbf{U}$ contains in any case all acts of order m with $m \leq g$ there are at least 2 non-trivial non-isomorphic irreducible acts, e. g. ${}_G G$ and ${}_G N$ with $N = \{xH \mid x \in G\}$ and G operating on N in the usual way. Indeed ${}_G G$ and ${}_G N$ are not isomorphic, even not for $|N| = g$, since one has for every $a \in G$ in the first case

$$\forall x \in G \ (xa = a \Leftrightarrow x = e),$$

in the second however

$$\forall x \in G \ (xH = H \Leftrightarrow x \in H)$$

thus x is not uniquely determined because $|H| > 1$. As $\mathbf{U}$ contains at least one reducible act th. 1 b therefore gives the existence of a sublattice of $\mathcal{t}$ of order 2^3, which shows that $\mathcal{t}$ is not of the asserted form.

The conditions in theorem 2 can be slightly weakened if one is concerned with one direction only:

Corollary 3: Let G be a group of prime order and let $\mathbf{U}$ - as a category of G-acts - contain

a) an act with two maximal orbits: t has the form

b) no act with two maximal orbits but a reducible act with exactly one maximal orbit: t is a pentagon.

c) no reducible act with a maximal orbit: t is a Boolean lattice of order 4 at the most. In more detail: let $\mathbf{U}$ contain an act with

1. a maximal orbit and
   α) a non-trivial reset: $t \in \mathbf{B}_4$.
   β) no non-trivial reset: $t \in \mathbf{B}_2$.

2. no maximal orbit and
   α) a non-trivial reset: $t \in \mathbf{B}_2$.
   β) no non-trivial reset: $t \in \mathbf{B}_1$.

In the cases a), b) and c1α) the diagrams for t are as in theorem 2 and otherwise:

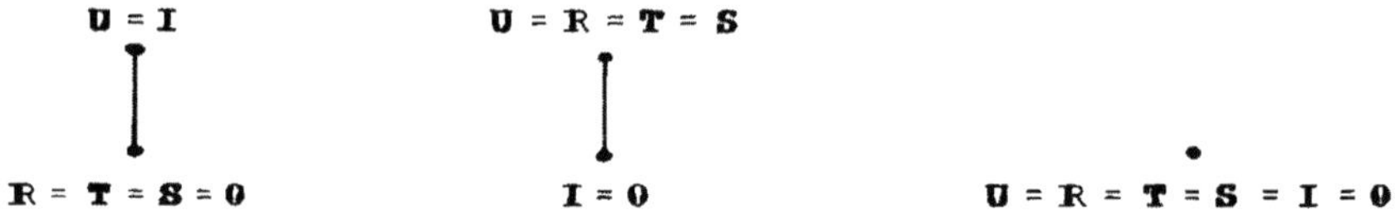

c1β)   c2α)   c2β)

The proof is similar to that of th. 2 or straight forward.

REFERENCES

[1] Amin, I.A., Wiegandt, R.: Torsion and torsionfree classes of acts. Contributions to General Algebra 2, Proc. Klagenfurt Conf. 1982. Wien, Stuttgart, 1983; 19 - 34.

[2] Lex, W.: Lattices of Torsion Theories for Semi-Automata. To appear in the Proc. of the Chico Conf. on Semigroups in 1986. D. Reidel, 1987.

[3] -, Wiegandt, R.: Torsion theory for acts. Studia Sci. Math. Hungar., 16(1981), 263 - 280.

[4] Veldsman, S., Wiegandt, R.: On the existence and non-existence of complementary radical and semisimple classes. Quaest. Math., 7(1984), 213 - 224.

Relativized Star-Free Expressions, First-Order Logic, and a Concatenation Game

D. Lippert, W. Thomas

Lehrstuhl für Informatik II, Büchel 29-31, D-5100 Aachen

Abstract : Star-free expressions with an additional constant for some fixed language are considered. In contrast to the well-known equivalence between star-free expressions and first-order logic (over finite orderings), it is shown here that in the relativized version star-free expressions are strictly weaker than the corresponding first-order formulas. For the proof, a concatenation game is introduced which captures the expressive power of the relativized star-free expressions.

1. Introduction and Statement of Result

A language $L \subseteq A^*$ over a finite alphabet A is called *star-free* if it can be constructed from finite sets of words using boolean operations and the concatenation product. The star-free languages form a natural and important proper subclass of the class of regular languages (which is generated when the star operation is added). Interesting connections to semigroup theory and to mathematical logic have been established by two fundamental theorems of Schützenberger and McNaughton. By Schützenberger's Theorem ([Sch 65]), a regular language is star-free iff its syntactic monoid contains only trivial groups; and McNaughton's Theorem ([McNP 71]) states that the star-free languages are precisely those languages that are definable in first-order logic (using an appropriate signature). In recent years these connections between formal language theory, semigroup theory and logic (also concerning systems of temporal logic) have been further worked out, especially in the context of a structure theory of star-free languages (cf. [CBrz 71], [Str 81], [P 84], [Th 82], [Th 84]).

The present paper is concerned with the relation between star-free expressions (denoting star-free languages) and first-order formulas. If an alphabet $A = \{a_1, \ldots, a_k\}$ is given, the *star-free expressions* over A are constructed in the standard way from constants $\emptyset, \varepsilon, a_1, \ldots, a_k$ (denoting the languages $\emptyset, \{\varepsilon\}, \{a_1\}, \ldots, \{a_k\}$, respectively), using the symbols $\sim$ (for complement w.r.t. A^*), $\cup$ (for union) and $\cdot$ (for concatenation). We also admit $\cap$ (for intersection) and write A^* instead of $\sim \emptyset$. The language determined in this way by the star-free expression e is denoted $L(e)$; sometimes we shall say that a word w "satisfies" e if $w \in L(e)$.

For the description of properties of words by *first-order formulas* we view any nonempty word w of length n as a finite linearly ordered model of the form $(\{1, \ldots, n\}, <, P_1, \ldots, P_k)$ where $P_i \subseteq \{1, \ldots, n\}$ is defined by : $m \in P_i$ iff the m-th letter of w is a_i. If a sentence φ of the corresponding first-order language is true in such a "word model" that represents the word w, we write $w \models \varphi$. In order to include the empty word as a possible model, we adopt the convention that sentences $\exists x \varphi$ are false and sentences $\forall x \varphi$ are true in the empty model.

Let us consider an example: The language $a_1{}^* a_2 a_1{}^*$, which is denoted over $A = \{a_1, a_2\}$ by the star-free expression

$$e_1 = \sim(A^* \cdot a_2 \cdot A^*) \cdot a_2 \cdot \sim(A^* \cdot a_2 \cdot A^*)$$

is defined by the first-order sentence

$$\varphi_1 = \exists x\big(P_2 x \wedge \forall y((y < x \vee x < y) \rightarrow P_1 y)\big).$$

Sometimes it is convenient to allow further (definable) symbols in order to facilitate the formulation of properties. For this purpose we use the constants *min* and *max* for the first resp. last position of a word, and the successor function *succ* and predecessor function *pred* over the ordering $<$ (where $succ(max) = max$ and $pred(min) = min$). For example, the language $a_1 a_1 a_2{}^*$, defined by the star-free expression

$$e_2 = a_1 \cdot a_1 \cdot \sim(A^* \cdot a_1 \cdot A^*)$$

can be conveniently represented in this extendend first-order language by the sentence

$$\varphi_2 = P_1 min \wedge (min < succ(min)) \wedge P_1 succ(min) \wedge \forall y(succ(min) < y \rightarrow P_2 y).$$

McNaughton's Theorem (stated in [McNP 71, Ch.10] in a somewhat different terminology) says that a language $L \subseteq A^*$ is star-free (i.e. represented by some star-free expression) iff $L = L(\varphi)$ for some first-order sentence φ.

It turned out that the connection between star-free expressions and first-order formulas is much tighter than indicated by McNaughton's result. This closer relationship is visible when comparing standard classifications of star-free expressions and first-order formulas. A star-free expression is of *dot depth n* if it involves n levels of concatenation (alternating with boolean operations). Formally, a star-free expression over $A = \{a_1, \ldots, a_k\}$ is of dot depth 0 if it is a boolean combination of $\emptyset$, ε, $a_1, \ldots, a_k$, and a star-free expression is of dot depth $n+1$ if it is a boolean combination of expressions $e_1 \cdot \ldots \cdot e_m$ $(m > 2)$ where each e_i is of dot depth n. The language classes $B_0, B_1, \ldots$, where $L \in B_n$ iff L is defined by a star-free expression of dot depth n form the *Brzozowski hierarchy* ([CBrz 71]). This hierarchy was shown to be strict in [BrzK 78], [Str 81], [Th 84]. On the other hand, one can classify first-order sentences (assumed to be transformed into prenex normal form) by their *quantifier alternation depth*, i.e. the number of quantifier blocks consisting of existential or of universal quantifiers in their quantifier prefix. By the main result of [Th 82], a language $L \subseteq A^*$ belongs to the n-th level of the Brzozowski hierarchy iff it is definable by a first-order sentence of quantifier alternation depth n (in the extended signature with *min*, *max* etc.), provided $n > 0$. (As an illustration consider the above expression e_1, which is of dot depth 2, and the corresponding sentence φ_1 whose prenex normal form is

$$\exists x \forall y(P_2 x \wedge (y < x \vee x < y \rightarrow P_1 y)).$$

This sentence is of quantifier alternation depth 2.) The correspondence between dot depth and quantifier alternation depth can be kept also for the restricted signature (without *min*, *max* etc.). In this case one considers the product operation $L \cdot a \cdot L'$ (where $a \in A$) instead of the concatenation product $L \cdot L'$ and otherwise proceeds as before. The resulting concatenation depth hierarchy of languages is the *Straubing hierarchy* of [Str 81]; in [PP 86] it is shown that for $(n > 0)$ its n-th level contains precisely the languages definable by first-order sentences of quantifier alternation depth n in the restricted signature.

The main point in these equivalence results is the correspondence between the star-free operations $\cup$, $\sim$, $\cdot$ and the logical connectives $\vee$, $\neg$, $\exists$. In view of this close relationship one may be led to think that the star-free expressions and first-order sentences are more or less "the same". The present paper shows that this impression is misleading: We prove that in a slightly more general context star-free expressions are strictly weaker than first-order formulas. In particular, this will clarify an essential difference between the dot for concatenation product and the existential quantifier (a difference which is hidden when we refer to "pure" star-free expressions).

The generalization to be considered here is given by the "relativization" of star-free expressions, where we adjoin a constant L for some fixed (usually non-star-free) language to the constants $\emptyset$, ε, $a_1, \ldots, a_k$. We call the resulting expressions *star-free L-expressions*. The corresponding extension of the first-order language consists in allowing additional atomic formulas of the form $[x,y] \in L$, meaning that "the segment from position x to position y is in L". We call sentences of this extended language *first-order L-sentences*. For instance, taking the language $L = b(aa)^*b$ over $A = \{a,b\}$, we may define $b((aa)^*b)^*$ by the star-free L-expression

$$b \cdot A^* \cap A^*b \cap \sim (A^* \cdot (ba^*b \cap \sim L) \cdot A^*)$$

where a^* abbreviates $\sim (A^* \cdot b \cdot A^*)$. A corresponding first-order L-sentence (using the predicate symbols P_a and P_b) is the following :

$$P_b min \wedge P_b max \wedge \ \forall x \forall y (x < y \wedge P_b x \wedge P_b y \wedge \forall z (x < z \wedge z < y \rightarrow P_a z) \ \rightarrow \ [x,y] \in L).$$

We denote the classes of languages definable by star-free L-expressions and first-order L-sentences by $SF(L)$ and $FO(L)$, respectively. By induction over the star-free L-expressions it is easy to verify that $SF(L) \subseteq FO(L)$. (Namely, formulas corresponding to $\emptyset$, ε, a_i, L are, respectively, $\exists x\ x < x$, $\forall x\ x < x$, $min = max \wedge P_i min$, $[min, max] \in L$. Furthermore, complement and union are simply definable using $\neg$ and $\vee$. Finally, given L and L' defined by φ and φ', $L \cdot L'$ is defined by $\exists x(\tilde{\varphi}(min,x) \wedge \tilde{\varphi}'(succ(x), max))$ where $\tilde{\varphi}(min,x)$ stands for the relativization of φ to the elements $\leq x$ and $\tilde{\varphi}'(succ(x), max)$ for the relativization to the elements $\geq succ(x)$.)

The main result of this paper says that the converse inclusion fails in general :

Theorem 1 There is a regular language L_0 such that $SF(L_0)$ is strictly contained in $FO(L_0)$.

The example language L_0 and the general proof strategy will be given in section 2. A crucial step in the proof is the introduction of a "concatenation game" which will be a useful tool in verifying that two given words cannot be distinguished by certain star-free L-expressions. This game may be considered as a suitable modification of the Ehrenfeucht-Fraissé game of first-order logic, capturing the expressive power of star-free expressions instead of first-order formulas. The game and its application in the desired proof will be presented in section 3. Finally, in section 4 we show (in Theorem 2) that the game is indeed appropriate for characterizing indistinguishability by star-free L-expressions.

It should be mentioned that the motivation for studying relativized star-free expressions arose in the investigation of certain systems of *interval temporal logic*, in particular the propositional interval logic of Moszkowski ([MoMa 84], [Mo 83]) used for the specification and verification of hardware. The formulas of this logic are very similar to the relativized star-free expressions; they contain variables for "intervals" that have to be interpreted by words in given languages. A complete comparison between first-order logic, Moszkowski's interval logic, and star-free expressions is carried out in [Lip 86]. It is shown there that (with respect to expressive power) Moszkowski's system is located strictly between star-free expressions and first-order logic.

2. The Example Language L_0

In the sequel we work with the alphabet $A = \{a, b\}$. The language L_0 required in Theorem 1 is given by the following regular expression :

$$b(aa)^*b(aa)^*b((aa)^*ab)^*(aa)^*b(aa)^*b.$$

Let us call a segment of the form ba^nb a *section*; we say that this section is even (resp. odd) if n is even (resp. odd). So a word in L_0 starts with two even sections (overlapping by one b), continues by odd sections, and ends again with two even sections. If we indicate an even section by |——| and a sequence of odd sections by |~~~~| then a word in L_0 has the form

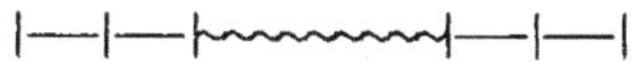 .

It follows from the characterization of star-free languages in terms of group-free monoids (or in terms of the "noncounting property", cf.[McNP 71]) that L_0 is not star-free.

Our task is to exhibit a language L_1 that belongs to $FO(L_0)$ but not to $SF(L_0)$. We let L_1 consist of all words which are built up from two L_0 words, overlapping by one section :

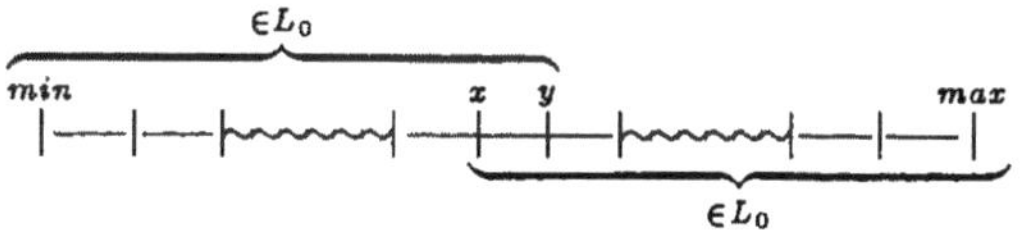

We give a formal definition of L_1 by presenting a first-order L_0-sentence defining L_1 :

$$\exists x \exists y([min, y] \in L_0 \wedge [x, max] \in L_0 \wedge x < y \wedge \forall z(x < z \wedge z < y \rightarrow P_a z)).$$

(Note that the segment from x to y has to be an even section by the requirement that $[min, y] \in L_0$ and $[x, max] \in L_0$.) So we have $L_1 \in FO(L_0)$, and it remains to show $L_1 \notin SF(L_0)$.

Before entering the proof it may be instructive to describe informally the essential defect of the concatenation dot as compared to the existential quantifier. In a formula such as $\exists x \exists y \varphi(x, y)$ we may have several independent clauses in $\varphi(x, y)$ involving the element x or y. This is well illustrated in the above formula defining L_1. However, in a star-free expression the reference to a "concatenation position" x is much more restricted since this reference is realized only by the concatenation dot, and this dot appears only once for each position where concatenation takes place. In the first-order description of the example language L_1, however, the positions x and y are both referenced twice : as a start, resp. end, of an L_0-segment, and as a start, resp. end, of a segment in ba^*b.

For the formal proof of Theorem 1 we classify L_0-expressions by their "concatenation complexity". Since the following definitions do not depend on the specific choice of L_0 we work in the sequel with an arbitrary language L instead. However, for technical convenience we assume that (over the alphabet $A = \{a_1, \ldots, a_k\}$)

(L) $\qquad L$ is nonempty and disjoint from $\{\varepsilon\}, \{a_1\}, \ldots, \{a_k\}$,

a condition which is clearly satisfied by L_0.

We define sets $E_0^L, E_1^L, \ldots$ of star-free L-expressions by :

E_0^L = set of boolean combinations of $\emptyset, \varepsilon, a_1, \ldots, a_k, L$,

E_{n+1}^L = set of boolean combinations of L-expressions $e \cdot e'$ with $e, e' \in E_n^L$.

Note that the sets E_n^L exhaust the set of all star-free L-expressions and that any language defined by an expression e in E_n^L is also defined by an expression in E_{n+1}^L (consider $e \cdot \varepsilon$). We say that two words w, w' are *n-equivalent* (and write $w \equiv_n^L w'$) if they are not distinguishable in E_n^L, i.e.

$$\forall e \in E_n^L : \quad w \in L(e) \Leftrightarrow w' \in L(e).$$

Our aim is to find for every $n > 0$ two words w_n , w'_n such that

$$(*) \qquad w_n \in L_1, \; w'_n \notin L_1, \; w_n \equiv_n w'_n.$$

This clearly suffices to verify $L_1 \notin SF(L_0)$. (Namely, if $L_1 \in SF(L_0)$, we could conclude for some n and some $e \in E_n$ that $L_1 = L(e)$. Thus *any* two words $w \in L_1, w' \notin L_1$ would be distinguished by $e \in E_n$; and hence we would not have $w \equiv_n w'$, contradicting (*).)

The desired words w_n, w'_n are easily defined. Using the picture representation as above, they are of the form

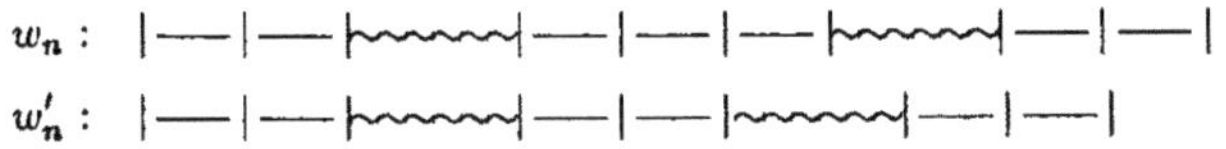

where in all cases an even section, indicated by |—|, is given by $ba^{2^{n+1}}b$ and a sequence of odd sections, indicated ⊢~~~⊣ , by $(ba^{2^{n+1}+1})^{2^{n+1}}b$.

Obviously we have $w_n \in L_1$ an $w'_n \notin L_1$. To complete the proof of the main result we have to show $w_n \equiv_n w'_n$. For this purpose it is convenient to use a game theoretical characterization of the relation $\equiv_n$. The next section introduces the underlying "concatenation game".

3. The Concatenation Game $\mathbf{C}_n^L(w, w')$

The game $C_n^L(w, w')$ is played by two players, called I and II, on the words w and w'. A play (or "match") of the game consists of n moves, each of them leading from a given configuration to a new configuration. A configuration after i moves ($i = 0, \ldots, n$) is given by a decomposition of both words w and w' into $i+1$ segments :

$$w = u_0 \ldots u_i \quad , \quad w' = u'_0 \ldots u'_i.$$

Assume $i < n$. In the $(i+1)$-th move, player I chooses some u_j (or u'_j) and splits it into two segments; player II has then to split the corresponding segment u'_j (or u_j) of the other word into two segments. A new configuration of the form

$$w = v_0 \dots v_{i+1} \quad , \quad w' = v'_0 \dots v'_{i+1}$$

will result. After n moves (starting with w and w' as initial configuration) we have reached a decomposition of the form

$$w = w_0 \dots w_n \quad , \quad w' = w'_0 \dots w'_n .$$

Player II has won this play if the correspondence $w_j \mapsto w'_j$ respects membership in $\{\varepsilon\}$, $\{a_1\},\dots,$ $\{a_k\}$, and L, i.e. we have, for $j = 0,\dots,n$,

$$\begin{aligned} w_j &= \varepsilon \quad \text{iff} \quad w'_j = \varepsilon, \\ w_j &= a \quad \text{iff} \quad w'_j = a \quad \text{for } a \in A, \\ w_j &\in L \quad \text{iff} \quad w'_j \in L \quad (\text{“}L\text{-clause”}). \end{aligned}$$

We say that *II wins* $C_n^L(w,w')$ if II has a winning strategy to win any play of $C_n^L(w,w')$. (It is not necessary here to define "winning strategy" formally, e.g. by introducing an appropriate function on configurations.) If II wins $C_n^L(w,w')$ we shall write $w \sim_n^L w'$. If in the winning condition for II the clause concerning the language L is dropped we speak of the game $C_n(w,w')$ and the relation $\sim_n$. In section 4 we shall show that $w \equiv_n^L w'$ holds iff $w \sim_n^L w'$. In the present section we note some elementary facts about the concatenation game and show $w_n \sim_n^{L_0} w'_n$ for the words w_n, w'_n defined in section 2.

As a first example consider $C_2(a^7, a^8)$. Player II does not win this game : Namely, player I can split a^8 into two segments a^4 in his first move. Player II will then create, while splitting a^7, at least one segment a^m with $m \leq 3$. Asume w.l.o.g. that this a^m is his first segment. Then I splits his first segment a^4 into two segments a^2 and will thus have created three segments none of which is ε or a. Now II, having to split a^m $(m \leq 3)$, has no possibility to achieve the same : any splitting of aaa or some shorter word into two pieces will generate some segment ε or a (and hence lead to a violation of the winning condition).

It should be clear from this discussion that player II will not run into such difficulties in the game $C_2(a^8, a^9)$; here II has "enough space" to react to any choice of player I. In general, with j moves ahead, II should simply guarantee that corresponding segments of the two given words are either identical or both of length $\geq 2^{j+1}$. Clearly this condition can be preserved during a play if it is true for the initial configuration. Hence we have

$$a^i \sim_n a^j \quad \text{if} \quad i,j \geq 2^{n+1}$$

(however, this relation does not hold if $i < 2^{n+1}$, $j \geq 2^{n+1}$).

A repetition of the above argument shows that on any two words of the form

$$(+) \qquad ba^{m_1}ba^{m_2}b\dots ba^{m_r}b \quad \text{with} \quad m_i \geq 2^{n+1},\ r \geq 2^{n+1},$$

player II wins the n move game (where we ignore the clause concerning L). The winning strategy just requires to play according to the above strategy *within* given sections, and to ensure that the number of full sections in corresponding segments of a configuration agree in the sense that, with j moves ahead, these numbers should be $\geq 2^{j+1}$ or else coincide. For brevity we call this strategy th "2^j-strategy".

If the L-clause in the winning condition is admitted then this strategy may of course fail : Consider the case $L = L_0$. Given a word w of the form (+) in L_0 and another such word w' outside L_0, player I may simply generate the trivial decompositions w, $w\varepsilon$, $w\varepsilon\varepsilon, \ldots$ whereupon II clearly has to react by generating w', $w'\varepsilon$, $w'\varepsilon\varepsilon, \ldots$ and thus will lose, since in the end the first segments will violate the L_0-clause.

However, this situation will obviously not arise if the given words w and w' do not contain any subsegments in L_0. In particular, we can conclude that for the words w_n and w'_n defined above

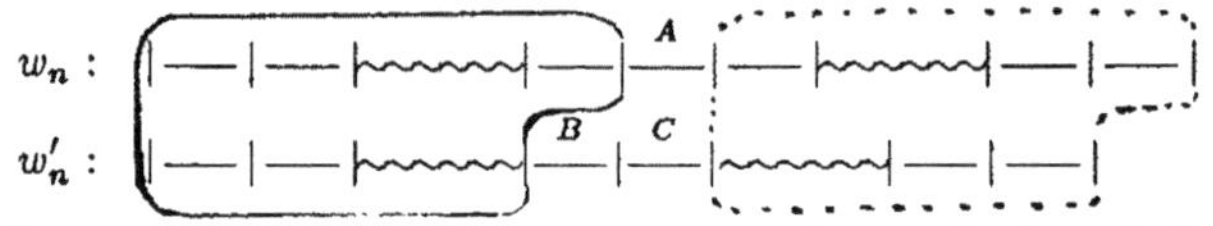

player II will win the n-move game on the two segments of w_n and w'_n encircled by the solid line even in the case that the L_0-clause is included in the winning condition. The same holds for the two segments encircled by the dotted line since again no L_0-word appears in these segments.

Based on these observations, it is easy to formulate a strategy which is a winning strategy for II in $C_n^{L_0}(w_n, w'_n)$. The strategy depends on the fact that in both w_n and w'_n there are only two segments that belong to L_0; moreover, creating one L_0-segment during a play of the game will automatically split (and thus destroy) the other L_0-segment. As a consequence, inside this latter segment player II can forget about the L_0-clause.

We describe the strategy for II by distinguishing the following cases after the first splitting proposed by I in his first move :

(1) I chose his first cut in the segments of w_n and w'_n encircled by the solid line (as given by the above picture), i.e. I destroyed either in w_n or in w'_n the left L_0-segment, but kept the right L_0-segment intact. Then II chooses his cuts in the rest of the play according to the 2^j-strategy on the segments w_n and w'_n encircled by the solid line. The two segments on the right of the solid line are identical; there player II simply copies all the splittings effected by I ("identity strategy").

(2) I chose his first cut in the segments enclosed by the dotted line of the above figure. This case is handled symmetrically to case (1).

(3) I chose his first cut outside the encircled segments in the above figure, i.e. destroyed both L_0-segments of the splitted word.

   (a) If the cut of player I occurs in section A, II reacts in the rest of the play by playing the 2^j-strategy on the two segments encircled by the solid line, and by playing the identity strategy on the two segments remaining on the right (which are the same).

   (b) If the cut of player I occurs in section B, II plays in the same fashion as in (a), by the 2^j-strategy on the left encircled parts and the identity strategy on the parts remaining on the right.

   (c) If the cut of player I occurs in section C, then a strategy defined symmetrically to (b) applies : II plays the 2^j-strategy on the parts encircled by the dotted line and by the identity strategy on the parts remaining to the left.

In all cases we obtain a winning strategy for player II. In particular, the L_0-clause in the winning condition will be met since on segments where it may turn out to be relevant, II uses the identity strategy. Hence we have $w_n \sim_n w'_n$ as desired.

4. Adequacy of the game

The purpose of this section is to show that the relations $\equiv_n^L$ and $\sim_n^L$ coincide :

Theorem 2 For any language $L \subseteq A^*$, satisfying condition (L) of section 2, any words $w,\ w' \in A^*$, and $n \geq 0$, we have

$$w \equiv_n^L w' \quad \text{iff} \quad w \sim_n^L w' \quad .$$

As a corollary of the proof we will also obtain that $w \equiv_n w'$ holds iff $w \sim_n w'$.

The *proof of Theorem 2* proceeds by induction on n. For $n = 0$ we deal with E_0^L-expressions and the 0-move game. Clearly the winning condition in $C_0^L(w, w')$ just says that w and w' satisfy the same expressions in E_0^L.

For the induction step we first state a simple property of $\sim_{n+1}^L$. In order to formulate it succinctly we write $\exists u_1 u_2 = u \ldots$ to mean $\exists u_1 \exists u_2 (u_1 u_2 = u \wedge \ldots)$ and $\forall u_1 u_2 = u \ldots$ to mean $\forall u_1 \forall u_2 (u_1 u_2 = u \rightarrow \ldots)$. We have

$$(*) \quad w \sim_{n+1}^L w' \quad \text{iff} \quad \begin{array}{l} (1)\ \forall\ w_1 w_2 = w\ \exists\ w'_1 w'_2 = w' : w_1 \sim_n^L w'_1 \wedge w_2 \sim_n^L w'_2 \\ \text{and}\ \ (2)\ \forall\ w'_1 w'_2 = w'\ \exists\ w_1 w_2 = w : w_1 \sim_n^L w'_1 \wedge w_2 \sim_n^L w'_2. \end{array}$$

(1) and (2) just formalize the condition that II wins the $(n+1)$-move game by reacting to I in his first move in such a way that he will win the remaining n-move games on the two generated pairs of words.

The inductive proof of the equality between $\equiv_n^L$ and $\sim_n^L$ will be completed if we have shown the analogue of $(*)$ for $\equiv_{n+1}^L$:

$$(**) \quad w \equiv_{n+1}^L w' \quad \text{iff} \quad \begin{array}{l} (1)\ \forall\ w_1 w_2 = w\ \exists\ w'_1 w'_2 = w' : w_1 \equiv_n^L w'_1 \wedge w_2 \equiv_n^L w'_2 \\ \text{and}\ \ (2)\ \forall\ w'_1 w'_2 = w'\ \exists\ w_1 w_2 = w : w_1 \equiv_n^L w'_1 \wedge w_2 \equiv_n^L w'_2. \end{array}$$

For the proof of $(**)$ it is useful to introduce a normal form of star-free L-expressions, similar to the distributive normal form of first-order formulas. Expressions in normal form will be disjunctions of expressions which we call *n-types*. The set of n-types will be denoted T_n (we drop the dependence of L in this notation). The sets T_n are defined inductively as follows :

- T_0 consists of the expressions $\varepsilon,\ a_1, \ldots, a_k,\ L$, and $\sim(\varepsilon \cup a_1 \cup \ldots \cup a_k \cup L)$.
- T_{n+1} consists of all satisfiable expressions of the form

$$\bigcap_{(e,e') \in T} e \cdot e' \cap \bigcap_{(e,e') \in \overline{T}} \sim(e \cdot e')$$

where $T \subseteq T_n \times T_n$ and $\overline{T} = (T_n \times T_n) - T$. (An expression e is satisfiable if $L(e) \neq \emptyset$.)

Thus an $(n+1)$-type determines the ways in which a word can be split into two segments satisfying two given n-types; the possibilities of splitting are captured by the set T. The following Lemma summarizes some basic facts about n-types :

Lemma For every $n \geq 0$:

(a) Each n-type is an expression in E_n^L.

(b) Each word satisfies exactly one n-type.

(c) ("Distributive normal form") Each satisfiable expression in E_n^L is equivalent to a disjunction of n-types.

Proof Part(a) is obvious from the definition of n-types. Part (b) is true for $n = 0$ since by the assumption (L) on L the expressions of T_0 define a partition of A^*. For the induction step of (b) note that the possibilities in which a given word can be split into segments determine a set T and hence the unique n-type satisfied by w. Part (c) is also shown by induction on n. The case $n = 0$ is easy using disjunctive normal form for expressions in E_0^L. Consider now an expression e in E_{n+1}^L, which is a Boolean combination of expressions $(e_1 \cdot e_2)$ with $e_1, e_2 \in E_n^L$. By the induction hypothesis, each of e_1, e_2 is equivalent to a disjunction of expressions $e' \cdot e''$ with e', e'' in T_n. Hence by disjunctive normal form, e can be written as a disjunction of expressions

$$\bigcap_{(e',e'')\in S_1} e' \cdot e'' \cap \bigcap_{(e',e'')\in S_2} \sim (e' \cdot e'')$$

where S_1, $S_2 \subseteq T_n \times T_n$ and (by satisfiability of the given expression e) $S_1 \cap S_2 = \emptyset$. But such an expression is equivalent to the disjunction

$$\bigcup_{S_1 \subseteq T,\ T \cap S_2 = \emptyset} \Big(\bigcap_{(e',e'')\in T} e' \cdot e'' \cap \bigcap_{(e',e'')\in \overline{T}} \sim (e' \cdot e'') \Big) .$$

Altogether we obtain thus a representation of e as a disjunction of n-types, as was to be shown.

As a consequence of the Lemma we note that the following conditions are equivalent :

(i) $w \equiv_n^L w'$

(ii) w and w' satisfy the same n-types

(iii) w and w' satisfy the same disjunctions of n-types

(iv) w and w' satisfy the same expressions in E_n^L.

For the proof note that (i) implies (ii) by part (a), (b) of the Lemma, and that (iii) implies (iv) by part (c) of the Lemma. Implications (ii)$\Rightarrow$(iii) and (iv)$\Rightarrow$(i) are obvious.

Using the equivalence between (i) and (ii) above we can now show the claim (**). Namely, we have that $w \equiv_{n+1}^L w'$ holds iff for some $T \subseteq T_n \times T_n$, w and w' satisfy $e' \cdot e''$ for all $(e', e'') \in T$, and w and w' do not satisfy $e' \cdot e''$ for all $(e', e'') \in (T_n \times T_n) - T$.

Hence for any splitting of w into two segments w_1 and w_2, we have $w_1 \in L(e')$ and $w_2 \in L(e'')$ for some $(e', e'') \in T$. Since $w' \in L(e' \cdot e'')$ it follows that also w' splits into two segments $w_1' \in L(e')$ and $w_2' \in L(e'')$. Because e', e'' are n-types we know $w_1 \equiv_n^L w_1'$ and $w_2 \equiv_n^L w_2'$. Starting with a splitting of w' we argue similarly. Hence we obtain

$$\forall w_1 w_2 = w\ \exists w_1' w_2' = w' : w_1 \equiv_n^L w_1' \wedge w_2 \equiv_n^L w_2' \quad ,$$
$$\forall w_1' w_2' = w'\ \exists w_1 w_2 = w : w_1 \equiv_n^L w_1' \wedge w_2 \equiv_n^L w_2' \quad ,$$

i.e. the right hand side of (**).

Conversely, if $w \equiv_{n+1}^L w'$ does not hold then w and w' satisfy different $(n+1)$-types, i.e. for some $(e', e'') \in T_n \times T_n$, say w satisfies $e' \cdot e''$ but w' does not. In this case there is some splitting of w into $w_1 \in L(e')$, $w_2 \in L(e'')$ such that *no* splitting of w' into segments w_1', w_2' exists with $w_1' \in L(e')$, $w_2' \in L(e'')$, i.e. with $w_1 \equiv_n^L w_1'$, $w_2 \equiv_n^L w_2'$. If w' satisfies $e' \cdot e''$ but w does not, we obtain in the analogous way that the right hand side of (**) does not hold. —Thus we have shown (**) and completed the proof of Theorem 2.

It is an easy task to generalize the above proof to the case where constants for several fixed languages $L_1, \ldots, L_k$ are allowed. If these languages satisfy condition (L) and are pairwise disjoint, then the proof carries over directly, otherwise one uses a refinement of 0-types involving all possible intersections of the languages L_i, $\sim L_i$, $\{a_i\}$, $\sim\{a_i\}$, $\{\varepsilon\}$, $\sim\{\varepsilon\}$.

By restricting to "pure" star-free expressions and the game without the L-clause in the winning condition we obtain from the above proof also the equivalence between $w \equiv_n w'$ and $w \sim_n w'$. This can be used to verify that certain languages are not star-free, without invoking the algebraic characterization of star-free sets. For instance, the fact that we have $a^i \sim a^j$ provided $i, j \geq 2^{n+1}$ (shown in section 3) may be used to show that the language $(aa)^*$ is not star-free : Namely, from that fact (and the equality $\equiv_n = \sim_n$) we see that for any n there are $w_n \in (aa)^*$ and $w_n' \notin (aa)^*$ with $w_n \equiv w_n'$, e.g. $w_n = a^{2^{n+1}}$, $w_n' = a^{2^{n+1}+1}$. Hence for any n, $(aa)^*$ is not definable by an expression in E_n and thus not star-free.

References

[BrzK 78] J.A. Brzozowski, R. Knast, The dot depth hierarchy of star-free languages is infinite, J.Comput.System Sci. 16 (1978), 37-55.

[CBrz 71] R.S. Cohen, J.A. Brzozowski, Dot depth of star-free events, J.Comput.System Sci. 5 (1971), 1-16.

[Lip 86] D. Lippert, Ausdrucksstärke der Intervall-Temporallogik: Eine Untersuchung mit spieltheoretischen Methoden. Diplomarbeit, RWTH Aachen 1986.

[MoMa 84] B. Moszkowski, Z. Manna, Reasoning in interval temporal logic, In: Logic of Programs (E. Clarke, D. Kozen, Eds.), Springer Lecture Notes in Computer Science 164 (1984), 371-384.

[Mo 83] B.C. Moszkowski, Reasoning about digital circuits, PhD Dissertation, Stanford University 1983.

[McNP 71] R. McNaughton, S. Papert, Counter-Free Automata, MIT Press, Cambridge, Mass. 1971.

[P 84] J.E. Pin, Variétés de langages formels, Masson, Paris 1984.

[PP 86] D. Perrin, J.E. Pin, First-order logic and star-free sets, J.Comput. System Sci. 32 (1986), 393-406.

[Sch 65] M.P. Schützenberger, On monoids having only trivial subgroups, Inf. Contr. 8 (1965), 190-194.

[Str 81] H. Straubing, A generalization of the Schützenberger product, Theor. Comput. Sci. 25 (1982), 107-110.

[Th 82] W. Thomas, Classifying regular events in symbolic logic, J. Comput. System Sci. 25 (1982), 360-376.

[Th 84] W. Thomas, An application of the Ehrenfeucht-Fraissé game in formal language theory, Mem. Soc. Math. France 16 (1984), 11-21.

SEMIGROUP EXTENSIONS OF PARTIAL GROUPOIDS

E.S. Ljapin
Department of Mathematics
Leningrad Pedagogical Institute
Leningrad, U.S.S.R.

For the theory of semigroups as well as for the theory of partial groupoids the situations when a partial operation can be extended to a complete associative operation are of substantial interest (see, for example, [1]).

This paper is devoted to an investigation of possibilities of such extensions. In §2 it is shown that this extension problem can be reduced in a sense to the problem of an internal extension. Using results of §2 and the conditions obtained in [2], it is possible to find another form for the criterion from §2 connected with the properties of extensions of partial transformations.

§1

1.1 For a partial operation on a set P, we will be using a multiplicative notation. Let θ be a partial operation on a set P and let $a,b \in P$. If the result of θ for (a,b) is defined and equals $c \in P$, we write $a \cdot b = c(\theta)$ (omitting θ when it is clear which operation is being considered). In the case when the result of θ for (a,b) is not defined, we write $a \cdot b = \emptyset(\theta)$.

A set P together with a partial operation θ defined on it called a partial groupoid and will be denoted by P^{θ}.

If for all $a,b \in P$, $a \cdot b \neq \emptyset$, then θ is called a complete operation, or simply an operation.

1.2 A partial operation θ on P is called weakly 3-associative if

$$(\forall a,b,c \in P)[(a\cdot b)\cdot c \neq \emptyset \ \& \ a\cdot(b\cdot c) \neq \emptyset \Longrightarrow (a\cdot b)\cdot c = a\cdot(b\cdot c)(\theta)].$$

1.3 The problem of how the general idea of associativity can be realized in the theory of partial operations was considered in various publications (see, for example, [3], [4], [5], [6]). Within the scope of the present work no variants of the definition of associativity other than the one given in 1.2 will be needed.

1.4 Given two partial groupoids $P_1^{\theta_1}$ and $P_2^{\theta_2}$ with $P_1 \subset P_2$, we will say that θ_2 is an extension of θ_1, or that θ_2 contains θ_1, and will write $\theta_1 \subset \theta_2$, if for all $a,b,c \in P_1$,

$$a\cdot b = c(\theta_1) \Longrightarrow a\cdot b = c(\theta_2).$$

If $\theta_1 \neq \theta_2$, then this extension is called proper.

It is obvious that the extension relation is an ordering. In what follows, when considering partial operations, we always look upon them from the point of view of this order relation (in particular, when talking about a maximal element in some collection of partial operations).

If for $P_1^{\theta_1}$ and $P_2^{\theta_2}$ we have $P_1 \subset P_2$ and $\theta_1 \subset \theta_2$, then the partial groupoid $P_2^{\theta_2}$ is said to be an extension of the partial groupoid $P_1^{\theta_1}$. If in this situation $P_1 = P_2$, then the extension is called internal. If for all $a, b, c \in P_1$,

$$a \cdot b = c(\theta_2) \Longrightarrow a \cdot b = c(\theta_1),$$

then the extension is called external.

1.5. It is easy to see that whenever θ_2 is an extension of θ_1, there exists a unique partial operation θ' satisfying $\theta_1 \subset \theta' \subset \theta_2$ such that θ' is an internal extension of θ_1 and θ_2 is an external extension of θ'.

1.6. If a partial operation θ on P is complete and associative, it is called a semigroup operation, since P^θ is a semigroup.

If a semigroup operation θ_2 is an extension of a partial operation θ_1, then we say that θ_2 is a semigroup extension for θ_1. Similarly to that $P_2^{\theta_2}$ is called a semigroup extension of the partial groupoid $P_1^{\theta_1}$.

1.7. Let P^θ be a partial groupoid. A subset M of P is called active (with respect to θ) if for all $a \in P$ and $z \in M$, $z \cdot a \neq \emptyset$ and $a \cdot z \neq \emptyset$.

An element m of P is called active if a singleton $\{m\}$ is an active subset of P.

1.8. A nonempty subset $J \subset P$ is called an ideal of a partial groupoid P^θ if

$$(\forall a \in P, z \in J)[(a \cdot z \neq \emptyset \Longrightarrow a \cdot z \in J) \ \& \ (z \cdot a \neq \emptyset \Longrightarrow z \cdot a \in J)].$$

1.9. It is obvious that the set P itself is always an ideal of a partial groupoid P^θ. It will be an active ideal only if θ is a complete operation.

1.10. An element e of a partial groupoid P^θ is called an identity if for every $a \in P$, $a \cdot e = e \cdot a = a(\theta)$.

An identity is an active element.

1.11. An element z of a partial groupoid P^θ is called a left zero, if for every $a \in P$, $z \cdot a = z(\theta)$.

The set of all left zeros of P^θ will be denoted by $L(P^\theta)$, or simply by L, if it is clear which partial groupoid is being discussed.

1.12. Let us record the following properties of $L(P^\theta)$.

(1) If θ' is an internal extension of θ, then $L(P^{\theta}) \subset L(P^{\theta'})$.

(2) $L(P^{\theta})$ is a subset of P closed with respect to θ.

(3) $L(P^{\theta})$ satisfies the identity $\xi_1 \cdot \xi_2 = \xi_1$.

(4) If $M \subset L(P^{\theta})$ and M is an ideal of P^{θ}, then $M = L(P^{\theta})$ and M is a universally minimal ideal of P^{θ}.

The properties (1), (2), and (3) are obvious. Let us show that (4) holds.

Let J be an arbitrary ideal of a partial groupoid P^{θ}. For any $z \in M$ and $a \in J$ we have $z = z \cdot a(\theta)$. However J is an ideal and hence $z \cdot a \in J$. Therefore $z \in J$, and thus $M \subset J$.

If $x \in L(P^{\theta})$, then for any $z \in M$, $x \cdot z = x(\theta)$. Since M is an ideal, we have $x \cdot z \in M$. It follows that $x \in M$ so that $L(P^{\theta}) \subset M$, i.e. $L(P^{\theta}) = M$.

1.13. It has been already noted (see, for example, [3], [7]) that in the theory of partial operations there are several (non-equivalent) different notions corresponding to the idea of a homomorphism, and each of these notions has a right for an independent existence.

In the present work we use only one of them.

For partial groupoids ${P_1}^{\theta_1}$ and ${P_2}^{\theta_2}$ a mapping $\phi: P_1 \to P_2$ is called a homomorphism if for all a, b, $c \in P_1$,

$$a \cdot b = c(\theta_1) \Longrightarrow (\phi a) \cdot (\phi b) = \phi c(\theta_2).$$

1.14. For a partial groupoid P^{θ} an equivalence ε on the set P is called a congruence if for all a, a', b, $b' \in P$,

$$a \sim a'(\varepsilon) \ \& \ b \sim b'(\varepsilon) \ \& \ a \cdot b \neq \emptyset(\theta) \ \& \ a' \cdot b' \neq \emptyset(\theta) \Longrightarrow$$
$$\Longrightarrow a \cdot b \sim a' \cdot b'(\varepsilon).$$

1.15. To any homomorphism ϕ of a partial groupoid ${P_1}^{\theta_1}$ into a partial groupoid ${P_2}^{\theta_2}$ there corresponds an equivalence ε_{ϕ} on the set P_1 called the kernel of ϕ such that for any a, $b \in P_1$,

$$a \sim b(\varepsilon_{\phi}) \Longleftrightarrow \phi a = \phi b.$$

1.16. It is easy to see that the kernel of a homomorphism is a congruence.

On the other hand, if ε is a congruence on a partial groupoid P^{θ}, then there exists a surjective homomorphism of a partial groupoid P^{θ} for which ε is the kernel. To prove this it is sufficient to take a partial groupoid whose set of elements, Q, is the collection of all ε-classes and whose partial operation is defined as follows: for any A, B, $C \in Q$, $A \cdot B = C$ if there exist $a \in A$, $b \in B$, and $c \in C$, such that $a \cdot b = c(\theta)$. A mapping $\phi: P \to Q$ for which ϕa is an ε-class containing $a \in P$ is, obviously, the desired homomorphism.

1.17. It should be kept in mind that in the theory of partial operations a bijective homomorphism need not be an isomorphism. In order for it to be an isomorphism the following condition should be satisfied:

$$a \cdot b = \emptyset(\theta_1) \Longrightarrow (\phi a)\cdot(\phi b) = \emptyset(\theta_2).$$

1.18. Let ϕ be a homomorphism of a partial groupoid $P_1{}^{\theta_1}$ into a partial groupoid $P_2{}^{\theta_2}$. Then we have the following naturally defined partial operation $\phi\theta_1$ on P_2:

$$(\forall a_2,b_2,c_2 \in P_2)[a_2 \cdot b_2 = c_2(\phi\theta_1) \Longleftrightarrow$$

$$\Longleftrightarrow (\exists a_1,b_1,c_1 \in P_1)(\phi a_1 = a_2 \ \& \ \phi b_1 = b_2 \ \& \ \phi c_1 = c_2 \ \& \ a_1 \cdot b_1 = c_1(\theta_1))]$$

It is obvious that θ_2 is an extension of $\phi\theta_1$ and ϕ is a surjective homomorphism of $P_1{}^{\theta_1}$ onto $(\phi P_1)^{\phi\theta_1}$. It is an isomorphism if and only if ϕ is injective

1.19. The notion of a dense embedding appeared first in the theory of semigroups [9]. It was used many times and further studied by a number of authors (see, for example, [10], [11]). Later it was extended to the theory of partial operations [8].

For a partial groupoid P^θ, a nonempty subset $A \subset P$ is called dense (or densely embedded) with respect to the class of all homomorphisms if for any non-injective homomorphism ϕ of P^θ, the restriction of ϕ to A is also non-injective.

1.20. Another possibility to introduce the notion of 1.19 follows from 1.16. An equivalence ε on P is called diagonal if for any $a,b \in P$, $a \sim b(\varepsilon)$ only if $a = b$.

A subset A of a partial groupoid P^θ is dense in P^θ with respect to the class of all homomorphisms if and only if for any non-diagonal congruence on P^Θ its restriction to A is also non-diagonal.

1.21. The set of all transformations of some set Ω is denoted by T_Ω. An associative operation σ of superposition is defined on T_Ω in the usual way. For any $P \subset T_\Omega$, the operation σ defines a partial operation on P which is the restriction of σ to P. We denote this partial operation by σ_P, or simply by σ if it is clear which set P is being discussed. It is obvious that σ_P is a complete operation if and only if P is a subsemigroup of the semigroup T_Ω.

1.23. The role of transformation semigroups in the problem concerning semigroup extensions of partial groupoids is well known.

<u>Lemma</u>. A partial groupoid P^θ has a semigroup extension if and only if there exists an injective homomorphism of P^θ into the semigroup T_Σ where Σ is any set of cardinality $|P|$ if P is infinite, and Σ is

countable if P is finite.

Proof. 1) Let $\phi : P \to T_\Sigma$ be an injective homomorphism of P^θ into T_Σ. According to (1.18) there is a partial operation $\phi\theta$ on T_Σ for which σ_Σ is a semigroup extension. Since ϕ is injective, the partial groupoids P^θ and $(\phi P)^{\phi\theta}$ are isomorphic. From the fact that $(\phi P)^{\phi\theta}$ has a semigroup extension it follows that P^θ also has a semigroup extension.

2) Now let P^θ have a semigroup extension H. Let H' be a subsemigroup of H generated by the set P. If P is infinite, then H' is equipotent with P. If P is finite, then H' is either finite, or countable. It is well known that there exists an injective homomorphism of H' into T_Σ. Its restriction to P is a desired injective homomorphism of P^θ into T_Σ.

1.24. Among various elements of the semigroup T_Ω(1.21) a special role is played by the so-called constant transformations. Let $\gamma \in \Omega$. A transformation c_γ for which $c_\gamma \alpha = \gamma$ for every $\alpha \in \Omega$, is called a constant transformation corresponding to γ. The set of all constant transformations is denoted for a while by C_Ω.

It is easy to see that for all $x \in T_\Omega$ and $z \in C_\Omega$, we have $z \cdot x = z \in C_\Omega$ and $x \cdot z \in C_\Omega$.

It follows that in any partial groupoid $Q^{\sigma Q}$, where $Q \subset T_\Omega$, if $C_\Omega \cap Q \neq \emptyset$, we have $L(Q^{\sigma Q}) = C_\Omega \cap Q$ and this set is an ideal in $Q^{\sigma Q}$. According to 1.12 it is a universally minimal ideal.

If $C_\Omega \subset Q$, then the ideal C_Ω is active.

§2.

2.1. Let P^θ be a partial groupoid. We will say that its internal extension $P^{\theta'}$ has property 2.1 if $P^{\theta'}$ satisfies the following conditions:

(1) θ' is weakly 3-associative,

(2) θ' does not have proper weakly 3-associative internal extensions,

(3) $L(P^\theta)$ is an active ideal in $P^{\theta'}$.

Let us note that if (3) is satisfied, then $L(P^\theta) \subset L(P^{\theta'})$. Thus according to 1.12(4), $L(P^{\theta'}) = L(P^\theta)$ and $L(P^\theta)$ is a universally minimal ideal in $P^{\theta'}$.

2.2. Lemma. If a partial groupoid P^θ satisfies the following conditions:

(1) θ is weakly 3-associative,

(2) θ has an identity e,

(3) $L(P^\theta) \neq \emptyset$

(4) $(\forall a \in P\setminus(L \cup e))(\forall z \in L)\ a \cdot z = \emptyset(\theta)$,

(5) $(\forall a, b \in P\setminus L)\ a \cdot b \,\bar{\in}\, L$,

then P^{θ} has an internal extension with property 2.1.

Proof. Let us consider a partial operation θ_0 which is an internal extension of θ; θ_0 is obtained from θ by adding the following relations:

$$(\forall a \in P \setminus L)(\forall z \in L)\, a \cdot z = z(\theta_0).$$

These relations do not contradict relations that define θ since P^{θ} satisfies (4).

Let us show that θ_0 is weakly 3-associative. Let $a, b, c \in P$ and let

$$u = (a \cdot b) \cdot c, \ v = a \cdot (b \cdot c) \quad (\theta_0).$$

If none of these elements belongs to L, then $u = v$ follows from the weak 3-associativity of θ. If $a \in L$, then $u = a$ and $v = a$. If $a \bar{\in} L$ but $b \in L$, then $u = b$ and $v = b$. If $a, b \bar{\in} L$ but $c \in L$, then according to (5), $a \cdot b(\theta_0) = a \cdot b(\theta)$ and hence $u = c$. Similarly $v = a \cdot (b \cdot c) = a \cdot c = c(\theta_0)$.

Let us consider the set Θ of all internal weakly 3-associative extensions θ' of θ_0 for which L is an active ideal in $P^{\theta'}$. This set is nonempty, since it contains θ_0, and is (partially) ordered by extension. Let $\{\theta_i\}_{i \in I}$ be a chain in Θ. It is easy to see that $\hat{\theta} = \bigcup_{i \in I} \theta_i$ is a partial operation on P which is a weakly 3-associative internal extension of θ_0. Since L is an active ideal with respect to θ_0, it will also be an active ideal with respect to $\hat{\theta}$ which is an internal extension of θ_0. Therefore by Zorn's Lemma Θ contains maximal elements.

Let $\tilde{\theta}$ be a maximal element of Θ. Then $\tilde{\theta}$ satisfies conditions 2.1(1) and 2.1(3). Let us show that $\tilde{\theta}$ also satisfies the condition 2.1(2). Let $\bar{\theta}$ be a weakly 3-associative internal extension of $\tilde{\theta}$. For any $z \in L$ and $x \in P$, L contains elements $x \cdot z(\tilde{\theta})$ and $z \cdot x(\tilde{\theta})$. Since $\theta_0 \subset \tilde{\theta} \subset \bar{\theta}$, elements $x \cdot z(\bar{\theta})$ and $z \cdot x(\bar{\theta})$ also belong to L. Therefore $\bar{\theta} \in \Theta$. However $\tilde{\theta} \subset \bar{\theta}$ and $\tilde{\theta}$ is a maximal element in Θ. Thus $\bar{\theta} = \tilde{\theta}$. It follows that $\tilde{\theta}$ satisfies the condition 2.1(2).

2.3. It should be noted that generally speaking the internal extension of θ constructed in 2.2 is not unique among internal extensions of θ having property 2.1.

2.4. For a partial groupoid P^{θ}, let $\xi_P = \max\{|P|, \aleph_0\}$.

2.5. Let P^{θ} be a partial groupoid. Let us construct a new partial groupoid $\bar{P}^{\bar{\theta}}$ which is an external extension of P^{θ}. Set $\bar{P} = P \cup Q \cup \{e\}$ where all three components on the right are pairwise disjoint and $|Q| = \xi_P$. Now let $\bar{\theta}$ be an external extension of θ defined by the following additional relations:

$$(\forall x \in P)(\forall z \in Q)\ x \cdot z = z, \ e \cdot x = x \cdot e = x(\bar{\theta}).$$

It is obvious that e is the identity of $\bar{P}^{\theta}$ and $L(\bar{P}^{\theta}) = Q$. It is also clear that if $P_1^{\theta_1}$ and $P_2^{\theta_2}$ are isomorphic partial groupoids, then $\bar{P}_1^{\theta_1}$ and $\bar{P}_2^{\theta_2}$ are isomorphic as well.

2.6. Let P^{θ} be a weakly 3-associative partial groupoid. Then it is easy to see that the partial groupoid $\bar{P}^{\theta}$ constructed in 2.5 satisfies conditions 2.2 (1) through (5).

2.7. It is clear that if a partial groupoid has a semigroup extension, then it is weakly 3-associative. The converse, in general, is not true (see, for example, [4]).

2.8. <u>Theorem</u>. A weakly 3-associative partial groupoid P^{θ} has a semigroup extension if and only if among internal extensions of the partial groupoid $\bar{P}^{\theta}$ (2.5) satisfying the condition 2.1 (such extensions exist due to 2.2 and 2.6) there exists one in which $L(\bar{P}^{\theta}) = Q$ is a dense subset with respect to the class of all homomorphisms (1.19, 1.20).

<u>Proof</u>. 1) Let θ^* be an internal extension of $\bar{\theta}$ satisfying all the properties mentioned in the formulation of the theorem. Let us show that the partial groupoid $\bar{P}^{\theta^*}$ satisfies the conditions of Theorem 3.1 [8] (with some change in terminology).

Conditions (1), (2), 3.1 [8] are satisfied because, according to the assumption, $\bar{P}^{\theta}$ has property 2.1. Since $Q = L(\bar{P}^{\theta})$, it follows from 2.1 that $Q = L(\bar{P}^{\theta^*})$ and Q is an active universally minimal ideal in $\bar{P}^{\theta^*}$.

The identity $\xi_1\xi_2 = \xi_1$ holds in Q with respect to $\bar{\theta}$. Since $\bar{\theta} \subset \theta^*$, this identity holds also with respect to θ^*. According to 3.2 [8], it follows that conditions (3), (4), 3.1 [8] are satisfied in $\bar{P}^{\theta^*}$.

Since $L(\bar{P}^{\theta}) = L(\bar{P}^{\theta^*}) = Q$, it follows that Q is a dense subset in $\bar{P}^{\theta^*}$ with respect to the class of all homomorphisms (1.19, 1.20), i.e., the condition (5) 3.1 [8] is satisfied.

According to 3.1 [8] $\bar{P}^{\theta^*}$ is isomorphic to some partial groupoid of transformations, R^{σ}, where $R \subset T_{\Omega}$. Since T_{Ω} is a semigroup extension of R^{σ} and R^{σ} is isomorphic to $\bar{P}^{\theta^*}$, it follows that $\bar{P}^{\theta^*}$ also has a semigroup extension. However $\bar{P}^{\theta^*}$ is an extension of P^{θ}. Thus P^{θ} must have a semigroup extension as well.

2) Suppose that P^{θ} has a semigroup extension. According to 1.23 there exists an injective homomorphism ϕ of P^{θ} into the semigroup T_{Σ} where $|\Sigma| = \xi_P$ (2.4). Then by 1.18, P^{θ} is isomorphic to $(\phi P)^{\phi\theta}$ where $\phi\theta \subset \sigma$. Let $\Omega = \Sigma \cup \{o_1, o_2\}$ where $o_1 \neq o_2$ and $o_1 o_2 \bar{\varepsilon} \Sigma$. Then $|\Omega| = |\Sigma|$. Let ψ be a mapping from T_{Σ} to T_{Ω} such that for any $u \in T_{\Sigma}$, $(\psi u)\alpha = u\alpha$ for every $\alpha \in \Sigma$ and $(\psi u)o_i = o_1 (i=1,2)$. It is evident that ψ is an injective homomorphism. Therefore,

according to 1.18, $\psi\phi$ is an isomorphism of P^{θ} onto $(\psi\phi P)^{\psi\phi\theta}$ where $\psi\phi\theta \subset \sigma$.

Since the properties of internal extensions listed in the formulation of the theorem are invariant under isomorphisms, the argument given in the last paragraph shows that it is sufficient for us to prove that there exists a desired internal extension of $\bar{P}^{\bar{\theta}}$ for such P^{θ} that $P \subset T_{\Omega}$ and $\theta \subset \sigma$ where $|\Omega| = \xi_P$. Herein $\Omega \ni o_1, o_2$ and for every $x \in P$ we have $x\alpha \in \Omega\backslash\{o_1,o_2\}$ for any $\alpha \in \Omega\backslash\{o_1,o_2\}$, and $xo_i = c_1(i=1,2)$.

Let Q be the set of all constant transformations of the set Ω and let e denote the identical transformation of Ω. Set $\bar{P} = P \cup Q \cup \{e\}$. (It is clear that P, Q and $\{e\}$ are pairwise disjoint sets.) Let us define a partial operation $\bar{\theta}$ on $\bar{P}$ as follows: if $a,b,c \in P$ and $a \cdot b = c(\theta)$, then set $a \cdot b = c(\bar{\theta})$; for all $z \in Q$ and $x \in P$, set $z \cdot x = z(\bar{\theta})$; finally, for all $x \in \bar{P}$, set $x \cdot e = e \cdot x = x(\bar{\theta})$. It is easy to see that $\sigma_{\bar{P}}$ is an internal extension of $\bar{\theta}$ and $\bar{\theta}$ is an external extension of θ of the type 2.5. Since θ is weakly 3-associative, then according to 2.5, 2.6, $L(\bar{P}^{\bar{\theta}}) = Q$ and $\bar{P}^{\bar{\theta}}$ satisfies conditions 2.2(1), (2), (3), (4), and (5). To finish the proof, we have to show that $\sigma_{\bar{P}}$ satisfies the condition 2.1 (with respect to $\bar{\theta}$) and Q is a dense subset in $\bar{P}^{\sigma_{\bar{P}}}$ with respect to the class of all homomorphisms.

Since $\sigma_{\bar{P}} \subset \sigma_{\Omega}$ and σ_{Ω} is a complete associative operation, $\sigma_{\bar{P}}$ is weakly 3-associative. Now $\bar{P}^{\sigma_{\bar{P}}}$ is a partial groupoid of transformations in the sense of 1.3, 1.10 [8]. Since $Q \subset \bar{P}$, it follows that this partial groupoid is complete with respect to constant transformations. Therefore, according to 3.1 [8], $\bar{P}^{\sigma_{\bar{P}}}$ has no proper internal weakly 3-associative extensions. From the definition of constant transformations, it follows that $L(\bar{P}^{\bar{\theta}}) = Q$ and $L(\bar{P}^{\sigma_{\bar{P}}}) = Q$. It is also clear that for all $x \in \bar{P}$ and $z \in Q$, $z \cdot x = z \in Q$ and $x \cdot z \in Q$. This means that Q is an active ideal in $\bar{P}^{\sigma_{\bar{P}}}$. Since $\bar{P}^{\sigma_{\bar{P}}}$ has property 2.1 with respect to $\bar{P}^{\bar{\theta}}$, then $Q = L(\bar{P}^{\bar{\theta}})$ is a universally minimal active ideal in $\bar{P}^{\sigma_{\bar{P}}}$. Therefore, according to 3.1 [8], Q is a dense subset of $\bar{P}^{\sigma_{\bar{P}}}$ with respect to the class of all homomorphisms.

§3.

3.1. Throughout this section M will stand for an arbitrary fixed nonempty set and e for some fixed element of this set.

Let F_M be the set of all partial transformations of M. It is well known that with respect to composition, F_M is a semigroup. For any $\alpha \in F_M$, let Π_{α} denote the domain of α. If $x \bar{\in} \Pi_{\alpha}$, then we will write $\alpha x = \emptyset$. The identical transformation of the set M will be denoted by E_M. For any $M' \subset M$, let $\alpha M'$ be the set of all elements $a \in M$ such that for some $x \in \Pi_{\alpha}$, $\alpha x = a$.

3.2. We will say that a set of partial transformations $H \subset F_M$ satisfies property 3.2 with respect to e if: (1) $E_M \in H$ (2) $He = M$; (3) $(\forall \alpha \in H)\ e \in \Pi_\alpha$; (4) $(\forall \alpha, \beta \in H)(\alpha \neq \beta \Longrightarrow \alpha e \neq \beta e)$. The set of all subsets of F_M that satisfy property 3.2 with respect to $e \in M$, will be denoted by $\Gamma_e = \Gamma_e(M)$.

3.3 The set of all partial operations on the set M with respect to which e is the identity will be denoted by $\Sigma_e = \Sigma_e(M)$.

3.4. Let $e \in M$. For any partial operation θ on M and any subset H of F_M, set $(\theta, H) \in \lambda_e$ if for all a, b, $c \in M$

$$a \cdot b = c(\theta) \Longleftrightarrow (\exists \alpha \in H)(\alpha e = a \& \alpha b = c).$$

It was proved in [2] that λ_e establsihes a 1-1 correspondence between Σ_e and Γ_e. If $(\theta, H) \in \lambda_e$, we will also write $\theta \underset{\lambda_e}{\longleftrightarrow} H$. For any $\theta \in \Sigma_e$, denote by H_θ such an element of Γ_e for which $\theta \underset{\lambda_e}{\longleftrightarrow} H_\theta$.

3.5 It should be noted that according to 3.2, for a given $a \in M$, a partial transformation α in the formula of 3.4 is uniquely defined.

3.6. For any elements u and v of a partial groupoid M^θ, we can define the following two subsets of M:

$$[u/v]_\ell = \{x \in M \mid v \cdot x = u\},\ [u/v]_r = \{x \in M \mid x \cdot v = u\}.$$

3.7. For any $\theta \in \Sigma_e$ and any u, $v \in M$, we have

$$[u/v]_\ell = \{x \in M \mid (\exists \gamma \in H_\theta)(\gamma e = v \& \gamma x = u)\},$$

$$[u/v]_r = \{x \in M\ (\exists \delta \in H_\theta)(\delta e = x \& \delta v = u)\}.$$

3.8 Lemma. Let $\theta \in \Sigma_e, \alpha, \beta \in H_\theta$, a, b, $x \in M$ where $x = \eta_x e$ (such $\eta_x \in H_\theta$ exists for any $x \in M$ and is unique). Then in M^θ we have

$$x \in [a/b]_r \cap [\beta e/\alpha e]_\ell \Longleftrightarrow \alpha x = \beta e \ \&\ \eta_x b = a.$$

Proof. 1) Let $x \in [a/b]_r \cap [\beta e/\alpha e]_\ell$. According to 3.7, there exist $\gamma, \delta \in H_\theta$ such that $\gamma e = \alpha e$, $\gamma x = \beta e$, $x = \delta e$, $a = \delta b$. Since $\gamma e = \alpha e$ and $\eta_x e = \delta e$, we have $\gamma = \alpha$ and $\delta = \eta_x$. Hence $\alpha x = \beta e$ and $\eta_x b = a$.

2) Let $\alpha x = \beta e$ and $\eta_x b = a$. According to 3.7, $\eta_x b = a$ and $\eta_x e = x$ imply $x \cdot b = a$, i.e. $x \in [a/b]_r$. Similarly, $\alpha e = (\alpha e)$ and $\alpha x = \beta e$ imply that $x \in [\beta e/\alpha e]_\ell$.

3.9. Lemma 3.8 clarifies the role of the following

Theorem. A partial operation $\theta \in \Sigma_e$ is weakly 3-associative iff for any $\alpha, \beta \in H_\theta$ and for all $a \in \Pi_\alpha$ and $b \in \Pi_\beta$,

$$[a/b]_r \cap [\beta e/\alpha e]_\ell \neq \emptyset \Longrightarrow \alpha a = \beta b.$$

Proof. 1) Let θ be weakly 3-associative. Suppose that $z \in [a/b]_r \cap [\beta e/\alpha e]_\ell$, i.e. $a = z \cdot b$ and $\beta e = (\alpha e) \cdot z$. Denote αa by a' and βb by b'. Then $a' = (\alpha e)\cdot a = (\alpha e)\cdot(z\cdot b)$ and $b' = (\beta e)\cdot b = [(\alpha e)\cdot z]\cdot b$. It follows that $a' = b'$.

2) Suppose that M^θ satisfies the condition given in the formulation of the theorem. Suppose also that for some $a,b,c,d,d' \in M$, we have $a\cdot(b\cdot c) = d$ and $(a\cdot b)\cdot c = d'$. Since $a\cdot b \neq \emptyset$, there exists $\alpha \in H_\theta$ such that $\alpha e = a$. Denote $b\cdot c$ by t. Since $d = a\cdot t$, there exists $\beta \in H_\theta$ such that $\beta e = a$ and $\beta t = d$. We have $\alpha e = \beta e$ whence $\alpha = \beta$. Now $d' = (a\cdot b)\cdot c = [(\alpha e)\cdot b]\cdot c$. It follows that there exists $\gamma \in H_\theta$ such that $\gamma e = (\alpha e)\cdot b$ and $\gamma c = d'$. Since $b \in [t/c]_r \cap [\gamma e/\alpha e]_\ell$ and $\alpha t = \beta t \neq \emptyset$, $\gamma c = d'$, we have $\alpha t = \gamma c$. Finally, $d = a\cdot t$ implies $d = \beta t = \alpha t$. Therefore $d = d'$, i.e. θ is weakly 3-associative.

3.10. For any $H, H' \subset F_M$, set $H \leq H'(\mu)$ iff for any $\alpha \in H$, there exists $\alpha' \in H'$ such that $\Pi_\alpha \subset \Pi_{\alpha'}$ and for each $x \in \Pi_\alpha$, $\alpha x = \alpha' x$. It is clear that the relation μ is reflexive and transitive.

3.11. In general, μ is not an order relation. It can be seen from the following example in which M is the set of all natural numbers. For any $k = 1,2,3,\ldots$, define $\alpha_k \in F_M$ as follows: $\Pi_{\alpha_k} = \{1,2,\ldots,k\}$ and $\alpha_k x = x$ for each $x \in \{1,2,\ldots,k\}$. It is obvious that for $H = \{\alpha_1,\alpha_3,\alpha_5,\ldots\}$ and $H' = \{\alpha_2,\alpha_4,\alpha_6,\ldots\}$, we have $H \leq H'(\mu)$ and $H' \leq H(\mu)$ though $H \neq H'$.

However it may happen that the restriction of μ to a certain set of subsets of F_M is an order relation.

3.12. Theorem. Let $\theta, \theta' \in \Sigma_e$. Then θ' is an extension of θ iff $H_\theta \leq H_{\theta'}(\mu)$.

Proof. 1) Let $\theta \subset \theta'$ and $\alpha \in H_\theta$. Take $b \in \Pi_\alpha$ and set $c = \alpha b$. Then $c = (\alpha e)\cdot b(\theta)$. Since $\theta \subset \theta'$, $c = (\alpha e)\cdot b(\theta')$. It follows that there exists $\alpha' \in H_{\theta'}$ such that $b \in \Pi_{\alpha'}$ and $\alpha' b = c = \alpha b$. Thus $H_\theta \leq H_{\theta'}(\mu)$.

2) Let $H_\theta \leq H_{\theta'}(\mu)$ and $a\cdot b = c(\theta)$ for some $a,b,c \in M$. Then there exists $\alpha \in H_\theta$ such that $b \in \Pi_\alpha$ and $\alpha e = a$, $\alpha b = c$. Since $H_\theta \leq H_{\theta'}(\mu)$, for this α there exists $\alpha' \in H_{\theta'}$ such that $\Pi_\alpha \subset \Pi_{\alpha'}$ and $\alpha x = \alpha' x$ for any $x \in \Pi_\alpha$. Therefore $\alpha' e = a$ and $\alpha' b = c$ which means that $a\cdot b = c(\theta')$.

3.13. Corollary. The restriction of μ to Γ_e is an order relation.

3.14. Lemma. Let $\theta \in \Gamma_e$. Then the partial groupoid M^θ has left zeros iff H_θ contains at least one constant transformation of M(1.24).

Proof. 1) Let $z \in M$ be a left zero in M^θ. Then $z\cdot z = z$ and therefore there exists $\alpha \in H_\theta$ such that $\alpha e = z$ and $\alpha z = z$. For any $a \in M$, $z\cdot a = z$ and hence there exists $\beta \in H_\theta$ such that $\beta e = z$, $\beta a = z$. Since $\alpha e = \beta e$, $\alpha = \beta$. Thus for any $a \in M$, $a = z$. In other words,

$\alpha \in H_\theta$ is a constant transformation corresponding to z.

2) Let $\xi_t \in H_\theta$ be the constant transformation of M corresponding to an element $t \in M$. For any $a \in M$, $\xi_t a = t = \xi_t e$. Since $\xi_t \in H_\theta$, it means that $t \cdot a = t(\theta)$. Thus t is a left zero of M^θ.

3.15. Theorem. Let M^θ be a partial groupoid with the identity e, and let $\emptyset \neq J \subset M$. Then J is an ideal of M^θ iff for any $\alpha \in H_\theta$:

(1) $\alpha J \subset J$, and

(2) $\alpha e \in J \Longrightarrow \alpha\Pi_\alpha \subset J$.

Furthermore, an ideal J will be active (1.7) iff for any $\alpha \in H_\theta$:

(3) $J \subset \Pi_\alpha$, and

(4) $\alpha e \in J \Longrightarrow \Pi_\alpha = M$.

Proof. 1) Let J be an ideal of M^θ. Take $\alpha \in H_\theta$, $z \in J \cap \Pi_\alpha$. Set $\alpha z = u$ and $\alpha e = v$. Then $v \cdot z = u$. Since J is an ideal, $u \in J$. Hence $\alpha J \subset J$.

Now assume that $\alpha e \in J$ and $x \in \Pi_\alpha$. Then $(\alpha e) \cdot x = \alpha x$. Since J is an ideal, we have $\alpha x \in J$, i.e. $\alpha\Pi_\alpha \subset J$.

2) Suppose that J satisfies (1) and (2). Assume that $a \cdot z = b$. for some $a, b \in M$ and $z \in J$. Then there exists $\alpha \in H_\theta$ such that $\alpha e = a$ and $\alpha z = b$. By (1) it follows that $b = \alpha z \in \alpha J \subset J$. On the other hand, if $c = z \cdot a$, then for some $\alpha \in H_\theta$ we have $\alpha e = z$, $a \in \Pi_\alpha$ and $\alpha a = c$. Since $\alpha e = z \in J$, then by (2), $c = \alpha a \in J$. Thus J is an ideal in M^θ.

3) Now suppose that J is an active ideal. For any $z \in J$ and $\alpha \in H_\theta$, set $\alpha e = u$ and $u \cdot z = z'$. Then there exists $\beta \in H_\theta$ such that $\beta e = u$ and $\beta z = z'$. Since $\alpha e = \beta e$, we have $\alpha = \beta$. Therefore $z \in \Pi_\beta = \Pi_\alpha$ so that $J \subset \Pi_\alpha$.

To prove (4), take any $a \in M$ and assume that $u = \alpha e \in J$. Since J is an active ideal, $u \cdot a \in J$. Then there exists $\gamma \in H_\theta$ such that $a \in \Pi_\gamma$, $\gamma e = u$ and $\gamma a = u \cdot a$. Since $\alpha e = \gamma e$, we have $\gamma = \alpha$. Therefore $a \in \Pi_\alpha$, i.e. $\Pi_\alpha = M$.

4) Now suppose that an ideal J satisfies conditions (3) and (4), i.e. all four conditions (1), (2), (3), and (4) are satisfied for J. For any $a \in M$, there exists $\alpha \in H_\theta$ such that $\alpha e = a$. Take any $z \in J$. Then $z \in \Pi_\alpha$. Denote αz by b. Since $\alpha e = a$ and $\alpha z = b$, we have $a \cdot z = b$, i.e. $a \cdot z \neq \emptyset$. There exists $\beta \in H_\theta$ such that $\beta e = z$. By (4) we get $\Pi_\beta = M$. It follows that $\beta a = c$ for some $c \in M$. Since $\beta e = z$ and $\beta a = c$, we have $z \cdot a = c$, i.e. $z \cdot a \neq \emptyset$. Thus J is active.

3.16. Theorem. Let M^θ be a partial groupoid with the identity e and ρ an equivalence relation on M. Then ρ is a congruence

(1.14) iff for all $\alpha, \beta \in H_\theta$ and for any $a \in \Pi_\alpha$, $b \in \Pi_\beta$,

$$\alpha e \sim \beta e(\rho) \ \& \ a \sim b(\rho) \Longrightarrow \alpha a \sim \beta b(\rho).$$

Proof. 1) Let ρ be a congruence, $\alpha e = u$, $\alpha a = a'$, $\beta e = v$, $\beta b = b'$, and $u \sim v(\rho)$, $a \sim b(\rho)$. Then $u \cdot a = a'$ and $v \cdot b = b'$. Since ρ is a congruence and $u \cdot a \neq \emptyset$, $v \cdot b \neq \emptyset$, we have $u \cdot a \sim v \cdot b(\rho)$, i.e. $a' \sim b'(\rho)$.

2) Suppose that ρ satisfies the condition given in the formulation of the theorem. Let $a \sim a'(\rho)$ and $b \sim b'(\rho)$ for some a, a', b, $b' \in M$. If $a \cdot b = c$, $a' \cdot b' = c'(\theta)$, then there exist $\alpha, \beta \in H_\theta$ such that $b \in \Pi_\alpha$, $b' \in \Pi_{\beta'}$, $\alpha e = a$, $\alpha b = c$, $\beta e = a'$, and $\beta b' = c'$. Since $\alpha e \sim \beta e(\rho)$ and $b \sim b'(\rho)$, then according to our assumption, $\alpha b \sim \beta b'(\rho)$, i.e. $a \cdot b \sim a' \cdot b'(\rho)$.

§4.

4.1. In 2.8 we obtained the criterion for a partial groupoid to have a semigroup extension. Using the relationship between partial operations and sets of partial transformations on a set, studied in §3, we can give a new form to this criterion.

4.2. In 2.5 for a partial groupoid P^θ we constructed another partial groupoid $\bar{P}^{\bar{\theta}}$. It had the identity e and $L(\bar{P}^{\bar{\theta}}) = Q$. According to 3.4, $\bar{\theta} \underset{\lambda e}{\longleftrightarrow} H_{\bar{\theta}}$, $H_{\bar{\theta}} \in \Gamma_e(\bar{P}^{\bar{\theta}})$. In what follows we will be considering such sets $G \in \Gamma_e(\bar{P}^{\bar{\theta}})$ that $H_{\bar{\theta}} \leq G(\mu)$.

4.3. We say that G has property 4.3 if for all $\alpha, \beta \in G$ and all $a \in \Pi_\alpha$, $b \in \Pi_\beta$, we have $\alpha a = \beta b$ whenever there exists $x \in \bar{P}$ such that $\alpha x = \beta e$ and $\eta_x b = a$ (since $G \in \Gamma_e(\bar{P}^{\bar{\theta}})$, there exists a unique partial transformation $\eta_x \in G$ such that $x = \eta_x e$).

4.4. Let us say that G has property 4.4 if for any $G' \in \Gamma_e(\bar{P}^{\bar{\theta}})$ having property 4.3, $G \leq G'(\mu)$ implies $G = G'$.

4.5. We say that G has property 4.5 if for any $\alpha \in G$, the following conditions are satisfied for $Q = L(\bar{P}^{\bar{\theta}})$:

(1) $\alpha Q \subset Q$; (2) $\alpha e \in Q \Longrightarrow \alpha \Pi_\alpha \subset A$; (3) $Q \subset \Pi_\alpha$; (4) $\alpha e \in Q \Longrightarrow \Pi_\alpha = \bar{P}$.

4.6. Denote by $\Gamma_e^*(\bar{P}^{\bar{\theta}})$ the set of all sets $G \in \Gamma_e(\bar{P}^{\bar{\theta}})$ such that $G \geq H_{\bar{\theta}}(\mu)$ and G has properties 4.3, 4.4, and 4.5.

4.7. Let $G \in \Gamma_e^*(\bar{P}^{\bar{\theta}})$. According to 3.4 there exists a unique partial operation $\tau \in \Sigma_e(\bar{P}^{\bar{\theta}})$ such that $\tau \underset{\lambda e}{\longleftrightarrow} G = H_\tau$. Furthermore, by 3.8, 3.9 and 3.12, τ is a weakly 3-associative internal extension of $\bar{\theta}$ which has no proper weakly 3-associative internal extensions. By 3.15 $Q = L(\bar{P}^{\bar{\theta}})$ is an active ideal of $\bar{P}^\tau$. Thus $\bar{P}^\tau$, being an internal extension of $\bar{P}^{\bar{\theta}}$, has property 2.1 with respect to $\bar{P}^{\bar{\theta}}$.

4.8. Now let $\bar{P}^{\tau}$ be an internal extension of $\bar{P}^{\bar{\theta}}$ having property 2.1. It follows from 3.8, 3.9, 3.12, and 3.15 that $H_{\tau} \in \Gamma_{e}^{*}(\bar{P}^{\bar{\theta}})$.

4.9. We say that $G \subset F_{\bar{P}}$ has property 4.9 if for any nondiagonal equivalence ρ on $\bar{P}$ which satisfies

$$\alpha e \sim \beta e(\rho) \ \&\ a \sim b(\rho) \Longrightarrow \alpha a \sim \beta b(\rho)$$

for all $\alpha, \beta \in G$ and $a \in \pi\alpha$, $b \in \pi\beta$,

the restriction of ρ to $Q = L(\bar{P}^{\bar{\theta}})$ is not the diagonal on Q.

4.10. Taking into account 1.20 and 3.16 and using 4.7 and 4.8, we can present Theorem 2.8 in the following form:

Theorem. A weakly 3-associative partial groupoid P^{θ} has a semigroup extension iff there exists $G \in \Gamma_{e}^{*}(\bar{P}^{\bar{\theta}})$ which has property 4.9.

4.11. In conclusion note that the class $\Gamma_{e}^{*}(\bar{P}^{\bar{\theta}})$ which appears in the formulation of Theorem 4.10 is not empty. Indeed, according to 2.6, $\bar{P}^{\bar{\theta}}$ satisfies conditions 2.2(1), (2), (3), (4), and (5). Therefore, by 2.2, there exists an internal extension $\bar{P}^{\nu}$ which has property 2.1. Then, according to 4.8, $H_{\nu} \in \Gamma_{e}^{*}(\bar{P}^{\bar{\theta}})$.

REFERENCES

1. E.S. Ljapin, Problems of the theory of semigroup extensions of partial groupoids, "Contemporary Algebra", I, Leningrad, 1974, 130-145 (in Russian).
2. E.S. Ljapin, On internal extensions of partial operations to complete associative ones, "Izv. Vysš. Učebn. Zaved. Matematika," No. 7(242), 1982, 40-44 (in Russian).
3. V.V. Rozen, Partial operations in ordered sets, Saratov Univ. Press, 1973. (in Russian)
4. E.S. Ljapin, An abstract characterization of partial groupoids of words with synonyms, "Algebraic Theory of Semigroups. Colloq. Math. Soc. János Bolyai", 20, 1976, 341-356.
5. E.S. Ljapin, Partielle Operationen in der Theorie der Halbgruppen, "Lecture Notes in Mathematics", 855, 1978, 33-48.
6. E.S. Ljapin, Weak associativity of partial operations, "Semigroup Varieties and Endomorphism Semigroups", Leningrad, 1979, 95-112 (in Russian).
7. G. Grätzer, Universal algebra, D. Van Nostrand Company, 1968.
8. E. S. Ljapin, An abstract characterization of one class of partial groupoids of transformations, "Izv. Vysš. Učebn. Zaved. Matematika", No. 6(241), 1982, 30-36 (in Russian).
9. E. S. Ljapin, Associative systems of all partial transformations, Dokl. Akad. Nauk SSSR", 88, 1953, 13-15 (in Russian).
10. L. M. Gluskin, On some dense embeddings, "Matem. Sb.", 61 1963, 175-206 (in Russian).
11. L. M. Gluskin, On dense extensions, "Trudy Moskov. Mat. Obshch.", 129, 1973, 119-131 (in Russian).

ON GROUPS HAVING FINITE MONADIC CHURCH-ROSSER PRESENTATIONS

Klaus Madlener
Friedrich Otto

Fachbereich Informatik
Universität Kaiserslautern
Postfach 3049
6750 Kaiserslautern
West Germany

Introduction

One of the standard ways of describing groups is by way of presentations involving generators and defining relations [cf.,e.g.,17]. It is therefore only natural to investigate the relationship between algebraic properties of groups and combinatorial properties of their presentations. Examples of investigations of this type are those for groups with one defining relation and those for small cancellation groups.

The first investigations resulting in algebraic characterizations were concerned with regular and context-free group languages. If $<\Sigma;L>$ is a group-presentation of a group G, then the set of all words presenting the identity of G is called the group language of this presentation. Anisimov [1] proved that a finitely generated group G has a presentation $<\Sigma;L>$ the group language of which is a regular language if and only if the group G is finite, and that in this situation every finite presentation of G has a regular group language. Muller and Schupp [18] proved that a finitely generated group G has a presentation with a context-free group language if and only if G is virtually free and accessible, and that in this situation all finite presentations of G give context-free group languages. Due to a recent result by Dunwoody [10] the mere technical condition of accessibility can be dropped from the Muller-Schupp result, i.e., a finitely generated group has context-free group language if and only if G is virtually free. An important subclass of this class of groups has been investigated by Haring-Smith [13]. He proved that a group G has a finitely generated presentation the reduced word problem of which is a simple language if and only if G is isomorphic to a free product of a free group of finite rank and finitely many finite groups. However, the property of having a reduced word problem that is a simple language is not an invariant of finitely generated presentations.

Since the class of groups is a subclass of the class of monoids, groups can also be described through monoid-presentations of the form $(\Sigma;T)$. Here Σ is an alphabet, and T is a Thue system on Σ. A Thue system can be interpreted as a string-rewriting system in that the process of substituting an occurrence of a left-hand side of a rule of T by an occurrence of the corresponding right-hand side defines a rewriting relation on Σ^*. Of particular interest are those Thue systems for which this process of rewriting defines unique normal forms for the elements of the monoid M_T presented by $(\Sigma;T)$. See e.g. [6] for an overview concerning systems of this form.

A monoid presentation $(\Sigma;T)$ will be called a Church-Rosser presentation if T is length-reducing, and T defines unique normal forms. Since many decision problems can be solved very efficiently for finite Church-Rosser presentations, it is only natural to ask which class of groups does allow presentations of this form.

The first result in this area was obtained by Cochet [8] who proved that a group can be described by a finite special Church-Rosser presentation if and only if G is isomorphic to a free product of finitely many finite or infinite cyclic groups. Gilman [11] conjectured that a group has a finite monadic Church-Rosser presentation if and only if G is isomorphic to a free product of a free group of finite rank and finitely many finite groups, which happens to be exactly the class of groups considered by Haring-Smith [13].

This conjecture is still open. Here we present what is known about it in some detail, pointing out the limitations of the proof techniques used so far to solve restricted cases. In particular, we are dealing with the following classes of groups:

- C_o, those groups that are presented by finite monadic Church-Rosser presentations which provide inverses of length one for all the generators,
- C_1, those groups that are presented by finite two-monadic Church-Rosser presentations,
- C_2, those groups that are presented by finite monadic Church-Rosser presentations, and
- C_3, those groups that have context-free group languages.

We shall see that $C_o = C_1 \subseteq C_2 \subseteq C_3$, and that C_o coincides with the class of groups that are isomorphic to the free product of a free group of finite rank and finitely many finite groups. In addition, we prove a few new results about groups that have finite Church-Rosser presentations.

1. Definitions

An alphabet Σ is a finite set the elements of which are called letters. Then Σ^* denotes the free monoid generated by Σ with identity 1. The length of a word $w \in \Sigma^*$ is denoted by $|w|$, the identity of words is written as $\equiv$, and the concatenation of words u and v is simply written as uv. In addition, superscripts are often used to abbreviate words, e.g., w^3 stands for the word www.

A Thue system T over Σ is a subset of $\Sigma^* \times \Sigma^*$, the elements of which are called (rewrite) rules. Let $\mathrm{dom}(T) = \{\ell \mid \exists r \in \Sigma^*: (\ell,r) \in T\}$ and $\mathrm{range}(T) = \{r \mid \exists \ell \in \Sigma^*: (\ell,r) \in T\}$ denote the sets of left-hand sides and right-hand sides of rules of T, respectively. T is called

- length-reducing if $|\ell| > |r|$ for each rule $(\ell,r) \in T$,
- monadic if it is length-reducing and $\mathrm{range}(T) \subseteq \Sigma \cup \{1\}$,
- two-monadic if it is monadic and $\mathrm{dom}(T) \subseteq \Sigma^2$, and
- special if it is length-reducing and $\mathrm{range}(T) = \{1\}$.

The Thue system T induces a congruence $\overset{*}{\underset{T}{\leftrightarrow}}$ on Σ^*, the Thue congruence generated by T, which is the reflexive transitive closure of the single-step derivation relation $\underset{T}{\leftrightarrow}$. For $w \in \Sigma^*$, the congruence class $\{z \in \Sigma^* \mid z \overset{*}{\underset{T}{\leftrightarrow}} w\}$ is denoted by $[w]_T$, and the factor monoid $\Sigma^*/\overset{*}{\underset{T}{\leftrightarrow}}$ of the free monoid Σ^* modulo the congruence $\overset{*}{\underset{T}{\leftrightarrow}}$ is denoted by M_T.

If a monoid M is isomorphic to M_T $(M \cong M_T)$, then the ordered pair $(\Sigma;T)$ is called a (monoid) presentation of M with generators Σ and defining relations T.

The Thue reduction $\overset{*}{\underset{T}{\rightarrow}}$ defined by T is the reflexive transitive closure of the relation $\underset{T}{\rightarrow}$, which is defined as follows: $u \underset{T}{\rightarrow} v$ if and only if $u \underset{T}{\leftrightarrow} v$ and $|u| > |v|$. Obviously, this relation is Noetherian, i.e., there does not exist an infinite chain of the form $u_1 \underset{T}{\rightarrow} u_2 \underset{T}{\rightarrow} \cdots$ If $u \overset{*}{\underset{T}{\rightarrow}} v$ one says that u reduces to v, u is an ancestor of v, and v is a descendant of u (modulo T). If there is no v such that $u \underset{T}{\rightarrow} v$, then u is called irreducible, otherwise it is reducible (modulo T). IRR(T) denotes the set of all irreducible words (modulo T).

The Thue system T is called Church-Rosser if every two congruent words have a common descendant. Thus, in a Church-Rosser Thue system no two distinct irreducible words are congruent, and so the set IRR(T) is a set of representatives for the monoid M_T. If in addition to being Church-Rosser the Thue system T is finite, then there exists an algorithm that on input a word $u \in \Sigma^*$ computes the irreducible descendant $\hat{u}$ of u in time $c \cdot |u|$ for some constant c depending only on T [5]. In particular, this means that the word problem for the monoid M_T is decidable in linear time.

Two Thue systems T_1 and T_2 over Σ are equivalent, if the congruences $\overset{*}{\underset{T_1}{\leftrightarrow}}$ and $\overset{*}{\underset{T_2}{\leftrightarrow}}$ coincide, and a Thue system T is called reduced (or normalized) if $\ell, r \in IRR(T-\{(\ell,r)\})$ hold for every rule $(\ell,r) \in T$. It has been noticed independently by several authors [3,15] that for each finite Church-Rosser Thue system T, there exists an equivalent one that is finite, Church-Rosser, and reduced, and that can be constructed effectively from T. In addition, if T is special or monadic, then the reduced Thue system is also special or monadic, respectively.

Finally, if a reduced Church-Rosser Thue system T contains a rule of the form (a,1) for some $a \in \Sigma$, then the monoid M_T is also presented by $(\Sigma-\{a\};\ T-\{(a,1)\})$, and $T' := T-\{(a,1)\}$ is again a reduced Church-Rosser Thue system. So in the following we will restrict our attention to Thue systems that are finite, reduced, and Church-Rosser, and that do not contain any rule of the form $(a,1) \in \Sigma\times\{1\}$. If T is a Thue system of this form then the ordered pair $(\Sigma;T)$ is called a finite Church-Rosser presentation.

Actually we are interested in finite Church-Rosser presentations $(\Sigma;T)$ such that the monoid M_T is a group. Obviously, M_T is a group if and only if, for each letter $a \in \Sigma$, there exists a word $u_a \in \Sigma^*$ such that $au_a \overset{*}{\underset{T}{\rightarrow}} 1$. Although it is undecidable in general, whether a given finite monoid-presentation actually presents a group, this problem is decidable for finite Church-Rosser presentations [21].

2. Groups having finite two-monadic Church-Rosser presentations

In combinatorial group theory groups are usually presented by what we call group presentations. A group presentation is an ordered pair $<\Sigma;L>$, where Σ is an alphabet and L is a subset of $\underline{\Sigma}^*$. Here $\bar{\Sigma}$ is an alphabet in 1-to-1 correspondence to Σ, this correspondence being given through a function $\bar{}: \Sigma \to \bar{\Sigma}$, $\Sigma \cap \bar{\Sigma} = \emptyset$, and $\underline{\Sigma} := \Sigma \cup \bar{\Sigma}$. The group presented by $<\Sigma;L>$ is M_{T_L}, where T_L is the following special Thue system on $\underline{\Sigma}$:

$$T_L := \{(w,1) \mid w \in L \cup \{a\bar{a}, \bar{a}a \mid a \in \Sigma\}\}.$$

Observe that in a group presentation each generator $a \in \Sigma$ has an inverse $\bar{a} \in \bar{\Sigma}$ of length one. So it is only natural to first consider finite monadic Church-Rosser presentations which provide inverses of length one for their generators. As it turns out presentations of this form are already two-monadic [3]. However, they only form a proper subclass of the class of all finite two-monadic Church-

Rosser presentations of groups as shown by the following example.

<u>Example 2.1</u>. Let $\Sigma = \{a,b,c\}$ and $T = \{(a^2,1),(b^2,1),(ab,c),(ac,b),(cb,a)\}$. Then $(\Sigma;T)$ is a finite two-monadic Church-Rosser presentation of the group $\mathbb{Z}_2 * \mathbb{Z}_2$. However, the generator c does not have an inverse of length one. □

If a group G is finite or free of finite rank, then G has obviously a finite monadic Church-Rosser presentation providing inverses of length one for its generators. If G_1 and G_2 have presentations of this form, then so does the free product $G_1 * G_2$. In fact, no other groups have presentations of this form as shown by Avenhaus and Madlener.

<u>Theorem 2.2</u> [3]. A group G has a finite monadic Church-Rosser presentation that provides inverses of length one for all its generators if and only if G is isomorphic to a free product of a free group of finite rank and finitely many finite groups.

This result was proved by showing that whenever $(\Sigma;T)$ is a finite monadic Church-Rosser presentation such that each generator $a \in \Sigma$ has an inverse of length one, then the following set M is finite:

$$M := \{w \in \Sigma^* \mid w \xleftrightarrow[T]{*} 1, \text{ but no proper factor } u \text{ of } w \text{ satisfies } u \xleftrightarrow[T]{*} 1\}.$$

From the presentation $(\Sigma;T)$ one can easily obtain a group presentation $\langle \Sigma;L \rangle$ of the same group M_T such that the set

$$M_0 := \{w \in \Sigma^* \mid w = 1 \text{ in } M_T, \text{ but no proper factor of } w \text{ is equal to } 1 \text{ in } M_T\}$$

is finite. Now the above result follows from a result of Haring-Smith.

<u>Theorem 2.3</u> [13]. Let $\langle \Sigma;L \rangle$ be a group presentation such that the set M_0 is finite. Then the group G presented by $\langle \Sigma;L \rangle$ is isomorphic to a free product of a free group of finite rank and finitely many finite groups.

Unfortunately Haring-Smith's result does not apply to finite two-monadic Church-Rosser presentations of groups in general.

<u>Example 2.1</u> (continued). For all $m > 1$, $c^m a c^m a \xleftrightarrow[T]{*} 1$, but no proper factor of $c^m a c^m a$ is congruent to 1. Hence, for the presentation $(\Sigma;T)$ the set M is infinite. □

Introducing the missing inverses of length one does not help in general, since when a letter $\bar{c}$ and rules $(c\bar{c},1)$, $(\bar{c}c,1)$ are added in Example 2.1, then the resulting Thue system cannot be completed to yield a finite Church-Rosser Thue system presenting the group $\mathbb{Z}_2 * \mathbb{Z}_2$. However, one can actually get rid of of all those generators that do not have an inverse of length one.

Definition 2.4. Let $(\Sigma;T)$ be a finite two-monadic Church-Rosser presentation such that M_T is a group. For $a \in \Sigma$, define $\lambda(a) := \min\{|u| \mid au \overset{*}{\underset{T}{\leftrightarrow}} 1\}$, and for $i \in \mathbb{N}$, let $\Sigma_i := \{a \in \Sigma \mid \lambda(a) = i\}$.

According to our definitions T does not contain a rule of the form $(a,1)$, $a \in \Sigma$, and so $\Sigma_0 = \emptyset$. As it turns out $\Sigma_{i+1} \neq \emptyset$ implies $\Sigma_i \neq \emptyset$, and hence, $\Sigma = \bigcup_{i=1}^{m} \Sigma_i$, where $m = |\Sigma|$. Further, for each $i > 1$ and each letter $a \in \Sigma_i$, $\Sigma^i \cap \{u \mid au \overset{*}{\underset{T}{\leftrightarrow}} 1\}$ consists of a single irreducible word $u_a \in \Sigma_1^i$ implying that there is a single word $v_a \in \Sigma_1^i$ such that $a \overset{*}{\underset{T}{\leftrightarrow}} v_a$. In particular, M_T is already generated by Σ_1. Obviously, Σ_1 is exactly the subset of generators from Σ that have inverses of length one. Now the following theorem reduces the two-monadic case to the result of Avenhaus and Madlener.

Theorem 2.5 [4]. Let $(\Sigma;T)$ be a finite two-monadic Church-Rosser presentation of a group G, let $\Sigma_1 := \{a \in \Sigma \mid \exists b \in \Sigma: ab \overset{*}{\underset{T}{\leftrightarrow}} 1\}$, and let $T_1 := T \cap (\Sigma_1^2 \times (\Sigma_1 \cup \{1\}))$. Then $(\Sigma_1;T_1)$ is a finite monadic Church-Rosser presentation of G that provides inverses of length one for its generators.

Thus, we have the following extension of Theorem 2.2.

Corollary 2.6 [4]. A group G has a finite two-monadic Church-Rosser presentation if and only if G is isomorphic to a free product of a free group of finite rank and finitely many finite groups.

Although finite two-monadic Church-Rosser presentations are sufficient to present all free products of free groups of finite rank by finitely many finite groups, these presentations are very inefficient. Since each finite two-monadic Church-Rosser presentation $(\Sigma;T)$ of a group G contains a finite monadic Church-Rosser presentation $(\Sigma_1;T_1)$ of the same group such that each generator $a \in \Sigma_1$ has an inverse of length one, it is sufficient to consider presentations of the latter form.

Theorem 2.7. Let $(\Sigma;T)$ be a finite monadic Church-Rosser presentation of a finite group G such that each generator $a \in \Sigma$ has an inverse of length one. Then $|\Sigma| = |G|-1$, and T is the non-trivial part of a multiplication table for G.

To prove this theorem the following two observations are used.

<u>Lemma 2.8</u>. Let $(\Sigma;T)$ be as in Theorem 2.7, and let $a \in \Sigma$. Then for each $n \geq 1$, there is a word $u_n \in \Sigma \cup \{1\}$ such that $a^n \xrightarrow[T]{*} u_n$.
<u>Proof</u>. Since the group G presented by $(\Sigma;T)$ is finite, the order $m := \mathrm{ord}_G(a)$ of a in G in finite, i.e., $a^m \xrightarrow[T]{*} 1$. Since T is two-monadic, this implies that $(a^2,u_2) \in T$ for some word $u_2 \in \Sigma \cup \{1\}$. Proceeding by induction on n we see that $u_n a \xrightarrow[T]{*} u_{n+1} \xleftarrow[T]{*} au_n$ for some words $u_n \in \Sigma \cup \{1\}$ and $u_{n+1} \in \mathrm{IRR}(T)$. Thus, either $|u_{n+1}| < |u_n| + |a|$ implying $u_{n+1} \in \Sigma \cup \{1\}$ or $u_n a \equiv u_{n+1} \equiv au_n$, i.e., $u_n \in \{a\}^*$. Since $u_n \in \Sigma \cup \{1\}$, and since a^2 is reducible, this means that $u_n \equiv 1$ and $u_{n+1} \equiv a$. □

<u>Lemma 2.9</u>. Let $(\Sigma;T)$ be as in Theorem 2.7. Then $\mathrm{IRR}(T) = \Sigma \cup \{1\}$.
<u>Proof</u>. Since T is two-monadic, $\mathrm{dom}(T) \subseteq \Sigma^2$, and so $\Sigma \cup \{1\} \subseteq \mathrm{IRR}(T)$. To prove the converse inclusion it is sufficient to show that, for all $a,b \in \Sigma$, $ab \in \Sigma^2$ is reducible. Let $a,b \in \Sigma$. If $a \equiv b$, then $ab \equiv a^2$ is reducible by Lemma 2.8. So let us assume that $a \not\equiv b$, and that ab is irreducible.

Since G is finite, there exists a minimal integer $n \geq 1$ such that $(ab)^n \xrightarrow[T]{*} 1$. Then also $(ba)^n \xrightarrow[T]{*} 1$, and since $ab \in \mathrm{IRR}(T)$, this implies $n \geq 2$ and $(ba,c) \in T$ for some $c \in \Sigma$. Thus, $(ba)^{n-1} \xrightarrow[T]{*} c^{n-1} \xrightarrow[T]{*} u_{n-1} \in \Sigma$ by Lemma 2.8, which in turn yields that $(ab)^n \xrightarrow[T]{*} au_{n-1}b \xrightarrow[T]{*} 1$, i.e., $(au_{n-1},\bar b) \in T$ or $(u_{n-1}b,\bar a) \in T$, where $\bar a$ ($\bar b$) denotes the inverse of length one of a (b). In particular, this gives that $u_{n-1} \xleftrightarrow[T]{*} \bar a\bar b \xleftrightarrow[T]{*} \bar c$, and so $u_{n-1} \equiv \bar c$. By Lemma 2.8 $(c^2,d) \in T$ for some $d \in \Sigma \cup \{1\}$.

Assume that $d \equiv 1$. Then $\bar c \xleftrightarrow[T]{*} c$, which yields $c \equiv \bar c \equiv u_{n-1}$. Now $(cb,\bar a) \in T$, and so $a \xleftrightarrow[T]{*} \bar b\bar c \equiv \bar b c$, i.e., $(\bar b c,a) \in T$. Thus, $ab \underset{T}{\leftarrow} \bar b cb \underset{T}{\rightarrow} \bar b\bar a$. Since ab is irreducible, we obtain $ab \equiv \bar b\bar a$, i.e., $a \equiv \bar b$ and $b \equiv \bar a$. Hence, $ab \equiv a\bar a \underset{T}{\rightarrow} 1$ contradicting the choice of ab.

Thus, $d \in \Sigma$, and so $(\bar c d,c) \in T$. Now $\bar b d \underset{T}{\leftarrow} a\bar c d \underset{T}{\rightarrow} ac$, and since $a \not\equiv \bar b$, $(ac,f) \in T$ implying $(\bar c\bar a,\bar f) \in T$ for some $f \in \Sigma \cup \{1\}$. Hence, $\bar b\bar a \underset{T}{\leftarrow} a\bar c\bar a \underset{T}{\rightarrow} a\bar f$. Since $\bar b \not\equiv a$, this implies that $(\bar b\bar a,g) \in T$ for some $g \in \Sigma \cup \{1\}$, and hence, that $(ab,\bar g) \in T$. ↯

Thus, ab is in fact reducible, i.e., $\mathrm{IRR}(T) = \Sigma \cup \{1\}$. □

It remains to prove Theorem 2.7. So let $(\Sigma;T)$ be a finite monadic Church-Rosser presentation of a finite group G such that each generator $a \in \Sigma$ has an inverse of length one. Since there is a one-to-one correspondence between the irreducible words in IRR(T) and the elements of G, Lemma 2.9 implies that $|\Sigma| = |G|-1$. Further, since each word of length two is reducible, T contains a rule of the form (ab,c), $c \in \Sigma \cup \{1\}$, for each pair of letters $a,b \in \Sigma$. Thus, T is in fact the non-trivial part of a multiplication table for G.

Observe that once we drop the condition that each generator must have an inverse of length one, we get finite monadic Church-Rosser presentations that are much more succinct, as shown by the obvious presentations for cyclic groups of finite order.

3. On Gilman's conjecture

Gilman conjectured that Corollary 2.6 remains valid when the restriction 'two-monadic' is relaxed to 'monadic'. This conjecture is still open. Here we present the current knowledge concerning it.

Example 3.1. Let $\Sigma = \{a,b,c,d\}$, and $T = \{(abc,1),(ca,d),(db,1),(bd,1)\}$. Then $(\Sigma;T)$ is a finite monadic Church-Rosser presentation that is not two-monadic. As can be checked easily the monoid M_T it presents is a group. Using the definitions of the previous section we obtain $\Sigma_1 = \{b,d\}$ and $\Sigma_2 = \{a,c\}$. Now $\Sigma_1^* \cap \mathrm{IRR}(T) = \{b\}^* \cup \{d\}^*$, and each word $w \in \Sigma_1^*$ reduces to an irreducible word $\hat{w} \in \Sigma_1^*$. Thus, $a,c \in \Sigma_2$ are not congruent to any words from Σ_1^*, i.e., Σ_1 does not generate M_T. □

Thus, the technique used to prove the result for two-monadic presentations does not work in the general case. In fact, it does not work for any finite monadic Church-Rosser presentation that is not two-monadic, as we see from the following result.

Theorem 3.2. Let $(\Sigma;T)$ be a finite monadic Church-Rosser presentation such that the monoid M_T is a group, and let Σ_1 be the set of generators that have an inverse of length one. Then the following two statements are equivalent:

(i) Σ_1 generates the monoid M_T;

(ii) T is two-monadic.

Proof. That (ii) implies (i) is essentially Theorem 2.5. Thus, it remains to verify the converse implication. So let $(\Sigma;T)$ be a finite monadic Church-Rosser presentation of a group, let $\Sigma_1 := \{a \in \Sigma \mid \exists b \in \Sigma\colon ab \underset{T}{=} 1\}$, and let $\langle\Sigma_1\rangle_{M_T}$ denote the submonoid of M_T generated by Σ_1, i.e., $\langle\Sigma_1\rangle_{M_T} = \{w \in \Sigma^* \mid \exists u \in \Sigma_1^*\colon u \overset{*}{\underset{T}{\leftrightarrow}} w\}$. Finally, let $T_2 := T \cap (\Sigma^2 \times (\Sigma\cup\{1\}))$, i.e., T_2 is the subsystem of T containing all two-monadic rules. Then T_2 is a finite two-monadic reduced Thue system on Σ, and it can be checked easily that T_2 is also Church-Rosser.

Assume that Σ_1 generates the monoid M_T, i.e., $<\Sigma_1>_{M_T} = \Sigma^*$. Since $\Sigma \subseteq \mathrm{IRR}(T)$, this implies that for each letter $a \in \Sigma$, there exists a non-empty word $u_a \in \Sigma_1^*$ such that $u_a \overset{*}{\underset{T}{\leftrightarrow}} a$. We must show that T itself is already two-monadic, i.e., $T = T_2$. To this end we need the following two claims.

<u>Claim 1</u>. Let $a_1, a_2, \ldots, a_m \in \Sigma$ such that $\ell = a_1a_2\ldots a_m \in \mathrm{dom}(T)$. If $m \geq 3$, then $a_1 \notin \Sigma_1$ and $a_m \notin \Sigma_1$.

<u>Proof</u>. Assume that $a_1 \in \Sigma_1$. Then there is a letter $\bar{a}_1 \in \Sigma_1$ such that $(\bar{a}_1a_1, 1) \in T$. Let $(\ell, r) \in T$ be the rule of T with left-hand side ℓ. Since T is monadic, $|r| \leq 1$. Now $a_2\ldots a_m \underset{T}{\leftarrow} \bar{a}_1a_1a_2\ldots a_m \equiv \bar{a}_1\ell \underset{T}{\rightarrow} \bar{a}_1r$, and since $m \geq 3$ and T is reduced, this implies $m = 3$, $r \in \Sigma$, and $a_2a_3 \equiv \bar{a}_1r$. Thus, $\ell \equiv a_1a_2a_3 \equiv a_1\bar{a}_1a_3$ contradicting the fact that T is reduced. Hence, $a_1 \notin \Sigma_1$, and analogously, $a_m \notin \Sigma_1$. □

<u>Claim 2</u>. If $T_2 \neq T$, then there exists a rule $(\ell, r) \in T - T_2$ such that, for some shortest word $z \in \Sigma_1^*$ satisfying $z \overset{*}{\underset{T}{\leftrightarrow}} \ell$, we already have $z \overset{*}{\underset{T_2}{\rightarrow}} \ell$.

<u>Proof</u>. Let $(\ell', r') \in T - T_2$, and let $z' \in \Sigma_1^*$ be a shortest word such that $z' \overset{*}{\underset{T}{\leftrightarrow}} \ell'$, i.e., $z' \equiv z_o \underset{T}{\leftrightarrow} z_1 \underset{T}{\leftrightarrow} \ldots \underset{T}{\leftrightarrow} z_n \equiv \ell'$. By Claim 1 $\ell' \notin \Sigma_1^*$, and hence, $n \geq 1$. If $z' \overset{*}{\underset{T_2}{\rightarrow}} \ell'$ holds, then Claim 2 is satisfied with (ℓ', r') for (ℓ, r) and z' for z. So assume that $z' \overset{*}{\underset{T_2}{\not\rightarrow}} \ell'$, and let $i := \min\{j \mid z_j \underset{T_2}{\not\rightarrow} z_{j+1}\}$. Then $z_i \equiv x\ell y$ and $z_{i+1} \equiv xry$ for some rule $(\ell, r) \in T - T_2$. Again by Claim 1 $\ell \notin \Sigma_1^*$ implying $i \geq 1$. According to the choice of index i $z' \equiv z_o \overset{*}{\underset{T_2}{\rightarrow}} z_i \equiv x\ell y$, and so $z' = uzv$, where $u \overset{*}{\underset{T_2}{\rightarrow}} x$, $z \overset{*}{\underset{T_2}{\rightarrow}} \ell$, and $v \overset{*}{\underset{T_2}{\rightarrow}} y$. Since $z' \in \Sigma_1^*$ is a shortest word from Σ_1^* satisfying $z' \overset{*}{\underset{T}{\leftrightarrow}} \ell'$, z is a shortest word from Σ_1^* satisfying $z \overset{*}{\underset{T}{\leftrightarrow}} \ell$. Thus, Claim 2 holds with (ℓ, r) and z chosen as above. □

Assume that $T \neq T_2$, and let $(\ell, r) \in T - T_2$ be a rule such that for some shortest word $z \in \Sigma_1^*$ satisfying $z \overset{*}{\underset{T}{\leftrightarrow}} \ell$, we already have $z \overset{*}{\underset{T_2}{\rightarrow}} \ell$. Let $\ell \equiv a_1a_2\ldots a_m$, $a_1, a_2, \ldots, a_m \in \Sigma$. Then $m \geq 3$, and $a_1, a_m \notin \Sigma_1$. Since T_2 is monadic, $z \overset{*}{\underset{T_2}{\rightarrow}} \ell = a_1a_2\ldots a_m$ implies that z can be factored as $z \equiv z_1z_2\ldots z_m$, where $z_i \overset{*}{\underset{T_2}{\rightarrow}} a_i$ for all i. Let $z_1 \equiv b_1b_2\ldots b_k$, $b_1, b_2, \ldots, b_k \in \Sigma_1$. Then $k \geq 2$, since $z_1 \in \Sigma_1^*$ and $a_1 \notin \Sigma_1$. We may assume that the rule $(\ell, r) \in T - T_2$ and the word $z \in \Sigma_1^*$ have been chosen such that in addition to the above properties $|z_1|$ is minimal. We distinguish two cases: $k = 2$ and $k \geq 3$.

(i) <u>k = 2</u>: Then $b_1b_2\ldots b_k \equiv b_1b_2 \xrightarrow[T_2]{} a_1$. Since $b_1 \in \Sigma_1$, there is a letter $\bar{b}_1 \in \Sigma_1$ such that $(\bar{b}_1b_1,1) \in T_2$, and $(\bar{b}_1a_1,b_2) \in T_2$. Thus, $b_2a_2\ldots a_m \xleftarrow[T]{} \bar{b}_1a_1a_2\ldots a_m \xrightarrow[T]{} \bar{b}_1r$.

If $r \equiv 1$, we obtain $b_2a_2\ldots a_m \xrightarrow[T]{*} \bar{b}_1$, and so $a_2\ldots a_m \xleftrightarrow[T]{*} \bar{b}_2\bar{b}_1$. Since T is reduced, $a_2\ldots a_m$ is irreducible implying that $m = 3$ and $a_2a_3 \equiv \bar{b}_2\bar{b}_1$. But $b_1 \in \Sigma_1$, while $a_3 \equiv a_m \notin \Sigma_1$ $\lightning$.

Hence, $r \in \Sigma$. If $\bar{b}_1r$ is irreducible, then $b_2a_2\ldots a_m \xrightarrow[T]{*} \bar{b}_1r$, and so $b_2a_2\ldots a_i \xrightarrow[T]{*} \bar{b}_1$ and $a_{i+1}\ldots a_m \xrightarrow[T]{*} r$ for some index i. Since $a_2\ldots a_m \in IRR(T)$, this implies that $i = m-1$, i.e., $a_m \equiv r$. Thus, $a_1a_2\ldots a_m \equiv \ell \xrightarrow[T]{} r \equiv a_m$, which in turn gives $a_1a_2\ldots a_{m-1} \xleftrightarrow[T]{*} 1$ $\lightning$.

Hence, $(\bar{b}_1r,d) \in T_2$ for some $d \in \Sigma \cup \{1\}$, i.e., $b_2a_2\ldots a_m \xrightarrow[T]{*} d$, and so $a_2\ldots a_m \xleftrightarrow[T]{*} \bar{b}_2d$. Since $m > 3$ and $a_2\ldots a_m \in IRR(T)$, we conclude that $m = 3$ and $a_2a_3 \equiv \bar{b}_2d$. Thus, $(\bar{b}_1r,a_3) \in T_2$ implying $(b_1a_3,r) \in T_2$. Hence, $b_1a_3 \xleftrightarrow[T]{} r \xleftrightarrow[T]{} \ell \equiv a_1a_2a_3$, which yields that $b_1 \xleftrightarrow[T]{*} a_1a_2$, i.e., $(a_1a_2,b_1) \in T_2$. $\lightning$

(ii) <u>k > 3</u>: Then $b_1b_2\ldots b_{k-1}b_k \xrightarrow[T_2]{*} a_1$. Since $b_1 \in \Sigma_1$, there is a letter $\bar{b}_1 \in \Sigma_1$ such that $(\bar{b}_1b_1,1) \in T_2$, and so $b_2\ldots b_k \xleftrightarrow[T]{*} \bar{b}_1a_1$.

If $\bar{b}_1a_1$ is irreducible, then $b_2\ldots b_i \xrightarrow[T]{*} \bar{b}_1$ and $b_{i+1}\ldots b_k \xrightarrow[T]{*} a_1$ for some index $i > 2$. Let $z' :\equiv b_{i+1}\ldots b_kz_2\ldots z_m \in \Sigma_1^*$. Then $|z'| < |z|$ and $z' \xrightarrow[T]{*} a_1a_2\ldots a_m \equiv \ell$ contradicting the choice of z.

Hence, $(\bar{b}_1a_1,d) \in T_2$ for some $d \in \Sigma \cup \{1\}$, and since $a_1 \notin \Sigma_1$, we actually have $d \in \Sigma$. Now $da_2\ldots a_m \xleftarrow[T]{} \bar{b}_1a_1a_2\ldots a_m \xrightarrow[T]{} \bar{b}_1r$, and since $m > 3$, this means that $da_2\ldots a_m$ is reducible. Thus, $(da_2\ldots a_i,f) \in T$ for some $i > 2$ and some $f \in \Sigma \cup \{1\}$. Since $(\bar{b}_1a_1,d) \in T_2$, also $(b_1d,a_1) \in T_2$, and so $a_1a_2\ldots a_i \xleftarrow[T]{}$ $b_1da_2\ldots a_i \xrightarrow[T]{} b_1f$. Now $i > 2$, and $a_1a_2\ldots a_j$ is irreducible for $j < m$. Thus, we either have $i = m$ or $i = 2$ and $a_1a_2 \equiv b_1f$. However, $b_1 \in \Sigma_1$, while $a_1 \notin \Sigma_1$, i.e., $a_1a_2 \not\equiv b_1f$. Hence, $i = m$, and so $(da_2\ldots a_m,f) \in T$, i.e., $(da_2\ldots a_m,f)$ is another rule from $T-T_2$.

We have $b_2b_3\ldots b_k \xleftarrow[T_2]{} \bar{b}_1b_1b_2\ldots b_k \xrightarrow[T_2]{*} \bar{b}_1a_1 \xrightarrow[T_2]{} d$. Since T_2 is also Church-Rosser, we obtain $b_2b_3\ldots b_k \xrightarrow[T_2]{*} d$. Let $u :\equiv$ $b_2b_3\ldots b_k$, and assume that there exists a word $v \in \Sigma_1^*$ such that $|v| < k-1$ and $v \xrightarrow[T]{*} d$. Then $|b_1v| < k$ and $b_1v \xrightarrow[T]{*} b_1d \xrightarrow[T]{} a_1$ contradicting the choice of z. Thus, the rule $(da_2\ldots a_m,f) \in T-T_2$ and the word $uz_2\ldots z_m$ satisfy the assumptions made for (ℓ,r) and z, but $|u| < k = |z_1|$. $\lightning$

This completes the proof of Theorem 3.2. □

Thus, if the technique developed for the two-monadic case should work at all for the monadic case, then the partitioning of the set Σ of generators must be done differently. One obvious way is the following.

Definition 3.3. Let $(\Sigma;T)$ be a finite monadic Church-Rosser presentation such that M_T is a group. For $a \in \Sigma$, define $i(a) := \min\{k \mid \exists u_a \in \Sigma^*: au_a \xrightarrow[T]{k} 1\}$, and let $\Sigma_j' := \{a \in \Sigma \mid i(a) = j\}$.

Notice that for two-monadic presentations this definition coincides with the one given in Section 2. However, in general this new definition does not solve our problems.

Example 3.1 (continued). $\Sigma_1' = \{a,b,d\}$ and $\Sigma_2' = \{c\}$. Since each word $w \in \Sigma_1'^*$ reduces to an irreducible word $\hat{w} \in \Sigma_1'^*$, $c \in \Sigma_2'$ is not congruent to any word from $\Sigma_1'^*$, i.e., Σ_1' does not generate M_T. □

It seems as if the monadic case cannot successfully be attacked at all using this technique. At least until now no way of partitioning the given set of generators has been found that would yield a result paralleling Theorem 2.5.

So far only negative results concerning Gilman's conjecture have been presented. Fortunately, there also are a few positive ones that are worth mentioning.

First of all if only sets of generators of cardinality two are considered, then we obtain the following characterization.

Theorem 3.4. A group G has a finite monadic Church-Rosser presentation with at most two generators if and only if G is either a finite or infinite cyclic group or if G is isomorphic to the free product of two cyclic groups at least one of which is finite.

This result can be proved by a detailed case analysis that we will not present here.

Let $(\Sigma;T)$ be a finite monadic Church-Rosser presentation of a group G. Since $[1]_T = \{w \in \Sigma^* \mid w \xleftrightarrow[T]{*} 1\}$ is a context-free language [7], this means that G is a context-free group [18]. Thus, by the characterization theorem for context-free groups of Muller and Schupp [18] G is virtually free, i.e., G has a free subgroup of finite index. If in addition G is torsion-free, then G is already a free group [18]. Thus the only interesting case that remains open is the one in which G does contain elements of infinite order and non-trivial elements of finite order. Notice that it is effectively decidable whether a group presented through a finite monadic

Church-Rosser presentation is torsion-free [20].

4. On groups having finite Church-Rosser presentations

Here we present a few interesting new results concerning groups that can be defined by finite Church-Rosser presentations. All these results are based on the observation that in a finite Church-Rosser presentation the commuting of elements can only be expressed rather poorly. The first result in this area is due to Avenhaus, Book, and Squier.

Theorem 4.1 [2]. Let G be an infinite abelian group. Then G has a finite Church-Rosser presentation if and only if G is isomorphic to $\mathbb{Z}$.

This result was then generalized by Diekert.

Theorem 4.2 [9]. Let G be an infinite group that has an abelian subgroup of finite index. Then G has a finite Church-Rosser presentation if and only if G is isomorphic to either $\mathbb{Z}$ or to the free product $\mathbb{Z}_2 * \mathbb{Z}_2$.

So we see that the property of being abelian or of containing an abelian subgroup of finite index is in fact very restrictive when dealing with groups presented by finite Church-Rosser presentations. What can be said about the groups obtainable in this way when this restriction is somewhat weakened ? In a forthcoming paper the authors derive the following result.

Theorem 4.3 [16]. Let G be a group that has a finite Church-Rosser presentation. Then for each element u of G, if u has infinite order in G, then the centralizer $C_G(u)$ of u in G is isomorphic to $\mathbb{Z}$.

From this theorem one easily obtains the following consequences.

Corollary 4.4. Let G be a group that has a finite Church-Rosser presentation.

(a) Every abelian subgroup S of G that contains an element of infinite order is isomorphic to $\mathbb{Z}$.
(b) Every finitely generated abelian subgroup of G is either finite or isomorphic to $\mathbb{Z}$.
(c) If the center C of G is non-trivial, then G is either finite or isomorphic to $\mathbb{Z}$.

(d) If G contains a non-trivial normal subgroup that is finite, then G itself is finite.

Observe that Theorem 4.3 implies that whenever two elements of G commute, then either both have finite or both have infinite order. Since the problem of characterizing those groups that have finite Church-Rosser presentations has not yet been solved, the above results can at least help to check that a given group does not have a presentation of this form. Note that for a finite Church-Rosser presentation $(\Sigma;T)$, if T is non-monadic, then $[u]_T$ might not even be recursive [19]. Nevertheless, if the monoid M_T is a group, it can be shown using arguments like those of 4.7 that it is in fact a context-free group.

Examples 4.5. (a) Let $G = F_2 \times \mathbb{Z}_2$, i.e., G is the direct product of the free group of rank 2 and the cyclic group of order 2. Then G is infinite, but $\mathbb{Z}_2$ is a finite normal subgroup of G.
(b) Let G_1 be given through the presentation $(\Sigma;S_1)$, where $\Sigma = \{a,b\}$ and $S_1 = \{(abba,1)\}$. Then G_1 is infinite, but not isomorphic to $\mathbb{Z}$. However, its center is non-trivial, since $1 \not\overset{*}{\leftrightarrow}_{S_1} a^2 \in C_{G_1}$.
(c) For $n > 2$, let G_n be given through the presentation $(\Sigma;S_n)$, where $\Sigma = \{a,b\}$ and $S_n = \{((ab)^n ba,1)\}$. Then G_n contains a subgroup that is isomorphic to $\mathbb{Z}(\frac{1}{n}) := \{p \cdot n^q \mid p,q \in \mathbb{Z}\}$ [14]. This subgroup is abelian, and it clearly contains elements of infinite order. However, it is not finitely generated, and so it is not isomorphic to $\mathbb{Z}$.
(d) Greendlinger's group Gr is presented by $<\Sigma;L>$, where $\Sigma = \{a,b,c\}$ and $L = \{abc\bar{a}\bar{b}\bar{c}\}$. The subgroup of Gr that is generated by ab and $c\bar{a}$ is free abelian of rank 2 [12].

Thus, none of these groups can be described by a finite Church-Rosser presentation.

At the end of Section 3 we mentioned that it is decidable whether or not a group given through a finite monadic Church-Rosser presentation is torsion-free. To conclude this paper we want to deal with the corresponding problem for groups presented by finite Church-Rosser presentations in general.

Since the set of irreducible words modulo a finite Thue system is a regular language, and since for a group given through a finite Church-Rosser Thue system this set is a set of representatives, the following is fairly straightforward.

Lemma 4.6. Let G be a group that has a finite Church-Rosser presentation. Then G is infinite if and only if G contains an element of infinite order.

Thus, for a group G of this form it is decidable whether or not G contains elements of infinite order. That the order of a given element can be determined effectively is our next result.

Theorem 4.7. The following task can be solved effectively:

INSTANCE: A finite Church-Rosser presentation $(\Sigma;T)$ presenting a group G, and a word $u \in \Sigma^*$.

TASK: Determine the order of u in G !

Proof. The order of u in G is finite if and only if $u^n \overset{*}{\underset{T}{\to}} 1$ for some integer $n \geq 1$. Now let $u \in \Sigma^*$ be irreducible, and for $m \geq 1$, let u_m denote the irreducible descendant of u^m. Further, let $\lambda := \max\{|\ell| \mid \ell \in \text{dom}(T)\}$ and $\mu(u) := 2\cdot(|u|+|u^{-1}|)\cdot(\lambda-1)+\lambda$, where $u^{-1} \in \Sigma^*$ denotes the irreducible word presenting the inverse of u in G. Our algorithm for determining the order of u in G will be based on the following claim.

Claim. The element u has infinite order in G if and only if there exists an integer $p \geq 1$ such that $|u_{p+1}| > |u_p| \geq \mu(u)$.

Proof. If u has infinite order in G, then $u_i \neq u_j$ for all $i \neq j$. Thus, there must clearly be an integer $p \geq 1$ satisfying the above condition. So assume conversely that $p \geq 1$ satisfies $|u_{p+1}| > |u_p| \geq \mu(u)$.

Since T is Church-Rosser, we have $uu_p \overset{*}{\underset{T}{\to}} u_{p+1}$, and since G is a group, we also have $u^{-1}u_{p+1} \overset{*}{\underset{T}{\to}} u_p$. Thus, $|u_p| - |u^{-1}| \leq |u_{p+1}| \leq |u_p| + |u|$ implying that $0 \leq |uu_p| - |u_{p+1}| \leq |u| + |u^{-1}|$. Hence, whenever $uu_p \overset{i}{\underset{T}{\to}} u_{p+1}$, then $i \leq |u|+|u^{-1}|$. Analogously, $u_pu \overset{j}{\underset{T}{\to}} u_{p+1}$ also implies $j \leq |u| + |u^{-1}|$.

Now $uu_p \overset{i}{\underset{T}{\to}} u_{p+1} \overset{j}{\underset{T}{\leftarrow}} u_pu$, and so we have the following factorizations: $u_p \equiv xt \equiv sz$ and $u_{p+1} \equiv vt \equiv sw$, where $ux \overset{i}{\underset{T}{\to}} v$ and $zu \overset{j}{\underset{T}{\to}} w$. Since $u, u_p \in \text{IRR}(T)$, and since $i,j \leq |u| + |u^{-1}|$, we conclude that $|x|,|z| \leq (|u| + |u^{-1}|)\cdot(\lambda-1)$. By choice of p this yields that $t \equiv yz$ and $s \equiv xy$ for some word $y \in \Sigma^*$ of length $|y| \geq \lambda$. Thus, we have the following situation:

$$u_p \equiv xyz,$$
$$u_{p+1} \equiv vyz \equiv xyw,$$
where $ux \overset{*}{\underset{T}{\to}} v$ and $zu \overset{*}{\underset{T}{\to}} w$.

Since G is a group, and $u \overset{*}{\underset{T}{\not\to}} 1$, $v \neq x$ and $w \neq z$. Further, $|u_{p+1}| > |u_p|$ implies that $|v| > |x|$ and $|w| > |z|$, i.e., $v \equiv xx_1$ and $w \equiv z_1z$ for some non-empty words $x_1, z_1 \in \Sigma^*$. Hence, $u_{p+1} \equiv xx_1yz \equiv xyz_1z$, which yields $x_1y \equiv yz_1$.

It remains to prove that $u_{p+k} \equiv xx_1^kyz$ for all $k \geq 1$. This is done by induction on k. Since $u_{p+1} \equiv xx_1yz$, it remains to prove that $u_{p+k} \equiv xx_1^k yz$ implies $u_{p+k+1} \equiv xx_1^{k+1} yz$. Now $uu_{p+k} \equiv uxx_1^k yz \xrightarrow[T]{*} vx_1^k yz \equiv xx_1^{k+1} yz \equiv xx_1^k yz_1 z$. The word $xx_1^k y$ is a factor of u_{p+k}, and hence, it is irreducible. The word $yz_1 z$ is a factor of u_{p+1}, and hence, it is also irreducible. Since $|y| > \lambda$, this means that the word $xx_1^{k+1} yz \equiv xx_1^k yz_1 z$ is irreducible, i.e., $u_{p+k+1} \equiv xx_1^{k+1} yz$. Thus, for all $k \geq 1$, $u^{p+k} \not\xrightarrow[T]{*} 1$, and so u has infinite order in G. □

To determine the order of u in G keep on computing the irreducible descendants u_i of u^i. If u has finite order, then eventually an integer n will be reached such that $u_n \equiv 1$; the first such integer is the order of u. If u has infinite order, then we will eventually reach an integer p such that $|u_{p+1}| > |u_p| \geq \mu(u)$. □

Notice that we have actually proved that the set $\Delta_T^*(\{u\}^*) \cap \mathrm{IRR}(T)$ of irreducible descendants of powers of u is a regular set, once $(\Sigma;T)$ is a finite Church-Rosser presentation defining a group. Moreover, given the word u, a regular expression describing this set can be constructed effectively.

Now how can we check whether or not a group G given through a finite Church-Rosser presentation $(\Sigma;T)$ is torsion-free ? Fortunately, we can determine a finite set of words that are reasonable candidates for elements of finite order, and it will turn out that G contains non-trivial elements of finite order if and only if one of these candidates has finite order.

Define a mapping $\hat{}: \Sigma^* \to \Sigma^*$ through $\hat{u} := \min\{v \in \Sigma^* \mid u \sim v\}$, where $\sim$ denotes the conjugacy relation with respect to G, and the minimum is taken with respect to the following ordering:

$$x < y \text{ if and only if } |x| < |y| \text{ or } |x| = |y| \text{ and } x <_{lex} y.$$

Here $<_{lex}$ denotes the lexicographic ordering on Σ^*. Observe that for each $u \in \Sigma^*$, $u \sim \hat{u}$, and $\hat{u}$ is irreducible.

<u>Lemma 4.8</u>. Let $(\Sigma;T)$ be a finite Church-Rosser presentation describing a group G, and let $u \in \Sigma^*$. If u has finite order, then there exists a word $v \in \Sigma^*$ and a rule $(\ell,r) \in T$ such that

- $v \sim u$,
- $|v| = |\hat{u}|$, and
- $\ell \equiv vz$ for some non-empty word $z \in \Sigma^*$.

<u>Proof</u>. Let $u \in \Sigma^*$, and assume that u has finite order $m \geq 2$. Then $1 \not\equiv \hat{u}$, and $\hat{u}^m \xrightarrow[T]{*} 1$. Thus, there is a rule $(\ell,r) \in T$ that is applicable to $\hat{u}^m$, i.e., $\hat{u}^m \equiv x\ell y$, where $x \equiv u_1$, $\hat{u} \equiv u_1u_2 \equiv u_3u_4$, $\ell \equiv u_2\hat{u}^p u_3$,

and $y \equiv u_4 \hat{u}^{m-p-2}$ for some integer $p > 0$. If $u_2 u_3 \equiv 1$, then $p > 1$, and hence $v := \hat{u}$ satisfies our claim.

So assume that $u_2 u_3 \not\equiv 1$, and consider $v := u_2 u_1$. Then $v \sim \hat{u} \sim u$, and so $v \sim u$ and $|v| = |\hat{u}|$. In particular, v is irreducible due to the choice of $\hat{u}$.

We have $\ell \equiv u_2 \hat{u}^p u_3 \equiv u_2 (u_1 u_2)^p u_3$. If $|\ell| < |\hat{u}|$, then $p = 0$ and $|u_3| < |u_1|$, i.e., $u_1 \equiv u_3 u_5$ for some word $u_5 \in \Sigma^*$. Thus, $\hat{u} \equiv u_1 u_2 \equiv u_3 u_5 u_2$, $\ell \equiv u_2 u_3$, and $v \equiv u_2 u_1 \equiv u_2 u_3 u_5$ contradicting the fact that v is irreducible. Hence, $|\ell| > |\hat{u}|$, i.e., $p > 0$ or $|u_3| > |u_1|$. In any case this means that v is an initial factor of ℓ. □

Thus, if a group G is presented by a finite Church-Rosser presentation $(\Sigma;T)$, then G contains an element of finite order $m \geq 2$ if and only if an initial factor v of a left-hand side of a rule of T has order m. In particular, the initial factors of the words in dom(T) provide a test set for determining whether or not G contains any non-trivial elements of finite order. Together with Theorem 4.7 this gives our final result.

Theorem 4.9. The following problem is effectively decidable:
INSTANCE: A finite Church-Rosser presentation $(\Sigma;T)$ defining a group G.
QUESTION: Is the group G torsion-free ?

Although Church-Rosser presentations are more general than monadic Church-Rosser ones, it appears as if the groups presented by the former and the groups presented by the latter have many algebraic and algorithmic properties in common. In fact, until now no example of a group G is known such that G has a finite Church-Rosser presentation, but G does not have a finite monadic Church-Rosser presentation.

References

1. A.V. Anisimov; Group languages; <u>Cybernetics</u> 7 (1971), 594-601.
2. J. Avenhaus, R.V. Book, C. Squier; On expressing commutativity by finite Church-Rosser presentations: a note on commutative monoids; <u>RAIRO Inf. Théorique</u> 18 (1984), 47-52.
3. J. Avenhaus, K. Madlener; On groups defined by monadic Thue systems; <u>Coll. Math. Soc. János Bolyai</u>: Algebra, Combinatorics and Logic in Computer Science 42 (1983), 63-71.
4. J. Avenhaus, K. Madlener, F. Otto; Groups presented by finite two-monadic Church-Rosser Thue systems; <u>Trans. Amer. Math. Soc.</u>, to appear.

5. R.V. Book; Confluent and other types of Thue systems; J. Assoc. Comput. Mach. 29 (1982), 171-182.
6. R.V. Book; Thue systems as rewriting systems; in J.P. Jouannaud (ed.), Rewriting Techniques and Applications, Lect. Notes Comput. Sci. 202 (1985), 63-94.
7. R.V. Book, M. Jantzen, C. Wrathall; Monadic Thue systems; Theoret. Comput. Sci. 19 (1982), 231-251.
8. Y. Cochet; Church-Rosser congruences on free semigroups; Coll. Math. Soc. János Bolyai: Algebraic Theory of Semigroups 20 (1976), 51-60.
9. V. Diekert; Some remarks on Church-Rosser Thue presentations; in Proc. of STACS 87, Passau, Lect. Notes Comp. Sci. 247 (1987), 272-285.
10. M.J. Dunwoody; The accessibility of finitely presented groups; Inventiones Mathematicae 81 (1985), 449-457.
11. R.H. Gilman; Computations with rational subsets of confluent groups; Proceedings of EUROSAM 84, Lect. Notes Comput. Sci. 174 (1984), 207-212.
12. M. Greendlinger; Problem of conjugacy and coincidence with the anticenter in group theory; Siberian Math. J. 7 (1966), 626-640.
13. R.H. Haring-Smith; Groups and simple languages; Trans. Amer. Math. Soc. 279 (1983), 337-356.
14. M. Jantzen; Thue systems and the Church-Rosser property; Proceedings of MFCS 84, Lect. Notes Comput. Sci. 176 (1984), 80-95.
15. D. Kapur, P. Narendran; The Knuth-Bendix completion procedure and Thue systems; SIAM J. Comput. 14 (1985), 1052-1072.
16. K. Madlener, F. Otto; Commutativity in groups presented by finite Church-Rosser Thue systems; RAIRO Inf. Theorique, to appear.
17. W. Magnus, A. Karrass, D. Solitar; Combinatorial Group Theory; 2nd. rev. ed., Dover Publ., New York, 1976.
18. D.E. Muller, P.E. Schupp; Groups, the theory of ends, and context-free languages; J. Comput. System Sci. 26 (1983), 295-310.
19. C. O'Dunlaing; Infinite regular Thue systems; Theoret. Comput. Sci. 25 (1983), 171-192.
20. F. Otto; Elements of finite order for finite monadic Church-Rosser Thue systems; Trans. Amer. Math. Soc. 291 (1985), 629-637.
21. F. Otto; On deciding whether a monoid is a free monoid or is a group; Acta Inf. 23 (1986), 99-110.

Automated Theorem Proving applied to the Theory of Semigroups

by

R. B. McFadden
Department of Mathematical Sciences
Northern Illinois University
DeKalb, IL 60115

ABSTRACT

An automated reasoning program may be used in many different ways to assist with research in mathematics and related fields. The program used here for research in semigroup theory is ITP (Interactive Theorem Prover), designed at Argonne National Laboratory and at Northern Illinois University. It is a general purpose program, flexible in that it may call upon one inference rule or another, with choice dependent upon a given task in a given environment. Each step is available for scrutiny by the user; runs may be stopped at any time, and changes may be made as the user dictates.

ITP has been used to generate several types of semigroups, and to provide detailed analyses of Green's relations on these examples. It has been used to establish connections between R, L and D, and between these relations and regularity. All the proofs generated are readily accessible, and in some cases exhibit novel features. These results are preliminary to the planned project of equipping ITP with sufficient knowledge of semigroup theory to enable it to prove theorems in more depth. A third use has been to use the defining axioms for a local semilattice to generate some facts about free such objects. The results, while elementary, are encouraging.

1. INTRODUCTION

One of the first uses of an automated reasoning program in the theory of semigroups was to answer the following question, posed by I. Kaplansky [10]:

Does there exist a finite semigroup that admits a nontrivial antiautomorphism but that does not admit any nontrivial involution?

Given an affirmative answer, clearly one would like to know the order of the smallest and, for example, how many nonisomorphic semigroups there are of that order.

What makes these, and similar, questions amenable to solution by a reasoning program? This question, naturally, depends on the nature of the program itself. There are many such programs; for an account of some of them see [2]. The one used for the results of this paper is ITP (Interactive Theorem Prover), developed at N.I.U and A.N.L. As a general purpose reasoning program, ITP is capable of carrying out a wide variety of computations in first order logic. The questions addressed in this paper were all answered by use of the following three operations:

1. Hyperresolution is an inference rule that matches facts against hypotheses in an "if-then" statement, and generates the conclusion with variables properly instantiated.

More formally, hyperresolution considers simultaneously a clause N that contains at least one negative literal, and a set of clauses A_j, each of which contains only positive literals; N and the A_j are assumed pairwise to have no variables in common. When successful, the rule yields a clause B containing only positive literals, obtained by finding a Most General Unifier that simultaneously unifies one positive literal in each of the A_j with a distinct negative literal in N, and taking the 'or' of all literals that do not participate in the unification.

In examples, let us use ′ for negation and | for *or*, and let us denote constants by letters from the beginning of the alphabet or by upper case letters, variables by letters from the end of the alphabet, subscripted if necessary. Then from

P ′ (x) | Q ′ (y) | R(x,y)
P(a)
Q(b)

a one-step application of hyperresolution yields

R(a,b);

from

P ′ (a) | Q(b)
P(a) | R(c)

it yields

Q(b) | R(c).

2. Paramodulation is a generalization of equality substitution. Applied to a pair of clauses of which at least one contains a positive equality literal, it yields a clause in which the equality substitution corresponding to the equality literal has occurred.

More formally:

Given a clause C containing a term t, and a clause D containing an equality literal $\alpha = \beta$, where t unifies with α under substitution σ, paramodulation derives the clause C_1 with t replaced by $\beta\sigma$ (and other literals from D added).

For example, from

C: EQUAL(sum(y,minus(y)),0)   t = sum(y,minus(y))
D: EQUAL(sum(0,x),x) α = sum(0,x), β = x

paramodulation yields

C_1: EQUAL(minus(0),0)

using σ: (y <--- 0, x <--- minus(0)).

3. Demodulation rewrites generated conclusions according to a set of (directed) equations.

More formally:

Given a clause C containing a term t, and a unit equality clause of the form $\alpha = \beta$, where t is an instance of $\alpha(t = \alpha\sigma)$, demodulation replaces C by C_1 obtained from C by replacing all occurrences of t by $t_1 = \beta\sigma$.

For example, applied to

P(f(a,a)) and f(x,x) = g(x)

demodulation

adds P(g(a))
deletes P(f(a,a)).

4. Subsumption checks new, demodulated facts to eliminate duplicates.

For example, the fact

le(m(X,m(Y,U)),m(X,m(Y,U)))

is subsumed by

le(m(x1,m(x2,U)),m(x1,m(x2,U))).

For a detailed explanation of these operations, and others, as well as descriptions of the use of ITP in several different areas of science, see [9]. For a description of several other automated reasoning programs and examples of their use see [2].

2. GENERATION OF EXAMPLES

Returning to the antiautomorphism problem, we state our associativity property as

$$m(m(x,y),z) = m(x,m(y,z))$$

and rely on this to take care of duplicates like those in the above example on subsumption. One can prove that the order of the antiautomorphism h, say, must be a multiple of four, so we start with the free semigroup on four generators b, c, d, e, make h permute these cyclically, and extend h to the whole semigroup. Denoting the involution by j, we use the clauses:

EQUAL(h(m(x,y)),m(h(y),h(x)))
EQUAL(j(m(x,y)),m(j(y),j(x)))
EQUAL(j(j(x)),x).

Appropriate relations on the generators are bbc = dde and the relation that says that all products of four or more elements are equal (so assuring finiteness):

EQUAL(m(b,m(b,c)),m(d,m(d,e)))
EQUAL(m(x,m(y,m(w,z))),m(b,m(b,m(b,b))))
EQUAL(h(b),c).

Completing the definition of h in the obvious way, ITP was used with these clauses. Its first run generated a finite semigroup (order 85) with a nontrivial antiautomorphism. But it has many involutions, arising from even permutations of the generators. The strategy now was to block involutions in such a way that h remained well-defined, but the involutions did not. The first semigroup to be generated which satisfies the requirements has order 83. Adding more relations and applying paramodulation yielded the smallest possible. It has order seven, and there are four such nonisomorphic semigroups of that order.

Two very appealing and useful features of ITP are, first, its ability to impose complicated conditions on the structure concerned, and, second, the fact that the user can stop a run and modify the conditions in light of results generated so far. As an illustration of this, consider a locally inverse semigroup S. Here the natural ordering is compatible with multiplication, and S may even have more than one compatible ordering imposed on it. If $\leq$ is a compatible ordering on S and S contains an idempotent maximum in this ordering, then the structure of S is known [1]. As an experiment, it was decided to ust ITP to construct an example of such a semigroup. For regularity we used:

if Inv(x,y) then m(x,m(y,x)) = x;
if Inv(x,y) then m(y,m(x,y)) = y;
if (m(x,m(y,x) = x) & (m(y,m(x,y)) = y) then Inv(x,y).

(This form of input is just as acceptable as the EQUAL clauses above; note that the implications must be given both ways.) Idempotents are defined by

if Idem(x) then m(x,x) =x

with its converse, and the natural ordering by

if Idem(x) & Idem(y) & le(x,y) then (m(x,y) = x);
if Idem(x) & Idem(y) & le(x,y) then (m(y,x) = x);
if Idem(x) & Idem(y) & (m(x,y) = x) & (m(y,x) = x) then le(x,y);

This was generalized in the obvious way for $\leq$, and then S was given two elements; an idempotent U and a non-idempotent X with $X < U$ together with the fact that x = xUx for all elements x of S. The first run generated five elements, namely U, X, UX, XU, UXU, Y(=XX). Then associativity took over and production of useful facts ceased. However, it was known to the users that $X^2 = X^3$. ITP was stopped and this information added to the input. Renaming some of the elements already generated as H = UXU, F = XU, G = UX, not only did the next run generate a nine element example complete with multiplication table and ordering, it also described Green's relations (more on this later) completely on the example. But there was even more information available, much more. Because of the "if-then" nature of many of the clauses, the program was able to describe the implications of various equalities of products. Thus if Y = UYU then Y = XF = GX. (The output is not as easily readible as this, but one soon learns to spot the worthwhile clauses.) Again, for example, if Inv(U,G) then U = H. Altogether, the program provided detail as complete as requested of four different examples, all non-orthodox; one of order nine, two of order six, and one of order five, the smallest possible. True, their existence had been known before, but certainly not by this method.

3. PROVING THEOREMS

Continuing with locally inverse semigroups, ITP was used to reprove some elementary theorems. For example, with just a little modification of the clauses above, it proved quickly the (known) fact that if e and f are idempotents with e $\leq$ f then ef, fe, fef, efe are all idempotent, and if S is naturally ordered in the sense that $x\omega y$ implies that x $\leq$ y, then efe = e. The technique is to provide the axioms in the form of facts and clauses, then to include a clause which denies the result sought. ITP then succeeds in providing a proof precisely when it deduces the null clause. What is particularly attractive about this procedure is that the program has the capability to establish a logfile of its derived clauses, including their logical relationships (only non-subsumed clauses are retained). This file may be consulted for the derivation of any clause; the one most sought is that of the null clause, for its derivation forms a complete proof of the result. The relationships are in the form of the particular operations chosen, and must be translated into mathematical terms.

Many such elementary results have been proved using the program. Some require no intervention from the user. Others require the blocking of certain logical paths because the program will pursue all paths equally unless the user weights certain objects such as constants; an elaborate weighting system is built into ITP. Still others require additional facts to generate more useful information, as the generation on the nine element example mentioned above required the knowledge that $X^2 = X^3$. In general, it is not productive to set up a system of axioms, state a complicated theorem, and just let the program run. Each alternative at a given level must be examined by the program before it can move to the next level, with the result that it can run out of steam if care is not taken. To obviate this, and to remove as much as possible the *ad hoc* nature of the user's intervention, an alternative approach is being tried.

Think of ITP as an automated assistant. Given large amounts of facts, it will sift them quickly and accurately. Given a set of axioms and a theorem reasonably accessible, it may be able to prove the theorem. The problem is, how does one assess what results are reasonably accessible? And, how does one avoid the ad hoc approach so temptingly near when one can add tidbits of information on the fly?

One way to proceed is simply to formulate the axioms for a given system and run ITP on those for a short time. Even complicated systems of axioms may be used, and this approach is not without merit. Consider, for example, the axioms for the biordered set of a locally inverse semigroup, or a local semilattice, as it is called in [6]. They are:

$$x \wedge x = x, (x \wedge y) \wedge (x \wedge z) = (x \wedge y) \wedge z,$$

$$(x \wedge y) \wedge ((x \wedge z) \wedge (x \wedge u)) = ((x \wedge y) \wedge (x \wedge z)) \wedge (x \wedge u),$$

together with their duals. These are ideal fodder for ITP when changed to the (m(x,x) = x) format, and the program quickly and easily deduces results from the axioms even without any further modification. Many are consequences of idempotency, particularly at first, but not all of them are. In [8], Schein proved that the meet operation used is associative if and only if the set E(m) is a normal band. Normality is in fact a consequence of the axioms. In a very short run, ITP proved that:

$$\text{(for any x, y, z, w in E)}, \ m(m(m(x,y),z),w) = m(m(m(x,z),y),w),$$

precisely the condition required. This was a new result. While it can be deduced without too much trouble from Nambooripad's axioms [7], ITP deduced it without ever having heard of those axioms.

Again, ITP was used to generate from a few given products a small example of a local semilattice. It produced a complete multiplication table, and then deduced that for two of the elements F and H, say, of the example,

$$m(F,x) = m(H,x) \text{ for any element x of E.}$$

Unknown to the program was the fact that F and H were R-related, and the equality just mentioned was an instance of the fact that if e and f are R-related idempotents in a regular semigroup then the sandwich sets S(g,e) and S(g,f) coincide. The meet operation in the local semilattice may be defined precisely in terms of the sandwich sets, all of which are singletons.

A second approach to the proof aspect of things is to consider the program as a hardworking graduate assistant. Teach it as lemmas the basic results of a branch of semigroup theory, then have it try to prove some theorems. This too is being tried, using the Moore method of instruction to build up the user's knowledge of how to formulate questions for ITP to consider. Already the program has proved most of the elementary facts about Green's relations and regularity as stated in Chapter 2 of [3]. Finiteness results present a difficulty not yet overcome. But consider, for example, how ITP proved that if one element of a D-class is regular then every element of that class is regular. Like several such proofs, this one offers just enough variation on the usual treatment to be intriguing. Using only closure and associativity, the program was given the facts that aLcRb, that a is regular via aXa =a, but that b is not regular via (m(b,m(y,b))=b)'. ITP used paramodulation to deduce that

m(c,x) = m(c,m(X,m(a,x))),

as follows. Paramodulating from the fact that

EQUAL(m(c,m(X,a)),m(U,a))

into the axiom

m(*m*(*x*1,*x*2) ,*x*3)=*m*(*x*1,*m*(*x*2,*x*3))

(the spaces are provided in the output itself as an aid to clarity) yields

m(*m*(*U*,*a*),*x*1)=*m*(*c*, *m*(*m*(*X*,*a*),*x*1)).

A couple of applications of associativity follow, then paramodulation from

m(c,m(X,m(a,x1))) = m(c,x1)

into

m(c,m(X,m(V,b))) = m(c,Z)

using aZ = Vb yields the null clause, hence the result.

Another useful application of the program is the use of rewrite rules. By a set of rewrite rules we mean a set of ordered pairs of expressions, written *lhs* ---> *rhs*. There is a similarity relation between *lhs* and *rhs*, usually of the form equality, inequality, implication . If, by ignoring the order of the pairs and allowing rewriting with *rhs* ---> *lhs* as well as *lhs* ---> *rhs*, one expression can be rewritten into another, then the two expressions are said to be similar with respect to the set of rewrite rules [2].

Taking paramodulation as the rewriting rule of inference is especially useful in solving word problems. Consider the situation with a single relation, say

baabbabaa = a

over an alphabet A = {a, b}. Given a word w over A with w having b as initial letter, can w be rewritten as w = au for some word u over A [4]? Via paramodulation, ITP solved this problem with the following proof; the numbers of the statements are taken from the output.

4. baabbabaa = a
5. w = baaaa
6. (w = ax)'
10. wx = baaaax
13. ax = baabbabaax
15. aabbabaa = baaaba
29. aabbaa = baaaba
30. baabax = aabbaax
49. w = aabwbbabaa

Both 10 and 13 are derived from the input (no variable clauses) using associativity several times and paramodulation once. To derive 15, start with 13 in the form

$$ax = baaabba \quad baax$$

and paramodulate into it from 4, using x = abbabaa. To derive 29, start with 13 in the form

$$ax = baabba \quad baax$$

and paramodulate into it from 15 using x = abbaa. Again associativity implies 30 with a single application of paramodulation. For 49, start with 30 in the form

$$baaa \quad bax = aabbaax,$$

take x = aabbabaa and paramodulate from 4. Finally, paramodulate from 10 into the result in the form

$$baaaa = aab \quad baaaabbabaa$$

with x = bbabaa, then use 5 and one more application of paramodulation. Clearly 6 and 49 yield 50, which is the null clause, hence the result.

This is a promising start on what is in general a hard problem. A good more has been done on the word problem for a single relation of type a = v for v in A , and the results will appear elswhere.

Finally, since this paper was delivered, ITP has been used to calculate the free inverse semigroup F2B2 on two generators in the variety determined by the five element Brandt semigroup. A complete description of F2B2 and its semilattice of idempotents have been calculated [5].

REFERENCES

[1] Blyth, T. S. and R. B. McFadden, Naturally ordered regular semigroups with a greatest idempotent, Proc. Roy. Soc. Edinburgh, 91A,107-122,1981.

[2] Bundy, A. The Computer Modelling of Mathematical Reasoning, Academic Press, 1983.

[3] Howie, J. M., An Introduction to Semigroup Theory, Academic Press, 1976.

[4] Lallement, G., Some algorithmic problems in semigroup presentations, This volume, 1986.

[5] Lusk, E., McFadden, R., Using autoomated reasoning tools: A study of the Semigroup F2B2, Semigroup Forum (36), 1987, 75-88.

[6] Meakin, J., The free local semilattice on a set, to appear.

[7] Nambooripad, K. S. S., Structure of regular semigroups, Mem. Amer. Math. Soc. 224(1979).

[8] Schein, B. M., Pseudo-semilattices and pseudo-lattices, Izv. Vyss, Ucebn, Zaved. Mat. 2(117)(1972), 81-94 (in Russian).

[9] Wos, L., R. Overbeek, E. Lusk, J. Boyle, Automated Reasoning, Prentice Hall, 1984.

[10] Winker, S. K., L. Wos, E. L. Lusk, Semigroups, Antiautomorphisms, and Involutions: A Computer Solution to an Open Problem, I, Mathematics of Computation, 37(156), 533-545.

SUBDIRECTLY IRREDUCIBLE WE-2 SEMIGROUPS WITH GLOBALLY IDEMPOTENT CORE

Attila Nagy
Tecnical University of Budapest
Transport Engineering Faculty
Department of Mathematics
1111 Budapest, Müegyetem rkp. 9.
Hungary

It is a very natural thing to study subdirect products of semigroups and the subdirectly irreducible semigroups. In the literature there are a lot of theorems about the exponential semigroups and their generalizations, for example, the WE-m semigroups which make possible to describe the subdirectly irreducible WE-2 semigroups with globally idempotent core.

The object of this paper is to give a short summary of the mentioned theorems and prove that a semigroup is a subdirectly irreducible WE-2 semigroup with globally idempotent core if and only if it is isomorphic to either G or G^0 or a non-trivial subdirectly irreducible band where G is a non-trivial subgroup of a quasicyclic group. This result reduces the problem of finding all subdirectly irreducible WE-2 semigroups with globally idempotent core to the problem of finding all non-trivial subdirectly irreducible bands.

As known, a subsemigroup S of the direct product of the family S_i, $i\in I$, of semigroups is called a <u>subdirect product</u> of semigroups S_i such that $\pi_i(S) = S_i$, for all $i\in I$ where π_i denotes the projection homomorphism of the direct product $\times S_i$ on S_i.

We say that a semigroup S is <u>decomposable</u> as a subdirect product of a family S_i, $i\in I$, of semigroups if S is isomorphic to the subdirect product of S_i, $i\in I$.

Finally, a semigroup S is <u>subdirectly irreducible</u> if whenever S is written as a subdirect product of a family of semigroups S_i, $i\in I$, then, for at least one $j\in I$, the projection homomorphism π_j maps S onto S_j isomorphically.

THEOREM 1. <u>A non-trivial semigroup is subdirectly irreducible if and only if it has a least proper congruence</u>.

The least non-empty ideal of a semigroup S (if it exists) is called the _kernel_ of S. The kernel of a semigroup with zero is trivial. We call an ideal a _non-trivial_ ideal if it contains at least two elements. The least non-trivial ideal of a semigroup (if it exists) is called the _core_ of S. If K is the core of a semigroup S, then either $K^2 = K$ or $K^2 = \{0\}$. In the first case we call K _globally idempotent_, in the second case K is called _nilpotent_.

THEOREM 2. (Corollary 3.3.1 of [6]). _Every non-trivial subdirectly irreducible semigroup has a core._

As known, a _homogroup_ is a semigroup which contains a kernel that is a group.

THEOREM 3. (Theorem 4.1 of [7]). _Every subdirectly irreducible homogroup without zero is a group._

THEOREM 4. (Theorem 3.6 of [7]). _Semigroups_ S _and_ S^1 (S and S^0) _are simultaneously subdirectly irreducible or reducible._

THEOREM 5. (Theorem 5.1 of [7]). _An abelian group is subdirectly irreducible if and only if it is a subgroup of a quasicyclic group, that is, if it is a_ p^{∞}_-group or a cyclic group of order_ p^n, _where_ p _is a prime and_ n _is a non-negative integer._

DEFINITION 6. ([8]). _A semigroup_ S _is called exponential if, for every_ $x,y \in S$ _and every positive integer_ n, _the following equality_ holds: $(xy)^n = x^n y^n$.

DEFINITION 7. ([2]). _A semigroup_ S _is called a weakly exponential semigroup if, for every elements_ x,y _in_ S _and every positive integer_ m, _there is a positive integer_ k _such that_ $(xy)^{m+k} = x^m y^m (xy)^k = (xy)^k x^m y^m$.

Evidently, the weakly exponential semigroup is a common generalization of the exponential semigroup and the nil semigroup.

Following Nordahl [4], semigroups satisfying the identity $(xy)^m = x^m y^m$ ($m \geq 2$) are called E-m _semigroups_.

If S is a semigroup, then let WE(S) denote the set of all positive integers m which satisfy the following condition. For every x,y in S, there is a positive integer k such that $(xy)^{m+k} = x^m y^m (xy)^k = (xy)^k x^m y^m$.

THEOREM 8. (Theorem 1 of [3]). For every semigroup S, WE(S) is a subsemigroup of the multiplicative semigroup of all positive integers.

DEFINITION 9. ([3]). A semigroup S will be called a WE-m semigroup if $m \in WE(S)$, supposing $m \geq 2$.

THEOREM 10. (Theorem 2 of [3]). Every WE-m semigroup is a semilattice of archimedean WE-m semigroups.

THEOREM 11. (Theorem 3 of [3]). A 0-simple WE-m semigroup is an E-m completely simple semigroup with 0-adjoined.

THEOREM 12. (Theorem 5 of [3]). A semigroup is a WE-m archimedean semigroup with idempotents if and only if it is a retract extension of an E-m completely simple semigroup by a nil semigroup.

THEOREM 13. A semigroup is a WE-2 archimedean semigroup with idempotents if and only if it is a retract extension of a direct product of a rectangular band and an abelian group by a nil semigroup.

DEFINITION 14. Let K be an ideal of a semigroup S. We say that K is a dense ideal of S if the equality relation on S is the only congruence on S whose restriction to K is the equality relation on K. If this is the case, then we say that S is a dense extension of K. We say that S is a maximal dense extension of K if S is a dense extension of K and maximal under inclusion.

THEOREM 15. (Lemma 3.7 of [8]). Every weakly reductive semigroup K has a maximal dense extension and, identifying the inner part of the translational hull $\Omega(K)$ of K with K, any maximal dense extension V of K is isomorphic to $\Omega(K)$, that is, there is an isomorphism of V onto $\Omega(K)$ leaving the elements of K fixed.

THEOREM 16. A semigroup S is a subdirectly irreducible WE-2 semigroup with globally idempotent core if and only if it satisfies one of the following conditions.

(1) S is isomorphic to either G or G^0, where G is a non-trivial subgroup of a quasicyclic group.

(2) S is a non-trivial subdirectly irreducible band.

Proof. Let S be a subdirectly irreducible WE-2 semigroup with globally idempotent core K. Then K is a 0-simple semigroup. Assume that S has no zero element. As K is a WE-2 semigroup, we have, by Theorem 11, that K is a completely simple E-2 semigroup. Then K is isomorphic with a Rees IxJ matrix semigroup M(G;I,J;P) over a commutative group G with sandwich matrix P where every element of P is the identity of G (K is a rectangular abelian group). Since S is subdirectly irreducible, K is a dense ideal of S. As K is completely simple, it is weakly reductive. Then, by Theorem 15, S can be embedded into the translational hull $\Omega(K)$ of K. So we may assume that $S\subseteq\Omega(K)$.

Let Map(A,B) denote the set of all transformations from a set A to a set B. Write T_A instead of Map(A,A). Then, by [5],

$$\Omega(K)\cong\{[h(\cdot),\bar{g}(\cdot),g(\cdot),(\cdot)f]\in T_I\times Map(I,G)\times Map(J,G)\times T_J:\ g(j)p_{(j)f,i} = p_{j,h(i)}\bar{g}(i),\ \text{for all } i\in I \text{ and } j\in J\}.$$

The product in $\Omega(K)$ is given by

$$[h_1(\cdot),\bar{g}_1(\cdot),g_1(\cdot),(\cdot)f_1][h_2(\cdot),\bar{g}_2(\cdot),g_2(\cdot),(\cdot)f_2] =$$
$$= [h_1 o h_2(\cdot),\bar{g}_1(h_2(\cdot))\bar{g}_2(\cdot),g_1(\cdot)g_2((\cdot)f_1),(\cdot)f_1 o f_2] \text{ (see [5])}.$$

Since $p_{ji} = e$, the identity of G, we get $g(j) = \bar{g}(i)$, for all $i\in I$ and $j\in J$. Thus g and $\bar{g}$ are constant mappings.

Define the following relations on K.

$$r_1 = \{((i,g,m),(j,h,n))\in K\times K:\ i = j\},$$
$$r_2 = \{((i,g,m),(j,h,n))\in K\times K:\ m = n\},$$
$$r_3 = \{((i,g,m),(j,h,n))\in K\times K:\ g = h\}.$$

It can be easily verified that r_t, t = 1,2,3, are congruences on K.

Let $\bar{r}_t = \{(a,b)\in S\times S:\ a,b\in K \text{ and } (xay,xby)\in r_t \text{ for all } x,y\in S^1 \text{ or else } a = b\}$, t = 1,2,3.

The reader can easily verify that $\bar{r}_t$, t = 1,2,3, are congruences on S. Let r_t^o denote the restriction of $\bar{r}_t$ to K, t = 1,2,3. We prove that $r_t^o = r_t$, t = 1,2,3. It is evident that $r_t^o\subseteq r_t$. We shall prove $r_t\subseteq r_t^o$. Let t = 1. Assume $(a,b)\in r_1$, $a,b\in K$. Let a = (i,g,m) and b = (j,h,n) for some $i,j\in I$, $m,n\in J$ and $g,h\in G$. Let $x,y\in S^1\subseteq\Omega(K)$ be arbitrary elements such that

$$x = [h_1(\cdot),\bar{g}_1(\cdot),g_1(\cdot),(\cdot)f_1]$$

and

$$y = [h_2(\cdot),\bar{g}_2(\cdot),g_2(\cdot),(\cdot)f_2].$$

Then

$$xay = [h_1(\cdot),\bar{g}_1(\cdot),g_1(\cdot),(\cdot)f_1](i,g,m)[h_2(\cdot),\bar{g}_2(\cdot),g_2(\cdot).(\cdot)f_2]$$
$$= (h_1(i),\bar{g}_1(i)g,m)[h_2(\cdot),\bar{g}_2(\cdot),g_2(\cdot),(\cdot)f_2]$$
$$= (h_1(i),\bar{g}_1(i)gg_2(m),(m)f_2)$$

and, similarly,

$$xby = (h_1(j),\bar{g}_1(j)hg_2(n),(n)f_2).$$

Since $(a,b)\varepsilon r_1$, that is $i = j$, it follows that $(xay,xby)\varepsilon r_1$. This implies that $(a,b)\varepsilon \bar{r}_1$ which means that $(a,b)\varepsilon r_1^0$. Thus $r_1 \subseteq r_1^0$, and so $r_1 = r_1^0$.

The equality of r_2 and r_2^0 can be proved in a similar way.

Let $t = 3$. Assume $(a,b)\varepsilon r_3$, $a,b\varepsilon K$. Let x and y be arbitrary elements of S^1. Using the above expression for a, b, x and y, we have

$$xay = (h_1(i),\bar{g}_1(i)gg_2(m),(m)f_2)$$

and

$$xby = (h_1(j),\bar{g}_1(j)hg_2(n),(n)f_2).$$

Since $(a,b)\varepsilon r_3$, that is $g = h$, we get $(xay,xby)\varepsilon\bar{r}_3$ as $\bar{g}_1(i) = \bar{g}_1(j) = g_2(m) = g_2(n)$. Thus $r_3 \subseteq r_3^0$, that is, $r_3 = r_3^0$.

By Theorem 13, K is a direct product of the left zero semigroup K/r_1, the right zero semigroup K/r_2 and the abelian group K/r_3. Thus $r_1 \cap r_2 \cap r_3 = id$ on K (id is the identity relation).

Let $\bar{r} = \bar{r}_1 \cap \bar{r}_2 \cap \bar{r}_3$. Denoting the restriction of $\bar{r}$ to K by r^0, we have $r^0 = r_1^0 \cap r_2^0 \cap r_3^0 = r_1 \cap r_2 \cap r_3 = id$ on K. Since K is a dense ideal, it follows that $\bar{r} = id$ on S. Since S is subdirectly irreducible, we get either $\bar{r}_1 = id$ or $\bar{r}_2 = id$ or $\bar{r}_3 = id$ on S, that is K is either a left zero semigroup or a right zero semigroup or an abelian group.

Consider the case when K is an abelian group. Then S is a homogroup and so, by Theorem 3, S is a subdirectly irreducible abelian group. By Theorem 5, it follows that S is a subgroup of a p^∞-group where p is a prime. So (1) is satisfied.

Assume that $K = M(G;I,J;P)$ is a right zero semigroup. Then it is left reductive. Let $z = \{(a,b)\varepsilon SxS\colon ka = kb, \text{ for all } k\varepsilon K\}$. It can be easily verified that z is a congruence on S. Since the restriction of z is the identity congruence on K, it follows that $z = id$ on S. Thus K is a left reductor of S (see [7]).

Since K is right zero, $|G| = |I| = 1$. Let $G = \{e\}$ and $I = \{f\}$. Then, for arbitrary elements $(f,e,m)\varepsilon K$ and $s = [h(\cdot),\bar{g}(\cdot),g(\cdot),(\cdot)f]\varepsilon S$ $sk = [h(\cdot),\bar{g}(\cdot),g(\cdot),(\cdot)f](f,e,m) = (h(f),\bar{g}(f)e,m) = (f,e,m) = k$.

Since S is a WE-2 semigroup, there is a positive integer v such that

$(ks)^{2+v} = (ks)^v k^2 s^2$. Since K is a right zero semigroup, the last expression can be replaced by $ks = (ks)ks^2 = k(sk)s^2 = k^2s^2 = ks^2$. From this last equation it follows that $s = s^2$, for every $s \in S$, as K is the left reductor of S. Thus S is a subdirectly irreducible band. So (2) is satisfied.

We can conclude a similar result if K is a left zero semigroup.

In the next, assume that S has a zero element. By Theorem 10, S is a semilattice of archimedean WE-2 semigroups. Let S_i denote the archimedean component of S containing the zero element 0 of S. There are two cases.

First assume $|S_i| = 1$. We show that the complement of the zero element in S is a subsemigroup. Let $a, b \in S$ with $a \neq 0$ and $b \neq 0$. If ab was 0 then the subset

$$X = \{x \in S : xb = 0\}$$

would be a non-trivial ideal in S, because S_i is a semilattice component. So $K \subseteq X$, that is $Kb = \{0\}$. We can verify, in a similar way, that the subset

$$Y = \{y \in S : Ky = \{0\}\}$$

is a non-trivial ideal of S. So $K \subseteq Y$, that is, $K^2 = \{0\}$, contradicting the assumption that K is a globally idempotent core. Thus $S - \{0\}$ is a semigroup.

Assume $|S_i| \geq 2$. Then $K \subseteq S_i$. Since K is globally idempotent, it is 0-simple. Then, by Theorem 11, K is a completely simple semigroup with 0-adjoined. Thus $k^n \neq 0$, for every $k \in K$ $(k \neq 0)$ and every positive integer n. But this contradicts the fact that S_i is an archimedean semigroup.

By the previous results and Theorem 4, $S - \{0\}$ is a subdirectly irreducible WE-2 semigroup without zero. If $S - \{0\}$ is a one-element semigroup, then S satisfies (2). If $S - \{0\}$ has at least two elements, then, using Theorem 4 and results proved for subdirectly irreducible WE-2 semigroups without zero, S is isomorphic to either G^0 or a subdirectly irreducible band, where G is a non-trivial subgroup of a quasicyclic group.

Thus we have proved that every subdirectly irreducible WE-2 semigroup with idempotent core satisfies either (1) or (2).

Since semigroups satisfying either (1) or (2) are subdirectly irreducible WE-2 semigroups with globally idempotent core, the theorem is proved.

REFERENCES

[1] Clifford, A.H. and G.B. Preston, The algebraic theory of semigroups, Amer. Math. Soc. Providence R.I., 1961 (Vol. I.)

[2] Nagy, A., Weakly exponential semigroups, Semigroup Forum, Vol. 28 (1984) 291-302.

[3] Nagy, A., WE-m semigroups, Semigroup Forum, Vol. 32(1985) 241-250

[4] Nordahl, T., Semigroups satisfying $(xy)^m = x^m y^m$, Semigroup Forum Vol. 8(1974) 332-346

[5] Petrich, M., The translational hull in semigroups and rings, Semigroup Forum 1(4)(1970) 283-360

[6] Petrich, M., Introduction to semigroups, Merrill, Columbus, 1973

[7] Schein, B.M., Homomorphisms and subdirect decomposition of semigroups, Pacific Journal of Math. Vol. 17. No 3(1966) 529-547

[8] Tamura, T. and J. Shafer, On exponential semigroups I., Proc. of the Japan Academy, Vol. 48, No 2(1972)

COMMUTATIVE MONOID RINGS WITH KRULL DIMENSION

Jan Okniński

Institute of Mathematics,
University of Warsaw,
00-901 Warsaw, Poland.

It is known that for any monoid S and any ring with unity A the monoid ring A[S] is artinian if and only if A is artinian and S is finite, [15]. Some conclusive results concerning related finiteness conditions of semigroup rings have been obtained in [8], [9]. On the other hand, our knowledge about noncommutative noetherian monoid rings is very meagre except the case of polycyclic by finite groups (c.f. [11]). In the commutative case it was shown in [3] that the monoid ring A[S] is noetherian if and only if A is noetherian and S is finitely generated. Some particular subclasses of commutative noetherian monoid rings have been extensively studied (c.f. [3],[6]).

While the theory of rings having Krull dimension in the sense of [4] is mainly motivated by noetherian rings, group rings of this type have been investigated in [10],[13],[14].

On the other hand the behaviour of the classical Krull dimension (defined by chains of prime ideals) in the class of commutative cancellative monoid rings is described in [3].

In this paper A will be a commutative ring with unity and S - a commutative monoid. We get a description of the classical Krull dimension of A[S] generalizing the above mentioned result concerning the cancellative case. Further, some necessary conditions for A[S] to have Krull dimension are given and a complete characterization of this dimension is proved in the case where S satisfies a.c.c. on principal ideals. The latter extends the description of noetherian A[S], ([3], Theorem 7.7).

The Krull dimension and the classical Krull dimension will be denoted by Kdim and clKdim respectively. If I is an ideal of $A[S]$ and $\sim_I$ denotes the congruence of S given by $a\sim_I b$ if and only if $a-b \in I$, then we write S_I for the semigroup $S/\sim_I$. For a subset X of S, by $\langle X\rangle$ we mean the subsemigroup generated by X. If S has a zero element, then it is denoted by θ and $A_0[S]$ is the contracted semigroup ring. Let η be the least semilattice congruence in S and let $\varphi: S \to S/\eta$ be the natural epimorphism (c.f. [1], §4.3). An archimedean component S_0 of S will be called minimal (least) nonzero if $S_0 \neq \theta$ and $\varphi(a) \nless \varphi(S_0)$ $(\varphi(a) \geq \varphi(S_0))$ for any $\theta \neq a \in S$ with regard to the natural order of S/η.

It is well known that, if T is a cancellative semigroup, then the torsion free rank of the group of fractions of T is equal to the supremum of the cardinalities of finite algebraically independent subsets of T. This allows us to define the rank of T (c.f. [3], p. 165). We will show that the notion of rank may be extended, in a reasonable way, to arbitrary semigroups. While for any semigroup S there exists a least cancellative congruence ρ in S, this cannot be used to define the rank of S. In fact, if S has a zero element, then ρ kills the whole semigroup. This suggests the following definition. If T is a 0-cancellative semigroup (i.e. T is cancellative possibly with θ adjoined), then let $T'=T\setminus\{\theta\}$ if $\theta \in T$ and $T'=T$ otherwise. Further, let $\bar{T}$ denote the group of fractions of T'. By the rank of S (rk S) we mean the supremum of torsion free ranks of all groups $\bar{T}$ arising from 0-cancellative homomorphic images T of S. Let Rk $S=\sup\{\text{rk}\, T \mid T$ - a free subsemigroup of $S\}$. The following result is crucial for applications of the rank.

Theorem 1. rk$S$ = Rk S.

Proof. Since rk T = Rk T for any cancellative semigroup T, the inequality rk $S \leq$ Rk S is an easy consequence of the definitions. Thus, it is enough to show that if $T=\langle a_1,\dots,a_n\rangle$ is a free subsemigroup of S with free generators $a_1,\dots,a_n$ then rk $S \geq n$. Let $\sim$ be the least separative congruence in S. From the description of $\sim$, ([1], Theorem 4.14), it follows that the image of T under the natural epimorphism $S \to S/\sim$ is isomorphic to T. Since rk $S \geq$ rk $S/\sim$, then we may assume that S is separative. It is easily seen that the semigroup $a_1\dots a_n T$ is contained in an archimedean component S_0 of S. Put $J = \{s \in S \mid \varphi(S_0) \nleq \varphi(s) S/\eta\}$ where

η is the least semilattice congruence in S and $\varphi: S \to S/\eta$ is the natural epimorphism. Then J is empty or it is an ideal of S (consisting exactly of archimedean components of S lying "not above" S_o). If $J \neq \emptyset$, than $\mathrm{rk}\, S \geq \mathrm{rk}\, S/J$ and so we may assume that $J = \{\Theta\}$ or $J = \emptyset$. Hence S_o is the least nonzero archimedean component of S. Since S_o is cancellative, ([1], Theorem 4.16), $\Theta \notin S_o$. Moreover, if $\Theta \in S$, then $S = \tilde{S} \setminus \{\Theta\}$ is a semigroup.
Let ρ be the least cancellative congruence in $\tilde{S}$. Then S/ρ or $\tilde{S}/\rho \cup \{\Theta\}$ if $\Theta \in S$, is a 0-cancellative image of S. If $\mathrm{rk}\, S < n$, then there exist $a,b \in T$, $a \neq b$, such that $a\rho b$ which means that $za=zb$ for some $z \in \tilde{S}$. Hence $zab=zba=zaa$ and $za \in S_o$ since $\varphi(za)=\varphi(z)\varphi(a)=\varphi(a)$ (note that $\varphi(z) \geq \varphi(a)$). This contradicts the fact that S_o is cancellative and proves the inequality $\mathrm{rk}\, S \geq n$.

Corollary 1. The following conditions are equivalent:

1) $\mathrm{rk}\, S = \infty$,
2) S has a free subsemigroup of infinite rank.

Proof. 2) $\Rightarrow$ 1) follows from Theorem 1.
1) $\Rightarrow$ 2) If 2) does not hold, then there exists a finite subset $\{a_1,\dots,a_n\} \subseteq S$ which is maximal among algebraically independent subsets of S. This easily implies that $\mathrm{rk}\, T \leq n$ for any 0-cancellative image T of S and hence $\mathrm{rk}\, S \leq n$.

The rank appears useful when describing the classical Krull dimension of commutative monoid rings. Our result extends that proved by Arnold and Gilmer in the cancellative case ([3], Theorem 21.4).

Theorem 2

1) $\mathrm{clKdim}\, A[S] = \sup\{\mathrm{clKdim}\, A[\bar{S}_P] \mid P$ - a prime ideal of $A[S]\}$,
2) $\mathrm{clKdim}\, A[S] < \infty$ if and only if $\mathrm{clKdim}\, A < \infty$ and $\mathrm{rk}\, S < \infty$, and in this case $\mathrm{clKdim}\, A[S] = \mathrm{clKdim}\, A[x_1,\dots,x_{\mathrm{rk}S}]$.

Proof. We know that $\mathrm{clKdim}\, A[S] = \sup\{\mathrm{clKdim}\, A[S]/_P \mid P$ - a prime ideal of $A[S]\}$. Let P be a prime ideal of $A[S]$. Since we have the natural epimorphisms $A[S] \to A[S_P] \to A[S]/_P$, then $\mathrm{clKdim}\, A[S] \geq \mathrm{clKdim}\, A[S_P] \geq \mathrm{clKdim}\, A[S]/_P$. Now S_P embeds into a domain and so it is a 0-cancellative semigroup. Then $A[S_P] = A[S_P']$ or $A[S_P] \simeq A[S_P'] \oplus A$ and $\mathrm{clKdim}\, A[S_P] = \mathrm{clKdim}\, A[S_P'] = \mathrm{clKdim}\, A[\bar{S}_P]$,

([3], Theorem 21.4). This establishes 1). The second assertion is an easy consequence of 1), [3], Theorem 17.1, and the well known bounds for the classical Krull dimension of polynomial rings ([2], Corollary 30.3).

The next result follows from Theorem 2, 2) and [2], Theorem 30.5.

Corollary 2. Assume that A is noetherian. Then $\text{clKdim}\, A[S] = \text{clKdim}\, A + \text{rk}\, S$.

Now, we pass to the Krull dimension in the sense of [4].

Lemma 1. Let I be an ideal of S. If $A[S]$ has Krull dimension, then the number of principal ideals of S which are maximal among principal ideals of S contained in I is finite.

Proof. Suppose that $a_1S, a_2S, \ldots$ are distinct maximal principal ideals of S contained in I. If I_j is the ideal of nongenerators of a_jS, then put $P = S/\bigcup_{j=1}^{\infty} I_j$. Since $a_iS \cap a_jS \subseteq I_j$ for any i,j, $i \neq j$, then P has an infinite set of mutually Θ-disjoint ideals. This yields an infinite set of independent ideals of the rings $A_o[P]$. Since $A_o[P]$ has Krull dimension, this contradicts the finiteness of the Goldie dimension of $A_o[P]$, (c.f. [4], Proposition 1.4).

Proposition 1. Assume that $A[S]$ has Krull dimension. Then

1) A has Krull dimension,
2) S has finitely many archimedean components,
3) $\text{rk}\, S < \infty$,
4) if $\text{rk}\, S > 0$, then A is noetherian,
5) if $\text{rk}\, S = 0$, then S is finite.

Proof. 1) is immediate.
2) Let η be the least semilattice congruence in S. Then $A[S/\eta]$ has Krull dimension. If S/η is infinite, then $A[S/\eta]$ has an infinite set of orthogonal idempotents. This contradicts the finiteness of the uniform dimension of $A[S/\eta]$ (c.f. [4], Proposition 1.4). Hence S/η is finite which is equivalent to 2).
3) Suppose that $\text{rk}\, S = \infty$. By 2) and Corollary 1 there exists an infinite algebraically independent subset of S contained in an

archimedean component of S. This implies, as in the proof of Theorem 1, that S has a 0-cancellative homomorphic image T with $\text{rk } T = \infty$. Since $A[T']$ has Krull dimension, then the group ring $A[\bar{T}]$, being a localization of $A[T']$, also has Krull dimension. This, in view of [14], (Proposition 2.4), contradicts the fact that $\text{rk } \bar{T} = \infty$.
4) Since $A[\bar{T}]$ has Krull dimension for a 0-cancellative homomorphic image T of S, $\text{rk } T > 0$, then the assertion follows from [14], Corollary 3.2.
5) By Theorem 1, S is periodic. For any $e=e^2 \in S$ we have $A[S] = A[eS] \oplus A_0[S/_{eS}]$ and the rings $A[eS], A[S/_{eS}]$ have Krull dimension. Since S has finitely many idempotents, then by repeating this procedure we can assume that S has no nontrivial idempotents. This means that $S=S_1 \cup S_0$ where S_1 is the group of units of S and S_0 is a nilideal of S or S_0 is empty. Now $A[S_1]$ is a homomorphic image of $A[S]$ and so it has Krull dimension. The periodicity of S_1 and [14] then imply that S_1 is finite. If $S_0 \neq \emptyset$, then $A_0[S_0]$ is a nilideal of $A_0[S]$ and from [4], Theorem 5.1, it follows that $A_0[S_0]$ is nilpotent. Thus $S_0^n=\theta$ for some $n \geq 1$ and it is enough to show that $S_0/_{S_0^2}$ is finite. Passing to the semigroup ring $A[S/_{S_0^2}]$ we can assume that $S_0^2=\theta$. Since for any $\theta \neq a \in S_0$ the ideal $aS = aS_1 \cup \{\theta\}$ is finite and it is a maximal principal ideal of S contained in S_0, then Lemma 1 implies that S_0 is finite which completes the proof.

The following result is a direct consequence of [5], Theorem 5.3 and [4], Lemma 1.1.

Lemma 2. Let $A \subseteq B$ be rings with the same unity such that B is a finitely generated A-module. Then A has Krull dimension if and only if B has Krull dimension and, in this case, $\text{Kdim } A = \text{K dim } B$.

Corollary 3. Assume that $\text{rk } S = 0$. Then $A[S]$ has Krull dimension if and only if A has Krull dimension and S is finite. Moreover, in this case, $\text{Kdim } A[S] = \text{Kdim } A$.

Proof. The assertion follows from Proposition 1,5) and from Lemma 2.

Let us notice that, if $\text{Kdim } A[S]$ exists, then in view of [4], Corollary 7.5, we get as in the proof of Theorem 2 $\text{Kdim } A[S] = \sup\{\text{Kdim } A[T'] \mid T$ - a 0-cancellative image of $S\} = \sup\{\text{Kdim } A[\bar{S}_P] \mid P$

- a prime ideal of $A[S]\}$.

In view of Proposition 1 and Corollary 3 we may further assume that A is noetherian. Thus, if S is finitely generated then $A[S]$ is noetherian and so it has Krull dimension, ([4], Proposition 1.3). The case of (commutative) group rings might suggest that the converse is true (c.f. [14]).

Example 1. Let $S=\{\theta,1,y,x_1,x_2,\dots\}$ with the multiplication defined by the rules: $x_ix_j=\theta$ for $i,j\geq 1$, $yx_1=x_1y=\theta$, $yx_i=x_iy=x_{i-1}$ for $i>1$. Plainly $\operatorname{rk} S = 1$ and S is not finitely generated. We will show that $\operatorname{Kdim} K[S] = 1$ for any field K. The subset $I=S\setminus\{1,y\}$ is an ideal of S, and $K[S]/_{K[I]} \simeq K[\langle y,1\rangle]$ is a ring with Krull dimension 1. Hence, it has Krull dimension 1 as a $K[S]$-module. Let $0\neq J\subseteq K_o[I]$ be an ideal of $K_o[S]$. Then for any $a=\sum_{i=1}^{n} k_ix_i\in J$, $k_n\neq 0$, we have $ay,ay^2,\dots,yy^{n-1}\in J$. This easily implies that $x_1,\dots,x_n\in J$. Hence $J=K_o[\{x_1,\dots,x_m,\theta\}]$ for some $m\geq 1$ or $J=K_o[I]$. This means that $\operatorname{Kdim} K_o[I]_{K_o[S]} = 0$ and so $K[I]$ is also artinian as a $K[S]$-module. By [4], Lemma 1.1, $\operatorname{Kdim} K[S]=1$. Observe that $x_1S\subseteq x_2S\subseteq\dots$, i.e. S does not satisfy the a.c.c. on principal ideals.

An analogues example may be also constructed in the cancellative case. However, in this situation, the rank of S must exceed 1. If $S=\langle x\rangle\times\langle y,y^{-1}\rangle\cup\langle y^{-1}\rangle\cup\{1\}$ is a subsemigroup of the torsion free group of rank 2 with generators x,y, then S is not finitely generated and one can check that $\operatorname{Kdim} K[S]=2$ for any field K.

It is known that any (not necessarily commutative) group G satisfies the a.c.c. on normal subgroups whenever $\operatorname{Kdim} A[G]$ exists, [10]. On the other hand, Example 1 shows that the a.c.c. on principal ideals of S is not a consequence of the existence of $\operatorname{Kdim} A[S]$. We will show that a complete characterization of the Krull dimension of $A[S]$ may be given when additionally assuming this finiteness condition on S.

Lemma 3. Let T be a semigroup satisfying the a.c.c. on principal ideals and let I be an ideal of T which is archimedean and has no non-zero idempotents. If the semigroup S/I^2 is finitely generated, then

S is also finitely generated.

Proof. Let $T/I^2 = \langle \bar{x}_1,\dots,\bar{x}_n \rangle$ where $\bar{x}_i$ denotes the image of some $x_i \in T$ under the natural homomorphism $T \to T/I^2$. For any $a \in I^2$, $a \neq 0$, we have $a=b\,c$, $b,c \in I$. Suppose that $aI^1 = bI^1$. Then $a=axc$ for some $x \in I^1$. If $\theta \in T$, then I is a nilsemigroup and so $xc \in I$ is nilpotent, a contradiction. Thus, assume that $\theta \notin T$. Then I has no idempotents. If $\sim$ is the least separative congruence in T, then $I/\sim$ is cancellative. This implies that $xc \sim (xc)^2$. From the description of $\sim$ it then follows that xc is a periodic element, a contradiction. Thus, $aI^1 \subsetneqq bI^1$ and similarly $aI^1 \subsetneqq cI^1$. Since S satisfies a.c.c. on principal ideals, then as in [12], Theorem 1, we show that I satisfies a.c.c. on principal ideals. Hence $a=b_1\dots b_m$ for some $b_i \in I$ generating maximal principal ideals b_iI^1 of I. Now $b_i \notin I^2$ which means that $b_i \in \langle x_1,\dots,x_n \rangle$. This implies that $I^2 \subseteq \langle x_1,\dots,x_n \rangle$ and so $T = \langle x_1,\dots,x_n \rangle$.

Theorem 3. Assume that S satisfies the a.c.c. on principal ideals. If $A[S]$ has Krull dimension, then S is finitely generated. Moreover, in this case:

1) $\operatorname{Kdim} A[S] = \operatorname{Kdim} A + \operatorname{rk} S$,
2) if $\operatorname{Kdim} A < \infty$, then $\operatorname{Kdim} A[S] = \operatorname{clKdim} A[S] < \infty$.

Proof. To prove that S is finitely generated we will proceed by induction on the number n_S of nonzero archimedean components of S. If $n_S=1$, then S is a group, possibly with zero, since the component S_1 of the identity is a group. Thus $A[S] = A[S_1]$ or $A[S] \simeq A[S_1] \oplus A$ and S is finitely generated by [14]. Now assume that $n_S > 1$. Since distinct idempotents of S belong to distinct archimedean components of S, ([1], Theorem 4.12), and the semigroups eS, S/eS inherit the hypotheses on S, then we can assume as in Proposition 1,5) that S has no nontrivial idempotents. Let S_o be a minimal nonzero archimedean component of S. Put $\tilde{S}_o = S_o \cup \{\theta\}$ if $\theta \in S$ and $\tilde{S}_o = S_o$ otherwise. Then $\tilde{S}_o$ is an ideal of S and $n_{S/\tilde{S}_o} < n_S$. Since $A[S/\tilde{S}_o]$ has Krull dimension, then by the induction hypothesis $S/\tilde{S}_o$ is finitely generated. If there exists another minimal nonzero archimedean component T_o of S, then similarly $S/\tilde{T}_o$ is finitely generated. Moreover $T_o \subseteq$ $\subseteq S \setminus S_o \cup \{\theta\}$. Hence, in this case, S is finitely generated (as generators of S we can take the union of the sets of generators of S modulo $\tilde{S}_0$ and those of S modulo $\tilde{T}_0$). Thus, we can assume that S_0 is the least nonzero archimedean component of S. If $\theta \in S \setminus S_0$, then $P = S \setminus \{\theta\}$ is a semigroup and $A[P] \simeq A[S]/A[\theta]$ has Krull dimension. This allows to assume that $\tilde{S}_0 = S_0$. Since, by Lemma 1, there are finitely many maximal principal ideals of S contained in S_0, then the a.c.c. on principal ideals in S yields $S_0 = \bigcup_{j=1}^{m} b_j S$ for some $b_j \in S_0$. This easily implies that the semigroup $S/_{S_0^2}$ is finitely generated (the generators arise from $b_1,\dots,b_m$ and from the generators of S/S_0). Hence S is finitely generated by Lemma 3.

As we have noticed $\operatorname{Kdim} A[S] = \sup\{\operatorname{Kdim} A[T'] \mid T$-a 0-cancellative image of $S\}$. By Corollary 3, Proposition 1 and Theorem 2 we can assume that A is noetherian. Observe that any T' inherits the hypotheses on S and so it is finitely generated by the first part of the proof. Since from [13], Theorem 2.5, it follows that $\operatorname{Kdim} A[\bar{T}] = \operatorname{Kdim} A + \operatorname{rk} \bar{T}$ then to establish 1) it is enough to prove that $\operatorname{Kdim} A[T'] = \operatorname{Kdim} A[\bar{T}]$ for any 0-cancellative image T of S. The inequality $\geq$ follows from the fact that $A[\bar{T}]$ is a localization of $A[T']$. From Theorem 1 it follows that there exist $a_1,\dots,a_n \in T'$, $n = \operatorname{rk} T'$, such that the subgroup H generated by $a_1,\dots,a_n$ has finite index in $\bar{T}$.

Then the extensions $A[H] \subseteq A[\bar{T}]$ and $A[H \cap T'] \subseteq A[T']$ satisfy the hypotheses of Lemma 2 and so $\operatorname{Kdim} A[H] = \operatorname{Kdim} A[\bar{T}]$ and $\operatorname{Kdim} A[H \cap T'] = \operatorname{Kdim} A[T']$. Thus, since H is the group of fractions of $H \cap T'$, then we can assume that $H = \bar{T}$. Now [7] implies that there exist $t_1,\dots,t_n \in A[T']$ such that $A[T']$ is a finitely generated $A[t_1,\dots,t_n]$-module. Thus, again by Lemma 2, $\operatorname{Kdim} A[T'] = \operatorname{Kdim} A[t_1,\dots,t_n]$, the latter not exceeding $\operatorname{Kdim} A + n$, ([4], Theorem 9.2). On the other hand $\operatorname{Kdim} A + n = \operatorname{K dim} A + \operatorname{rk} \bar{T} = \operatorname{Kdim} A[\bar{T}]$ which establishes the inequality $\operatorname{Kdim} A[T'] \leq \operatorname{Kdim} A[\bar{T}]$, and hence 1). If $\operatorname{Kdim} A < \infty$, then by 1) $\operatorname{Kdim} A[S] < \infty$. Now 2) is a consequence of [4], Proposition 7.8.

Theorem 3 generalizes the nontrivial implication in the characterization of noetherian monoid rings ([3], Theorem 7.7) and provides an alternative proof of this result.

We close with a conjecture concerning the case where S need not be finitely generated. Let L(S) be the lattice of congruences of S. Since the Krull dimension makes sense in the class of lattices, then we can define Kdim S as the Krull dimension of L(S). Plainly, if Kdim A[S] exists then Kdim S exists.

Conjecture. Kdim A[S] exists if and only if Kdim A, Kdim S exist and, in this case, rk S is finite and Kdim A[S] = Kdim A + rk S.

The following example shows that a subsemigroup of a finitely generated semigroup need not have Krull dimension.

Example 2. Let $S = xyT \cup \{1\}$ where T is the free semigroup of rank 2 with generators x,y and with 1 adjoined. Then S has infinitely many maximal principal ideals which easily implies that Kdim S does not exist.

References

1. Clifford A.H., Preston G.B. The algebraic theory of semigroups, Math. Surveys of the Amer. Math., Soc. 7, Providence, 1961.
2. Gilmer R., Multiplicative ideal theory, Marcel Dekker, New York, 1972.
3. Gilmer R., Commutative semigroup rings, Chicago Lect. in Math., Chicago, 1984.
4. Gordon R., Robson J.C., Krull dimension, Memoirs of the Amer. Math. Soc. 133, Providence, 1973.
5. Lemonnier B., Dimension de Krull et codeviation. Application au theoreme d'Eakin, Comm. Algebra 6(1978), 1647-1665.
6. Matsuda R., Notes on noetherian semigroup rings, Bull. Fac. Sci. Ibaraki Univ. 15(1983), 9-16.
7. Moh T.T., On a normalization lemma for integers and an application of four colors theorem, Houston J.Math. 5(1979), 119-123.
8. Okniński J., When is the semigroup ring perfect, Proc. Amer. Math. Soc. 89(1983), 49-51.
9. Okniński J., Semilocal semigroup rings, Glasgow Math. J. 25(1984), 37-44.
10. Park J.K., Skew group rings with Krull dimension, Math. J. Okayama Univ. 25(1983), 75-80.
11. Passman D.S., Group rings of polycyclic groups, Group theory: essays for Philip Hall, London Math. Soc., 1984.
12. Saito T., Note on minimal conditions for principal ideals of a semigroup, Math. Japon. 13(1968), 95-104.
13. Smith P.F., On the dimension of group rings, Proc. London Math. Soc. 25(1972), 288-302.
14. Woods S.M., Existence of Krull dimension in group rings, J. London Math. Soc. 9(1975), 406-410.
15. Zelmanov E.I. Semigroup algebras with identities, Sib. Math. J. 18(1977), 787-798.

Languages induced by certain homomorphisms of a free monoid

Mario Petrich and Gabriel Thierrin[1]

UNIVERSITY OF WESTERN ONTARIO
LONDON, CANADA

1. Introduction and summary. For a given nonempty set X, any subset of X^*, the free monoid on X, is called a language over the alphabet X. In the extensive theory of (formal) languages there are many ways of creating new languages L based on properties of words in L, or on properties of the syntactic monoid of L, to mention only a few. Another way is to introduce and study various families of languages based on some general property of languages.

We offer here new types of languages, namely those subsets of X^* which are saturated by congruences induced by some special homomorphisms of X^*. Some of these homomorphisms are among the familiar ones producing free objects in certain varieties of monoids, whereas others are new. For each of these homomorphisms we perform a similar analysis with some variations depending on the case at hand.

We now summarize the main ideas of the paper. Let X be a nonempty set and X^* be the free monoid on X. To every equivalence relation τ on X^* one may associate the family of all languages over X which are saturated by τ. Some interesting families of languages may be obtained in this way if τ is restricted to be a fully invariant congruence or, more precisely, the congruence induced on X^* by a homomorphism of X^* onto a relatively free monoid over X. . A language belonging to such a family has the property of being closed under certain transformations of the words which reflect the identities valid in the corresponding homomorphic image. By varying these transformations, we obtain further languages by the requirement that they be closed under all such transformations. With these ideas as background, we consider here several examples of such transformations and briefly study the languages so obtained.

A considerable part of the paper is devoted to proving that various congruences (induced by the homomorphisms studied here) are syntactic.

[1]This research was supported in part by Grant #7877 of the Natural Sciences and Engineering Research Council of Canada.

2. Terminology and notation

Let M be a monoid and let A be a nonempty subset of M. The relation P_A defined on M by aP_Ab if ($xay \in A$ if and only if $xby \in A$) ($x,y \in M$) is a congruence, called the principal congruence defined by A. The set A is a union of classes of P_A and $P_A \subseteq P_C$ for every class C of P_A. A subset D of M is said to be disjunctive if P_D is the identity relation. If φ is the canonical homomorphism of M onto the quotient monoid M/P_A, then $\varphi(A)$ is a disjunctive subset of M/P_A.

Let $X = \{a_1,a_2,\ldots,a_k\}$ be a finite alphabet and let X^* be the free monoid generated by X. Elements of X^* are called words over X. The length of a word w is the number of occurrences in w of the letters of X and is denoted by lg(w). The empty word is denoted by 1. A language L over X is any subset of X^*. The principal congruence defined by the language L is also called the syntactic congruence of L and the quotient monoid $syn(L) = X^*/P_L$ is the syntactic monoid of L. A congruence ρ defined on X^* is said to be syntactic if there is a language L over X such that $\rho = P_L$. A monoid M is called a syntactic monoid if there exists a language L over X such that M is isomorphic to X^*/P_L. A monoid is syntactic if and only if it contains a disjunctive set and it is finitely generated.

Let $\mathbb{N}$ be the monoid of non-negative integers under addition and $\mathbb{N}^k$ be the direct product of k copies of $\mathbb{N}$. For any mapping θ, we denote by $\overline{\theta}$ the equivalence relation induced by θ.

The above introduced X will be fixed throughout the paper. The finiteness of the alphabet is merely a notational convenience; indeed, some of our results can be easily modified to extend to the case of an arbitrary alphabet while others are generally valid without any change. All our languages will be over the alphabet X.

3. Commutative languages

For our discussion, we will need the following concept.

Definition 3.1. Two words over X are rearrangements of each other if each can be obtained from the other one by a permutation of its variables. A congruence ρ on a semigroup S is commutative if S/ρ is commutative. The mapping Ψ defined by $\Psi : w \longrightarrow (p_1,p_2,\ldots,p_k)$

$(w \in X^*)$ where p_i is the number of occurrences of a_i in w for $1 \le i \le k$, is a homomorphism $\psi : X^* \longrightarrow \mathbb{N}^k$ called the <u>Parikh mapping</u>.

<u>Lemma</u> 3.2. The congruence $\bar{\psi}$ on X^* induced by the Parikh mapping is the least commutative congruence on X^*.

<u>Proof</u>. Let ρ be a commutative congruence on X^* and let $\psi(u) = \psi(v) = (p_1, p_2, \ldots, p_k)$. Let $w = a_1^{p_1} a_2^{p_2} \ldots a_k^{p_k}$; then $u\rho w$ and $w\rho v$ so that $u\rho v$. Hence $\bar{\psi} \subseteq \rho$.

The following result can essentially be found in the literature.

<u>Theorem</u> 3.3. The following conditions on a language L over X are equivalent.

(i) $\psi^{-1}\psi(L) = L$.

(ii) L is closed under rearrangements of words.

(iii) L saturates the least commutative congruence $\bar{\psi}$ on X^*.

(iv) syn(L) is commutative.

<u>Proof</u>. (i) <u>implies</u> (ii). Let $u \in L$ and v be a rearrangement of u. Hence $\psi(u) = \psi(v)$ and thus $v \in \psi^{-1}\psi(L) = L$, as required.

(ii) <u>implies</u> (iii). Let $u \in L$ and $u\bar{\psi}v$. Then $\psi(u) = \psi(v)$ which means that v is a rearrangement of u. Hence $v \in L$ so L saturates $\bar{\psi}$.

(iii) <u>implies</u> (iv). By maximality of P_L, we obtain $\bar{\psi} \subseteq P_L$ and thus $syn(L) = X^*/P_L$ is commutative.

(iv) <u>implies</u> (i). Let $u \in \psi^{-1}\psi(L)$ so that $\psi(u) = \psi(v)$ for some $v \in L$ and thus $u\bar{\psi}v$. Since syn(L) is commutative, we get $\bar{\psi} \subseteq P_L$ and hence uP_Lv. But then $v \in L$ implies $u \in L$. This proves that $\psi^{-1}\psi(L) \subseteq L$ and the opposite inclusion is trivial.

Note that languages satisfying condition (iv) above are usually referred to as <u>commutative languages</u>. Recall that a congruence ρ on a semigroup S is <u>fully invariant</u> if ρ is invariant under all endomorphisms of S. In order to prove that all fully invariant commutative congruences ρ on X^* are syntactic, we need some preparation.

Let i, p and k be positive integers. We will construct the free object on k generators in the variety $\mathcal{V}$ of commutative monoids satisfying the identity $x^i = x^{i+p}$. Let $C_{i,p}$ be the cyclic semigroup generated by z subject to the defining relation $z^i = z^{i+p}$. Let $C_{i,p}^1$ be the semigroup $C_{i,p}$ with an identity adjoined if $C_{i,p}$ is not a group, otherwise let $C^1_{i,p} = C_{i,p}$. Finally let $C = (C_{i,p})^k$, the direct product of k copies of $C^1_{i,p}$. Define a map $\kappa : X \longrightarrow C$ by

$$\kappa : a_j \longrightarrow (1,1,\ldots,z,1,\ldots,1) \qquad (j = 1,2,\ldots k)$$

where z is in the j-th position.

<u>Lemma</u> 3.4. (κ,C) is a free object in the variety $\mathcal{V} = [xy = yx,\ x^i = x^{i+p}]$ on X.

<u>Proof</u>. Let $M \in \mathcal{V}$ and let $\varphi : X \longrightarrow M$ be any function. We now define $\tau : C \longrightarrow M$ by

$$\tau : (z^{p_1},z^{p_2},\ldots,z^{p_k}) \longrightarrow (a_1\varphi)^{p_1}(a_2\varphi)^{p_2} \ldots (a_k\varphi)^{p_k},\ (1,1,\ldots,1) \longrightarrow 1.$$

In order to show that τ is single valued, assume that

$$(z^{p_1},z^{p_2},\ldots,z^{p_k}) = (z^{q_1},z^{q_2},\ldots,z^{q_k}).$$

Then for $j = 1,2,\ldots,k$, we have $z^{p_j} = z^{q_j}$, so either $p_j = q_j < i$ or $p_j,q_j \geq i$ and $p_j \equiv q_j \pmod p$. It follows that $(a_j\varphi)^{p_j} = (a_j\varphi)^{q_j}$ trivially in the first case and in the second case since M satisfies the identity $x^i = x^{i+p}$. Hence τ is single valued.

Now τ is a homomorphism since M is commutative and it obviously maps the identity onto the identity. The diagram

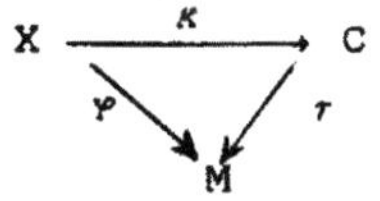

obviously commutes. The uniqueness of τ follows from the easily checked fact that $\kappa(X)$ generates C.

<u>Lemma</u> 3.5. Let M be a monoid such that $S = M\backslash\{1\}$ is a finite cyclic semigroup. Then M is a syntactic monoid.

<u>Proof</u>. The monoid M is of the form $M = \{1,a,a^2,\ldots,a^i,a^{i+1},\ldots,a^{i+p-1}\}$ where $p > 0$, $a^i = a^{i+p}$ and $G = \{a^i,\ldots,a^{i+p-1}\}$ is a cyclic group. If $p = 1$, then $G = \{a^i\}$ is evidently a disjunctive subset of M. Suppose now that $p > 1$. We will show that the subset $D = \{a,a^2,\ldots,a^i\}$ is disjunctive. Let R be the syntactic congruence defined by D and suppose that xRy.

If $x,y \in G$, then because G is a group and $G \cap D = \{a^i\}$, we have $x = y$. If $x = 1$ or $y = 1$, then $x = y$ because $\{1\}$ is a class of R. Let $x,y \in D$. If $x \neq y$, then $x = a^m$, $y = a^n$ with, say, $m < n \leq i$. Hence $xa^{i-n+1} = a^{m+i-n+1} = a^k \in D$ because $k \leq i$,

$ya^{i-n+1} = a^{n+i-n+1} = a^{i+1} \notin D$ because $a^i \neq a^{i+1}$. It follows then that xRy, a contradiction. Thus $x = y$. If $x \in D$, $y \notin D$, then we have also a contradiction because $1\cdot x\cdot 1 \in D$ and $1\cdot y\cdot 1 \notin D$.

__Theorem__ 3.6. Every fully invariant commutative congruence on X^* is syntactic.

__Proof__. It was proved in ([3], Proposition 5.1) that $\overline{\mathcal{P}}$, the least commutative congruence on X^*, is syntactic. In view of ([1], Theorem 1), the only other fully invariant commutative non-universal congruences on X^* are the least congruences ρ satisfying the condition $x^i \rho x^{i+p}$ for some fixed $i,p > 0$ and all $x \in X^*$. For such a congruence ρ, we have that X^*/ρ is a free object in the variety $\mathcal{V} = [xy = yx,\ x^i = x^{i+p}]$ on k generators.

By Lemma 3.4, we have $X^*/\rho \cong C$ where $C = \left\{C^1_{i,p}\right\}^k$, and $C_{i,p}$ is a cyclic semigroup of index i and period p. Now Lemma 3.5 yields that each $C^1_{i,p}$ has a disjunctive subset. It was proved in [5] that if each semigroup S_i, $i = 1,2,\ldots,n$, has a disjunctive subset, so does their direct product $\prod_{i=1}^{n} S_i$. Applying this to C, we obtain that C has a disjunctive subset D. Lifting D to X^* by means of ρ, we get that ρ is syntactic.

With the above notation, we further introduce

$$\mathcal{P}_{i,p} : w \longrightarrow (z^{q_1}, z^{q_2}, \ldots, z^{q_k}) \quad (w \in X^*)$$

where for $j = 1,2,\ldots,k$, q_j is the greatest non-negative integer satisfying $0 \leq q_j < i + p$, $q_j \equiv \#_{a_j}(u) \pmod p$, where $\#_{a_j}(u)$ stands for the number of occurrences of a_j in u.

A congruence ρ on X^* such that $X^*/\rho \in \mathcal{V}$ is called a $\mathcal{V}$-__congruence__.

__Lemma__ 3.7. The mapping $\mathcal{P}_{i,p}$ is a homomorphism of X^* onto C which induces the least $\mathcal{V}$-congruence on X^*, where $\mathcal{V} = [xy = yx,\ x^i = x^{i+p}]$.

__Proof__. Let

$$\mathcal{P}_{i,p}(u) = (z^{q_1}, z^{q_2}, \ldots, z^{q_k}), \qquad \mathcal{P}_{i,p}(v) = (z^{r_1}, z^{r_2}, \ldots, z^{r_k})$$

so that

$$\begin{aligned}\mathcal{P}_{i,p}(u)\mathcal{P}_{i,p}(v) &= (z^{q_1}, z^{q_2}, \ldots, z^{q_k})(z^{r_1}, z^{r_2}, \ldots, z^{r_k}) \\ &= (z^{q_1+r_1}, z^{q_2+r_2}, \ldots, z^{q_k+r_k}). \qquad (1)\end{aligned}$$

On the other hand,

$$\psi_{i,p}(uv) = (z^{s_1}, z^{s_2}, \ldots, z^{s_k}) \qquad (2)$$

where, for $j = 1,2,\ldots,k$,

$$0 \leq q_j < i + p, \quad q_j \equiv \#_{a_j}(u) \pmod p,$$

$$0 \leq r_j < i + p, \quad r_j \equiv \#_{a_j}(v) \pmod p,$$

$$0 \leq s_j < i + p, \quad s_j \equiv \#_{a_j}(uv) \pmod p,$$

and q_j, r_j, s_j are greatest with these properties. Further, $0 \leq q_j + r_j - tp < i + p$ for some non-negative integer t such that $q_j + r_j - (t-1)p \geq i + p$, and

$$q_j + r_j - tp \equiv \#_{a_j}(u) + \#_{a_j}(v) = \#_{a_j}(uv) \equiv s_j \pmod p.$$

Since also $0 \leq s_j < i + p$, and both $q_j + r_j - tp$ and s_j are greatest with these properties, we conclude that $q_j + r_j - tp = s_j$. Now $z^{q_j+r_j} = z^{q_j+r_j-tp} = z^{s_j}$. By (1) and (2), we have that $\psi_{i,p}$ is a homomorphism. It is evident that $\psi_{i,p}$ maps X^* onto C.

Since $C \in \mathcal{V}$, it follows that $\gamma_{i,p} \subseteq \bar{\psi}_{i,p}$ where $\gamma_{i,p}$ is the least $\mathcal{V}$-congruence on X^* and $\bar{\psi}_{i,p}$ is the congruence on X^* induced by $\psi_{i,p}$. We have the commutative diagram

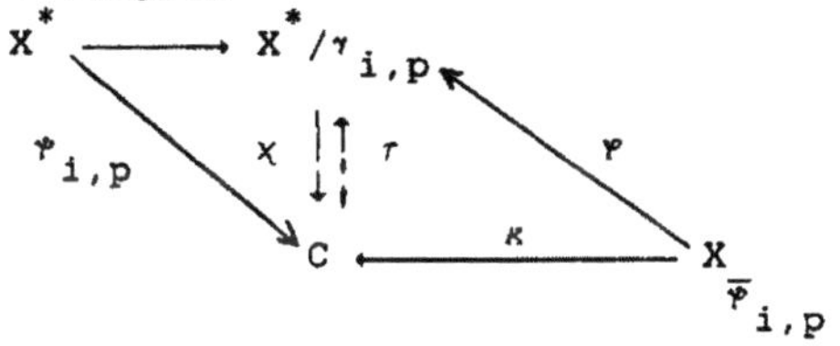

where χ is the unique homomorphism induced by $\gamma_{i,p} \subseteq \bar{\psi}_{i,p}$ and $\varphi : a_j \longrightarrow a_j\gamma_{i,p}$ for $j = 1,2,\ldots,k$. Since C is free, there exists a unique homomorphism τ making the right triangular diagram above commute. For $j = 1,2,\ldots,k$, we have

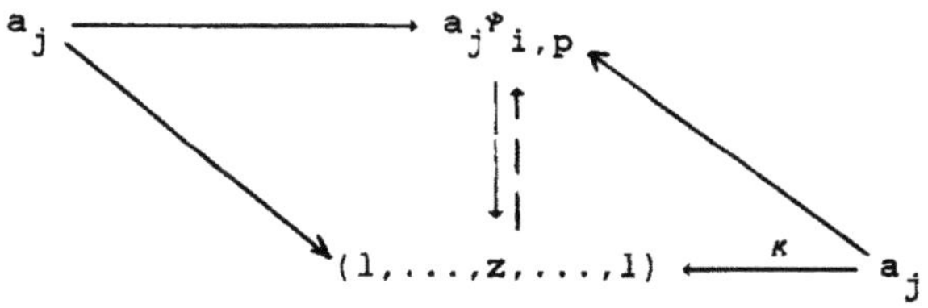

which implies that the mappings χ and τ are mutually inverse. But then

$\gamma_{i,j} = \bar{\varphi}_{i,p}$ as asserted..

Since X^* is free, $\varphi_{i,p}$ is the unique extension of the mapping $a_j \longrightarrow (1,1,\ldots,z,1,\ldots)$, with z in the j-th position, for $j = 1,2,\ldots,k$, to a homomorphism of X^* onto C.

A result analogous to that in Theorem 3.3 can also be established for the function $\varphi_{i,p}$.

4. Powerful languages

Definition 4.1. Let $\mathcal{F}$ denote the set of all families of subsets of X under the multiplication $\{H_i\}_{i\in I} \cdot \{K_j\}_{j\in J} = \{H_i \cup K_j\}_{i\in I, j\in J}$. For each $w \in X^*$, let c(w) denote the content (or alphabet) of w, namely the set of all variables occurring in w.

It is well known that the mapping $\chi : w \longrightarrow c(w)$ is a homomorphism of X^* onto the monoid of all subsets of X under union which induces the least semilattice congruence on X^*. Let $\mathcal{L}$ denote the set of all languages on X. As a consequence of this, we deduce the following simple result.

Proposition 4.2. The mapping $\chi : L \longrightarrow \{c(w) \mid w \in L\}$ $(L \in \mathcal{L})$ is a homomorphism of $\mathcal{L}$ onto $\mathcal{F}$.

In describing the languages induced on X^* by the homomorphism χ, it is convenient to introduce the following concept.

Definition 4.3. A language L over X is powerful if it satisfies

(i) L is commutative;

(ii) $uv^2 \in L \Leftrightarrow uv \in L \quad (u,v \in X^*)$.

We are now ready for the desired characterization.

Theorem 4.4. The following conditions on a language L over X are equivalent.

(i) $L = \chi^{-1}\chi(L)$.

(ii) L is powerful.

(iii) L saturates the least semilattice congruence $\bar{\chi}$ on X^*.

(iv) syn(L) is a semilattice.

Proof. (i) implies (ii). Let $w \in L$ and let w' be a rearrangement of w. Then $c(w) = c(w')$ so that $w' \in \chi^{-1}\chi(L) = L$ and L is commutative. Next let $uv^2 \in L$. Then $c(uv) = c(uv^2)$ which implies that $uv \in \chi^{-1}\chi(L) = L$. Similarly $uv \in L$ implies $uv^2 \in L$. Consequently

L is powerful.

(ii) implies (iii). Let $u, v \in X^*$ be such that $u \in L$ and $c(u) = c(v) = \{x_1, x_2, \ldots, x_n\}$. Let p_i be the number of occurrences of x_i in u for $1 \le i \le n$. By commutativity of L, we then have $v = x_i^{p_1} x_2^{p_2} \ldots x_n^{p_n} \in L$. If p_n is even, property (ii) gives that $x_1^{p_1} \ldots x_n^{\frac{p_n}{2}} \in L$ and if $p_n > 1$ is odd, then $x_1^{p_1} \ldots x_n^{\frac{p_n - 1}{2}} \in L$. Continuing this procedure, we obtain that $x_1^{p_1} \ldots x_{n-1}^{p_{n-1}} x_n \in L$. By commutativity, we also have $x_n x_1^{p_1} \ldots x_{n-1}^{p_{n-1}} \in L$ and we may apply the same procedure to get $x_n x_1^{p_1} \ldots x_{n-1} \in L$ so that $x_{n-1} x_n x_1^{p_1} \ldots x_{n-2}^{p_{n-2}} \in L$. Obvious induction then implies that $x_1 x_2 \ldots x_n \in L$.

Let q_i be the number of occurrences of x_i in v for $1 \le i \le n$. Reversing the procedure applied above, we may show that $x_1^{q_1} x_2^{q_2} \ldots x_n^{q_n} \in L$. Commutativity of L now gives that $v \in L$.

Statement (iii) now follows from the well-known characterization of the least semilattice congruence $\bar{\chi}$ on X : $u \bar{\chi} v \Leftrightarrow c(u) = c(v)$.

(iii) implies (iv). By maximality of P_L, we have $\bar{\chi} \subseteq P_L$ and thus syn(L) is a semilattice.

(iv) implies (i). We always have $L \subseteq \chi^{-1}\chi(L)$. Let $u \in \chi^{-1}\chi(L)$ so that $\chi(u) = \chi(v)$ for some $v \in L$ which implies that $c(u) = c(v)$. The hypothesis implies that L is commutative by Theorem 3.3 and hence that it is powerful. By the implication "(ii) ⇒ (iii)", we then get that $u \in L$. Therefore $L = \chi^{-1}\chi(L)$.

In order to characterize all languages over X which induce the least semilattice congruence $\bar{\chi}$ on X^*, we will first determine all disjunctive subsets of $X^*/\bar{\chi}$ which we may think of as the set of all subsets of X under set-theoretic union.

We write Y for the set of all subsets of X under union. Also let

$$\bar{X} = \{X \setminus \{a\} \mid a \in X\}.$$

Lemma 4.5. Let $\mathcal{D} \subseteq Y$. Then $\mathcal{D}$ is disjunctive if and only if either $x \in \mathcal{D}$ and $\bar{X} \cap \mathcal{D} = \emptyset$ or $X \notin \mathcal{D}$ and $\bar{X} \subseteq \mathcal{D}$.

Proof. Assume first that $X, X\setminus\{a\} \in \mathcal{D}$ for some $a \in X$. Then for any subset P of X, we have $P \cup X = X$, $P \cup (X\setminus\{a\}) = X$ or $X\setminus\{a\}$ so that

$XP_{\mathcal{D}}X\backslash\{a\}$ and $\mathcal{D}$ is not disjunctive. The direct part of the lemma now follows by contrapositive since the complement of a disjunctive set is disjunctive.

Conversely, assume that either $X \in \mathcal{D}$ and $\overline{X} \cap \mathcal{D} = \emptyset$ or $X \notin \mathcal{D}$ and $\overline{X} \subseteq \mathcal{D}$. Let $A,B \in Y$ be such that $A \neq B$. By symmetry, we may assume that $a \in A\backslash B$ for some $a \in X$. It follows that $(X\backslash\{a\}) \cup A = X$, $(X\backslash\{a\}) \cup B = X\backslash\{a\}$. The hypothesis now implies that $AP_{\mathcal{D}}B$ and $\mathcal{D}$ is disjunctive.

Proposition 4.6. Let $A_1, A_2, \ldots, A_n$ be a family of distinct subsets of X satisfying:

$$\text{either } |A_1| = k \text{ and } |A_i| < k - 1 \text{ for } 1 < i \leq n,$$

$$\text{or } |A_1| = |A_2| = \ldots = |A_k| = k - 1 \text{ and } |A_i| < k - 1 \text{ for } k < i \leq n.$$

Then $L = \{w \in X^* \mid c(w) \in \{A_1, A_2, \ldots, A_n\}\}$ has the property $P_L = \overline{X}$. Conversely, every language L such that $P_L = \overline{X}$ can be so obtained.

Proof. This follows directly from the lemma by lifting disjunctive subsets from Y to X^*.

It was observed in [2] that every finite semilattice has a disjunctive subset. By lifting this disjunctive subset to X^*, we obtain that every semilattice congruence is syntactic.

5. Transformations preserving the order of first occurrences

Here we consider languages L closed under the transformations preserving the order of the first occurrence of variables of a word in L.

Notation 5.1. For each $w \in X^*$, let $i(w)$, the initial of w, be the word obtained from w by retaining only the first occurrence of each variable in w. Let $\Phi = \{w \in X^* \mid i(w) = w\}$ under the multiplication $u \cdot v = i(uv)$.

Recall that a semigroup S is a left regular band if it satisfies the identities $x^2 = x$, $axa = ax$. It is well known that Φ is a free left regular band on X (qua monoid) and that the mapping $\varphi : w \longrightarrow i(w)$ $(w \in X^*)$ is a homomorphism of X^* onto Φ which induces the least left regular band congruence $\overline{\varphi}$ on X^*.

We can perform an analysis of this case analogous to that in the preceding two sections. We limit ourselves to the few most interesting highlights.

Theorem 5.2. The following conditions on a language L over X are equivalent.

(i) $L = \varphi^{-1}\varphi(L)$.

(ii) L is closed under the transformations α which preserve the order of the first occurrence of variables in a word (i.e., $i(w) = i(\alpha(w))$).

(iii) L saturates the least left regular band congruence $\bar{\varphi}$ on X^*.

(iv) syn(L) is a left regular band.

Proof. (i) implies (ii). Let $u \in L$ and $v \in X^*$ be a word such that the order of occurrences of variables in u equals that in v. It follows that $i(u) = i(v)$ and thus $\varphi(u) = \varphi(v)$ so that $v \in \varphi^{-1}\varphi(L) = L$, as required.

(ii) implies (iii). Let $u \in L$ and $v \in X^*$ be such that $u\bar{\varphi}v$. Then $i(u) = i(v)$ and hence u and v have identical order of first occurrences of variables. The hypothesis now implies that $v \in L$. Hence L saturates $\bar{\varphi}$.

(iii) implies (iv). We must show that $xay \in L \Leftrightarrow xa^2y \in L$, $xauay \in L \Leftrightarrow xauy \in L$ for all $a,u,x,y \in X^*$. This follows directly from the hypothesis since $X^*/\bar{\varphi}$ is a left regular band.

(iv) implies (i). We always have $L \subseteq \varphi^{-1}\varphi(L)$. Let $u \in \varphi^{-1}\varphi(L)$ so that $\varphi(u) = \varphi(v)$ for some $v \in L$. This means that $i(u) = i(v)$ which evidently implies that $\bar{u} = \bar{v}$ in the syntactic monoid syn(L). Since $v \in L$ and the syntactic congruence of L saturates L, we obtain that $u \in L$. Therefore $L = \varphi^{-1}\varphi(L)$.

For $w \in \Phi$, we have $w = a_{i_1}a_{i_2} \ldots a_{i_n}$ where the a_{i_j} are distinct and $\{i_1,i_2,\ldots,i_n\} \subseteq \{1,2,\ldots,k\}$. In fact, $w = a_{j_{1\sigma}}a_{j_{2\sigma}} \ldots a_{j_{n\sigma}}$ where $j_1 < j_2 < \ldots < j_n$ and σ is a permutation of the set $\{j_1,j_2,\ldots,j_n\}$. We define the *parity* of w, par(w), to be the parity of the permutation σ. We also define the *parity* of a natural number n, par(n), to be its usual parity of being even or odd.

Lemma 5.3. (C.M. Reis) Let $M = \{w \in \Phi \mid lg(w) = k\}$. If u,v are distinct elements of M of the same parity, then there exists $w \in \Phi$ such that $par(wu) \neq par(wv)$.

Proof. First assume that there exists $a \in X$ such that $u = xay$

and $v = waz$ for some $x,y,z \in \Phi$ such that $par(lg(x)) \neq par(lg(w))$. Then $au = axy$ and $av = awz$. Now axy is obtained by performing $lg(x)$ transpositions on u whereas awz is obtained by performing $lg(w)$ transpositions on v. Since $par(lg(x)) \neq par(lg(w))$, it follows that $par(au) \neq par(av)$.

In the contrary case, we have the following setting: for every $a \in X$, a occurs in an even (odd) position in u if and only if a occurs in an even (odd) position in v. Since $u \neq v$, there exist distinct letters a and b such that $u = xay$ and $v = xbz$ for some $x,y,z \in \Phi$. But then b occurs in y and a occurs in z which shows that $u = xapbq$, $v = xbras$ for some $p,q,r,s \in \Phi$. By assumption, we have $par(lg(x)) = par(lg(xbr))$ and $par(lg(xap)) = par(lg(x))$. Let $m = lg(x)$, $n = lg(xbr)$ and $\ell = lg(xap)$. Note that m,n and ℓ have the same parity. Now $(ab)u = abxpq$ and $(ab)v = abxrs$, where abxpq is obtained from u by performing $m + \ell - 1$ transpositions whereas abxrs is obtained from v by performing $m + n$ transpositions. Since $par(m + n) \neq par(m + \ell - 1)$, we conclude that $par(abu) \neq par(abv)$.

Lemma 5.4. The monoid Φ has a disjunctive subset.

Proof. Let M be as in Lemma 5.3. Thus M consists of all words of the form $a_{1\sigma}a_{2\sigma}\ldots a_{k\sigma}$ where σ is a permutation of the set $\{1,2,\ldots,k\}$. Let

$$L = \{a_{1\sigma}a_{2\sigma}\ldots a_{k\sigma} \mid \sigma \text{ is an even permutation}\}.$$

We will show that L is a disjunctive subset of Φ. To this end, let $u,v \in \Phi$, $u \neq v$. Then one of the following cases occurs.

Case 1: $u = wau'$, $v = wbv'$ for some $w,u',v' \in \Phi$, $a,b \in X$, $a \neq b$. Let $p \in M$. Then $up \neq vp$ and $up,vp \in M$. If $par(up) \neq par(vp)$, then either $up \in L$, $vp \notin L$ or $up \notin L$, $vp \in L$. If $par(up) = par(vp)$, then by Lemma 5.3, there exists $q \in \Phi$ such that $par(qup) \neq par(qvp)$, and hence either $qup \in L$, $qvp \notin L$ or $qup \notin L$, $qvp \in L$. Therefore, in either case uP_Lv.

Case 2: $u = vu'$ for some $u' \in \Phi$. Hence $u' \neq 1$; let a be the first letter occurring in u'. We may assume that $k > 1$; for otherwise there is nothing to prove. Let $b \in X$, $b \neq a$, and $p \in M$. Then $ubp = vu'bp \neq vbp$ and $ubp,vbp \in M$. Considering the parity of ubp and vbp, the same type of argument as above shows that uP_Lv.

The remaining case $v = uv'$ for some $v' \in \Phi$ is symmetric to Case 2. Therefore, L is a disjunctive subset of Φ (even in the case $k = 1$).

As we mentioned in the introduction, all our results extend to the case of an arbitrary cardinality of the set X. In certain cases, the arguments presented hold in the general case while in some others the difference is merely notational. The preceding proof represents a notable exception.

We now present the needed argument when X is infinite. First we linearly order X in any fashion. Relative to this order, we may speak, analogously as above, of the parity of a word. Hence let L be the set of all words in Φ of even parity. Let $u, v \in \Phi$, $u \neq v$, and consider the same cases as in the above proof.

In Case 1, the only modification is the choice of p. In the present case, we take $p \in \Phi$ such that $c(up) = c(vp)$. The same argument as above shows that $u \not{R}_L v$.

Consider Case 2. Let $a \in X \setminus c(u)$. Then $uau = vu'au \neq vau$ and $c(uau) = c(vau)$. The same type of argument as in Case 1 in the above proof shows that again $u \not{R}_L v$.

The remaining case $v = uv'$ for $v' \in \Phi$ is again symmetric. Therefore L is a disjunctive subset of Φ...

<u>Corollary</u> 5.5. The congruence induced by Ψ is syntactic.

6. <u>The mappings</u> φ <u>and</u> Ψ

We now combine the homomorphism φ of Section 3 with the homomorphism Ψ of Section 5.

<u>Notation</u> 6.1. Let

$$Q = \{((p_1, p_2, \ldots, p_k), u) \in \mathbb{N}^k \times \Phi \mid p_i > 0 \Leftrightarrow a_i \in c(u)\}$$

with multiplication

$$((p_1, p_2, \ldots, p_k), u)((q_1, q_2, \ldots, q_k), v)$$
$$= ((p_1 + q_1, p_2 + q_2, \ldots, p_k + q_k), i(uv)).$$

<u>Proposition</u> 6.2. The mapping $\xi : w \longrightarrow (\varphi(w), \Psi(w))$ $(w \in X^*)$ is a homomorphism of X^* onto Q.

<u>Proof</u>. It is clear that ξ maps X^* into Q. The homomorphism property follows from that of φ and Ψ. If $((p_1, p_2, \ldots, p_k), u) \in Q$, then $u = a_{i_1} a_{i_2} \ldots a_{i_n}$, so for $w = a_{i_1}^{p_{i_1}} a_{i_2}^{p_{i_2}} \ldots a_{i_n}^{p_{i_n}}$ we obtain $\xi(w) = ((p_1, p_2, \ldots, p_k), u)$. Therefore ξ maps X^* onto Q.

<u>Theorem</u> 6.3. The following conditions on a language L over X are equivalent.

(i) $L = \xi^{-1}\xi(L)$.

(ii) L is closed under the transformations α such that $\alpha(w)$ is a rearrangement of w and $i(\alpha(w)) = i(w)$.

(iii) L saturates the least congruence κ on X^* for which X^*/κ satisfies the identity $xyx = x^2y$.

(iv) syn(L) satisfies the identity $xyx = x^2y$.

<u>Proof</u>. (i) <u>implies</u> (ii). Let $u \in L$, $v \in X^*$ and assume that v is a rearrangement of u and $i(u) = i(v)$. Hence $\psi(u) = \psi(v)$ and $\wp(u) = \wp(v)$ so that $\xi(u) = \xi(v)$. But then $v \in \xi^{-1}\xi(L) = L$, as required.

(ii) <u>implies</u> (iii). Let $a \in L$ and $a\kappa b$. There exists a sequence of elements $z_1, z_2, \ldots, z_n$ in X^* such that $a = z_1$, $z_n = b$ and z_{i+1} is obtained from z_i by an elementary transition relative to the identity $xyx = x^2y$ for $i = 1,2,\ldots,n-1$. These transitions are of the form $z_i = sut$, $z_{i+1} = svt$ where either $u = xyx$, $v = x^2y$ or $u = x^2y$, $v = xyx$ for some $x,y \in X^*$. It is clear that then z_{i+1} is a rearrangement of z_i and since $i(xyx) = i(x^2y)$ we also have $i(z_i) = i(z_{i+1})$, $i = 1,2,\ldots,n-1$. If now $a \in L$, the hypothesis on L implies that $z_2 \in L$, and continuing this procedure, we finally get $b \in L$. Therefore κ is saturated by L.

(iii) <u>implies</u> (iv). By maximality of P_L, we have $\kappa \subseteq P_L$ and thus syn(L) satisfies the identity $xyx = x^2y$.

(iv) <u>implies</u> (i). For $u \in X^*$, we first find a canonical word for u relative to rearrangements and initials. Indeed, let

$$\xi(u) = (p_1, p_2, \ldots, p_k; a_{i_1} a_{i_2} \ldots a_{i_n}).$$

Then $i(u) = a_{i_1} a_{i_2} \ldots a_{i_n}$ and we may form the word

$\bar{u} = a_{i_1}^{p_{i_1}} a_{i_2}^{p_{i_2}} \ldots a_{i_n}^{p_{i_n}}$ with the property that $\xi(u) = \xi(\bar{u})$.

We show next that $uP_L\bar{u}$ by induction on the cardinality of c(u). If $|c(u)| = 1$, then $u = a^n$ for some $a \in X$ and some positive integer n, whence $u = \bar{u}$. Assume that the statement is valid for all words u for which $|c(u)| < n$. Let $c(u) = n$. Then $u = va$ for some $a \in X$ and $v \in X^*$. If $a \notin c(v)$, then $\bar{u} = \bar{v}aP_Lva = u$ by the induction hypothesis since $|c(v)| = n - 1$. Assume that $a \in c(v)$. Then $\bar{v} = a_{i_1}^{k_1} a_{i_2}^{k_2} \dots a_{i_n}^{k_n}$ for some $a_{i_1}, \dots, a_{i_n} \in X$ and positive integers $k_1, \dots, k_n$. Hence $a = a_{i_p}$ for some $1 \leq p \leq n$. Thus

$$\bar{u} = a_{i_1}^{k_1} a_{i_2}^{k_2} \dots a_{i_p}^{k_p+1} \dots a_{i_n}^{k_n}$$

$$= a_{i_1}^{k_1} a_{i_2}^{k_2} \dots a_{i_p}^{k_p-1} a_{i_p}^{2} \left[a_{i_{p+1}}^{k_p+1} \dots a_{i_n}^{k_n} \right]$$

$$P_L\ a_{i_1}^{k_1} a_{i_2}^{k_2} \dots a_{i_p}^{k_p} a_{i_{p+1}}^{k_{p+1}} \dots a_{i_n}^{k_n} a_{i_p} = \bar{v} a_{i_p} P_L va = u,$$

as required.

Now let $u \in \xi^{-1}\xi(L)$. Then there exists $v \in L$ such that $\xi(u) = \xi(v)$. It follows that $\bar{u} = \bar{v}$ and thus $uP_L\bar{u} = \bar{v}P_L v$. Now since $v \in L$, we also have $u \in L$. Consequently $\xi^{-1}\xi(L) \subseteq L$ and the opposite inclusion is trivial.

7. Transformations preserving the position of first occurrences

We now consider languages L closed under the transformations preserving the position of the first occurrence of variables of a word in L. In order to do this, we must consider the length of a word w. It is clear that the mapping $\lambda : w \longrightarrow lg(w)$ $(w \in X)$ is a homomorphism of X^* onto the infinite cyclic monoid $\mathbb{N}$.

Notation 7.1. Let

$$\Pi = \{(p;s_1,s_2,\dots,s_k) \mid p \geq s_i \geq 0,\ s_i = s_j > 0 \Rightarrow i = j,$$
$$s_r = 1 \text{ for some } 1 \leq r \leq k,\ s_1 = s_2 = \dots s_k = 0 \Rightarrow p = 0\}$$

with multiplication

$$(p;s_1,\dots,s_k)(q;t_1,\dots,t_k) = (p + q;[s_1,t_1],\dots,[s_k,t_k])$$

where

$$[s_i,t_i] = \begin{cases} s_i & \text{if } s_i > 0, \\ p + t_i & \text{if } s_i = 0, t_i > 0, \\ 0 & \text{if } s_i = t_i = 0. \end{cases}$$

One can verify directly that Π is a monoid. This will actually follow from the next proposition.

Proposition 7.2. Define a mapping π by
$\pi : w \longrightarrow (p;s_1,s_2,\ldots,s_k)$ $(w \in X^*)$ where p is the length of w, s_i is the position of the first occurrence of a_i in w if a_i occurs in w and $s_i = 0$ otherwise. Then π is a homomorphism of X^* onto Π.

Proof. That π maps X^* into Π is verified easily. The homomorphism property follows immediately from the definition of the multiplication in Π. Let $t = (p;s_1,s_2,\ldots,s_k) \in \Pi$. If all s_i are equal to 0, then p = 0 and $\pi(\emptyset) = t$. In the contrary case, let $1 = s_{i_1} < s_{i_2} < \ldots < s_{i_q}$ be the set of all nonzero s_i's and let

$$w = a_{i_1}^{s_{i_2}-1} a_{i_2}^{s_{i_3}-s_{i_2}} \ldots a_{i_{q-1}}^{s_{i_q}-s_{i_{q-1}}} a_{i_q}^{p-s_{i_q}+1} .$$

Simple reflection shows that $\pi(w) = t$ in this case. Therefore π maps X^* onto Π.

The congruence induced on X^* by π is generally not fully invariant, so the homomorphism π is not associated with a variety. We thus have here a very restricted analogue of Theorems 3.3, 4.4 and 5.2 as follows.

Proposition 7.3. A language L has the property that $L = \pi^{-1}\pi(L)$ if and only if L is closed under the transformations which preserve the length of the words and the position of the first occurrence of each variable in a word.

Proof. The argument here is similar to those in the proofs of theorems cited above and is omitted.

In order to gain some inkling into the structure of the monoid Π, we will embed it into a semidirect product of some familiar monoids. Toward this end we need some preparation.

Lemma 7.4. Let $\mathbb{N}^0 = \{0,1,\ldots\}$ with multiplication

$$m \circ n = \begin{cases} m & \text{if } m > 0 \\ n & \text{if } m = 0. \end{cases}$$

Then $\mathbb{N}^0$ is a left zero semigroup with an identity adjoined.

Proof. Clearly $\mathbb{N}^0\backslash\{0\}$ is a left zero semigroup and 0 is an identity of $\mathbb{N}^0$.

<u>Lemma</u> 7.5. Let $\mathbb{N}$ act on $\mathbb{N}^0$ by

$$m \cdot n = \begin{cases} m + n & \text{if } n > 0 \\ 0 & \text{if } n = 0. \end{cases}$$

Then

$$m \cdot (n \circ p) = (m \cdot n) \circ (m \cdot p), \tag{3}$$

$$(m + n) \cdot p = m \cdot (n \cdot p), \tag{4}$$

$$0 \cdot n = n, \quad n \cdot 0 = 0, \tag{5}$$

$$m \cdot n = m \cdot p \Rightarrow n = p, \tag{6}$$

for all $m,n,p, \in \mathbb{N}$.

<u>Proof</u>. On the one hand,

$$m \cdot (n \circ p) = \begin{cases} m + (n \circ p) & \text{if } n \circ p > 0 \\ 0 & \text{if } n \circ p = 0 \end{cases}$$

$$= \begin{cases} m + n & \text{if } n > 0 \\ m + p & \text{if } n = 0,\ p > 0 \\ 0 & \text{if } n = p = 0 \end{cases}$$

and on the other hand,

$$(m \cdot n) \circ (m \cdot p) = \begin{cases} m \cdot n & \text{if } m \cdot n > 0 \\ m \cdot p & \text{if } m \cdot n = 0 \end{cases}$$

$$= \begin{cases} m + n & \text{if } n > 0 \\ m + p & \text{if } n = 0,\ p > 0 \\ 0 & \text{if } n = p = 0 \end{cases}$$

which proves (3). Furthermore,

$$(m + n) \cdot p = \begin{cases} m + n + p & \text{if } p > 0 \\ 0 & \text{if } p = 0 \end{cases}$$

$$m \cdot (n \cdot p) = \begin{cases} m + (n \cdot p) & \text{if } n \cdot p > 0 \\ 0 & \text{if } n \cdot p = 0 \end{cases}$$

$$= \begin{cases} m + m + p & \text{if } p > 0 \\ 0 & \text{if } p = 0 \end{cases}$$

which proves (4). Relations (5) clearly hold. Assume that $m \cdot n = m \cdot p$. If $n = 0$, then $m \cdot n = 0$ so $m \cdot p = 0$ whence either $p = 0$ or $m + p = 0$, so in either case $p = 0$. If $n > 0$, then $m \cdot n = m + n$ so $m \cdot p = m + p = m + n$ so that $n = p$. Therefore $n = p$ and (6) holds.

The preceding lemma yields that $\mathbb{N}$ acts on $\mathbb{N}^0$ by monomorphisms. Using this action, we may define a semidirect product of $\mathbb{N}^0$ and $\mathbb{N}$, where the latter's operation is the usual addition. More generally, we can do this with the direct product $(\mathbb{N}^0)^k$ of k-copies of $\mathbb{N}^0$ as follows.

<u>Lemma</u> 7.6. Let $\mathbb{N}$ act on $(\mathbb{N}^0)^k$ by

$$m \cdot (n_1, n_2, \ldots, n_k) = (m \cdot n_1, m \cdot n_2, \ldots, m \cdot n_k).$$

Then $\mathbb{N}$ acts on $(\mathbb{N}^0)^k$ by monomorphisms and we may define the semidirect product $(\mathbb{N}^0)^k \dot{\times} \mathbb{N}$.

<u>Proof</u>. The first assertion follows directly from Lemma 7.5. In view of this, the usual argument shows that the semidirect product $(\mathbb{N}^0)^k \dot{\times} \mathbb{N}$, defined on the set $\mathbb{N}^k \times \mathbb{N}$ with multiplication

$$((m_1, m_2, \ldots, m_k), n)((p_1, p_2, \ldots, p_k), q)$$
$$= ((m_1, m_2, \ldots, m_k) \circ n \cdot (p_1, p_2, \ldots, p_k), n + q),$$

is a monoid.

With these preparations, we may now prove the embedding result aluded to earlier.

<u>Proposition</u> 7.7. The mapping $\sigma : (p; s_1, s_2, \ldots, s_k) \longrightarrow ((s_1, s_2, \ldots, s_k), p)$ is an embedding of Π into $(\mathbb{N}^*)^k \dot{\times} \mathbb{N}$.

<u>Proof</u>. It suffices to check the homomorphism property. Indeed, on the one hand,

$$((p; s_1, s_2, \ldots, s_k)(q; t_1, t_2, \ldots, t_k))\sigma$$
$$= (p + q, [s_1, t_1], [s_2, t_2], \ldots, [s_k, t_k])\sigma$$
$$= (([s_1 t_1], [s_2, t_2], \ldots, [s_k, t_k]), p + q)$$

and on the other hand,

$$(p; s_1, s_2, \ldots, s_k)\sigma(q; t_1, t_2, \ldots, t_k)\sigma$$
$$= ((s_1, s_2, \ldots, s_k), p)((t_1, t_2, \ldots, t_k), q)$$
$$= ((s_1, s_2, \ldots, s_k \circ p \cdot (t_1, t_2, \ldots, t_k), p + q)$$
$$= ((s_1 \circ p \cdot t_1, s_2 \circ p \cdot t_2, \ldots, s_k \circ p \cdot t_k), p + q)$$

where

$$s_i \circ p \cdot t_i = \begin{cases} s_i & \text{if } s_i > 0 \\ p \cdot t_i & \text{if } s_i = o \end{cases} = \left\{ \begin{matrix} s_i & \text{if } s_i > 0 \\ p + t_i & \text{if } s_i = 0,\ t_i > 0 \\ 0 & \text{if } s_i = t_i = 0 \end{matrix} \right\} = [s_i, t_i]$$

which proves that σ is a homomorphism.

8. <u>The functions</u> φ <u>and</u> π

We now combine the homomorphism φ of Section 3 with a part of the homomorphism π of Section 7.

Notation 8.1. Let

$$P = \{(p_1,p_2,\ldots,p_k;s_1,s_2,\ldots,s_k) \mid s_i,p_i \geq 0,\ p_i = 0 \Leftrightarrow s_i = 0,\ s_i = s_j > 0 \Rightarrow i = j\}$$

with multiplication

$$(p_1,p_2,\ldots,p_k;s_1,s_2,\ldots,s_k)(q_1,q_2,\ldots,q_k;t_1,t_2,\ldots,t_k)$$
$$= (p_1 + q_1,p_2 + q_2,\ldots,p_k + q_k;[s_1,t_1],[s_2,t_2],\ldots,[s_k,t_k])$$

where $[s_i,t_i]$ was introduced in Notation 6.1 with $p = p_1 + p_2 + \ldots + p_k$.

One can verify directly that P is a monoid. This also follows easily from Proposition 7.2. Our last homomorphism can be given as follows.

Proposition 8.2. Define a mapping θ by

$$\theta : w \longrightarrow (p_1,p_2,\ldots,p_k;s_1,s_2,\ldots,s_k) \quad (w \in X^*)$$

where p_i is the number of occurrences of a_i in w and s_i is the position of the first occurrence of a_i in w if a_i occurs in w and $s_i = 0$ otherwise. Then θ is a homomorphism of X^* into P.

Proof. The straightforward verification is omitted.

In trying to establish whether θ maps X^* onto P there arise certain difficulties, so it appears that θ is not surjective. As an analogue of Proposition 7.3 we have the following simple result.

Proposition 8.3. A language L has the property that $L = \theta^{-1}\theta(L)$ if and only if L is closed under rearrangements of words which preserve the position of the first occurrence of each variable in a word.

Proof. The argument goes along the same lines as in the earlier cases and is omitted.

We also have an obvious analogue of Proposition 8.3 for the homomorphism λ introduced at the beginning of this section.

Proposition 8.4. A language L has the property that $L = \lambda^{-1}\lambda(L)$ if and only if L is closed under the transformations which preserve the length of words.

As in the preceding section, we may obtain an embedding into a semidirect product as follows.

Proposition 8.5. Let the direct product $\mathbb{N}^k$ act on the direct product $(\mathbb{N}^0)^k$ by

$$(m_1,m_2,\ldots,m_k) \cdot (n_1,n_2,\ldots,n_k) = (m \cdot n_1, m \cdot n_2,\ldots,m \cdot n_k)$$

where $m = m_1 + m_2 + \ldots + m_k$. Then $\mathbb{N}^k$ acts on $(\mathbb{N}^0)^k$ by monomorphisms so we may define the semidirect product $(\mathbb{N}^0)^k \dot{\times} \mathbb{N}^k$. Its multiplication can be written briefly as

$$((m_i),(n_i))((p_i),(q_i)) = ((m_i) \circ ((n_i) \cdot (p_i)),(n_i) + (q_i))$$

or, more explicitly as $((m_i \circ n \cdot p_i),(n_i + q_i))$ where $n = n_1 + n_2 + \ldots + n_k$. Moreover, the mapping

$$(p_1,p_2,\ldots,p_k;s_1,s_2,\ldots,s_k) \longrightarrow ((s_1,s_2,\ldots,s_k),(p_1,p_2,\ldots,p_k))$$

is an embedding of P into $(\mathbb{N}^0)^k \dot{\times} \mathbb{N}^k$.

Proof. These assertions either follow from the corresponding statements in the preceding section or are easy to verify.

Since $\mathbb{N}^0$ is a left zero semigroup with an identity adjoined, it is a left regular band, and thus so is the direct product $(\mathbb{N}^0)^k$. Therefore P is embedded above into a semidirect product of a left regular band and a commutative monoid.

9. The syntactic congruences $\bar{\xi}$, $\bar{\pi}$, and $\bar{\theta}$

In this section we will show that the congruences induced on X^* respectively by ξ, π and θ are all syntactic. After introducing the following concept we prove a general criterion on a congruence to be syntactic.

Definition 9.1. A congruence ρ on X^* is called a lg-congruence if each ρ-class contains only words of the same length.

Theorem 9.2. Let ρ be a lg-congruence on X^* such that $u^n \rho v u^{n-1}$ with $lg(u) = lg(v)$ and $n > 0$ implies $u \rho v$. Then ρ is a syntactic congruence.

Proof. Let $L_1,L_2,\ldots,L_n,\ldots$ be the classes of ρ ordered in the following way:

if L_i contains a word of length m and L_j a word of length n with $m < n$, then $L_i < L_j$

for the classes containing words of the same length, then use any order.

For every $n > 0$, let $E_n = (L_n)^{2^n}$; E_n is contained in a class of ρ. Let F_n be the class of ρ containing E_n and let $L = F_1 \cup F_2 \cup \ldots \cup F_n \cup \ldots$

Let $u,v \in L_i$ and $x,y \in X^*$. If $xuy \in L$, then $xuy \in F_n$ for some n. Since F_n is a class of ρ and ρ is a congruence, $xuy\rho xvy$ implies $xvy \in F_n$. Therefore $\rho \subseteq P_L$.

Conversely, let uP_Lv, $r = lg(u)$, $s = lg(v)$ and $k = s - r$. Suppose that $r \neq s$; we may assume that $k > 0$. There exists $x \in X^*$ such that $vx \in L$ where the words in L_n are of length $t > k$. Further, for some $y \in X^*$ we have $vxy \in E_n \subseteq F_n$. Since uP_Lv; we obtain $uxy \in L$. Now $lg(uxy) < lg(vxy)$ implies that $uxy \in F_m$ with $m < n$. It follows that $k = lg(vxy) - lg(uxy) \geq t^{2^n} - t^{2^m}$, contradicting the fact that $t > k$. Hence $lg(u) = lg(v)$. There exists $n > 0$ such that $u^n \in L$ and hence $u^n \in F_i$ for some i. This implies $vu^{n-1} \in F_i$, that is $u^n\rho vu^{n-1}$ and by hypothesis of the theorem, we have $u\rho v$. Therefore $P_L \subseteq \rho$ and the equality prevails.

__Corollary__ 9.3. The congruence induced by ξ is syntactic.

__Proof__. Since $\overline{\xi}$ is a lg-congruence, we have only to show that the second condition of Theorem 9.2 is satisfied. Suppose that $\xi(u^n) = \xi(vu^{n-1})$. The words u^n and vu^{n-1} are Parikh equivalent and hence u and v are also Parikh equivalent. From that it follows that $c(u) = c(v)$ and therefore $\xi(u) = \xi(v)$.

__Corollary__ 9.4. The congruences induced by π and θ are syntactic.

__Proof__ This follows from Theorem 9.2, because these congruences are all lg-congruences and it is immediate that they satisfy the second condition of this theorem.

REFERENCES

[1] T.J. Head, The varieties of commutative monoids, Nieuw Archief v. Wisk. 16(1968), 203-206.

[2] H. Jürgensen, Inf-Halbverbände als syntaktische Halbgruppen, Acta Math. Acad. Sci. Hung. 31(1978), 37-41.

[3] Y. Guo, H.J. Shyr, G. Thierrin, F-disjunctive languages, International J. of Computer Math. 18(1986), 219-237.

[4] H.J. Shyr, Free monoids and languages, Soochow Univ., Taipei, Taiwan, 1979.

[5] E. Valkema, Zur Charakterisierung formaler Sprachen durch Halbgruppen, Dissertation, University of Kiel, 1974.

Infima in the power set of free semigroups

G. Pollák
Mathematical Research Institute of
Hungarian Academy of Science
H-1053 Budapest, Reáltanoda u. 13-15.

The results of this paper were obtained as auxiliary ones while dealing with infinite independent systems of identities; in this field they find application repeatedly [1]. In the same time it seems to the author that they are of some interest for themselves, too. In particular, the type of ordering of infinite antichains called below "reducing", often gives non-trivial characterization to different posets; what follows can be an example to this situation.

Let X be an arbitrary alphabet, X^* the free monoid over X. We put $a \leqq b$ for $a,b \in X^*$ if a is a subword of b (by "subword" we always mean connected subword). This partial ordering induces a quasiorder on the set $\mathcal{P}_\infty(X^*)$ of infinite subsets of X^* by

Definition 1. If B,C are infinite subsets of X^* then C <u>can be reduced to</u> B (notation: $B \cdot\!\leqq C$) if $\forall b \in B \exists c \in C (b \leqq c)$.

One of our aims is to prove that $\langle \inf \mathcal{P}_\infty(X^*), \leqq \rangle$ exists (the infimum of the quasiordered set $\mathcal{Q}$ - quoset for short - being defined as the maximal subset J of A such that 1/ for every $A \in \mathcal{Q}$ there is a $B \in \mathcal{J}$ with $B \leqq A$ and 2/ comparable elements of $\mathcal{J}$ are equivalent, i.e. $A \leqq A'$ implies $A' \leqq A$ - and to find it. For this sake we have to find the infimum of another quoset.

Let $X^\omega = \{u : u = x_1 x_2 \ldots\}$ be the set of all infinite words (of type ω) over X. Denote the set of finite subwords of u by W(u), and put

$$u \leqq v \overset{\text{def}}{\Longleftrightarrow} W(u) \subseteq W(v).$$

Clearly, $\leqq$ is a quasiorder. Furthermore, for $a \in W(u)$, denote by $\nu(a,u)$ the maximum of the lengths $|b|$ of those subwords b of u which do not contain a, if this maximum exists, and put $\nu(a,u) = \infty$ otherwise. E.g. if a does occur in u, but only finitely many times, then $\nu(a,u) = \infty$.

Definition 2. An infinite word u is said to be <u>dense</u> if $\nu(a,u)$ is finite for every $a \in W(u)$.

Theorem 1. $\inf\langle X^\omega, \leqq\rangle$ consists of all dense words.

Proof. Let D denote the set of all dense words, and let $u \in X^\omega$. First we construct a dense word $u_d \leqq u$. As any pure periodical word $a^\omega = aaa\ldots$ is obviously dense, we can choose $u_d = a^\omega$ if u contains arbitrarily high powers of a. Now suppose that for every $a \in X^*$, some power $a^{n(a)}$ does not occur already in u. The same holds then of course for every $u' \leqq u$. Next we show that there is a $u' \leqq u$ and an $x \in X$ such that $\nu(x,u')$ is finite. This is clear if $|X| = 2$ ($|X| = 1$ is excluded by the assumption just made), because if $\nu(x,u) = \infty$ for one of the letters then arbitrarily high powers of the other letter are subwords of u. Suppose the assertion holds for $|X| < n$, and let $|X| = n$, $\nu(x,u) = \infty$ for some $x \in X$. This means that there are arbitrarily long subwords of u which belong to $(X\setminus\{x\})^*$; denote their set by W'(u). If $a = x_1\ldots x_{k-1} \in W'(u)$ is such that infinitely many words of W'(u) start with a (the empty word has this property) then there must be an $x_k \in X\setminus\{x\}$ such that ax_k has the same property; thus we can construct an infinite word $\bar{u} = x_1x_2\ldots \in (X\setminus\{x\})^\omega$ such that $\bar{u} \leqq u$. By assumption, $\nu(x_i,u')$ is finite for some $x_i \in (X\setminus\{x\}) \cap W(u)$ and some $u' \leqq \bar{u} \leqq u$. This proves the assertion.

Now suppose $\nu(a,u')$ is finite for some $u' \leqq u$ and some $a_1 \in W(u')$. Put $Y = \{a_1a_2 : |a_2| < \nu(a,u')\}$. It is easy to see that $u' \in Y^\omega$. By what has been just proved, there is an $a_1a_2 \in Y$ and an infinite word $u'' \leqq u'$ such that $\nu(a_1a_2,u'')$ is finite (in the alphabet Y, but this, of course, does not matter). Hence we can construct an infinite word $u_d = a_1a_2\ldots$ such that $\nu(a_1\ldots a_i,u_d)$ is finite for every i. However, it is easy to see that this implies $\nu(a,u_d)$ for every $a \in W(u_d)$, i.e. $u_d \in D$.

It remains to show that $u \leqq v$ ($u,v \in D$) implies $v \leqq u$. Suppose $u \leqq v$, $v \nleqq u$. Then there is an $a \in W(v)$ such that $a \notin W(u)$. However v has by definition only finitely many subwords which do not contain a; on the other hand, the infinite word u has infinitely many subwords. This contradiction proves the equivalence of u and v.

Finally, if u is dense and $u' \leqq u$ then u' obviously is dense wich proves the maximality of D.

Now we can prove

Theorem 2. $\inf\langle P_\infty(X^*), \leqq\rangle = \{M : M \subseteq P_\infty(W(u)),\ u \in \inf\langle X^\omega, \leqq\rangle\}$.

Proof. The proof can be achieved through two simple lemmas. In order to formulate them, denote by $P(u)$ the set of all finite prefixes of u.

Lemma 1. *For every* $B \in \mathcal{P}_\infty(X^*)$ *there is a* $u \in X^\omega$ *such that* $P(u) \leqq B$.

The assertion is almost obvious: the empty word is a prefix of all members of B, and if one has a word $x_1 \ldots x_k$ which is a prefix of infinitely many members of B, then one can find an $x_{k+1} \in X$ such that $x_1 \ldots x_{k+1}$ has the same property. The infinite word $u = x_1 x_2 \ldots$ satisfies the condition of the Lemma.

Lemma 2. If $u \in X^\omega$ *then* $P(u)$ *is equivalent to* $W(u)$. *If, moreover,* u *is dense, then* $W(u)$ *is equivalent with each of its infinite subsets.*

Indeed, on the one hand, $P(u) \subseteqq W(u)$, and on the other, every $a \in W(u)$ occurs in some prefix of u. The second assertion immediately follows from the definition of density.

Now by Lemmas 1 and 2 there is a $u \in X^\omega$ for every $B \in \mathcal{P}_\infty(X^*)$ such that $W(u) \leqq B$, and, by Theorem 1, this u can be chosen to be dense, which proves the theorem.

In the applications it is not so much $\inf \langle \mathcal{P}_\infty(X^*), \leqq \rangle$ that is of importance as the basis of a subset of $\mathcal{P}_\infty(X^*)$ the elements of which are called cover antichains (by a basis of a quoset Q we mean a set of pairwise incomparable elements of Q which contains for every $b \in Q$ an $a \leqq b$).

Definition 2. A subset B of X^* is said to *cover* $a \in X^*$ if a can be covered by subwords which belong to B, i.e. if for every occurrance of a letter x in a there is a $b \in B$ s.th. $a = a'ba''$ and the given occurrance of x lies in b. Notation: $B \succ a$.

Definition 3. $A (\subset X^*)$ is a cover antichain if no $a \in A$ can be covered by $A \setminus \{a\}$.

Theorem 3. *Let* $u = x_1 x_2 \ldots \in X^\omega$. *The set of all finite prefixes* $P(u)$ *of* u *contains an infinite cover antichain iff u is not (pure) periodical.*

Proof. Denote by p_k the prefix $x_1 \ldots x_k$ of u. Let u be periodical, p_k its minimal period, and suppose $K \subseteqq P(u)$. Every $p \in K$ is of the form $p = p_k^r p_i$, $0 \leqq i < k$. If $p_{i(1)}, \ldots, p_{i(s)}$ are all prefixes which occur here as second components, and $a_j = p_k^{r_j} p_{i(j)}$ is the shortest word in

K with the second component $p_{i(j)}$, then, clearly, $\{a_1,\dots,a_s\}$ covers every $p\in K$. Hence K is not a cover antichain.

Now suppose u is non-periodical. If u has a finite subword a which occurs in u only a finite number of times, then the minimal prefix p_n which contains all occurrances of a cannot occur any later in u, because else a had one more occurrance in u. Hence $\{p_n,p_{n+1},\dots,\}$ is an infinite cover antichain contained in P(u). So we can assume for the rest of the proof that (u is non-periodical and) every finite subword of u occurs in it infinitely many times. Next we show that this condition implies that whenever p is a finite subword of u, there are finite subwords a_1 and a_2, $a_1 \neq a_2$, of equal length such that pa_1 and pa_2 also are subwords of u.

Suppose not. By assumption, u can be factorized in the form u = =bpcpu' where (u' is infinite and) every prefix of u' is equal to the prefix of cpu', having the same length; in particular, u' = cpu". By induction we obtain that u' is periodical with period cp. However then either u is periodical or it is of the form u = au*, u ≠ 0, where u* is periodical with some period p' but the last letters of a and of p' are different. In this latter case, ap' occurs in u only once. In both cases we have a contradiction with one of the assumptions.

Now we are going to construct an infinite sequence $p_{f(1)},p_{f(2)},\dots$ of prefixes of increasing length with the following property: $\{p_{f(1)},\dots,p_{f(i-1)}\}$ does not cover $p_{f(i)}$, and besides, there is a subword a_i of u longer than $p_{f(i)}$ and such that none of $p_{f(1)},\dots$ $\dots,p_{f(i)}$ is a suffix of a_i. Let $p_{f(1)} = \emptyset$, $a_1 = x_1$. Suppose by induction that $\{p_{f(1)},\dots,p_{f(i)}\}$ and a_i already have the required property, and define $p_{f(i+1)}$ as follows. Choose b_1,b_2, $b_1 \neq b_2$ so that both a_ib_1 and a_ib_2 are subwords of u. Clearly, they can be supposed to be of the forms $b_1 = cx$, $b_2 = cy$, $x \neq y$, $x,y\in A$, and also it can be achieved, that a_ic have no occurrances of a_i but the explicit one. As a_i occurs in u infinitely many times, u also has subwords a and a' having a_ib_1 and a_ib_2, respectively, for prefixes, and a_i for suffix. Suppose a and a' are minimal with this property, i.e. a_i does not occur "inside" them. Then $q = \bar{q}aq' = \bar{\bar{q}}a'q''$. Let $\bar{q}a$ be shorter than $\bar{\bar{q}}a'$, and put $p_{f(i+1)} = \bar{q}a$, $a_{i+1} = \bar{\bar{q}}a'$. As a_i is a suffix of both a and a', none of $p_{f(1)},\dots,p_{f(i)}$ is a suffix of either of them, whence it follows in particular that $\{p_{f(1)},\dots,p_{f(i)}\}$ does not cover $p_{f(i+1)}$. Furthermore, $p_{f(i+1)}$ is not a suffix of a_{i+1}, either , because none of a and a' is a suffix of the other one. Thus, the construction can be carried out for i+1, too, and $\{p_{f(1)},\ldots\}$ is an infinite cover antichain, which completes the proof.

Now denote the set of all infinite cover antichains by $\mathcal{A}$. For constructing a basis of $\mathcal{A}$, we introduce a linear order $\lhd$ on X, and extend it lexicographically to X^{ω} and X* (if a is a prefix of b, we also put $a \lhd b$). Next we show:

Lemma 3. Among the infinite words equivalent to the dense word u there is a lexicographically first one.

Proof. Let u' be the word with the property that its prefix p_k of length k is the lexicographically first subword of this length in u. Such word exists: its first letter is the lexicographically first letter of u, and if p_k is determined, and x is the lexicographically first letter such that $p_k x \in W(u)$, then put $p_{k+1} = p_k x$. We claim that u' is equivalent with u (whence, by Lemma 2, it is dense, too). Indeed, $u' \leq u$ by definition. Conversely, if $a \in W(u)$ then a is a subword of $p_{\nu(a,u)+1}$. From the definition of u follows now that it satisfies the condition of the lemma.

Denote the set of non-periodical lexicographically first dense words by L. Define for $u \in L$ the set $A(u) = \{a_1, a_2, \ldots\} \subseteq P(u)$ as follows. Let x_1 be the first letter of u and put $a_1 = x_1$. If a_k is already defined choose $a_{k+1} = a'_{k+1}x$ $(x \in X)$ to be the shortest prefix of u such that a'_{k+1} is a prefix of some power of a_k but a_{k+1} is not, i.e. $a'_{k+1} = a_k^m c$, $a_k = cyb_k$, $x \neq y \in X$. We have:

Lemma 4. A(u) is a cover antichain.

Proof. As $a_1, a_2, \ldots$ all are prefixes of u, it is sufficient to prove that no prefix of u can be a nontrivial suffix of a_r. Suppose this holds for a_k; let c,x,y be as above. By assumption, cy is a prefix of a_k and hence of u, but cx is not. Since $u \in L$, this implies $cy \lhd cx$ (because by the construction of the elements of L each of the prefixes of u precedes any other subword lexicographically), i.e. $y \lhd x$. If $c = c'd$ then both dx and dy are subwords of u, and $dy \lhd dx$. This implies that no prefix of u can be a suffix of cx. If some prefix of u were a suffix of a_{k+1} but not of cx then a (shorter) prefix of u would be a suffix of a_k, besides nontrivial, contrary to the assumption.

Now we construct a basis of $\langle \mathcal{A}, \leq \rangle$.

Theorem 4. The set

$B=\{\{xa^{n_i}:\{n_i\}\in\mathcal{P}_\infty(\mathbb{N}), a\in X^*, x\in X$, x <u>different from the last letter of</u> $a\}\cup$

$\cup\{\{a^{n_i}x:\{n_i\}\in\mathcal{P}_\infty(\mathbb{N}), a\in X^*, x\in X$, x <u>different from the first letter of</u> $a\}\cup$

$\cup\ \{A(u):u\in L\}$

<u>is a basis of</u> $\langle\mathcal{A}, \leqq\rangle$.

<u>Proof.</u> Let $A\in\mathcal{A}$, and suppose first that A can be reduced to some cyclic monoid $\{b\}^*$ or, equivalently, that $W(b^\omega)\leqq A$. Denote the longest subword of $c\in A$ contained in $W(b^\omega)$ by $b(c)$. As A is a cover antichain, $A\cap W(b^\omega)$ is finite, so we can suppose it is empty, and, since X is finite, either $xb(c)\leqq c$ or $b(c)x\leqq c$ occurs infinitely many times with some $x\in X$. We can even achieve, replacing b if necessary by a word obtained from it by a cyclic permutation of its letters, that in the first case $b(c) = b^{n(c)}\ b'$, b' a prefix of b (in the second case $b(c) = b''.b^{n(c)}$, b'' a suffix of b). Then x is different from the last letter of b by the maximality of $b(c)$, and $\{xb^{n(c)}:c\in A\}\leqq A$ ($\{b^{n(c)}x:c\in A\}\leqq A$, resp.)

If, on the other hand, A cannot be replaced by any cyclic monoid, this means that every word b has a power which is not a subword of any element of A. By Theorem 2, A can be replaced by the set of finite subwords $W(u)$ of a dense word u which is then non-periodical. By Lemma 3, u can be chosen from L and then, by Lemma 4, $A(u)\leqq A$, q.e.d.

<u>Remark.</u> Theorem 1 implies that a finite word is avoidable on X (as defined in [1]) if and only if it is avoided by some non-periodical infinite dense word - or, equivalently, by an element of L.

References

[1] Bean, D.R. - Ehrenfeucht, A. - McNulty, G.F., Avoidable patterns in strings of symbols, <u>Pacific J. Math.</u> **85** (1979), 261-294.

[2] Pollák, G., A syntactic method in the theory of varieties of semigroups (to appear)

UPDATE ON THE PROBLEMS IN "INVERSE SEMIGROUPS" BY M. PETRICH

N.R. Reilly
Department of Mathematics and Statistics
Simon Fraser University
Burnaby, B.C. V5A 1S6
Canada

Petrich's book "Inverse semigroups" [2] appeared in 1984. As the preface indicates, the book was several years in the writing and the manuscript was extensively commented upon in the preliminary stages by several of the leading contributors in the field. Many unanswered questions arose in the preparation of the manuscript. Those that were not resolved at the time and which were considered interesting by the author appeared in the book as open problems. In all, there are fifty-five problems presented in the book.

However, work continued in many of the areas relating to these problems after the manuscript was sent to the printers and, not surprisingly, many of the problems have since been solved. The objective here is to report on the present status of the problems in Petrich's book. We shall only comment on those for which solutions or partial solutions have been obtained. In most cases we shall simply give an appropriate reference together with a brief description of the solution or a brief comment on the solution. In a few instances, where the solution does not appear elsewhere, we will provide a solution.

Naturally we adopt the notation and terminology of Petrich's book. We will provide no background or explanation of the problems, since that can be found in Petrich's book.

Of the fifty-five problems, fifteen have been solved completely or partly.

III.7.11(i) Investigate the classes of the congruence $\hat{\sigma}$ induced by the homomorphism $\rho \to \rho \vee \sigma$ (see 5.6). (For each $\rho \in C(S)$, we have $\rho \subseteq \pi_\rho \subseteq \rho \vee \sigma$, where all three congruences are $\hat{\sigma}$-related and the last one is the greatest element of its class.) Do $\hat{\sigma}$-classes have least elements?

Not every $\hat{\sigma}$-class need have a minimum member. Let N denote the natural numbers, Z_2 the additive group of integers modulo 2 and $S = Z_2 \times N$ with multiplication

$$(a,x)(b,y) = (a+b,\max(x,y)).$$

For each $n \in N$, let the relation ρ_n be defined on S by

$$(a,x)\rho_n(b,y) \Leftrightarrow \text{either } (a,x) = (b,y) \text{ or } x,y \geq n .$$

Then $\cap\rho_n$ is the identity congruence ε while $\sigma \vee \rho_n$ is the universal congruence ω, for all n, but $\sigma \vee \varepsilon = \sigma \neq \omega$. Thus the $\hat{\sigma}$-class of the ρ_n has no smallest element.

NOTE. In problem III.7.11(iii), the author informs me that the final line should read

... such that $e' \tau f'$ and $(e',f') \leq (e,f)$.

IV.4.16(i) In groups, Brandt semigroups, and bicyclic semigroups every congruence is induced by a transitive representation by one-to-one partial transformations on a set. Find more classes of inverse semigroups with this property.

In [3], Petrich and Rankin have obtained results along these lines for the class of completely semisimple inverse semigroups with zero. We require the following definition.

DEFINITION A semigroup S is a special extension of an ideal I if every idempotent separating congruence on I extends uniquely to S.

THEOREM 1.[3] Let S be an inverse semigroup with zero. Then S is such that

(i) S is completely semisimple with S/J finite,

(ii) all congruences are induced by transitive representations

if and only if

(a) there exists a (unique) principal series $0 = S_0 \subset S_1 \subset \ldots \subset S_n = S$ such that

(b) S_i/S_{i-1} is a Brandt semigroup

(c) S_{i+1}/S_{i-1} is a special extension of S_i/S_{i-1} by S_{i+1}/S_i .

Now Scheiblich [7] has described the form of all congruences on the symmetric inverse semigroup $I(X)$. When X is finite this shows that the conditions (a), (b) and (c) of Theorem 1 are satisfied by $I(X)$ so that we have

COROLLARY 2.[3] Let X be a finite set. Then all congruences on $I(X)$ are induced by transitive representations.

Žitomorskiǐ (see [8]) showed that any congruence ρ on a bisimple inverse semigroup S is completely determined by any of its classes. On account of this fact, Petrich and Rankin [3] were able to show that ρ is induced by the transitive representation by right translations of the right cosets of $(e\rho)\omega$, for any $e \in E_S$. Thus

COROLLARY 3. [3] Every congruence on a bisimple inverse semigroup is induced by a transitive representation.

VI.6.11(i) Let an inverse semigroup S be a normal extension of an inverse semigroup K. Is it true that $\theta(S:K)$ is the greatest congruence on S contained in the relation ρ defined by

$$a\rho b \Leftrightarrow aK = bK,\ Ka = Kb?$$

First note that, by Definition VI.2.3, $\theta(S:K)$ is a homomorphism. In place of $\theta(S:K)$ we should read the congruence $\rho(S:K)$, say, induced by $\theta(S:K)$. The answer is then "no".

Let S be a non-trivial group with identity e and let $K = \{e\}$. Then $\rho(S:K) = S \times S$ but $\rho = \varepsilon$, the identity congruence.

VII.3.7(i) Is the translational hull of an E-reflexive inverse semigroup E-reflexive?

A much more general result is true, as we will now show.

LEMMA 4. Let the inverse semigroup V be a dense extension of the inverse semigroup S. Let $\theta : S \to T$ be an epimorphism. Then θ extends uniquely to a homomorphism $\varphi : V \to \Omega(T)$. Moreover, $\xi = \theta \circ \theta^{-1}$ is idempotent separating (respectively, idempotent pure) if and only if $\eta = \varphi \circ \varphi^{-1}$ is idempotent separating (respectively, idempotent pure).

Proof. That θ has a unique extension φ is well known and its description is as follows.

For any $v \in V$, $t \in T$, let $s \in S$ be such that $t = s\theta$.

Then

$$v\varphi = (\lambda^v, \rho^v) \in \Omega(T) \quad \text{where} \quad \lambda^v t = (vs)\theta,\ t\rho^v = (sv)\theta.$$

In particular, for $u,v \in V$, $s \in S$,

$$(usv)\theta = (u\varphi)(s\varphi)(v\varphi) = (\lambda^u,\rho^u)(s\theta)(\lambda^v,\rho^v) = \lambda^u(s\theta)\rho^v .$$

Now suppose that ξ is idempotent separating and consider any two distinct idempotents $e,f \in V$. Since V is a dense extension of S, $eSe \neq fSf$. Without loss of generality we may assume that $eSe \nsubseteq fSf$. Hence we may assume that there exists $g \in E_S$ such that $g \in eSe \setminus fSf$. Then

$$g\theta = (ege)\theta = \lambda^e(g\theta)\rho^e.$$

Since ξ is idempotent separating $g\theta \notin (fSf)\theta = \lambda^f T\rho^f$. Thus $\lambda^e T\rho^e \not\subseteq \lambda^f T\rho^f$. Therefore $e\varphi \neq f\varphi$ and η is idempotent separating.

Now suppose that ξ is idempotent pure. Let $v \in V \setminus E_V$. Then there exists $g \in E_S$ such that $g\mathcal{R}gv \neq g$. Since g is an idempotent, gv cannot be an idempotent. Hence

$$(g\theta)\rho^v = (gv)\theta \notin E_T,$$

so that ρ^v is not an idempotent right translation. Hence $v\varphi$ is not an idempotent and therefore η is idempotent pure.

The reverse implications are obvious.

For any class C of inverse semigroups, let

$C_K = \{S : S$ is an inverse semigroup and there exists an idempotent pure convergence ρ on S with $S/\rho \in C\}$,

$C_T = \{S : S$ is an inverse semigroup and there exists an idempotent separating congruence ρ on S with $S/\rho \in C\}$.

<u>LEMMA</u> 5. If C is a class of inverse semigroups which is closed under dense extensions, then C_K and C_T are also closed under dense extensions.

<u>Proof</u>. Let $S \in C_K$ and V be a dense extension of S. Let ρ be an idempotent pure congruence on S with $S/\rho \in C$. Let $\theta : a \to a\rho$ be the natural homomorphism of S onto S/ρ. By Lemma 4, θ extends to a unique homomorphism $\varphi : V \to \Omega(S/\rho)$ and $\varphi \circ \varphi^{-1}$ is idempotent pure. By hypothesis, $V\varphi \in C$ and so, by the definition of C_K, $V \in C_K$. Thus C_K is closed under dense extensions. The proof that C_T is closed under dense extensions is entirely similar.

<u>THEOREM</u> 6. Let C be any of the classes that occur in the min-network (that is, the upper part of diagram III.8.9.). Then C is closed under dense extensions.

<u>Proof</u>. This follows by induction from Lemma 4 and 5.

Since the class of E-reflexive semigroups is one of the classes in the min-network and since the translational hull of an inverse semigroup is also a dense extension, it follows from Theorem 6 that, in particular, the answer to question VII.3.7(i) is positive.

VII.6.13(i) <u>Does the last property in</u> 6.12(v) <u>characterize the inversely well ordered semilattices</u>?

Yes. Clearly if Y is inversely well ordered then $E_S = Y$ implies that S is F-inverse. Suppose conversely, that Y is such that if $E_S = Y$, then S is F-inverse.

Suppose that I is an ideal in Y with no maximum element. Let $S = \bigcup_{\alpha \in Y} G_\alpha$ be the semilattice of groups where

$$G_\alpha = \begin{cases} \{0\} & \text{if } \alpha \notin I \\ Z_2 & \text{if } \alpha \in I \end{cases}$$

and the structural homomorphisms are the obvious monomorphisms. Then, for any $\alpha \in I$, $(1)_\alpha \sigma$ has no maximum element and so S is not F-inverse.

VIII.4.9(i) Establish an analogue of 4.4 for the following classes: semilattices, groups, Clifford semigroups.

An analogue already exists for groups; see, for example, [9]V.1.a. For semilattices it is easy to verify the following

THEOREM 7. Let Y be a semilattice and $A \subseteq Y$. Then A generates a free subsemilattice of Y if and only if for all $a_1,\dots,a_{n+1} \in A$

$$a_1 \dots a_{n+1} \neq a_1 \dots a_n.$$

X.4.12(ii) Let S be a bisimple inverse monoid and R be the $\mathcal{R}$-class of its identity. Is every congruence ρ on S uniquely determined by the sets

$$N = \{a \in S \mid a\mathcal{H}1, a\rho 1\},$$
$$Q = \{a \in S \mid a\mathcal{R}1, a^{-1}a\rho 1\}?$$

If so, characterize abstractly the pairs (N,Q).

No. Compare the minimal group congruence and the universal congruence on the bicyclic semigroup.

XII.1.12(i) Is every relatively free inverse semigroup completely semisimple?

Departing slightly from the notation in [2], for any variety $\mathcal{V}$, let $F\mathcal{V}_X$ denote the relatively free inverse semigroup in $\mathcal{V}$ on the set X.

THEOREM 8 [6] Let X be a non-empty set, $x,y \in X$ and $x \neq y$. Let

$$a = xyxy(xyxy)^{-1}xyx^{-1}y^{-1} \in F\mathcal{I}_X$$

and let $\mathcal{V}$ be the variety defined by the identity

$$aa^{-1} = a^2a^{-2} .$$

Then FV_X is not completely semisimple.

Let $\rho = \rho(V)$ be the fully invariant congruence on FI_X corresponding to V and $g = a\rho$. Then clearly $gg^{-1} = g^2g^{-2}$. Also

$$(g^{-1}g)(gg^{-1}) = (g^{-1}g)(gg^{-1})(g^{-1}g) = g^{-1}g^2g^{-2}g$$
$$= (g^{-1}g)(g^{-1}g) = g^{-1}g.$$

Hence $g^{-1}g \leq gg^{-1}$. The rest of the long and technical proof of Theorem 6 is devoted to showing that $g^{-1}g \neq gg^{-1}$. It then follows that the inverse subsemigroup of FV_X generated by g is isomorphic to the bycyclic semigroup. Thus FV_X is not completely semisimple.

XII.1.12(ii) Characterize varieties of inverse semigroups generated by their antigroups (equivalently, fundamental inverse semigroups).

There are only a few varieties V for which much is known about the detailed structure of FV_X. However, it is sometimes possible to establish general properties of FV_X such as being E-unitary, fundamental etc. This is the approach taken by Reilly and Trotter [6] and Trotter [10]. Since many of the results can be stated to cover other properties besides fundamental we will include those properties also when convenient.

The first observation provides a more specific characterization than that presented in the problem.

THEOREM 9.[6] Let S be an inverse semigroup and $V = V(S)$, the variety generated by S. If S is combinatorial (respectively, completely semisimple or E-unitary) then FV_X is also combinatorial (respectively, completely semisimple or E-unitary). If S is fundamental and X is infinite then so is FV_X.

COROLLARY 10.[6] A variety V is generated by fundamental (respectively, combintorial, completely semisimple or E-unitary) inverse semigroups if and only if FV_X is fundamental (respectively, combinatorial, completely semisimple or E-unitary) for some (equivalently all) infinite sets X.

If C denotes the variety generated by the bicyclic semigroup (or, equivalently, the variety generated by the free inverse semigroup on a single generator) then one immediate consequence of Theorem 9 is that FC_X is combinatorial, completely semisimple and E-unitary.

The requirement in the final statement of Theorem 9 that X be infinite is indeed necessary since it does not always hold for finite sets X (see [6], Lemma 5.6).

Theorem 9 and its corollary shifts the focus of the problem to that of characterizing those varieties V for which FV_X is fundamental when X is infinite. One such characterization can be obtained by working with the corresponding fully invariant congruences. Recall that S is fundamental if and only if $\mu_S = \varepsilon$. This is equivalent to the condition:

$$a \in S,\ a^{-1}ea = eaa^{-1},\ \text{for all}\ \ e \in E_S \Rightarrow a \in E_S.$$

Trotter [10], used this characterization of fundamental to derive the next observation. For $a \in FI_X$, we denote by $c(a)$ the set of variables appearing in a.

LEMMA 11. [10] Let V be a variety of inverse semigroups. Then the following conditions are equivalent.

(i) FV_X if fundamental.

(ii) For all $a \in FI_X$, $z \in X \setminus c(a)$,

$$(a^{-1}zz^{-1}a,\ zz^{-1}aa^{-1}) \in \rho(V) \Rightarrow a \in \ker \rho(V).$$

The next result provides a useful way of deducing from the fact that one variety has a certain property that others also have it.

THEOREM 12. [6] Let V and W be varieties of inverse semigroups with $V \subseteq W$ and $\ker \rho(V) = \ker \rho(W)$. If FV_X is combinatorial fundamental or E-unitary, then so if FW_X. If FV_X is completely semisimple and combinatorial, then FW_X is completely semisimple.

The requirement that $\ker \rho(V) = \ker \rho(W)$ may seem difficult to verify. However, it is equivalent to saying that there is an idempotent pure congruence ρ on FW_X where X is infinite, such that $FW_X/\rho \in V$ and in some situations this can be seen to hold. In particular, if V is a variety of groups then this is equivalent to saying that FW_X is E-unitary over V or that W has E-unitary covers over V. With regard to the final statement of Theorem 12, there do exist V and W satisfying all the hypotheses except that FV_X is not combinatorial and such that FW_X is not completely semisimple (see [6], Remark 4.3).

Theorem 12 is useful for showing that whole intervals of varieties are such that their relatively free members have various properties as the following discussion indicates.

We will denote by G (respectively AG) the variety of groups (respectively abelian groups) and, for any variety of groups U we will denote by U^{max} the largest variety with E-unitary covers over U. Several alternative characterizations of U^{max} are given in [6], Theorem 3.7.

Theorem 12 can then be used to establish the following:

<u>THEOREM</u> 13. [6] For any variety V such that $C \subseteq V \subseteq AG^{max}$, FV_X is combinatorial, completely semisimple and E-unitary over AG.

The interval $[C, AG^{max}]$ is infinite. This and further information on varieties V with FV_X fundamental can be found in [6].

Trotter [10] used Lemma 11(ii) as motivation for a related concept. For any variety V, FV_X is said to be <u>completely fundamental</u> if for any $a \in FI_X$, $z \in X \setminus c(a)$, $(a^{-1}zz^{-1}a)\rho(V)$, $(zz^{-1}aa^{-1})\rho(V)$ comparable implies that $a \in \ker \rho(V)$.

Clearly, if FV_X is completely fundamental then it is fundamental but the converse does not hold ([10] Theorem 3.7.). However,

<u>LEMMA</u> 14. [10] If FV_X is completely semisimple and fundamental, then it is completely fundamental.

Varieties V for which FV_X is completely fundamental are quite plentiful. The following construction was introduced by Reilly [4].

Let G be a group, $M(G) = M^0(G, \{1\}, G; \Delta)$ and

$$N(G,G) = M(G) \cup G$$

be the extension of the combinatorial Brandt semigroup $M(G)$ by G^0 for which products in G and $M(G)$ are as given, while for $g \in G$, $(x,1,y) \in M(G)$,

$$g(x,1,y) = (xg^{-1},1,y), \quad (x,1,y)g = (x,1,yg), \quad g0 = 0 = 0g.$$

Then $N(G,G)$ is a fundamental inverse semigroup.

For any variety U of groups, let $U^{cf} = V(N(FU_X, FU_X))$.

<u>THEOREM</u> 15. [10] Let U be a non-trivial variety of groups. Then $V \in [U^{cf}, U^{max}]$ if and only if FV_X (for X infinite) is completely fundamental and E-unitary with FU_X as its maximum group homomorphic image.

XII.3.9(i) <u>If</u> $V \in L(I)$, <u>what can be said about</u> $(\rho(V))^{min}$, $(\rho(V))^{max}$?

One small observation can be made here. By definition ([2],III.4.6.), $(\rho(V))^{min} = (\rho(V) \cap L)^*$. Now, $\rho(V)$ and L are both fully invariant relations on FI_X and hence so also is $(\rho(V))^{min}$. But, in [1], Lemma 2.2, Pastijn and Trotter show that the congruence generated by a fully invariant relation on a semigroup S is also fully invariant. Hence, $(\rho(V))^{min}$ is a fully invariant congruence.

XII.4.20(ii) Construct a suitable isomorphic copy of the free strict inverse semigroup on a non-empty set.

A description of the free combinatorial strict inverse semigroup FB_X on a non-empty set X based on a solution of the word problem for FB_X is provided by Reilly in [5]. We will say that a word w in $Y = X \cup X^{-1}$ is in canonical form if $w = w_1 w_1^{-1} \ldots w_k w_k^{-1} w_{k+1}$ where

(i) the words w_i are in reduced form.

(ii) $w_{k+1} \in r(w) = \{w_i : i = 1,\ldots,k\}$ and $r(w)$ is closed under initial segments.

Every element in the free inverse semigroup can be written in canonical form in a manner which is unique to within the order of the factors $w_i w_i^{-1}$. Let $a \in Z = Y^+$, have canonical form $a = a_1 a_1^{-1} \ldots a_m a_m^{-1} g$ where

$$a_1 = a_{11} \ldots a_{1m_1} \quad \text{and} \quad g = g_1 \ldots g_r$$

in reduced form. Then we write

$$s(a) = \begin{cases} g_1 & \text{if } g \neq \emptyset \\ a_{11} & \text{if } g = \emptyset \end{cases} \qquad e(a) = \begin{cases} g_r^{-1} & \text{if } g \neq \emptyset \\ a_{11} & \text{if } g = \emptyset . \end{cases}$$

In addition, let

$$S(a) = \{a_i : i = 1,\ldots,m\} \quad \text{and} \quad Y(a) = \{y \in X \cup X^{-1} : y \text{ occurs in } a\}.$$

Define the relations γ_a and δ_a on $Y(a)$ by

$$x \,\gamma_a\, y \Leftrightarrow \begin{cases} \text{either (i)} & x,y \in S(a) \\ \text{or (ii)} & \text{there exists an element } u \in Z \cup \{0\} \\ & \text{such that either } ux^{-1}y \in S(a) \text{ or} \\ & uy^{-1}x \in S(a) \text{ where these elements} \\ & \text{are reduced as written} \end{cases}$$

δ_a = the equivalence relation on $Y(a)$ generated by γ_a.

These concepts enable us to formulate a solution to the word problem for FB_X.

THEOREM 16.[5] Let $a,b \in Z$ have canonical forms

$$a = a_1 a_1^{-1} \ldots a_m a_m^{-1} g \qquad b = b_1 b_1^{-1} \ldots b_n b_n^{-1} h$$

and let $\rho = \rho_B$. Then

(i) $a\rho = b\rho \Leftrightarrow \delta_a = \delta_b = \delta$, say, $(s(a),s(b)) \in \delta$ and $(e(a), e(b)) \in \delta$,

(ii) $a\rho \; R \; b\rho \Leftrightarrow \delta_a = \delta_b = \delta$, say, $(s(a),s(b)) \in \delta$,

(iii) $a\rho \; L \; b\rho \Leftrightarrow \delta_a = \delta_b = \delta$, say, $(e(a),e(b)) \in \delta$,

(iv) $a\rho \; \mathcal{D} \; b\rho \Leftrightarrow \delta_a = \delta_b$,

(v) $a\rho$ is an idempotent $\Leftrightarrow s(a) = e(a)$.

From this solution of the word problem, it is possible to derive a description of the free combinatorial strict inverse semigroup employing graphs in a manner modelled on the description of the free inverse semigroup obtained by Munn ([2],VIII.3).

A strict graph over a finite non-empty subset Q of X is a labelled digraph $\Gamma = (V,E,l)$ with V as the set of vertices, $E \subseteq V \times V$ as the set of edges, $R = Q \cup Q^{-1}$ as the full set of labels and $l(v_1,v_2)$ as the set of labels on the edge (v_1,v_2) for some function $l:E \longrightarrow 2^Y \setminus \{\emptyset\}$, such that

(i) Γ is connected,

(ii) each edge has at least one label from Y,

(iii) if $(v_1,v_2) \in E$ and $q \in l(v_1,v_2)$ then $(v_2,v_1) \in E$ and $q^{-1} \in l(v_2,v_1)$,

(iv) each element of Q appears exactly once as a label.

A birooted strict graph over $Q \subseteq X$ is a quintuple (V,E,l,s,e) where (V,E,l) is a strict graph over Q and $s,e \in V$. If $\Gamma_i = (V_i,E_i,l_i,s_i,e_i)$, $i = 1,2$, are two birooted strict graphs over $Q \subseteq X$, then an isomorphism $\theta:\Gamma_1 \longrightarrow \Gamma_2$ is a bijection of V_1 onto V_2 such that

(i) $(u,v) \in E_1 \Leftrightarrow (u\theta,v\theta) \in E_2$,

(ii) $(u,v) \in E_1 \Rightarrow l(u,v) = l(u\theta,v\theta)$,

(iii) $s_2 = s_1\theta$, $e_2 = e_1\theta$.

An isomorphism between strict graphs is defined similarly by dropping the conditions (ii) and (iii).

With each element of F we are going to associate an isomorphism class of strict graphs. For any $a \in Z$, let $\delta = \delta_a$ and define

$$V_a = Y(a)/\delta, E_a = \{(z\delta, z^{-1}\delta): z \in Y(a)\}$$

$$l_a(z\delta,z^{-1}\delta) = \{x \in Y(a): x\delta = z\delta, \; x^{-1}\delta = z^{-1}\delta\}$$

$$s_a = s(a)\delta \; , \; e_a = e(a)\delta .$$

LEMMA 17. $\Gamma_a = (V_a,E_a,l_a)$ is a strict graph over $c(a)$, the set of variables from x that appear in a, and $\Gamma_a^* = (V_a,E_a,l_a,s_a,e_a)$ is a birooted strict graph.

For any (birooted) strict graph Γ, let $[\Gamma]$ denote the isomorphism class of Γ. Theorem 16 and Lemma 17 now combine to give the following graphical interpretation of properties of elements of S.

THEOREM 18. [5] The mapping

$$\chi : a\rho \longrightarrow [\Gamma_a^*]$$

is a bijection of F onto the set of isomorphism classes of birooted strict graphs over finite non-empty subsets of X. Moreover,

(i) $a\rho = b\rho \Leftrightarrow [\Gamma_a^*] = [\Gamma_b^*]$,

(ii) $(a\rho, b\rho) \in R \Leftrightarrow [\Gamma_a] = [\Gamma_b]$ and $s_a = s_b$,

(iii) $(a\rho, b\rho) \in L \Leftrightarrow [\Gamma_a] = [\Gamma_b]$ and $e_a = e_b$,

(iv) $(a\rho, b\rho) \in D \Leftrightarrow [\Gamma_a] = [\Gamma_b]$,

(v) $a\rho$ is an idempotent if and only if $s_a = e_a$.

An algorithm for deriving $[\Gamma_{ab}^*]$ directly from $[\Gamma_a^*]$ and $[\Gamma_b^*]$ is provided in [5]. In addition, by counting the number of graphs involved, formulae are developed for the number of elements, idempotents and D-classes in the free combinatorial strict inverse semigroup on any finite number of generators.

XII.6.12(iii) For $n \geq 0$, let $M_n = \langle M_n \rangle$. Is it true that

$$C \cap (M_n \vee G) = M_n \vee AG \text{ ?}$$

(The inclusion $\supseteq$ is easy to establish.)

An affirmative answer is provided by Reilly and Trotter in [6], Corollary 3.11.

XII.6.12(v) Is there a variety in C which is not small?

Let Z denote the additive group of integers and $N_\infty = N(Z,Z)$ (where the construction $N(G,G)$, for a group G, is as for the discussion on Problem XII.1.12(ii)) and $V_\infty = V(N_\infty)$. It is shown by Reilly in [4], Theorem 7.7, that $V_\infty \subseteq C$ and that V_∞ is not small.

XII.7.13. Does there exist a cryptic variety which is not completely semisimple?

Let N_∞ and V_∞ be as in the solution to Problem XII.6.12(v). Again, it is shown in [4] that $V_\infty \subseteq C$. Thus, if the variety V contains C, then it contains V_∞ and therefore N_∞. But (see [4]), N_∞ is not cryptic. Thus any variety containing C is not cryptic. Consequently, any variety that is cryptic does not contain C and is therefore completely semisimple ([4], Corollary 5.13).

XIII.2.8(i) Does the class of E-unitary inverse semigroups have the weak amalgamation property?

Let C denote the bicyclic semigroup and σ the minimum group congruence on C. Let $S = C \cup C/\sigma$ with multiplication defined to be as given in C and C/σ and by

$$a(b\sigma) = (a\sigma)(b\sigma) = (a\sigma)b \qquad \text{for } a,b \in C$$

otherwise. Let $S_1 = S_2 = S$ but denote the idempotents of S_1 and S_2 by

$$E_{S_1} = \{e_0,e_1,e_2\ldots\},\ E_{S_2} = \{f_0,f_1,f_2,\ldots\}.$$

Let $U = \{e_0,e_1,e_2,e_3,\ldots\} \cup S/\sigma = \{f_0,f_1,f_3,f_4,\ldots\} \cup S/\sigma$

under the correspondence which is the identity on S/σ and identifies $e_0 \leftrightarrow f_0$, $e_1 \leftrightarrow f_1$, $e_2 \leftrightarrow f_3$, $e_3 \leftrightarrow f_4$ etc. Let $a \in S_1$, $b \in S_2$ be such that

$$aa^{-1} = e_0,\ a^{-1}a = e_1,\ bb^{-1} = f_0,\ b^{-1}b = f_1. \qquad (*)$$

Suppose that $[S_1,S_2;U]$ is embedded in the E-unitary inverse semigroup T. Let ω be the identity of C/σ. Then $\omega a = a\sigma = b\sigma = \omega b$ so that $(a,b) \in \sigma_T$. By (*), $a\mathcal{H}b$. Since T is E-unitary we must have $a = b$. However,

$$a^{-2}e_0a^2 = e_2 = f_3 \quad (\text{in } T)$$
$$\neq f_2 = b^{-2}f_0b^2 = b^{-2}e_0b^2 \quad (\text{in } T).$$

Thus $a^2 \neq b^2$ (in T). Hence $a \neq b$ (in T), a contradiction. Therefore $[S_1,S_2,U]$ is not embeddable in an E-unitary inverse semigroup (even weakly).

A similar argument shows that the class $\mathcal{E}$ of E-unitary inverse semigroups does not have the special amalgamation property. To see this, take S_1,S_2 as above and $U = E_{S_1} \cup C/\sigma = E_{S_2} \cup C/\sigma$. If $[S_1,S_2;U]$ is strongly embeddable in T, then $a \neq b$ (in T) and $a\mathcal{H}b$ but $\omega a = \omega b$.

REFERENCES

[1] F.J. Pastijn and P.G. Trotter, Lattices of completely regular semigroup varieties, (preprint).
[2] M. Petrich, Inverse Semigroups, Wiley, New York, 1984.
[3] M. Petrich and S. Rankin, Congruences induced by transitive representations of inverse semigroups, Glasgow Math. J. 29 (1987), 21-40.
[4] N.R. Reilly, Minimal non-cryptic varieties of inverse semigroups, Quart. J. Math. Oxford (2), 36 (1984), 467-487.
[5] N.R. Reilly, Free combinatorial strict inverse semigroups, J. London Math. Soc. (to appear).

[6] N.R. Reilly and P.G. Trotter, Properties of relatively free inverse semigroups, Transactions Amer. Math. Soc. 294 (1986), 243-262.
[7] H.E. Scheiblich, Concerning congruences on symmetric inverse semigroups, Czechoslovak Math. J., 23 (1973), 1-10.
[8] B.M. Schein, A remark concerning congruences of [0-] bisimple inverse semigroups, Semigroup Forum, 3 (1971), 80-83.
[9] E. Schenkman, Group Theory, Van Nostrand, New York, 1965.
[10] P.G. Trotter, Relatively free inverse semigroups, Quarterly J. Math. Oxford Ser(2) 37 (1986), 357-374.

MINIMAL CLANS: A CLASS OF ORDERED PARTIAL SEMIGROUPS INCLUDING BOOLEAN RINGS AND LATTICE-ORDERED GROUPS

Klaus D. Schmidt

Seminar für Statistik, Universität Mannheim, A 5
6800 Mannheim, West Germany

ABSTRACT

The present paper contains a rather comprehensive investigation of the properties of minimal clans - a new class of ordered partial semigroups which includes Boolean rings and lattice-ordered groups as special cases. It is shown that minimal clans preserve many properties of Boolean rings and lattice-ordered groups, that Boolean rings and lattice-ordered groups can be identified as minimal clans having, respectively, a minimal domain of addition or a maximal set of invertible elements, and that minimal clans in turn can be characterized in the classes of all symmetric clans, semiclans, and normal clans. Minimal clans are also compared with some other ordered algebraic structures with partial or complete addition that have been studied in the literature. An example of a minimal clan which is neither a Boolean ring nor a lattice-ordered group is provided by the collection of all fuzzy subsets of a given set.

1. INTRODUCTION

A few years after the fundamental papers by Stone [27] and Birkhoff [3] on generalized Boolean algebras and lattice-ordered groups had appeared, Birkhoff posed the following problem: "Is there a common abstraction which includes Boolean algebras (rings) and lattice-ordered groups as special cases?" [4; p. 233]; see also [5; p. 318].

From a purely algebraic point of view, Birkhoff's problem is certainly suggested by the observation that not only lattice-ordered groups but also generalized Boolean algebras are at the same time a group and a lattice. This is a consequence of Stone's result asserting the existence of a bijection between generalized Boolean algebras and rings with idempotent multiplication such that symmetric difference and intersection in the generalized Boolean algebra correspond, respectively, to addition and multiplication in the ring. In the terminology of Stone and Birkhoff, a ring with idempotent multiplication is said to be a Boolean ring, but in view of Stone's result it has also become customary to define a Boolean ring to be a generalized Boolean algebra, i.e. a relatively complemented distributive lattice having a least element. In the present paper we shall use this latter definition of a Boolean ring.

Another reason for studying Birkhoff's problem originates in the idea of developing a unified approach to certain aspects of measure and operator theory: Since a common abstraction of vector measures on a Boolean ring and of linear operators on a Riesz space has to be defined on a common abstraction of Boolean rings and lattice-ordered groups, solutions to Birkhoff's problem are indispensable for giving a unified approach e.g. to the Jordan decomposition of vector measures and linear operators; for a first result in this direction, see [23].

Several solutions to Birkhoff's problem or its commutative variant have been proposed in the literature, and it turned out that very different solutions are possible. This can already be seen from the early papers on this subject written by Swamy [28], Wyler [29], Nakano [17], and Rama Rao [20]. One way to classify the different algebraic structures solving Birkhoff's problem consists in distinguishing them according to the type of addition that has to be defined in Boolean rings in order to identify them as a special case of the algebraic structure in question.

A common abstraction of Boolean rings and lattice-ordered groups which is in the spirit of the previous remarks in connection with Stone's result was proposed by Rama Rao [20] who gave a characterization of the direct product of a ring with idempotent multiplication and a lattice-ordered group. In this algebraic structure the cancellation law holds, but order and addition need not be compatible since, by Stone's result, Boolean rings are included as a special case if in Boolean rings addition is defined to be the symmetric difference.

Common abstractions of Boolean rings and lattice-ordered groups which are ordered algebraic structures in the sense of Fuchs [15] were proposed by Swamy [28] in the commutative case and by Bosbach [6,7] in the general case. In these algebraic structures order and addition are compatible, but the cancellation law need not hold since in both of them Boolean rings are included as a special case if in Boolean rings addition is defined to be the union.

Some further common abstractions of Boolean rings and lattice-ordered groups were proposed by Wyler [29], Bosbach [8], and Billhardt [2] in the general case and by Schmidt [23,24] in the commutative case. The algebraic structures proposed by these authors can be subsumed under the notion of an ordered partial semigroup - an ordered algebraic structure in which only a partial addition is defined such that order and addition are compatible. Moreover, the cancellation law holds in the algebraic structures considered by Wyler [29], Bosbach [8], and Schmidt [24], and it can also be imposed on those considered by Billhardt [2] and Schmidt [23] without violating their property of being a common abstraction of Boolean rings and lattice-ordered groups. In each of these algebraic structures Boolean rings are included as a special case if in Boolean rings addition is defined only for disjoint elements to be the union or, equivalently, the symmetric difference.

A postulate of Rama Rao requires that a common abstraction of Boolean rings and lattice-ordered groups should, in order to be useful, possess "as much as possible of the richness of the structures common to both" Boolean rings and lattice-ordered groups [20; p. 411]. The choice of properties common to Boolean rings and lattice-ordered groups which are to be preserved by a common abstraction will, of course, to a great extent depend on the intended applications. The preceding discussion shows that a choice is possible, in particular, between the completeness of addition and the compatibility of order and addition together with the validity of the cancellation law. For the applications mentioned before, we feel that the second of these possibilities is the more convenient one.

We are thus led to considering ordered partial semigroups having the cancellation property. With regard to Rama Rao's postulate, we also require the lattice property, and imposing an additional condition, which is independent of the others and which is closely related to the existence of relative complements in Boolean rings and the modular law in commutative lattice-ordered groups, we arrive at the definition of a common abstraction of Boolean rings and lattice-ordered groups which we shall call a minimal clan.

The present paper contains a rather detailed discussion of the properties of minimal clans and their relation to other ordered algebraic structures with partial or complete addition that have been studied earlier. After giving some preliminary results on ordered partial semigroups (Section 2), we show that minimal clans preserve many properties of Boolean rings and lattice-ordered groups (Section 3), that Boolean rings and lattice-ordered groups can be identified as minimal clans having, respectively, a minimal domain of addition or a maximal set of invertible elements (Sections 4 and 5), and that minimal clans in turn can be characterized in the classes of all symmetric clans, semiclans, and normal clans, which were respectively introduced by Wyler [29], Bosbach [8], and Billhardt [2] (Sections 6 and 7). We also compare minimal clans with some ordered algebraic structures with complete addition which were introduced by Bosbach [6,7,9], Swamy [28], and Riesz [21] (Section 8), and we complete the discussion of minimal clans with a few remarks concerning their axioms, examples of minimal clans which are neither a Boolean ring nor a lattice-ordered group, and their relation to the Riesz D-semigroups introduced by Dinges [14] which originally raised our interest in ordered partial semigroups (Section 9).

I would like to thank Professors B. Bosbach, K. Keimel, and H.J. Weinert for several stimulating discussions on the subject of this paper.

2. ORDERED PARTIAL SEMIGROUPS

In this preliminary section, we introduce ordered partial semigroups and some related notions which will be needed in the sequel. With regard to the definitions of the algebraic structures we consider, we remark that a neutral element will always be assumed to exist and that the additive notation will be used throughout although commutativity will usually not be assumed. The additive notation has also been used by Birkhoff [3,4,5] in the case of lattice-ordered groups, and in the case of minimal clans it is also convenient with regard to the applications mentioned in the introduction.

A <u>partial groupoid</u> is a set $\mathbb{E}$ with a relation $S \subseteq \mathbb{E}\times\mathbb{E}$ and a map $+ : S \longrightarrow \mathbb{E}$ such that

- there exists an element $0 \in \mathbb{E}$ satisfying $(0,x) \in S$, $(x,0) \in S$, and $0+x = x = x+0$ for all $x \in \mathbb{E}$.

If $(\mathbb{E},S,+)$ is a partial groupoid, then the set S is said to be the <u>domain of addition</u>, the map $+ : S \longrightarrow \mathbb{E}$ is called <u>addition</u>, and the (unique) element $0 \in \mathbb{E}$ satisfying $(0,x) \in S$, $(x,0) \in S$, and $0+x = x = x+0$ for all $x \in \mathbb{E}$ is said to be the <u>zero element</u> of $\mathbb{E}$.
In the sequel, we shall frequently write

$$x+y \text{ has property (P)}$$

instead of the full statement

$$(x,y) \in S \text{ and } x+y \text{ has property (P)}.$$

This will considerably simplify the notation.

A partial groupoid $(\mathbb{E},S,+)$ is associative if

- for all $x, y, z \in \mathbb{E}$, $(x,y) \in S$ and $(x+y,z) \in S$ if and only if $(y,z) \in S$ and $(x,y+z) \in S$, and in this case $(x+y)+z = x+(y+z)$;

it is commutative if

- for all $x, y \in \mathbb{E}$, $(x,y) \in S$ if and only if $(y,x) \in S$, and in this case $x+y = y+x$;

it has the cancellation property if

- $x = y$ holds for all $x, y \in \mathbb{E}$ satisfying $u+x+v = u+y+v$ for some $u, v \in \mathbb{E}$;

and it has complete addition if

- $S = \mathbb{E}\times\mathbb{E}$.

A partial groupoid with complete addition will be said to be a groupoid and will be denoted by $(\mathbb{E},\mathbb{E}\times\mathbb{E},+)$, and a corresponding simplification in the case of complete addition will also be used for the special partial groupoids to be defined now.

A partial semigroup is a partial groupoid $(\mathbb{E},S,+)$ which is associative. If $(\mathbb{E},S,+)$ is a partial semigroup, then an element $x \in \mathbb{E}$ is invertible if there exist $u, v \in \mathbb{E}$ satisfying $u+x = 0 = x+v$, and this is equivalent to the condition that there exists some $x^* \in \mathbb{E}$ satisfying $x^*+x = 0 = x+x^*$. For an invertible element $x \in \mathbb{E}$, the (unique) element $x^* \in \mathbb{E}$ satisfying $x^*+x = 0 = x+x^*$ is said to be the inverse of x. The set of all invertible elements of $\mathbb{E}$ will be denoted by $\mathbb{E}_*$.

2.1. Lemma.

Let $(\mathbb{E},S,+)$ be a partial semigroup.

(a) If $x \in \mathbb{E}_*$, then $(x,y) \in S$ and $(y,x) \in S$ for all $y \in \mathbb{E}$.

(b) If $\mathbb{E}_* = \mathbb{E}$, then $S = \mathbb{E}\times\mathbb{E}$.

An ordered partial semigroup is a partial semigroup $(\mathbb{E},S,+)$ with an order relation $\leq$ such that

- $u+x+v \leq u+y+v$ holds for all $x, y \in \mathbb{E}$ satisfying $x \leq y$ and for all $u, v \in \mathbb{E}$ satisfying $(u,x) \in S$ and $(u+x,v) \in S$ as well as $(u,y) \in S$ and $(u+y,v) \in S$.

If $(\mathbb{E},S,+,\leq)$ is an ordered partial semigroup, then an element $x \in \mathbb{E}$ is positive if it satisfies $0 \leq x$, and two positive elements $x, y \in \mathbb{E}$ are disjoint if their greatest lower bound $x \wedge y$ exists and is equal to 0. The set of all positive elements of $\mathbb{E}$ will be denoted by $\mathbb{E}_+$, and the collection of all pairs of disjoint positive elements of $\mathbb{E}$ will be denoted by $\mathbb{E}^{\perp}\mathbb{E}$. Furthermore, for $x, z \in \mathbb{E}$ satisfying $x \leq z$, the set $[x,z] := \{ y \in \mathbb{E} \mid x \leq y \leq z \}$ is said to be the order interval with endpoints x and z.

An ordered partial semigroup $(\mathbb{E},S,+,\leq)$ is positive if

- $\mathbb{E}_+ = \mathbb{E}$;

it has the lattice property if

- $(\mathbb{E},\leq)$ is a lattice, i.e. any two elements $x, y \in \mathbb{E}$ have a least upper bound $x \vee y$ and a greatest lower bound $x \wedge y$;

it has the difference property if

- for all $x, y \in \mathbb{E}$ having a least upper bound $x \vee y$ and a greatest lower bound $x \wedge y$, there exist $u, v \in \mathbb{E}_+$ satisfying $u + x = x \vee y = x + v$ and $u + x \wedge y = y = x \wedge y + v$;

and it has property (M) if

- for all $x, y \in \mathbb{E}$ having a least upper bound $x \vee y$ and a greatest lower bound $x \wedge y$, $(x,y) \in S$ if and only if $(x \vee y, x \wedge y) \in S$, and in this case $x + y = x \vee y + x \wedge y$.

2.2. Lemma.

Let $(\mathbb{E},S,+,\leq)$ be an ordered partial semigroup.

(a) If $S \subseteq \mathbb{E}^{\perp}\mathbb{E}$, then $\mathbb{E}_+ = \mathbb{E}$.

(b) If $\mathbb{E}_+ = \mathbb{E}$, then $\mathbb{E}_* = \{0\}$.

2.3. Lemma.

Let $(\mathbb{E},S,+,\leq)$ be an ordered partial semigroup having the difference property.

(a) If $u \leq x$, $v \leq y$, and $(x,y) \in S$, then $(u,v) \in S$.

(b) If $u \leq x$ and $x \in \mathbb{E}_*$, then $u \in \mathbb{E}_*$.

The second assertion of Lemma 2.3 may be regarded as a converse of the following property of $\mathbb{E}_+$: If $x \leq w$ and $x \in \mathbb{E}_+$, then $w \in \mathbb{E}_+$.

A lattice-ordered partial semigroup is an ordered partial semigroup having the lattice property.

2.4. Lemma.

Let $(\mathbb{E},S,+,\leq)$ be a lattice-ordered partial semigroup having the difference property.

(a) $x \wedge 0 + x \vee 0 = x = x \vee 0 + x \wedge 0$ for all $x \in \mathbb{E}$.

(b) $\mathbb{E}_+ = \mathbb{E}$ if and only if $\mathbb{E}_* = \{0\}$.

(c) If $x \wedge y = 0$, then $x + y = x \vee y = y + x$.

(d) $\mathbb{E}^{\perp}\mathbb{E} \subseteq S$.

(e) If $S = \mathbb{E}^{\perp}\mathbb{E}$, then $(\mathbb{E},S,+,\leq)$ is commutative and positive.

Proof. Consider $x \in \mathbb{E}$. By the difference property, there exist $u, v \in \mathbb{E}$ satisfying $u + x = x \vee 0 = x + v$ and $u + x \wedge 0 = 0 = x \wedge 0 + v$.

The second identity yields $x\wedge 0 \in \mathbb{E}_*$ and $u = (x\wedge 0)^* = v$, and now the first identity gives $x = 0+x = (x\wedge 0 + u) + x = x\wedge 0 + (u+x) = x\wedge 0 + x\vee 0$ and, similarly, $x = x\vee 0 + x\wedge 0$. This proves (a).
(b) follows from (a) and Lemma 2.2, (c) and hence (d) is obvious from the difference property, and (e) follows from (c) and Lemma 2.2. □

For commutative lattice-ordered partial semigroups having the difference property, assertions (a) and (d) of Lemma 2.4 can be improved:

2.5. Lemma.
Let $(\mathbb{E},S,+,\leq)$ be a commutative lattice-ordered partial semigroup having the difference property.

(a) $(x,y) \in S$ if and only if $(x\vee y, x\wedge y) \in S$, and in this case $x+y = x\vee y + x\wedge y$.

(b) If $x\wedge y \in \mathbb{E}_*$, then $(x,y) \in S$ and $(y,x) \in S$.

Proof. Consider $x, y \in \mathbb{E}$ and choose $w \in \mathbb{E}_+$ satisfying $x\vee y = x+w$ and $w + x\wedge y = y$, by the difference property and commutativity. If either $(x,y) \in S$ or $(x\vee y, x\wedge y) \in S$, then all sums in the identity

$$x+y = x + (w + x\wedge y) = (x+w) + x\wedge y = x\vee y + x\wedge y$$

are defined. This proves (a).
(b) follows from (a) and Lemma 2.1. □

By Lemma 2.5, a lattice-ordered partial semigroup having the difference property is commutative if and only if it has property (M).

To conclude this section, we remark that none of the previous results involves the cancellation property.

3. MINIMAL CLANS

In this section, we introduce minimal clans and proceed with a detailed discussion of their properties. We generalize most of the results on commutative minimal clans proven in [24], and we also prove several additional results which even in the commutative case are new. From the results of this section it can be seen that minimal clans preserve to a large extent the properties of Boolean rings and lattice-ordered groups.

A minimal clan is a lattice-ordered partial semigroup having the cancellation property and the difference property. Throughout this section, let $(\mathbb{E},S,+,\leq)$ be a minimal clan.

The following results are immediate from the definition of minimal clans:

<u>3.1. Lemma.</u>
$x \in \mathbb{E}_*$ if and only if there exists some $w \in \mathbb{E}$ satisfying either $w + x = 0$ or $0 = x + w$.

<u>3.2. Lemma.</u>
For all $x, y \in \mathbb{E}$, there exist unique elements $u, v \in \mathbb{E}_+$ satisfying $u + x = x \vee y = x + v$ and $u + x \wedge y = y = x \wedge y + v$.

The following result describes two permanence properties of minimal clans:

<u>3.3. Theorem.</u>
(a) $(\mathbb{E}_*, \mathbb{E}_* \times \mathbb{E}_*, +, \leq)$ is a minimal clan.
(b) $(\mathbb{E}_+, (\mathbb{E}_+ \times \mathbb{E}_+) \cap S, +, \leq)$ is a minimal clan.

Proof. It is easy to see that $(\mathbb{E}_*, \mathbb{E}_* \times \mathbb{E}_*, +, \leq)$ is an ordered partial semigroup having the cancellation property. To see that it also has the difference property and the lattice property, consider $x, y \in \mathbb{E}_*$ and $u, v \in \mathbb{E}_+$ satisfying $u + x = x \vee y = x + v$ and $u + x \wedge y = y = x \wedge y + v$. By Lemma 2.3, we have $x \wedge y \in \mathbb{E}_*$, hence $u \in \mathbb{E}_*$ and $v \in \mathbb{E}_*$, and thus $x \vee y \in \mathbb{E}_*$. Therefore, $(\mathbb{E}_*, \mathbb{E}_* \times \mathbb{E}_*, +, \leq)$ is a minimal clan.
(b) is obvious. □

As a consequence of the difference property and the cancellation property, we obtain the following <u>order cancellation property</u>:

<u>3.4. Lemma.</u>
If $u + x + v \leq u + y + v$, then $x \leq y$.

<u>3.5. Corollary.</u>
If $u + x \wedge y = y = x \wedge y + v$ and $w + x \wedge y = x = x \wedge y + z$, then $u \wedge w = 0 = v \wedge z$.

Proof. By assumption, we have $u \wedge w + x \wedge y \leq x \wedge y$ and thus $u \wedge w \leq 0$, by Lemma 3.4, and from $x \wedge y \leq y = u + x \wedge y$ and $x \wedge y \leq x = w + x \wedge y$ we obtain $0 \leq u$ and $0 \leq w$, and thus $0 \leq u \wedge w$. This implies $u \wedge w = 0$, and a similar argument yields $v \wedge z = 0$. □

Corollary 3.5 will frequently be used in the sequel; in most cases, it will be applied in connection with Lemma 2.4.

Our next result is the <u>refinement property</u>:

3.6. Theorem.

If $x_1, x_2, \ldots, x_m \in \mathbb{E}_+$ and $y_1, y_2, \ldots, y_n \in \mathbb{E}_+$ are such that

$$\sum_{i=1}^{m} x_i = \sum_{j=1}^{n} y_j ,$$

then there exist $z_{ij} \in \mathbb{E}_+$ satisfying

$$x_i = \sum_{j=1}^{n} z_{ij}$$

for all $i \in \{1,2,\ldots,m\}$,

$$y_j = \sum_{i=1}^{m} z_{ij}$$

for all $j \in \{1,2,\ldots,n\}$, and

$$\left(\sum_{k=i+1}^{m} z_{kj} \right) \wedge \left(\sum_{l=j+1}^{n} z_{il} \right) = 0$$

for all $i \in \{1,2,\ldots,m-1\}$ and $j \in \{1,2,\ldots,n-1\}$.

Proof. The assertion is trivial in the case $m = 1$ or $n = 1$. Let us first consider the case $m = n = 2$. Define $z_{11} := x_1 \wedge y_1$, choose $z_{12} \in \mathbb{E}_+$ satisfying $x_1 \vee y_1 = y_1 + z_{12}$ and $x_1 = x_1 \wedge y_1 + z_{12}$, and choose $z_{21} \in \mathbb{E}_+$ satisfying $x_1 \vee y_1 = x_1 + z_{21}$ and $y_1 = x_1 \wedge y_1 + z_{21}$. Then we have $z_{11} \in \mathbb{E}_+$ as well as

$$x_1 = z_{11} + z_{12}$$

and

$$y_1 = z_{11} + z_{21} .$$

Now define $z := x_1 + x_2 = y_1 + y_2$. Then we have

$$z_{11} + z_{12} + z_{21} = x_1 + z_{21} = x_1 \vee y_1 \leq z ,$$

and we may choose $z_{22} \in \mathbb{E}_+$ satisfying

$$z_{11} + z_{12} + z_{21} + z_{22} = z = x_1 + x_2 = z_{11} + z_{12} + x_2 ,$$

which yields

$$x_2 = z_{21} + z_{22} .$$

Furthermore, Corollary 3.5 yields

$$z_{12} \wedge z_{21} = 0 ,$$

and this implies

$$z_{12} + z_{21} = z_{21} + z_{12} ,$$

by Lemma 2.4. Thus we have

$$z_{11} + z_{21} + z_{12} + z_{22} = z = y_1 + y_2 = z_{11} + z_{21} + y_2 ,$$

which yields

$$y_2 = z_{12} + z_{22} .$$

This proves the assertion in the case $m = n = 2$.

Consider now $m \in \{3,4,\dots\}$ and assume that the assertion is true for all $m', m'' \in \{1,2,\dots,m-1\}$ and some $n \in \{2,3,\dots\}$.
Letting $m' := m-1$ and

$$x_i' := \begin{cases} x_i & , \text{ if } i \in \{1,2,\dots,m'-1\} \\ x_{m-1} + x_m & , \text{ if } i = m' \end{cases}$$

we obtain

$$\sum_{i=1}^{m'} x_i' = \sum_{i=1}^{m} x_i = \sum_{j=1}^{n} y_j ,$$

and by hypothesis there exist $z_{ij}' \in \mathbb{E}_+$ satisfying

$$x_i' = \sum_{j=1}^{n} z_{ij}'$$

for all $i \in \{1,2,\dots,m'\}$,

$$y_j = \sum_{i=1}^{m'} z_{ij}'$$

for all $j \in \{1,2,\dots,n\}$, and

$$\left(\sum_{k=i+1}^{m'} z_{kj}' \right) \wedge \left(\sum_{l=j+1}^{n} z_{il}' \right) = 0$$

for all $i \in \{1,2,\dots,m'-1\}$ and $j \in \{1,2,\dots,n-1\}$.
Furthermore, letting $m'' := 2$ and $x_i'' := x_{m+i-2}$ for $i \in \{1,2\}$, we obtain

$$\sum_{i=1}^{2} x_i'' = x_{m'}' = \sum_{j=1}^{n} z_{m'j}' ,$$

and by hypothesis there exist $z_{ij}'' \in \mathbb{E}_+$ satisfying

$$x_i'' = \sum_{j=1}^{n} z_{ij}''$$

for all $i \in \{1,2\}$,

$$z_{m'j}' = \sum_{i=1}^{2} z_{ij}''$$

for all $j \in \{1,2,\dots,n\}$, and

$$z_{2j}'' \wedge \left(\sum_{l=j+1}^{n} z_{2l}'' \right) = 0$$

for all $j \in \{1,2,\dots,n-1\}$.

Now define, for all $i \in \{1,2,...,m\}$ and $j \in \{1,2,...,n\}$,

$$z_{ij} := \begin{cases} z'_{ij} , & \text{if } i \in \{1,2,...,m-2\} \\ z''_{1j} , & \text{if } i = m-1 \\ z''_{2j} , & \text{if } i = m . \end{cases}$$

This yields

$$x_i = \sum_{j=1}^{n} z_{ij}$$

for all $i \in \{1,2,...,m\}$,

$$y_j = \sum_{i=1}^{m} z_{ij}$$

for all $j \in \{1,2,...,n\}$, and

$$\left(\sum_{k=i+1}^{m} z_{kj} \right) \wedge \left(\sum_{l=j+1}^{n} z_{il} \right) = 0$$

for all $i \in \{1,2,...,m-1\}$ and $j \in \{1,2,...,n-1\}$.
Interchanging the role of m and n in the induction step completes the proof. □

A generalization of the refinement property to sums of arbitrary elements will be given below. We now prove the distributive laws:

3.7. Theorem.

(a) If $(x,y) \in S$ and $(x,z) \in S$,
then $x + y \vee z = (x+y) \vee (x+z)$ and $x + y \wedge z = (x+y) \wedge (x+z)$.

(b) If $(x,z) \in S$ and $(y,z) \in S$,
then $x \vee y + z = (x+z) \vee (y+z)$ and $x \wedge y + z = (x+z) \wedge (y+z)$.

Proof. Choose $v' \in \mathbb{E}_+$ satisfying

$$y \vee z = z + v' \quad \text{and} \quad y = y \wedge z + v' ,$$

choose $z' \in \mathbb{E}_+$ satisfying

$$y \vee z = y + z' \quad \text{and} \quad z = y \wedge z + z' ,$$

and define

$$u := x + y \wedge z .$$

Then we have

$$x + y = x + y \wedge z + v' = u + v'$$

and

$$x + z = x + y \wedge z + z' = u + z'$$

as well as

$$v' \wedge z' = 0 \quad ,$$

by Corollary 3.5. Choose now $v'' \in \mathbb{E}_+$ satisfying

$$(u+v') \vee (u+z') = u + v' + v'' \quad \text{and} \quad u + z' = (u+v') \wedge (u+z') + v''$$

and choose $z'' \in \mathbb{E}_+$ satisfying

$$(u+v') \vee (u+z') = u + z' + z'' \quad \text{and} \quad u + v' = (u+v') \wedge (u+z') + z'' \quad .$$

Then we have

$$v' + v'' = z' + z'' \quad ,$$

by the cancellation property, as well as

$$v'' \wedge z'' = 0 \quad ,$$

by Corollary 3.5. By the refinement property, there exist $z_{ij} \in \mathbb{E}_+$ satisfying

$$v' = z_{11} + z_{12} \quad \text{and} \quad v'' = z_{21} + z_{22}$$

as well as

$$z' = z_{11} + z_{21} \quad \text{and} \quad z'' = z_{12} + z_{22} \quad .$$

From

$$0 \leq z_{11} \leq v' \wedge z' = 0$$

and

$$0 \leq z_{22} \leq v'' \wedge z'' = 0$$

we obtain

$$z' = v'' \quad .$$

This yields

$$\begin{aligned} x + y \vee z &= x + (y + z') \\ &= (x + y) + z' \\ &= (u+v') + v'' \\ &= (u+v') \vee (u+z') \\ &= (x+y) \vee (x+z) \quad , \end{aligned}$$

and using the cancellation property we also obtain

$$x + y \wedge z = u = (u+v') \wedge (u+z') = (x+y) \wedge (x+z) \quad .$$

This proves (a).
The proof of (b) is similar and hence omitted. □

The following result shows that each minimal clan is a distributive lattice:

<u>3.8. Theorem.</u>

(a) $x \wedge (y \vee z) = (x \wedge y) \vee (x \wedge z)$.

(b) $x \vee (y \wedge z) = (x \vee y) \wedge (x \vee z)$.

Proof. Choose $v \in \mathbb{E}_+$ satisfying

$$x \vee (y \wedge z) = y \wedge z + v \quad \text{and} \quad x = x \wedge y \wedge z + v \quad ,$$

choose $v' \in \mathbb{E}_+$ satisfying

$$x \vee y = x + v' \quad \text{and} \quad y = x \wedge y + v' \quad ,$$

and choose $v'' \in \mathbb{E}_+$ satisfying

$$x \vee z = x + v'' \quad \text{and} \quad z = x \wedge z + v'' \quad .$$

Then we have

$$x \wedge y \wedge z + v \wedge v' \wedge v'' \leq x \wedge y \wedge z$$

and thus

$$v \wedge v' \wedge v'' = 0 \quad ,$$

by Lemma 3.4, and this yields

$$v + v' \wedge v'' = v \vee (v' \wedge v'') \quad ,$$

by Lemma 2.4. Using the distributive laws we obtain

$$\begin{aligned} (x \vee y) \wedge (x \vee z) &= (x+v') \wedge (x+v'') \\ &= x + v' \wedge v'' \\ &= x \wedge y \wedge z + v + v' \wedge v'' \\ &= x \wedge y \wedge z + v \vee (v' \wedge v'') \\ &= (x \wedge y \wedge z + v) \vee (x \wedge y \wedge z + v' \wedge v'') \\ &\leq x \vee ((x \wedge y + v') \wedge (x \wedge z + v'')) \\ &= x \vee (y \wedge z) \quad , \end{aligned}$$

and the converse inequality is obvious. This proves (b).
It is well-known that (a) is a consequence of (b); see [5; p.11]. □

Our next result is the <u>decomposition property</u>:

<u>3.9. Theorem.</u>
If

$$x \leq \sum_{i=1}^{n} y_i \quad ,$$

then there exist $x_i \in \mathbb{E}$ satisfying

$$x = \sum_{i=1}^{n} x_i$$

and

$$x_i \leq y_i$$

for all $i \in \{1,2,\ldots,n\}$.
Moreover, if $x \in \mathbb{E}_+$ and $y_1, y_2, \ldots, y_n \in \mathbb{E}_+$, then the x_i can be choosen such that $0 \leq x_i \leq x \wedge y_i$ holds for all $i \in \{1,2,\ldots,n\}$.

Proof. Let us first consider the case $n = 2$. Choose $u \in \mathbb{E}_+$ satisfying

$$u + y_2 = x \vee y_2 \quad \text{and} \quad u + x \wedge y_2 = x$$

and define

$$x_1 := u \wedge y_1 \quad .$$

Then we have

$$x_1 \leq y_1 .$$

Choose now $v \in \mathbb{E}_+$ satisfying

$$u = u \wedge y_1 + v = x_1 + v .$$

Then we have

$$x = u + x \wedge y_2 = x_1 + v + x \wedge y_2 .$$

Define

$$x_2 := v + x \wedge y_2 .$$

Then we have

$$x = x_1 + x_2 ,$$

hence

$$x_1 + x_2 = x \leq (u+y_2) \wedge (y_1+y_2) = u \wedge y_1 + y_2 = x_1 + y_2 ,$$

by the distributive laws, and thus

$$x_2 \leq y_2 ,$$

by the order cancellation property.
Moreover, if x, y_1, y_2 are positive, then the same is true for x_1 and x_2, and it is then clear that in this case $0 \leq x_i \leq x \wedge y_i$ holds for all $i \in \{1,2\}$. This proves the assertion in the case $n = 2$. The general case now follows by induction. □

<u>3.10. Corollary.</u>
If $x, y, z \in \mathbb{E}_+$ and $(y,z) \in S$, then $x \wedge (y+z) \leq x \wedge y + x \wedge z$.

<u>3.11. Corollary.</u>
If $x, y, z \in \mathbb{E}_+$ and $(y,z) \in S$, then $x \wedge (y+z) = x \wedge (x \wedge y+z) = x \wedge (x \wedge z+y)$.

Another consequence of the decomposition property is the following property of order intervals:

<u>3.12. Theorem.</u>
If $u \leq x$, $v \leq y$, and $(x,y) \in S$, then $[u,x] + [v,y] = [u+v,x+y]$.

Proof. Choose $z \in \mathbb{E}_+$ satisfying $x = u + z$, and choose $w \in \mathbb{E}_+$ satisfying $w + v = y$. Using the order cancellation property, we obtain

$$u + [0,z] = [u,u+z] = [u,x]$$

and

$$[0,w] + v = [v,w+v] = [v,y] ,$$

as well as

$$u + [0,z+w] + v = [u+v,u+z+w+v] = [u+v,x+y] .$$

It is therefore sufficient to prove

$$[0,z] + [0,w] = [0,z+w] ,$$

and the validity of this identity is obvious from the decomposition property. □

The following finite sum property extends the refinement property to sums of arbitrary elements:

3.13. Theorem.
If

$$\sum_{i=1}^{m} x_i = \sum_{j=1}^{n} y_j ,$$

then there exist $z_{ij} \in \mathbb{E}$ satisfying

$$x_i = \sum_{j=1}^{n} z_{ij}$$

for all $i \in \{1,2,\ldots,m\}$,

$$y_j = \sum_{i=1}^{m} z_{ij}$$

for all $j \in \{1,2,\ldots,n\}$, and

$$\left(\sum_{k=i+1}^{m} z_{kj} \right) \wedge \left(\sum_{l=j+1}^{n} z_{il} \right) = 0$$

for all $i \in \{1,2,\ldots,m-1\}$ and $j \in \{1,2,\ldots,n-1\}$.

Proof. Let us first consider the case $m = n = 2$. Choose $v \in \mathbb{E}_+$ satisfying $x_1 = x_1 \wedge y_1 + v$, choose $u \in \mathbb{E}_+$ satisfying $u + x_2 \wedge y_2 = x_2$, choose $z \in \mathbb{E}_+$ satisfying $y_1 = x_1 \wedge y_1 + z$, and choose $w \in \mathbb{E}_+$ satisfying $w + x_2 \wedge y_2 = y_2$. Then we have

$$v \wedge z = 0 = u \wedge w ,$$

by Corollary 3.5, and the identity $x_1 + x_2 = y_1 + y_2$ together with the cancellation property yields

$$v + u = z + w .$$

Using Corollary 3.11 we obtain

$$v = v \wedge (z+w) = v \wedge w = (v+u) \wedge w = w ,$$

and a similar argument yields

$$u = z .$$

Now define $z_{11} := x_1 \wedge y_1$, $z_{12} := v = w$, $z_{21} := u = z$, and $z_{22} := x_2 \wedge y_2$. Then we have

$$x_i = z_{i1} + z_{i2}$$

and

$$y_j = z_{1j} + z_{2j}$$

for all $i, j \in \{1,2\}$, as well as

$$z_{21} \wedge z_{12} = 0 .$$

This completes the proof in the case $m = n = 2$.

The general case now follows by induction as in the proof of Theorem 3.6. □

We now turn to the discussion of the Jordan decomposition. For $x \in \mathbb{E}$, define $x^+ := x \vee 0$, $x^- := (x \wedge 0)^*$, and $|x| := x^+ \vee x^-$. Note that x^- is well-defined, by Lemma 2.3, that x^+, x^-, and $|x|$ are positive, and that $|x| = 0$ is equivalent with $x = 0$.

3.14. Theorem.
$x^- + x = x^+ = x + x^-$ and $x^+ \wedge x^- = 0$.

Proof. By Lemma 2.4, we have $x \wedge 0 + x^+ = x = x^+ + x \wedge 0$, and this yields $x^+ = x^- + x$ and $x + x^- = x^+$. Furthermore, using the distributive laws, we obtain $x^+ \wedge x^- = (x+x^-) \wedge (0+x^-) = x \wedge 0 + x^- = 0$. □

3.15. Corollary.
$x^- + x^+ = |x| = x^+ + x^-$.

This follows from Theorem 3.14 and Lemma 2.4.

3.16. Corollary.
$x \leq y$ if and only if $x^+ \leq y^+$ and $y^- \leq x^-$.

Proof. If $x \leq y$ holds, then we have $x^+ \leq y^+$, and from $y \wedge 0 + y^- = 0 = x \wedge 0 + x^- \leq y \wedge 0 + x^-$ we obtain $y^- \leq x^-$, by the order cancellation property.
Conversely, if $x^+ \leq y^+$ and $y^- \leq x^-$ holds, then we have $x + y^- \leq x + x^- = x^+ \leq y^+ = y + y^-$, by Theorem 3.14, and thus $x \leq y$, by the order cancellation property. □

The following result concerns the uniqueness of the Jordan decomposition:

3.17. Theorem.
If $y \wedge z = 0$ and either $z + x = y$ or $y = x + z$, then $y = x^+$ and $z = x^-$.

Proof. Let us consider the case $y = x + z$. Since y and z are positive, we have $x^+ \leq y \leq x^- + y$ and thus

$$y = y \wedge (x^- + y) = y \wedge (x^- + x + z) = y \wedge (x^+ + z) = y \wedge x^+ = x^+ ,$$

by Theorem 3.14 and Corollary 3.11. Using Theorem 3.14 again, we obtain $x + z = y = x^+ = x + x^-$ and thus $z = x^-$, by the cancellation property. □

For invertible elements we obtain some further results related to the Jordan decomposition:

3.18. Theorem.

If $x \in \mathbb{E}_*$, then $(x^*)^+ = x^-$, $(x^*)^- = x^+$, and $|x^*| = x \vee x^* = |x|$.

Proof. Using the distributive laws and Theorem 3.14 we obtain

$$(x^*)^+ = x^* \vee 0 = (x^*+0) \vee (x^*+x) = x^* + 0 \vee x = x^* + x^+ = x^- ,$$

hence

$$(x^*)^- = (x^{**})^+ = x^+ ,$$

and thus

$$|x^*| = (x^*)^+ \vee (x^*)^- = x^- \vee x^+ = |x| .$$

Furthermore, we have

$$x \wedge x^* \le x^+ \wedge (x^*)^+ = x^+ \wedge x^- = 0 ,$$

and we may choose $v \in \mathbb{E}_+$ satisfying $x \vee x^* = x + v$ and $x^* = x \wedge x^* + v$. This yields

$$0 = x + x^* = x + x \wedge x^* + v \le x + v = x \vee x^*$$

and thus

$$|x| = x^+ \vee x^- = x^+ \vee (x^*)^+ = x \vee 0 \vee x^* \vee 0 = x \vee x^* ,$$

which completes the proof. □

3.19. Theorem.

If $x, y \in \mathbb{E}_*$ and $x + y = y + x$, then $|x+y| \le |x| + |y| = |y| + |x|$.

Proof. We first remark that the identity $x + y = y + x$ implies

$$x^* + y^* = (y+x)^* = (x+y)^* = y^* + x^*$$

and

$$x^* + y + x + y^* = x^* + x + y + y^* = 0 ,$$

which yields

$$x^* + y = (x+y^*)^* = y + x^* .$$

Thus we have

$$(x+y) \vee (x+y)^* = (x+y) \vee (x^*+y^*) \le x \vee x^* + y \vee y^* ,$$

and using the distributive laws we obtain

$$\begin{aligned} x \vee x^* + y \vee y^* &= (x+y) \vee (x+y^*) \vee (x^*+y) \vee (x^*+y^*) \\ &= (y+x) \vee (y+x^*) \vee (y^*+x) \vee (y^*+x^*) \\ &= y \vee y^* + x \vee x^* . \end{aligned}$$

Now the assertion follows from Theorem 3.18. □

We remark that the commutativity assumption cannot be omitted in the previous result. This is due to the fact that minimal clans generalize lattice-ordered groups, as will be shown in Section 5, and that for the latter commutativity is equivalent to the validity of the triangle inequality for arbitrary elements; see [5; p. 307].

We finally study the <u>torsion problem</u>. For $x \in \mathbb{E}$ and $n \in \mathbb{N}$ satisfying

$$\left(\sum_{i=1}^{m} x , x \right) \in S$$

for all $m \in \{1,2,\ldots,n-1\}$, define

$$nx := \sum_{i=1}^{n} x .$$

3.20. Theorem.
If nx is defined, then $(nx)^+ = nx^+$, $(nx)^- = nx^-$, and $|nx| = n|x|$.

Proof. By Theorem 3.14, we have

$$x + x + x^- + x^- = x + x^+ + x^- = x + x^- + x + x^- = x^+ + x^+$$

and, using Corollary 3.15, a similar argument yields

$$x^+ + x^+ + x^- + x^- = |x| + |x| .$$

Furthermore, from Corollary 3.11 and Theorem 3.14 we obtain

$$(x^+ + x^+) \wedge (x^- + x^-) = 0 .$$

By induction, this yields

$$nx^+ = nx + nx^-$$

and

$$n|x| = nx^+ + nx^- ,$$

as well as

$$(nx^+) \wedge (nx^-) = 0 .$$

Now Theorem 3.17 yields $nx^+ = (nx)^+$ and $nx^- = (nx)^-$, and from Corollary 3.15 we obtain $n|x| = |nx|$. □

3.21. Corollary.
If nx is defined and $nx \in \mathbb{E}_+$, then $x \in \mathbb{E}_+$.

Proof. By assumption and Theorems 3.14 and 3.20, we have $nx = (nx)^+ = nx + (nx)^- = nx + nx^-$, and thus $nx^- = 0$. This yields $0 \leq x^- \leq nx^- = 0$, hence $x^- = 0$, and thus $x = x^+ \in \mathbb{E}_+$. □

3.22. Corollary.
If nx is defined and $nx = 0$, then $x = 0$.

Proof. By Corollary 3.21, we have $0 \leq x \leq nx = 0$ and thus $x = 0$. □

This means that each minimal clan is torsion-free.

4. BOOLEAN RINGS

In this section, we compare minimal clans and Boolean rings which were introduced by Stone [26,27] under the name "generalized Boolean algebra".

A Boolean ring is a set $\mathbb{E}$ with an order relation $\leq$ such that

- $(\mathbb{E},\leq)$ is a distributive lattice;
- there exists an element $0 \in \mathbb{E}$ satisfying $0 \leq x$ for all $x \in \mathbb{E}$; and
- for all $x, z \in \mathbb{E}$ satisfying $x \leq z$, there exists some $w \in \mathbb{E}$ satisfying $x \wedge w = 0$ and $x \vee w = z$.

If $(\mathbb{E},\leq)$ is a Boolean ring, then the (unique) element $0 \in \mathbb{E}$ satisfying $0 \leq x$ for all $x \in \mathbb{E}$ is said to be the least element of $\mathbb{E}$, for $x, z \in \mathbb{E}$ satisfying $x \leq z$ the (unique) element $w \in \mathbb{E}$ satisfying $x \wedge w = 0$ and $x \vee w = z$ is said to be the relative complement of x in z , and two elements $x, y \in \mathbb{E}$ are disjoint if they satisfy $x \wedge y = 0$. As in the case of ordered partial semigroups, the collection of all pairs of disjoint elements of $\mathbb{E}$ will be denoted by $\mathbb{E}^{\perp}\mathbb{E}$.

4.1. Theorem.
Let $(\mathbb{E},\leq)$ be a Boolean ring with least element 0 .
Then $(\mathbb{E},\mathbb{E}^{\perp}\mathbb{E},\vee,\leq)$ is a positive commutative minimal clan with zero element 0 .

Proof. It is clear that $(\mathbb{E},\mathbb{E}^{\perp}\mathbb{E},\vee)$ is a commutative partial groupoid with zero element 0 . Moreover, for $x, y, z \in \mathbb{E}$ satisfying $x \wedge y = 0$ and $(x \vee y) \wedge z = 0$, we have $x \wedge z = 0$, and thus $y \wedge z = 0$ and $x \wedge (y \vee z) = (x \wedge y) \vee (x \wedge z) = 0$, as well as $(x \vee y) \vee z = x \vee (y \vee z)$. Therefore, $(\mathbb{E},\mathbb{E}^{\perp}\mathbb{E},\vee)$ is a positive commutative partial semigroup, and it is then clear that $(\mathbb{E},\mathbb{E}^{\perp}\mathbb{E},\vee,\leq)$ is a positive commutative lattice-ordered partial semigroup.
Furthermore, for $x, y, z \in \mathbb{E}$ satisfying $x \wedge z = 0$, $y \wedge z = 0$, and $x \vee z = y \vee z$, we have

$$x = x \wedge (x \vee z) = x \wedge (y \vee z) = (x \wedge y) \vee (x \wedge z) = x \wedge y ,$$

and a similar argument yields

$$y = x \wedge y .$$

Thus we have $x = y$. Therefore, $(\mathbb{E},\mathbb{E}^{\perp}\mathbb{E},\vee,\leq)$ has the cancellation property.
Finally, for all $x, y \in \mathbb{E}$, there exists some $v \in \mathbb{E}$ satisfying $x \wedge v = 0$ and $x \vee y = x \vee v$ as well as

$$y = (x \wedge v) \vee y = (x \vee y) \wedge (v \vee y) = (x \vee v) \wedge (y \vee v) = (x \wedge y) \vee v .$$

Therefore, $(\mathbb{E},\mathbb{E}^{\perp}\mathbb{E},\vee,\leq)$ has the difference property. □

4.2. Theorem.

Let $(\mathbb{E},S,+,\leq)$ be a minimal clan with zero element 0.
Then the following are equivalent:

(a) $(\mathbb{E},\leq)$ is a Boolean ring with least element 0.

(b) $x+y = x\vee y$ holds for all $x, y \in \mathbb{E}$ satisfying $(x,y) \in S$.

(c) $S \subseteq \mathbb{E}^{\perp}\mathbb{E}$.

(d) $S = \mathbb{E}^{\perp}\mathbb{E}$.

Proof. Suppose first that (a) holds. Consider $x, y \in \mathbb{E} = \mathbb{E}_+$ satisfying $(x,y) \in S$ and choose $v \in \mathbb{E}_+$ satisfying $x\vee y = x+v$ and $y = x\wedge y + v$. Then we have $0 \leq x\wedge y$, hence $v \leq y$, and thus $x\vee y \leq x+y$. By assumption, there exists some $w \in \mathbb{E} = \mathbb{E}_+$ satisfying $(x\vee y)\wedge w = 0$ and $(x\vee y)\vee w = x+y$. Then we have

$$x\vee y + w = (x\vee y)\vee w = x + y ,$$

by Lemma 2.4, as well as

$$x\wedge w = 0 = y\wedge w .$$

By the refinement property, there exist $z_{11}, z_{12}, z_{21}, z_{22} \in \mathbb{E}_+$ satisfying

$$x\vee y = z_{11} + z_{12} \quad \text{and} \quad w = z_{21} + z_{22}$$

as well as

$$x = z_{11} + z_{21} \quad \text{and} \quad y = z_{12} + z_{22} .$$

From

$$0 \leq z_{21} \leq x\wedge w = 0$$

and

$$0 \leq z_{22} \leq y\wedge w = 0$$

we obtain

$$w = z_{21} + z_{22} = 0$$

and thus

$$x\vee y = x + y .$$

Therefore, (a) implies (b).
By Lemma 3.2 and the cancellation property, (b) implies (c).
By Lemma 2.4, (c) implies (d).
Suppose now that (d) holds. By Theorem 3.8, $(\mathbb{E},\leq)$ is a distributive lattice. By Lemma 2.2, $0 \leq x$ holds for all $x \in \mathbb{E}$. Consider now $x, z \in \mathbb{E}$ satisfying $x \leq z$ and choose $w \in \mathbb{E}_+$ satisfying $z = x+w$, by the difference property. Then we have $(x,w) \in S$, hence $x\wedge w = 0$, and thus $x\vee w = x+w = z$, by Lemma 2.4. Therefore, (d) implies (a). □

In view of Lemma 2.4, the previous results may be summarized as follows:

4.3. Corollary.
Boolean rings are precisely the minimal clans having a minimal domain of addition.

It would be interesting to know whether each minimal clan $(\mathbb{E},S,+,\leq)$ contains a greatest Boolean ring. With regard to this problem, we remark that the assertion of [24; Corollary 3.3] is false since the ordered partial semigroup $(\mathbb{E}_+,\mathbb{E}^\perp\mathbb{E},+,\leq)$ need not have the difference property.

5. LATTICE-ORDERED GROUPS

In this section, we compare minimal clans and lattice-ordered groups which were introduced by Birkhoff [3]. Commutative or totally ordered lattice-ordered groups had been studied earlier; for references and detailed information on lattice-ordered groups, we refer to the books by Birkhoff [4,5], Fuchs [15], and Bigard, Keimel, and Wolfenstein [1].

A group is a semigroup in which each element is invertible, an ordered group is an ordered semigroup in which each element is invertible, and a lattice-ordered group is a lattice-ordered semigroup in which each element is invertible.

5.1. Theorem.
Each lattice-ordered group is a minimal clan.

Proof. Let $(\mathbb{E},\mathbb{E}\times\mathbb{E},+,\leq)$ be a lattice-ordered semigroup satisfying $\mathbb{E}_\star = \mathbb{E}$. Then $(\mathbb{E},\mathbb{E}\times\mathbb{E},+,\leq)$ has the cancellation property. Furthermore, using $\mathbb{E}_\star = \mathbb{E}$, it is not hard to see that the identities

$$w + x\wedge y + z = (w+x+z)\wedge(w+y+z)$$

and

$$y^\star\wedge x^\star = (x\vee y)^\star$$

hold for all $w, x, y, z \in \mathbb{E}$. Consider $x, y \in \mathbb{E}$. For $u := x\vee y + x^\star$, we have

$$u + x = x\vee y + x^\star + x = x\vee y$$

and

$$\begin{aligned} u + x\wedge y &= x\vee y + x^\star + x\wedge y + y^\star + y \\ &= x\vee y + (y^\star\wedge x^\star) + y \\ &= x\vee y + (x\vee y)^\star + y \\ &= y , \end{aligned}$$

and for $v := x^\star + x\vee y$ a similar argument yields $x\vee y = x+v$ and $y = x\wedge y + v$. Therefore, $(\mathbb{E},\mathbb{E}\times\mathbb{E},+,\leq)$ has the difference property. □

5.2. Theorem.

Let $(\mathbb{E},S,+,\leq)$ be a minimal clan.

Then the following are equivalent:

(a) $(\mathbb{E},S,+,\leq)$ is a lattice-ordered group.

(b) $\mathbb{E}_{+} \subseteq \mathbb{E}_{*}$.

(c) $\mathbb{E} = \mathbb{E}_{*}$.

Proof. The equivalence of (a) and (c) is obvious, and the equivalence of (b) and (c) follows from Lemmas 2.3 and 2.4. □

The previous results may be summarized as follows:

5.3. Corollary.

Lattice-ordered groups are precisely the minimal clans having a maximal set of invertible elements.

We finally remark that each minimal clan $(\mathbb{E},S,+,\leq)$ contains a greatest lattice-ordered group, namely $(\mathbb{E}_{*},\mathbb{E}_{*}\times\mathbb{E}_{*},+,\leq)$. This follows from Theorems 3.3 and 5.2.

6. SYMMETRIC CLANS

In this section, we compare minimal clans and symmetric clans which were introduced by Wyler [29]. Since symmetric clans are defined in terms of a partial subtraction and an induced partial addition, it is necessary to consider first a more general algebraic structure in which a partial addition can be shown to exist.

A preclan is a set $\mathbb{E}$ with an order relation $\leq$, a relation $T \subseteq \mathbb{E}\times\mathbb{E}$, and a map $- : T \longrightarrow \mathbb{E}$ such that

(C0) $(\mathbb{E},\leq)$ is a lattice;

(C1) $(x,y) \in T$ holds for all $x, y \in \mathbb{E}$ satisfying $x \leq y$;

(C2) for all $x, y, z \in \mathbb{E}$ satisfying $x \vee y \leq z$, $x \leq y$ if and only if $z - y \leq z - x$;

(C3) $y - x \leq z - x$ and $(z-x) - (y-x) = z - y$ holds for all $x, y, z \in \mathbb{E}$ satisfying $(x,y) \in T$, $(x,z) \in T$, and $y \leq z$; and

(C4) for all $x, y, z \in \mathbb{E}$ satisfying $y \leq z$ and $x \leq z - y$, there exists some $u \in \mathbb{E}$ satisfying $(u,z) \in T$ and $z - u = (z-y) - x$.

If $(\mathbb{E},\leq,T,-)$ is a preclan, then the set T is said to be the domain of subtraction, and the map $- : T \longrightarrow \mathbb{E}$ is called subtraction.

Let $(\mathbb{E},\leq,T,-)$ be a preclan. By [29; (2.4)], the equation $x - x = x$ has a unique solution, which will be denoted by 0. Furthermore, for all $x, y \in \mathbb{E}$ satisfying $x \leq y$, we have $0 = x - x \leq y - x$, by [29; (2.4)] and (C3). This means that subtraction maps the set

$$\mathbb{E}\overset{\leq}{-}\mathbb{E} \quad := \quad \{ (x,y) \in \mathbb{E}\times\mathbb{E} \mid x \leq y \}$$

into the set

$$\mathbb{E}_+ \quad := \quad \{ z \in \mathbb{E} \mid 0 \leq z \} \quad .$$

Finally, for all $u, x, y, z \in \mathbb{E}$ satisfying $y \leq z$, $x \leq z - y$, $(u,z) \in T$, and $z - u = (z-y) - x$, we have $z - z = 0 \leq z - u$, by [29; (2.4)] and the preceding remark, and this yields $(u,z) \in \mathbb{E}\overset{\leq}{-}\mathbb{E}$, by [29; (2.2)]. It is now easy to see that $(\mathbb{E},\leq,\mathbb{E}\overset{\leq}{-}\mathbb{E},-)$ is a preclan too.

For a preclan $(\mathbb{E},\leq,T,-)$, define a relation

$$\tilde{S} \quad := \quad \{ (x,y) \in \mathbb{E}\times\mathbb{E} \mid \text{there exists some } z \in \mathbb{E} \text{ satisfying } y \leq z \text{ and } x \leq z - y \}$$

and let

$$\tilde{+} \; : \; \tilde{S} \longrightarrow \mathbb{E}$$

denote the map associating with each pair $(x,y) \in \tilde{S}$ the unique element $u \in \mathbb{E}$ satisfying $z - u = (z-y) - x$ for all $z \in \mathbb{E}$ such that $y \leq z$ and $x \leq z - y$. Then the map $\tilde{+} : \tilde{S} \longrightarrow \mathbb{E}$ is well-defined [29; (2.5)], and it will be called <u>induced addition</u>. Subtraction and induced addition are related as follows: For all $x, y \in \mathbb{E}$ satisfying $x \leq y$, we have $(y-x,x) \in \tilde{S}$ and $(y-x) \tilde{+} x = y$, by [29; (3.1)].

A preclan $(\mathbb{E},\leq,T,-)$ is <u>symmetric</u> if

(C5) for all $x, y \in \mathbb{E}$ satisfying $x \leq y$, there exists some $v \in \mathbb{E}$ satisfying $(x,v) \in \tilde{S}$ and $x \tilde{+} v = y$,

and it is <u>commutative</u> if

(C6) for all $x, y \in \mathbb{E}$, $(x,y) \in \tilde{S}$ if and only if $(y,x) \in \tilde{S}$, and in this case $x \tilde{+} y = y \tilde{+} x$.

By [29; (4.1)], each commutative preclan is symmetric.

A <u>clan</u> is a preclan $(\mathbb{E},\leq,T,-)$ such that

(C7) $(x,y) \in \tilde{S}$, $(y,x) \in \tilde{S}$, and $x \tilde{+} y = x \vee y = y \tilde{+} x$ holds for all $x, y \in \mathbb{E}$ satisfying $x \wedge y = 0$.

In order to compare minimal clans and symmetric clans, we have to introduce an induced partial subtraction in minimal clans:

For a minimal clan $(\mathbb{E},S,+,\leq)$, define a relation

$$\tilde{T} \quad := \quad \{ (x,y) \in \mathbb{E}\times\mathbb{E} \mid x \leq y \}$$

and let

$$\tilde{-} : \tilde{T} \longrightarrow \mathbb{E}$$

denote the map associating with each pair $(x,y) \in \tilde{T}$ the unique element $u \in \mathbb{E}_+$ satisfying $u + x = y$. Then the map $\tilde{-} : \tilde{T} \longrightarrow \mathbb{E}$ is well-defined, by Lemma 3.2, and it will be called <u>induced subtraction</u>.

<u>6.1 Theorem.</u>
Let $(\mathbb{E},\leq,T,-)$ be a symmetric clan.
Then $(\mathbb{E},\tilde{S},\tilde{+},\leq)$ is a minimal clan satisfying $\tilde{\tilde{T}} \subseteq T$ and $y \mathbin{\tilde{\tilde{-}}} x = y - x$ for all $x, y \in \mathbb{E}$ satisfying $(x,y) \in \tilde{\tilde{T}}$.

Proof. It is easy to see from (C0) and [29; (3.1), (4.4), (3.3), and (3.2)] that $(\mathbb{E},\tilde{S},\tilde{+},\leq)$ is a lattice-ordered partial semigroup having the cancellation property.
Furthermore, for all $x, y \in \mathbb{E}$, there exists some $u \in \mathbb{E}_+$ satisfying

$$u = x \vee y - x = y - x \wedge y$$

and thus

$$u \mathbin{\tilde{+}} x = x \vee y \quad \text{and} \quad u \mathbin{\tilde{+}} x \wedge y = y ,$$

by [29; (5.5), (3.1), and (5.1)], and there also exists some $v \in \mathbb{E}_+$ satisfying

$$x \vee y = x \mathbin{\tilde{+}} v \quad \text{and} \quad y = x \wedge y \mathbin{\tilde{+}} v ,$$

by the duality principle for symmetric clans [29; (4.3)]. Therefore, $(\mathbb{E},\tilde{S},\tilde{+},\leq)$ has the difference property.
We have thus shown that $(\mathbb{E},\tilde{S},\tilde{+},\leq)$ is a minimal clan.
For the final assertion, we first note that (C1) yields $\tilde{\tilde{T}} = \mathbb{E} \overset{\leq}{-} \mathbb{E} \subseteq T$.
Furthermore, for all $x, y \in \mathbb{E}$ satisfying $(x,y) \in \tilde{\tilde{T}}$, we have

$$(y \mathbin{\tilde{\tilde{-}}} x) \mathbin{\tilde{+}} x = y = (y - x) \mathbin{\tilde{+}} x ,$$

by [29; (3.1)], and thus

$$y \mathbin{\tilde{\tilde{-}}} x = y - x ,$$

by the cancellation property. □

<u>6.2. Theorem.</u>
Let $(\mathbb{E},S,+,\leq)$ be a minimal clan.
Then $(\mathbb{E},\leq,\tilde{T},\tilde{-})$ is a symmetric clan satisfying $\tilde{\tilde{S}} = S$ and $x \mathbin{\tilde{\tilde{+}}} y = x + y$ for all $x, y \in \mathbb{E}$ satisfying $(x,y) \in \tilde{\tilde{S}}$.

Proof. Let us first prove that $(\mathbb{E},\leq,\tilde{T},\tilde{-})$ is a preclan.
Obviously, (C0) and (C1) are satisfied.
To verify (C2), consider $x, y, z \in \mathbb{E}$ satisfying $x \vee y \leq z$. Then we have $(x,z) \in \tilde{T}$ and $(y,z) \in \tilde{T}$. Define $u := z \mathbin{\tilde{-}} x$ and $v := z \mathbin{\tilde{-}} y$, which means $u + x = z = v + y$.
If $x \leq y$ holds, then there exists some $w \in \mathbb{E}_+$ satisfying $w + x = y$,

and we have $v+w+x = v+y = z = u+x$, hence $v+w = u$, and thus $z \tilde{-} y = v \leq v+w = u = z \tilde{-} x$.
Conversely, if $v = z \tilde{-} y \leq z \tilde{-} x = u$ holds, then there exists some $w \in \mathbb{E}_+$ satisfying $v+w = u$, and we have $v+w+x = u+x = z = v+y$, hence $w+x = y$, and thus $x \leq w+x = y$.
To verify (C3), consider $x, y, z \in \mathbb{E}$ satisfying $(x,y) \in \tilde{T}$, $(x,z) \in \tilde{T}$, and $y \leq z$. Then we have $(y,z) \in \tilde{T}$. Define $u := z \tilde{-} x$, $v := z \tilde{-} y$, and $w := y \tilde{-} x$, which means $w+x = y$ and $u+x = z = v+y$.
Then we have $v+w+x = v+y = z = u+x$, hence $v+w = u$, and thus $y \tilde{-} x = w \leq v+w = u = z \tilde{-} x$, which yields $y \tilde{-} x \leq z \tilde{-} x$ as well as $(z \tilde{-} x) \tilde{-} (y \tilde{-} x) = z \tilde{-} y$.
To verify (C4), consider $x, y, z \in \mathbb{E}$ satisfying $y \leq z$ and $x \leq z \tilde{-} y$. Then we have $(y,z) \in \tilde{T}$ and $(x, z \tilde{-} y) \in \tilde{T}$. Define $v := z \tilde{-} y$ and $w := (z \tilde{-} y) \tilde{-} x$, which means $v+y = z$ and $w+x = z \tilde{-} y = v$.
Then we have $w+x+y = v+y = z$, and thus $(x,y) \in S$, $(x+y, z) \in \tilde{T}$, and $z \tilde{-} (x+y) = w = (z \tilde{-} y) \tilde{-} x$.
We have thus shown that $(\mathbb{E}, \leq, \tilde{T}, \tilde{-})$ is a preclan.
We next prove that the induced addition in the preclan $(\mathbb{E}, \leq, \tilde{T}, \tilde{-})$ agrees with addition in the minimal clan $(\mathbb{E}, S, +, \leq)$. Consider $x, y \in \mathbb{E}$.
First, if $(x,y) \in \overset{\approx}{S}$, then there exists some $z \in \mathbb{E}$ satisfying $y \leq z$ and $x \leq z \tilde{-} y$. Define $w := (z \tilde{-} y) \tilde{-} x$, which means $w+x = z \tilde{-} y$. Then we have $w+x+y = z$, which yields $(x,y) \in S$ as well as

$$z \tilde{-} (x+y) = w = (z \tilde{-} y) \tilde{-} x .$$

In particular, it follows from the previous argument that the identity

$$z \tilde{-} (x+y) = (z \tilde{-} y) \tilde{-} x$$

holds for all $z \in \mathbb{E}$ satisfying $y \leq z$ and $x \leq z \tilde{-} y$, and this yields

$$x \overset{\approx}{+} y = x + y ,$$

by the definition of $x \overset{\approx}{+} y$.
Conversely, if $(x,y) \in S$, define $z := (x+y) \vee y$. Then we have

$$z = (x+y) \vee (0+y) = x \vee 0 + y ,$$

by the distributive laws, hence $y \leq z$ and $x \leq x \vee 0 = z \tilde{-} y$, and thus $(x,y) \in \overset{\approx}{S}$.
It now follows from the difference property of $(\mathbb{E}, S, +, \leq)$ and Lemma 2.4 that the preclan $(\mathbb{E}, \leq, \tilde{T}, \tilde{-})$ is even a symmetric clan. □

In view of (C1), the previous results may be summarized as follows:

<u>6.3. Corollary.</u>
Let $(\mathbb{E}, \leq)$ be a lattice.
Then there exists a bijection between the minimal clans $(\mathbb{E}, S, +, \leq)$ and the symmetric clans having a minimal domain of subtraction $(\mathbb{E}, \leq, T, -)$ such that $\tilde{T} = T$ and $\tilde{S} = S$.

Corollary 6.3 explains the name of minimal clans.

We conclude this section with a few remarks concerning the definition of symmetric clans. First, we observe that the induced addition in a symmetric clan $(\mathbb{E},\leq,T,-)$ is already determined by the properties of subtraction on the set $\mathbb{E}\overset{\leq}{-}\mathbb{E} \subseteq T$ and that $(\mathbb{E},\leq,\mathbb{E}\overset{\leq}{-}\mathbb{E},-)$ is a symmetric clan whenever this is the case for $(\mathbb{E},\leq,T,-)$. Furthermore, Wyler [29; p. 177] pointed out that the induced addition in a symmetric clan induces an extended subtraction which, however, does not extend addition in turn. In view of these remarks, it appears that the really important partial operation in symmetric clans is not subtraction but rather the induced addition, and that it is sufficient to consider symmetric clans having a minimal domain of subtraction. By Corollary 6.3, these are precisely the minimal clans. Since minimal clans are defined in terms of a single partial operation and since the domain of (induced) addition has also to be identified in the case of symmetric clans in order to verify axioms (C5) and (C7), it appears that the axioms of minimal clans are more natural and easier to check than those of symmetric clans.

7. SEMICLANS AND NORMAL CLANS

In this section, we compare minimal clans with semiclans and normal clans which were, respectively, introduced by Bosbach [8] and Billhardt [2].

Let us first consider semiclans:

A <u>semiclan</u> is a set $\mathbb{E}$ with a relation $S \subseteq \mathbb{E}\times\mathbb{E}$, a map $+ : S \longrightarrow \mathbb{E}$, and an order relation $\leq$ such that

(S0) $(\mathbb{E},\leq)$ is a lower semilattice, i.e. any two elements $x, y \in \mathbb{E}$ have a greatest lower bound $x \wedge y$;

(S1) for all $x, z \in \mathbb{E}$ satisfying $x \leq z$, there exist $u, v \in \mathbb{E}$ satisfying $u+x = z = x+v$;

(S2) $x = y$ holds for all $x, y \in \mathbb{E}$ satisfying either $u+x = u+y$ for some $u \in \mathbb{E}$ or $x+v = y+v$ for some $v \in \mathbb{E}$;

(S3) $(u+x)\wedge(u+y) = u + x\wedge y$ holds for all $u, x, y \in \mathbb{E}$ satisfying $(u,x) \in S$ and $(u,y) \in S$, and $(x+v)\wedge(y+v) = x\wedge y + v$ holds for all $v, x, y \in \mathbb{E}$ satisfying $(x,v) \in S$ and $(y,v) \in S$;

(S4) for all $x, y, z \in \mathbb{E}$, $(x,y) \in S$ and $(x+y,z) \in S$ if and only if $(y,z) \in S$ and $(x,y+z) \in S$, and in this case $(x+y)+z = x+(y+z)$; and

(S5) $x+y = x \vee y = y+x$ holds for all $x, y \in \mathbb{E}$ satisfying $x \wedge y + z = z$ for some $z \in \mathbb{E}$ and such that $x \vee y$ exists.

If $(\mathbb{E},S,+,\leq)$ is a semiclan, then there exists a (unique) element $0 \in \mathbb{E}$ satisfying $0+x = x = x+0$ for all $x \in \mathbb{E}$, by [8; p. 317]. Therefore, each semiclan is an ordered partial semigroup.

<u>7.1. Theorem.</u>

Let $(\mathbb{E},S,+,\leq)$ be an ordered partial semigroup.

Then the following are equivalent:

(a) $(\mathbb{E},S,+,\leq)$ is a minimal clan.

(b) $(\mathbb{E},S,+,\leq)$ is a semiclan having the lattice property.

Proof. Each minimal clan clearly satisfies axioms (S0), (S1), (S2), and (S4), and it also has the lattice property. Moreover, each minimal clan satisfies axioms (S3) and (S5), by Theorem 3.7 and Lemma 2.4. Conversely, each semiclan having the lattice property clearly is a lattice-ordered partial semigroup having the cancellation property, and it also has the difference property, by [8; (1.19) and (1.3), and the duals of these statements]. □

Let us now consider normal clans:

A <u>normal clan</u> is a set $\mathbb{E}$ with a relation $S \subseteq \mathbb{E}\times\mathbb{E}$, a map $+ : S \longrightarrow \mathbb{E}$, and an order relation $\leq$ such that

(N0) $(\mathbb{E},\leq)$ is a directed lower semilattice, i.e. $(\mathbb{E},\leq)$ is a lower semilattice and for all $x, y \in \mathbb{E}$ there exists some $z \in \mathbb{E}$ satisfying $x \leq z$ and $y \leq z$;

(N1) there exists an element $0 \in \mathbb{E}$ satisfying $0+x = x = x+0$ for all $x \in \mathbb{E}$;

(N2) for all $x, y, z \in \mathbb{E}$, $(x,y) \in S$ and $(x+y,z) \in S$ if and only if $(y,z) \in S$ and $(x,y+z) \in S$, and in this case $(x+y)+z = x+(y+z)$;

(N3) for all $x, z \in \mathbb{E}$ satisfying $x \leq z$, there exist $u, v \in \mathbb{E}$ satisfying $u+x = z = x+v$;

(N4) $(u+x+v)\wedge(u+y+v) = u + x\wedge y + v$ holds for all $u, v, x, y \in \mathbb{E}$ satisfying $(u,x) \in S$ and $(u+x,v) \in S$ as well as $(u,y) \in S$ and $(u+y,v) \in S$; and

(N5) for all $x_1, x_2, y_1, y_2 \in \mathbb{E}$ satisfying $x_1+x_2 = y_1+y_2$, there exist $z_{ij} \in \mathbb{E}$ satisfying $x_i = z_{i1}+z_{i2}$ and $y_j = z_{1j}+z_{2j}$ for all $i, j \in \{1,2\}$ as well as $z_{12}\wedge z_{21} = 0$.

Thus, each normal clan is an ordered partial semigroup.

We shall need the following two lemmas on normal clans:

7.2. Lemma.

Let $(\mathbb{E},S,+,\leq)$ be a normal clan.

(a) If $x \leq z$,
then there exist $u, v \in \mathbb{E}_+$ satisfying $u+x = z = x+v$.

(b) If $x_1, x_2, y_1, y_2 \in \mathbb{E}_+$ are such that $x_1+x_2 = y_1+y_2$,
then there exist $z_{ij} \in \mathbb{E}_+$ satisfying $x_i = z_{i1}+z_{i2}$ and $y_j = z_{1j}+z_{2j}$ for all $i, j \in \{1,2\}$ as well as $z_{12} \wedge z_{21} = 0$.

Proof. Consider first $x, z \in \mathbb{E}$ satisfying $x \leq z$. By (N3), there exists some $y \in \mathbb{E}$ satisfying $y+x = z$ and some $w \in \mathbb{E}$ satisfying $w+y\wedge 0 = 0$. Then we have

$$\begin{aligned} z &= y + x \\ &= y + w + y\wedge 0 + x \\ &= y + w + (y+x)\wedge(0+x) \\ &= y + w + z\wedge x \\ &= y + w + x \quad , \end{aligned}$$

by (N4), hence $(y,w) \in S$, and from

$$w + y\wedge 0 = 0 = y\wedge 0 + w$$

we obtain

$$(y+w)\wedge(0+w) = y\wedge 0 + w = 0 \quad ,$$

by (N4), and thus $y+w \in \mathbb{E}_+$. Now define $u := y+w$. Then we have $u \in \mathbb{E}_+$ and $u+x = z$, and in the same way it can be shown that there exists some $v \in \mathbb{E}_+$ satisfying $z = x+v$. This proves (a).

Consider now $x_1, x_2, y_1, y_2 \in \mathbb{E}_+$ satisfying $x_1+x_2 = y_1+y_2$. By (N5), there exist $z_{ij} \in \mathbb{E}$ satisfying $x_i = z_{i1}+z_{i2}$ and $y_j = z_{1j}+z_{2j}$ for all $i, j \in \{1,2\}$ as well as $z_{12}\wedge z_{21} = 0$. Thus we have $z_{12} \in \mathbb{E}_+$ and $z_{21} \in \mathbb{E}_+$. Using (N4), we obtain

$$z_{11} = z_{11} + z_{12}\wedge z_{21} = (z_{11}+z_{12})\wedge(z_{11}+z_{21}) = x_1\wedge y_1$$

and thus $z_{11} \in \mathbb{E}_+$, and in the same way we obtain $z_{22} \in \mathbb{E}_+$. This proves (b). □

7.3. Lemma.

Let $(\mathbb{E},S,+,\leq)$ be a normal clan having the cancellation property.

(a) $(\mathbb{E},S,+,\leq)$ has the lattice property.

(b) $(\mathbb{E},S,+,\leq)$ has the difference property.

Proof. Consider $x, y \in \mathbb{E}$. By (N0), there exists some $z \in \mathbb{E}$ satisfying $x \leq z$ and $y \leq z$. By Lemma 7.2, there exist $v, w \in \mathbb{E}_+$ satisfying

$$x + v = z = y + w \quad .$$

By (N5), there exist $z_{ij} \in \mathbb{E}$ satisfying

$$x = z_{11}+z_{12} \quad \text{and} \quad v = z_{21}+z_{22}$$

as well as

$$y = z_{11} + z_{21} \quad \text{and} \quad w = z_{12} + z_{22} \quad ,$$

and also

$$z_{12} \wedge z_{21} = 0 \quad .$$

Then we have

$$z_{22} = z_{12} \wedge z_{21} + z_{22} = (z_{12} + z_{22}) \wedge (z_{21} + z_{22}) = w \wedge v$$

and thus $z_{22} \in \mathbb{E}_+$. Now define

$$\tilde{z} := z_{11} + z_{12} + z_{21} = z_{11} + z_{21} + z_{12} \quad .$$

Since z_{12} , z_{21} , and z_{22} are positive, we have $x \leq \tilde{z}$, $y \leq \tilde{z}$, and $\tilde{z} \leq z$, and since the definition of $\tilde{z}$ is independent of the particular choice of the upper bound z , it follows that $\tilde{z}$ is the least upper bound of x and y . This proves (a).

Furthermore, letting $z := \tilde{z} = x \vee y$ in the proof of (a), we obtain $z_{22} = 0$ and thus $z_{21} = v$. This yields

$$x \vee y = z = x + v$$

and

$$\begin{aligned} y &= z_{11} + z_{21} \\ &= z_{11} + z_{12} \wedge z_{21} + z_{21} \\ &= (z_{11} + z_{12}) \wedge (z_{11} + z_{21}) + z_{21} \\ &= x \wedge y + v \quad , \end{aligned}$$

and in a similar way it can be shown that there exists some $u \in \mathbb{E}_+$ satisfying $u + x = x \vee y$ and $u + x \wedge y = y$. This proves (b). □

<u>7.4. Theorem.</u>

Let $(\mathbb{E}, S, +, \leq)$ be an ordered partial semigroup.

Then the following are equivalent:

(a) $(\mathbb{E}, S, +, \leq)$ is a minimal clan.

(b) $(\mathbb{E}, S, +, \leq)$ is a normal clan having the cancellation property.

Proof. Each minimal clan clearly satisfies axioms (N0), (N1), (N2), and (N3), and it also has the cancellation property. Furthermore, each minimal clan satisfies axioms (N4) and (N5), by Theorems 3.7 and 3.13. Conversely, each normal clan having the cancellation property clearly is an ordered partial semigroup having the cancellation property, and it also has the lattice property and the difference property, by Lemma 7.3. □

Since each semiclan has the cancellation property, the previous results may be summarized as follows:

7.5 Corollary.

Let $(\mathbb{E},S,+,\leq)$ be an ordered partial semigroup.
Then the following are equivalent:

(a) $(\mathbb{E},S,+,\leq)$ is a minimal clan.

(b) $(\mathbb{E},S,+,\leq)$ is a semiclan and a normal clan.

8. MINIMAL CLANS WITH COMPLETE ADDITION

In this section, we compare minimal clans with some ordered algebraic structures with complete addition which were introduced by Bosbach [6,7,9], Swamy [28], and Riesz [21]. We start with the general case, proceed with the commutative case, and conclude with the positive commutative case.

In [9], Bosbach introduced divisibility semiloops:

A divisibility semiloop is a groupoid $(\mathbb{E},\mathbb{E}\times\mathbb{E},+)$ with an order relation $\leq$ such that

- $(\mathbb{E},\mathbb{E}\times\mathbb{E},+)$ has the cancellation property;
- $(\mathbb{E},\leq)$ is a lower semilattice;
- $((u+x)+v)\wedge((u+y)+v) = (u+x\wedge y)+v$ holds for all $u, v, x, y \in \mathbb{E}$; and
- for all $x, y \in \mathbb{E}$ satisfying $w+x \leq y$ for some $w \in \mathbb{E}$, there exists some $u \in \mathbb{E}$ satisfying $u+x = y$,
  and for all $x, y \in \mathbb{E}$ satisfying $x+z \leq y$ for some $z \in \mathbb{E}$, there exists some $v \in \mathbb{E}$ satisfying $x+v = y$.

Thus, each associative divisibility semiloop is an ordered partial semigroup with complete addition.

8.1. Theorem.

Let $(\mathbb{E},S,+,\leq)$ be an ordered partial semigroup.
Then the following are equivalent:

(a) $(\mathbb{E},S,+,\leq)$ is a minimal clan with complete addition.

(b) $(\mathbb{E},S,+,\leq)$ is an associative divisibility semiloop.

Proof. The assertion follows from Theorem 3.7 and [9; Proposition 1.4 and its proof]. □

In [6,7], Bosbach introduced strong divisibility semigroups.

A strong divisibility semigroup is a semigroup $(\mathbb{E},\mathbb{E}\times\mathbb{E},+)$ with an order relation $\leq$ such that

- $(\mathbb{E},\leq)$ is a lower semilattice;
- $(u+x+v)\wedge(u+y+v) = u + x\wedge y + v$ holds for all $u, v, x, y \in \mathbb{E}$; and
- for all $x, y \in \mathbb{E}$, there exist $u, v \in \mathbb{E}$ satisfying $u + x\wedge y = x = x\wedge y + v$.

Thus, each strong divisibility semigroup is an ordered partial semigroup with complete addition.

8.2. Theorem.

Let $(\mathbb{E},S,+,\leq)$ be an ordered partial semigroup.
Then the following are equivalent:

(a) $(\mathbb{E},S,+,\leq)$ is a minimal clan with complete addition.

(b) $(\mathbb{E},S,+,\leq)$ is a semiclan with complete addition.

(c) $(\mathbb{E},S,+,\leq)$ is a normal clan with complete addition having the cancellation property.

(d) $(\mathbb{E},S,+,\leq)$ is a strong divisibility semigroup having the cancellation property.

Proof. It is clear from Theorem 7.4 that (a) and (c) are equivalent. Furthermore, (a) implies (b), by Theorem 7.1, and it is obvious from the definitions and the existence of a zero element in semiclans that (b) implies (d). Finally, it follows from [6; (2.7) and its proof] that (d) implies (a). □

As a consequence of Theorem 8.2, each semiclan with complete addition has the lattice property. Furthermore, Theorems 8.1 and 8.2 yield the following analogon of Corollary 7.5:

8.3. Corollary.

Let $(\mathbb{E},S,+,\leq)$ be an ordered partial semigroup.
Then the following are equivalent:

(a) $(\mathbb{E},S,+,\leq)$ is a minimal clan with complete addition.

(b) $(\mathbb{E},S,+,\leq)$ is a semiclan and a strong divisibility semigroup.

(c) $(\mathbb{E},S,+,\leq)$ is a divisibility semiloop and a strong divisibility semigroup.

In the terminology of Brehmer [11], a commutative strong divisibility semigroup is said to be a C_o-lattice. Brehmer also introduced C-lattices:

A C-lattice is a C_o-lattice $(\mathbb{E},\mathbb{E}\times\mathbb{E},+,\leq)$ such that

- for all $x, y \in \mathbb{E}$, there exists a least element $u \in \mathbb{E}_+$ satisfying $u + x\wedge y = y$.

The following result is immediate from Theorem 8.2:

8.4. Corollary.

Let $(\mathbb{E},S,+,\leq)$ be an ordered partial semigroup.
Then the following are equivalent:

(a) $(\mathbb{E},S,+,\leq)$ is a commutative minimal clan with complete addition.

(b) $(\mathbb{E},S,+,\leq)$ is a C_o-lattice having the cancellation property.

(c) $(\mathbb{E},S,+,\leq)$ is a C-lattice having the cancellation property.

Since C_o-lattices and commutative strong divisibility semigroups are the same, each C_o-lattice is a distributive lattice, by [6; (2.9) and (2.10)]. This contradicts a remark in [11; p. 26]. For C_o-lattices having the cancellation property, the property of being a distributive lattice also follows from Corollary 8.4 and Theorem 3.8.

A common abstraction of Boolean rings and lattice-ordered groups which is closely related to C-lattices was considered by Swamy [28] who introduced dually residuated lattice-ordered semigroups:

A dually residuated lattice-ordered semigroup is a commutative semigroup $(\mathbb{E},\mathbb{E}\times\mathbb{E},+)$ with an order relation $\leq$ such that

- $(\mathbb{E},\leq)$ is a lattice;
- $u+x\vee y = (u+x)\vee(u+y)$ and $u+x\wedge y = (u+x)\wedge(u+y)$ holds for all $u, x, y \in \mathbb{E}$;
- for all $x, y \in \mathbb{E}$, there exists a least element $u \in \mathbb{E}$ satisfying $y \leq u+x$;
- for all $x, y \in \mathbb{E}$, $u\vee 0+x \leq x\vee y$ holds for the least element $u \in \mathbb{E}$ satisfying $y \leq u+x$; and
- for all $x \in \mathbb{E}$, $0 \leq u$ holds for the least element $u \in \mathbb{E}$ satisfying $x \leq u+x$.

Thus, each dually residuated lattice-ordered semigroup is a commutative lattice-ordered partial semigroup with complete addition, and it follows from [28; Lemma 8] that it is even a C-lattice. Therefore, the following result is a consequence of Corollary 8.4:

8.5. Corollary.

Each dually residuated lattice-ordered semigroup having the cancellation property is a commutative minimal clan with complete addition.

On the other hand, if $(\mathbb{E},\leq)$ is a Boolean ring with least element 0, then $(\mathbb{E},\mathbb{E}\times\mathbb{E},\vee,\leq)$ is a dually residuated lattice-ordered semigroup with zero element 0, and it is obvious that also each commutative lattice-ordered group is a dually residuated lattice-ordered semigroup.

In [21], Riesz introduced fundamental domains:

A <u>fundamental domain</u> is a commutative semigroup $(\mathbb{E},\mathbb{E}\times\mathbb{E},+)$ with an order relation $\leq$ such that

- $(\mathbb{E},\mathbb{E}\times\mathbb{E},+)$ has the cancellation property;
- $x = y = 0$ holds for all $x, y \in \mathbb{E}$ satisfying $x + y = 0$;
- for all $x_1, x_2, y_1, y_2 \in \mathbb{E}$ satisfying $x_1 + x_2 = y_1 + y_2$, there exist $z_{ij} \in \mathbb{E}$ satisfying $x_i = z_{i1} + z_{i2}$ and $y_j = z_{1j} + z_{2j}$ for all $i, j \in \{1,2\}$; and
- for all $x, z \in \mathbb{E}$, $x \leq z$ holds if and only if there exists some $u \in \mathbb{E}$ satisfying $u + x = z$.

Thus, each fundamental domain is a positive commutative ordered partial semigroup with complete addition. Riesz also considered fundamental domains having the lattice property and property (M), and he pointed out that for such fundamental domains the refinement property can be omitted in the definition.

<u>8.6. Theorem.</u>

Let $(\mathbb{E},S,+,\leq)$ be an ordered partial semigroup.
Then the following are equivalent:

(a) $(\mathbb{E},S,+,\leq)$ is a positive commutative minimal clan with complete addition.

(b) $(\mathbb{E},S,+,\leq)$ is a fundamental domain having the lattice property and property (M).

Proof. If $(\mathbb{E},S,+,\leq)$ is a commutative minimal clan satisfying $S = \mathbb{E}\times\mathbb{E}$ and $\mathbb{E}_+ = \mathbb{E}$, then it is a fundamental domain having the lattice property and property (M), by Theorem 3.6 and Lemma 2.5. Conversely, if $(\mathbb{E},S,+,\leq)$ is a fundamental domain having the lattice property and property (M), then it is a lattice-ordered partial semigroup having the cancellation property. To see that it also has the difference property, consider $x, y \in \mathbb{E}$. Then we have

$$x + y = x\vee y + x\wedge y \quad .$$

Choose $u \in \mathbb{E} = \mathbb{E}_+$ satisfying

$$u + x\wedge y = y \quad .$$

Then we have

$$x + u + x\wedge y = x + y = x\vee y + x\wedge y$$

and thus

$$u + x = x + u = x\vee y \quad ,$$

which means that $(\mathbb{E},S,+,\leq)$ has the difference property. Thus we have shown that $(\mathbb{E},S,+,\leq)$ is a positive commutative minimal clan with complete addition. □

In his paper [21], Riesz gave several examples of fundamental domains having the lattice property and property (M). Actually, if $(\mathbb{E},\mathbb{E}\times\mathbb{E},+,\leq)$ is a commutative lattice-ordered group, then it follows from Theorems 5.1, 5.2, 3.3, and 8.6 that $(\mathbb{E}_+,\mathbb{E}_+\times\mathbb{E}_+,+,\leq)$ is a fundamental domain having the lattice property and property (M), and it can be seen from [25; Theorems 2.2 and 2.3] that in fact every fundamental domain having the lattice property and property (M) can be obtained in this way.

9. REMARKS

For partial groupoids it is possible to formulate several different associative laws which in the case of groupoids agree with the usual one. These different associative laws lead of course to different definitions of a "partial semigroup". The partial semigroups considered in this paper are the partial groupoids $(\mathbb{E},S,+)$ which satisfy the following associative law:

- for all $x, y, z \in \mathbb{E}$, $(x,y) \in S$ and $(x+y,z) \in S$ if and only if $(y,z) \in S$ and $(x,y+z) \in S$, and in this case $(x+y)+z = x+(y+z)$.

This associative law was also used by Bosbach [8] and Billhardt [2] in the definitions of semiclans and normal clans, and by Dinges [14] and Schmidt [23] in the definitions of Riesz D-semigroups and Riesz S-semigroups, which will be discussed at the end of this section. Furthermore, it can be seen from the subsequent result of Conrad [13] that this associative law is indeed a very natural one.

To formulate Conrad's result, we need the following construction:

For a partial groupoid $(\mathbb{E},S,+)$ and an element ∞ not belonging to $\mathbb{E}$, define $\mathbb{E}' := \mathbb{E}\cup\{\infty\}$ and a map $+' : \mathbb{E}'\times\mathbb{E}' \longrightarrow \mathbb{E}'$ by letting

$$x +' y := \begin{cases} x+y, & \text{if } x \in \mathbb{E},\ y \in \mathbb{E},\ \text{and } (x,y) \in S \\ \infty, & \text{otherwise.} \end{cases}$$

Then $(\mathbb{E}',\mathbb{E}'\times\mathbb{E}',+')$ is a groupoid.

9.1. Proposition.
Let $(\mathbb{E},S,+)$ be a partial groupoid.
Then the following are equivalent:
(a) $(\mathbb{E},S,+)$ is a partial semigroup.
(b) $(\mathbb{E}',\mathbb{E}'\times\mathbb{E}',+')$ is a semigroup.

For a proof of Conrad's result, see [13] or [12; Lemma 3.7].

In other words, the partial semigroups considered in the present paper are precisely the partial groupoids which are obtained from a semigroup with an infinity by discarding the infinity.

An associative law stronger than the above was proposed by Brandt [10] who required

- for all $x, y, z \in \mathbb{E}$, $(x,y) \in S$ and $(x+y,z) \in S$ if and only if $(y,z) \in S$ and $(x,y+z) \in S$, and in this case $(x+y)+z = x+(y+z)$, and
- $(x+y,z) \in S$, $(x,y+z) \in S$, and $(x+y)+z = x+(y+z)$ holds for all $x, y, z \in \mathbb{E}$ satisfying $(x,y) \in S$ and $(y,z) \in S$,

and an associative law weaker than the above was proposed by Schelp [22] who required

- for all $x, y, z \in \mathbb{E}$ satisfying $(x,y) \in S$ and $(y,z) \in S$, $(x+y,z) \in S$ if and only if $(x,y+z) \in S$, and in this case $(x+y)+z = x+(y+z)$.

For a further discussion of associative laws in partial groupoids, see Ljapin [16].

While the existence of a zero element is explicitly required in the definitions of normal clans and minimal clans, this is not the case for the definitions of symmetric clans and semiclans. The following result shows that there is also an equivalent definition of minimal clans in which the existence of a zero element is not explicitly required:

<u>9.2. Theorem.</u>

Let $\mathbb{E}$ be a set with a relation $S \subseteq \mathbb{E}\times\mathbb{E}$, a map $+ : S \longrightarrow \mathbb{E}$, and an order relation $\leq$.

Then the following are equivalent:

(a) $(\mathbb{E},S,+,\leq)$ is a minimal clan.

(b) $(\mathbb{E},S,+,\leq)$ satisfies the following axioms:

(M1) for all $x, y, z \in \mathbb{E}$, $(x,y) \in S$ and $(x+y,z) \in S$ if and only if $(y,z) \in S$ and $(x,y+z) \in S$, and in this case $(x+y)+z = x+(y+z)$;

(M2) $u+x+v \leq u+y+v$ holds for all $x, y \in \mathbb{E}$ satisfying $x \leq y$ and for all $u, v \in \mathbb{E}$ satisfying $(u,x) \in S$ and $(u+x,v) \in S$ as well as $(u,y) \in S$ and $(u+y,v) \in S$;

(M3) $x \leq y$ holds for all $x, y \in \mathbb{E}$ satisfying $u+x+v \leq u+y+v$ for some $u, v \in \mathbb{E}$;

(M4) $(\mathbb{E},\leq)$ is a lattice; and

(M5) for all $x, y \in \mathbb{E}$, there exist $u, v \in \mathbb{E}$ satisfying $u+x = x\vee y = x+v$ and $u+x\wedge y = y = x\wedge y+v$.

Proof. It is clear from the definition of minimal clans and Lemma 3.4 that (a) implies (b).
Suppose now that (b) holds. In order to prove that $(\mathbb{E},S,+,\leq)$ is a minimal clan, it is sufficient to show that there exists an element $0 \in \mathbb{E}$ satisfying $0+x = x = x+0$ for all $x \in \mathbb{E}$, and that for all $x, y \in \mathbb{E}$ the elements $u, v \in \mathbb{E}$ given by (M5) satisfy $0 \leq u$ and $0 \leq v$.
First, letting $x = y$ in (M5), we see that there exists an element $u_x \in \mathbb{E}$ satisfying $u_x + x = x$, and this element is unique since, for each $w \in \mathbb{E}$ satisfying $w+x = x$, we have

$$u_x + w + x = u_x + x = u_x + u_x + x ,$$

by (M1), and thus $w = u_x$, by (M3). Consider now $x, y \in \mathbb{E}$ and define $z := x \vee y$. By (M5), there exist $u', u'' \in \mathbb{E}$ satisfying

$$u' + z = z \vee x = z \quad \text{and} \quad u' + x = u' + z \wedge x = x$$

as well as

$$u'' + z = z \vee y = z \quad \text{and} \quad u'' + y = u'' + z \wedge y = y .$$

This yields

$$u_x = u' = u_z = u'' = u_y .$$

Since x and y are arbitrary, this means that there exists a unique element $u_0 \in \mathbb{E}$ satisfying $u_0 + x = x$ for all $x \in \mathbb{E}$, and in the same way it can be shown that there exists a unique element $v_0 \in \mathbb{E}$ satisfying $x = x + v_0$ for all $x \in \mathbb{E}$. This implies that there exists a unique element $0 \in \mathbb{E}$ satisfying $0+x = x = x+0$ for all $x \in \mathbb{E}$.
Second, for $x, y \in \mathbb{E}$ and $u, v \in \mathbb{E}$ satisfying

$$u + x = x \vee y = x + v \quad \text{and} \quad u + x \wedge y = y = x \wedge y + v ,$$

we have

$$0 + 0 + x = x \leq x \vee y = u + x = 0 + u + x$$

and thus $0 \leq u$, by (M3), and a similar argument yields $0 \leq v$.
Therefore, $(\mathbb{E},S,+,\leq)$ is a minimal clan. □

Although the set of axioms given in Theorem 9.2 is rather concise, we feel that the original definition of minimal clans is more transparent and perhaps easier to verify in concrete situations.

The following example shows that the difference property cannot be omitted in the definition of minimal clans:

<u>9.3. Example.</u>
Let $(\mathbb{H},\mathbb{H}\times\mathbb{H},+,\leq)$ be a lattice-ordered group containing at least two elements.
Let $\mathbb{E}$ denote the collection of all order intervals of $\mathbb{H}$, define $S := \mathbb{E}\times\mathbb{E}$, define a map $+ : S \longrightarrow \mathbb{E}$ by letting

$$[a,b] + [c,d] := [a+c,b+d] \quad ,$$

and define an order relation $\leq$ by letting

$$[a,b] \leq [c,d]$$

if and only if $a \leq c$ and $b \leq d$.

Then $(\mathbb{E},S,+,\leq)$ is a lattice-ordered semigroup having the cancellation property which does not have the difference property.

In fact, the axioms concerning addition are obviously satisfied, order and addition are compatible, and $(\mathbb{E},\leq)$ is a lattice since any two order intervals $[a,b]$ and $[c,d]$ possess a least upper bound, given by $[a\vee c,b\vee d]$, and a greatest lower bound, given by $[a\wedge c,b\wedge d]$.

Therefore, $(\mathbb{E},S,+,\leq)$ is a lattice-ordered (partial) semigroup having the cancellation property.

However, for $a \in \mathbb{H}\setminus\{0\}$, we have either $a^+ \in \mathbb{H}_+\setminus\{0\}$ or $a^- \in \mathbb{H}_+\setminus\{0\}$.

This yields $\mathbb{H}_+ \neq \{0\}$, and for $b \in \mathbb{H}_+\setminus\{0\}$ we have

$$[0,b] \leq [b,b] \quad ,$$

but there is no order interval $[c,d] \in \mathbb{E}$ satisfying

$$[c,d] + [0,b] = [b,b] \quad .$$

Therefore, $(\mathbb{E},S,+,\leq)$ does not have the difference property.

The lattice-ordered semigroups considered in Example 9.3 are of interest in interval mathematics. For further details on this subject, we refer to the papers by Nickel [18,19] and Schmidt [25].

In Sections 4 and 5 of this paper, we have characterized Boolean rings and lattice-ordered groups as minimal clans having a minimal domain of addition or a maximal set of invertible elements, respectively. The following example shows that there exist minimal clans which are neither a Boolean ring nor a lattice-ordered group:

<u>9.4. Example.</u>

Let Ω be an arbitrary set.

Let $\mathbb{E}$ denote the collection of all functions $\Omega \longrightarrow [0,1]$, define $S := \{ (x,y) \in \mathbb{E}\times\mathbb{E} \mid x(\omega) + y(\omega) \leq 1 \text{ for all } \omega \in \Omega \}$, define a map $+ : S \longrightarrow \mathbb{E}$ by letting

$$(x+y)(\omega) := x(\omega) + y(\omega)$$

for all $\omega \in \Omega$, and define an order relation $\leq$ by letting

$$x \leq y$$

if and only if $x(\omega) \leq y(\omega)$ holds for all $\omega \in \Omega$.

Then $(\mathbb{E},S,+,\leq)$ is a positive commutative minimal clan which is neither a Boolean ring nor a lattice-ordered group.

In fact, it is obvious that $(\mathbb{E},S,+,\leq)$ is a positive commutative minimal clan. Furthermore, for the function $x \in \mathbb{E}$ given by $x(\omega) := 1/2$

for all $\omega \in \Omega$, we have $(x,x) \in S$ and $x \wedge x = x \neq 0$. By Theorem 4.2, this implies that $(\mathbb{E},\leq)$ cannot be a Boolean ring, and from $\mathbb{E}_{*} = \{0\}$ and Theorem 5.2 we conclude that $(\mathbb{E},S,+,\leq)$ cannot be a lattice-ordered group.

The functions $\Omega \longrightarrow [0,1]$ considered in Example 9.4 are said to be the membership functions of the fuzzy subsets of Ω which were introduced by Zadeh [30]. For a fuzzy subset A of Ω , the membership function $\chi_A : \Omega \longrightarrow [0,1]$ indicates for each $\omega \in \Omega$ the degree to which the element ω belongs to A , just as the characteristic function $\chi_B : \Omega \longrightarrow \{0,1\}$ of an ordinary subset B of Ω indicates for each $\omega \in \Omega$ whether or not the element ω belongs to B . Usually, fuzzy sets and their membership functions are identified, so that we can briefly say that the fuzzy subsets of a given set form a minimal clan which is neither a Boolean ring nor a lattice-ordered group.

Another class of minimal clans which is neither a Boolean ring nor a lattice-ordered group is given by the following result:

<u>9.5. Theorem.</u>
Let $(\mathbb{E},S,+,\leq)$ be a minimal clan with complete addition containing at least two elements.
Then $(\mathbb{E}_+,\mathbb{E}_+\times\mathbb{E}_+,+,\leq)$ is a minimal clan which is neither a Boolean ring nor a lattice-ordered group.

Proof. By Theorem 3.3, $(\mathbb{E}_+,\mathbb{E}_+\times\mathbb{E}_+,+,\leq)$ is a minimal clan. Furthermore, for $x \in \mathbb{E}\setminus\{0\}$, we have either $x^+ \in \mathbb{E}_+\setminus\{0\}$ or $x^- \in \mathbb{E}_+\setminus\{0\}$. This yields $\mathbb{E}_+ \neq \{0\}$, and for $z \in \mathbb{E}_+\setminus\{0\}$ we have $z \wedge z = z \neq 0$. By Theorem 4.2, this implies that $(\mathbb{E},\leq)$ cannot be a Boolean ring.
Finally, we have $(\mathbb{E}_+)_+ = \mathbb{E}_+$ and thus $(\mathbb{E}_+)_* = \{0\}$, by Lemma 2.2, and it now follows from Theorem 5.2 that $(\mathbb{E}_+,\mathbb{E}_+\times\mathbb{E}_+,+,\leq)$ cannot be a lattice-ordered group. □

In particular, a nontrivial positive minimal clan with complete addition (and hence, by Theorem 8.6, a nontrivial fundamental domain having the lattice property and property (M)) cannot be a Boolean ring or a lattice-ordered group. Thus, applying Theorem 9.5 to lattice-ordered groups, we see that there exist minimal clans having a minimal set of invertible elements and a maximal domain of addition which fail to be a Boolean ring or a lattice-ordered group. This should be compared with Theorems 4.2 and 5.2.

A common abstraction of Boolean rings and fundamental domains was considered by Dinges [14] who introduced Riesz D-semigroups:

A <u>Riesz D-semigroup</u> is a commutative partial semigroup $(\mathbb{E},S,+)$ with an order relation $\leq$ such that

- $(\mathbb{E},S,+)$ has the cancellation property;
- $x = y = 0$ holds for all $x, y \in \mathbb{E}$ satisfying $x+y = 0$;
- for all $x_1, x_2, y_1, y_2 \in \mathbb{E}$ satisfying $x_1+x_2 = y_1+y_2$, there exist $z_{ij} \in \mathbb{E}$ satisfying $x_i = z_{i1}+z_{i2}$ and $y_j = z_{1j}+z_{2j}$ for all $i, j \in \{1,2\}$; and
- for all $x, z \in \mathbb{E}$, $x \leq z$ holds if and only if there exists some $u \in \mathbb{E}$ satisfying $u+x = z$.

Thus, each Riesz D-semigroup is a positive commutative ordered partial semigroup, and it is obvious from the definitions that an ordered partial semigroup is a fundamental domain if and only if it is a Riesz D-semigroup with complete addition. We have the following analogon of Theorem 8.6:

<u>9.6. Theorem.</u>
Let $(\mathbb{E},S,+,\leq)$ be an ordered partial semigroup.
Then the following are equivalent:

(a) $(\mathbb{E},S,+,\leq)$ is a positive commutative minimal clan.

(b) $(\mathbb{E},S,+,\leq)$ is a Riesz D-semigroup having the lattice property and property (M).

The proof is similar to that of Theorem 8.6.

In particular, if $(\mathbb{E},\leq)$ is a Boolean ring with least element 0 , then $(\mathbb{E},\mathbb{E}^{\perp}\mathbb{E},\vee,\leq)$ is a Riesz D-semigroup with zero element 0 , by Theorems 4.1 and 9.6. On the other hand, it is clear that a nontrivial Riesz D-semigroup cannot be a (commutative) lattice-ordered group.

The following common abstraction of Riesz D-semigroups and commutative lattice-ordered groups was introduced in [23]:

A <u>Riesz S-semigroup</u> is a commutative ordered partial semigroup $(\mathbb{E},S,+,\leq)$ such that

- for all $x_1, x_2, y_1, y_2 \in \mathbb{E}$ satisfying $x_1+x_2 = y_1+y_2$, there exist $z_{ij} \in \mathbb{E}$ satisfying $x_i = z_{i1}+z_{i2}$ and $y_j = z_{1j}+z_{2j}$ for all $i, j \in \{1,2\}$; and
- for all $x, z \in \mathbb{E}$ satisfying $x \leq z$, there exists some $u \in \mathbb{E}_+$ satisfying $u+x = y$.

It is obvious from the definitions that an ordered partial semigroup is a Riesz D-semigroup if and only if it is a positive Riesz S-semigroup having the cancellation property, and it follows from Theorem 3.6 that each commutative minimal clan (and hence, by Theorem 5.1, each commutative lattice-ordered group) is a Riesz S-semigroup.

We finally present two diagrams which visualize some of the relations existing between the different classes of ordered partial semigroups considered in this paper:

<u>Ordered Partial Semigroups - The General Case</u>

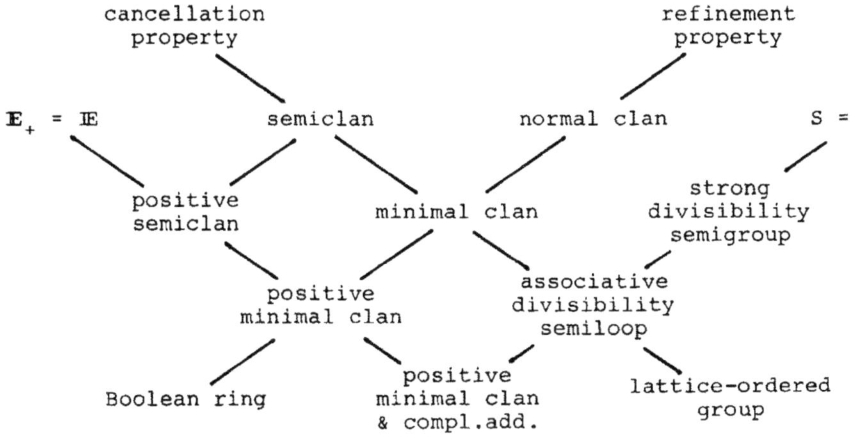

<u>Ordered Partial Semigroups - The Commutative Case</u>

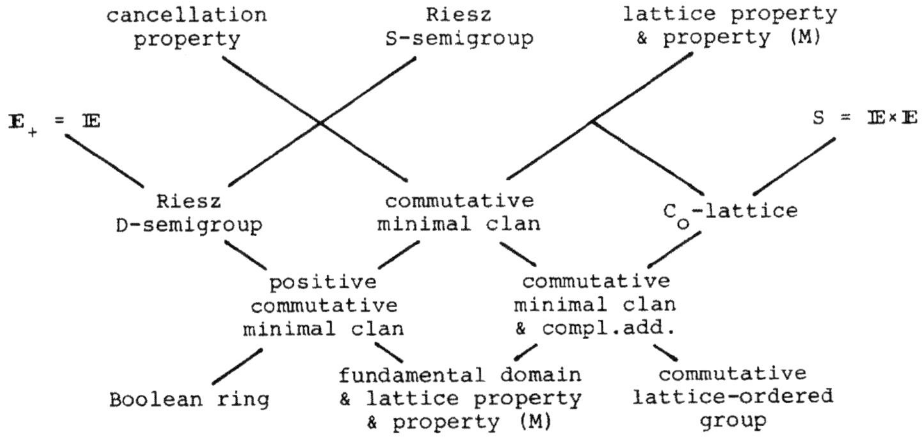

REFERENCES

[1] Bigard, A., Keimel, K., Wolfenstein, S.:
Groupes et Anneaux Réticulés.
Lecture Notes in Mathematics, vol. 608.
Berlin - Heidelberg - New York: Springer 1977.

[2] Billhardt, B.:
Zum Clan der normalen Teilbarkeitshalbgruppe.
Dissertation.
Kassel: Fachbereich Mathematik der Gesamthochschule Kassel 1981.

[3] Birkhoff, G.:
Lattice-ordered groups.
Ann. of Math. 43, 298-331 (1942).

[4] Birkhoff, G.:
Lattice Theory. (Second) Revised Edition.
Providence, Rhode Island: Amer. Math. Soc. 1948.

[5] Birkhoff, G.:
Lattice Theory. Third (New) Edition.
Providence, Rhode Island: Amer. Math. Soc. 1967.

[6] Bosbach, B.:
Zur Theorie der Teilbarkeitshalbgruppen.
Semigroup Forum 3, 1-30 (1971).

[7] Bosbach, B.:
Schwache Teilbarkeitshalbgruppen.
Semigroup Forum 12, 119-135 (1976).

[8] Bosbach, B.:
Concerning semiclans.
Arch. Math. 37, 316-324 (1981).

[9] Bosbach, B.:
Lattice ordered binary systems.
Mathematische Schriften Kassel, Preprint Nr. 4/84.
Kassel: Fachbereich Mathematik der Gesamthochschule Kassel 1984.

[10] Brandt, H.:
Über eine Verallgemeinerung des Gruppenbegriffes.
Math. Ann. 96, 360-366 (1927).

[11] Brehmer, S.:
Algebraic characterisation of measure and integral by the method of Caratheodory.
In: Proc. Conf. Topology and Measure (Zinnowitz 1974), Part 1, pp. 23-53.
Greifswald: Ernst-Moritz-Arndt-Universität 1978.

[12] Clifford, A.H., Preston, G.B.:
The Algebraic Theory of Semigroups, vol. I.
Providence, Rhode Island: Amer. Math. Soc. 1961.

[13] Conrad, P.:
Generalized semigroup rings.
J. Indian Math. Soc. (N.S.) 21, 73-95 (1957).

[14] Dinges, H.:
Zur Algebra der Maßtheorie.
Bull. Greek Math. Soc. 19, 25-97 (1978).

[15] Fuchs, L.:
Teilweise geordnete algebraische Strukturen.
Göttingen: Vandenhoeck & Ruprecht 1966.

[16] Ljapin, E.S.:
Partielle Operationen in der Theorie der Halbgruppen.
In: Semigroups.
Lecture Notes in Mathematics, vol. 855, pp. 33-48.
Berlin - Heidelberg - New York: Springer 1981.

[17] Nakano, T.:
Rings and partly ordered systems.
Math. Z. 99, 355-376 (1967).

[18] Nickel, K.:
Verbandstheoretische Grundlagen der Intervall-Mathematik.
In: Interval Mathematics.
Lecture Notes in Computer Science, vol. 29, pp. 251-262.
Berlin - Heidelberg - New York: Springer 1975.

[19] Nickel, K.:
Intervall-Mathematik.
Z. Angew. Math. Mech. 58, T72-T85 (1978).

[20] Rama Rao, V.V.:
On a common abstraction of Boolean rings and lattice ordered groups I.
Monatsh. Math. 73, 411-421 (1969).

[21] Riesz, F.:
Sur quelques notions fondamentales dans la théorie générale des opérations linéaires.
Ann. of Math. 41, 174-206 (1940).

[22] Schelp, R.H.:
A partial semigroup approach to partially ordered sets.
Proc. London Math. Soc. (3) 24, 46-58 (1972).

[23] Schmidt, K.D.:
A general Jordan decomposition.
Arch. Math. 38, 556-564 (1982).

[24] Schmidt, K.D.:
A common abstraction of Boolean rings and lattice ordered groups.
Comp. Math. 54, 51-62 (1985).

[25] Schmidt, K.D.:
Embedding theorems for cones and applications to classes of convex sets occurring in interval mathematics.
In: Interval Mathematics 1985.
Lecture Notes in Computer Science, vol. 212, pp. 159-173.
Berlin - Heidelberg - New York: Springer 1986.

[26] Stone, M.H.:
Postulates for Boolean algebras and generalized Boolean algebras.
Amer. J. Math. 57, 703-732 (1935).

[27] Stone, M.H.:
The theory of representations for Boolean algebras.
Trans. Amer. Math. Soc. 40, 37-111 (1936).

[28] Swamy, K.L.N.:
Dually residuated lattice ordered semigroups.
Math. Ann. 159, 105-114 (1965).

[29] Wyler, O.:
Clans.
Comp. Math. 17, 172-189 (1966).

[30] Zadeh, L.A.:
Fuzzy Sets.
Inform. Control 8, 338-353 (1965).

LES SYSTEMES ENTIERS D'EQUATIONS SUR UN ALPHABET DE 3 VARIABLES

Jean-Claude SPEHNER
Université de Haute Alsace,
Faculté des Sciences et Techniques
68093 MULHOUSE Cédex, FRANCE

Abstract.- If Σ^* is the free monoid over the alphabet Σ, a part S of $\Sigma^* x \Sigma^*$ is said to be a system of equations over the alphabet Σ ; the elements of Σ are the variables of S and each morphism ϕ of Σ^* into a free monoid A^* such that, $\forall$ (e,e') $\in$ S, $\phi(e) = \phi(e')$ is said to be a solution of S. If L is the code which generates the smallest free submonoid of A^* which contains $\phi(\Sigma^*)$, then card L is called the rank of ϕ. The greatest rank of a solution of S is called the rank of S. A system S is said to be entire if it admits a solution ϕ such that $S = \phi^{-1} \circ \phi$.

First we recall that all the equations of a entire system over a finite alphabet are generated by a finite graph.

In [Spe 78] we have shown that every non-commutative submonoid C^* of a free monoid such that card C = 3 admits a finite presentation (Σ,ρ) with card $\rho \leqslant 2$. Using these results and a notion of characteristic, we give a classification for all the entire systems of rank 2 over $\Sigma = \{x,y,z\}$.

Two systems of equations over the same alphabet are called equivalent if they have the same solutions. If T is a part of a system S which is equivalent to S and of minimal cardinality, then card T is called the dimension of S. We prove here the following inequality

dim(S) + rank(S) $\leqslant$ 3

An error of [Spe 78] is corrected in an appendix.

Résumé.- Nous rappelons d'abord que les équations d'un système entier S d'équations sur un alphabet fini sont générées par un graphe fini.

Dans [Spe 78] nous avons montré que tout sous-monoïde non commutatif C^* de monoïde libre tel que card C = 3 admet une présentation (Σ,ϕ) avec card $\rho \leqslant 2$. En introduisant une notion de caractéristique, nous en déduisons une classification de tous les systèmes entiers de rang 2 sur $\Sigma = \{x,y,z\}$.

Pour tout système S, le plus petit des cardinaux des parties de S équivalentes à S est appelé la dimension de S. Nous montrons que tout système entier S sur $\Sigma = \{x,y,z\}$ vérifie l'inégalité

$$\dim(S) + \mathrm{rang}(S) \leqslant 3 .$$

Une erreur de [Spe 78] est corrigée en annexe.

1.- La génération des équations d'un système entier d'équations

1.1. Notations et rappels.- (i) A^* est le monoïde libre d'alphabet A ($A \neq \phi$) ; 1 est l'élément neutre de A^* ; si $u \in A^*$, $|u|$ est la longueur du mot u et, $\forall\, a \in A$, $|u|_a$ est le nombre d'occurrences de la lettre a dans le mot u ; $\mathrm{alph}(u) = \{a \in A ; |u|_a > 0\}$.

Si $w = uv$, u [resp. v] est appelé facteur gauche [resp. droit] de w et est noté $u = wv^{-1}$ [resp. $v = u^{-1}w$].

Si $C \subset A^*$, C^* est le sous-monoïde de A^* engendré par C, $C^+ = C^* \setminus \{1\}$ et $\mathring{C} = C^+ \setminus (C^+)^2$ est la plus petite partie génératrice de C^*.

C est appelé un code lorsque le sous-monoïde C^* est libre et que $C = \mathring{C}$.

(ii) Un mot u de A^+ est dit imprimitif s'il existe $v \in A^+$ et $r > 1$ tels que $u = v^r$; dans le cas contraire il est dit primitif.

Pour tout mot w de A^+, il existe un unique mot primitif u tel que $w \in u^+$; ce mot u est appelé la racine primitive de w.

Si u est un mot primitif, l'égalité $uu = vuv'$ avec $v,v' \in A^*$ implique ou bien $v = 1$ ou bien $v' = 1$; on exprime cette propriété en disant qu'un mot primitif ne peut pas chevaucher son carré.

Deux mots u et u' de A^* sont dits conjugués s'il existe des mots u_1 et u_2 de A^* tels que $u = u_1u_2$ et $u' = u_2u_1$.

Si $uv = vu'$, v est appelé un facteur de conjugaison de (u,u').

u et u' sont conjugués si, et seulement si, (u,u') admet un facteur de conjugaison.

Si v est un facteur de conjugaison de (u,u') avec $u \neq 1$, il existe un couple unique (h_1,h_2) de mots de A^* et des entiers uniques $m > 0$ et $n \geqslant 0$ tels que $u = (h_1h_2)^m$, $u' = (h_2h_1)^m$ et $v = (h_1h_2)^n h_1$ avec h_1h_2 primitif d'après A. Lentin et M.P. Schützenberger [LS 67].

(iii) Σ^* étant le monoïde libre d'alphabet Σ, tout couple $E = (e,e')$ de $\Sigma^* \times \Sigma^*$ [resp. toute partie S non vide de $\Sigma^* \times \Sigma^*$] est appelé [e] équation [resp. système d'équations] sur l'alphabet Σ ou dans le monoïde libre Σ^* ; les lettres de Σ sont appelées les variables de l'équation E [resp. du système S].

Tout homomorphisme ϕ de Σ^* dans un monoïde libre A^* tel que $\phi(e) = \phi(e')$ [resp. $\phi(e) = \phi(e')$ $\forall\,(e,e') \in S$] est appelé une solution de l'équation E [resp. du système d'équations S].

Tout homomorphisme $\theta : A^* \to B^*$ de monoïdes libres et dit continu [non-erasing ou λ-free en anglais (λ désignant l'élément neutre de B^*)] si $1 \notin \theta(A)$ c'est-à-dire si $\theta^{-1}(1) = \{1\}$.

Si $\phi : \Sigma^* \to A^*$ et $\psi : \Sigma^* \to B^*$ sont 2 solutions de E [resp. S], on dit que ϕ divise ψ s'il existe un homomorphisme continu $\theta : A^* \to B^*$ tel que $\psi = \theta \circ \phi$; les solutions ϕ et ψ sont dites équivalentes si elles se divisent mutuellement.

Une solution ϕ de E [resp. S] est dite principale si toute solution de E [resp. S] qui divise ϕ est équivalente à ϕ.

(iv) Pour tout sous-monoïde M de A^*, le plus petit sous-monoïde libre de L de A^* contenant M est appelé l'enveloppe libre de M [Spe 75] (voir aussi [BPPR 79] et [Pec 84]) et le code $\overset{o}{L}$ est appelé noyau libre de M.

Si $\phi : \Sigma^* \to A^*$ est un homomorphisme de monoïdes libres, le cardinal du noyau libre de $\phi(\Sigma^*)$ est appelé rang de ϕ.

Le rang maximum des solutions d'une équation E [resp. d'un système d'équations S] est appelé rang de l'équation E [resp. du système S].

(v) Une équation E = (e,e') est dite équilibrée [resp. triviale] si, $\forall z \in \Sigma$, $|e|_z = |e'|_z$ [resp. e = e'].

(vi) Si $\rho \subset \Sigma^*x\Sigma^*$, un couple (v,v') de $\Sigma^*x\Sigma^*$ encore noté $v \to v'$ est appelé une ρ-transition élémentaire s'il existe $v_1, v_2 \in \Sigma^*$ et $(u,u') \in \rho$ tels que ou bien $v = v_1uv_2$ et $v' = v_1u'v_2$

ou bien $v = v_1u'v_2$ et $v' = v_1uv_2$.

Une suite $w = w_o \to w_1 \;\dots\;\dots \to w_n = w'$ est appelée une suite de ρ -transitions élémentaires de w à w' si, $\forall i \in \{0,\dots,n-1\}$, $w_i \to w_{i+1}$ est une ρ-transition élémentaire.

Si $c(\rho)$ est la plus petite congruence de Σ^* contenant ρ, $(w,w') \in c(\rho)$ si, et seulement si, il existe une suite de ρ-transitions élémentaires de w à w'.

(vii) Un couple (Σ,ρ) où $\rho \subset \Sigma^*x\Sigma^*$ est appelé une présentation d'un monoïde M si M est isomorphe au quotient du monoïde libre Σ^* d'alphabet Σ par la plus petite congruence $c(\rho)$ de Σ^* contenant ρ.

Le lecteur trouvera des compléments dans les livres de A. Lentin [Len 72] et de M. Lothaire [Lot 83] (voir aussi [Eil 74] et [Lal 79]).

Définition 1.2.- Un système d'équations S sur Σ est dit entier [Spe 85] s'il admet une solution ϕ telle que $S = \phi^{-1} \circ \phi = \{(e,e') \in \Sigma^*x\Sigma^* \; ; \; \phi(e) = \phi(e')\}$.

Définition 1.3.- Soient ϕ un homomorphisme continu du monoïde libre Σ^* d'alphabet Σ dans un monoïde libre A^* d'alphabet A, $C = \phi(\Sigma)$ et le système entier $S = \phi^{-1} \circ \phi$ d'équations sur Σ.

(i) Si les éléments $z \in \Sigma$, $\omega \in \Sigma^*$ et d, d' $\in A^*$ vérifient $\phi(z) = d\phi(\omega)d'$, $d\phi(\omega) \neq 1$ et $\phi(\omega)d' \neq 1$ le couple (d,d') est appelé un Σ-couple relatif à z, les couples (z,ω) et (ω,z) sont appelés les motifs produits par le Σ-couple (d,d') et $(d,d')_{(z,\omega)}$ est appelé un Σ-couple étiqueté.

En outre $(1,1)_{(1,1)}$ est considéré comme un Σ-couple étiqueté qui est dit trivial.

(ii) $\forall\, u \in A^*$, soit Fg(u) [resp. Fd(u)] l'ensemble des facteurs gauches [resp. droits] de u distincts de u et de 1.

Soient $Fg(C) = \bigcup_{u \in C} Fg(u)$, $Fd(u) = \bigcup_{u \in C} Fd(u)$ et $Fb(C) = Fg(C) \cap Fd(C)$.

(iii) Soit $\mathcal{G}(C)$ le multi-graphe orienté dont les sommets sont les éléments de $Fg(C) \cup Fd(C) \cup \{1\}$ et dont les arcs sont les Σ-couples étiquetés.

Le sous-multigraphe $\mathcal{L}(C)$ de $\mathcal{G}(C)$ qui est la composante fortement connexe du sommet 1 dans $\mathcal{G}(C)$ est appelé le graphe des équations de C^*.

(iv) Si $\sigma = ((d_0,d_1)_{(z_1,\omega_1)},(d_1,d_2)_{(z_2,\omega_2)},\ldots,(d_{r-1},d_r)_{(z_r,\omega_r)})$ est un chemin de $\mathcal{L}(C)$, d_0 [resp. d_r] est appelé origine [resp. extrémité] de σ et, si la longueur r de σ est paire [resp. impaire], le couple $(u,u') = (z_1\omega_2z_3 \ldots z_{r-1}\omega_r,\ \omega_1z_2\omega_3 \ldots \omega_{r-1}z_r)$ [resp. $(u,u') = (z_1\omega_2z_3 \ldots \omega_{r-1}z_r,\ \omega_1z_2\omega_3 \ldots z_{r-1}\omega_r)$] et son inverse (u',u) sont appelés les motifs produits par le chemin σ. Si r est pair,
$\alpha(z_1\omega_2z_3 \ldots z_{r-1}\omega_r)d_r = d_0\alpha(\omega_1z_2\omega_3 \ldots \omega_{r-1}z_r)$
et, si r est impair, $\alpha(z_1\omega_2z_3 \ldots \omega_{r-1}z_r) = d_0\alpha(\omega_1z_2\omega_3 \ldots z_{r-1}\omega_r)d_r$.

(v) Tout chemin $\sigma = ((d_0,d_1)_{(z_1,\omega_1)}, \ldots, (d_{r-1},d_r)_{(z_r,\omega_r)})$ tel que $d_0 = d_r = 1$ est appelé un circuit unitaire.

Théorème 1.4.- Soient ϕ un homomorphisme continu du monoïde libre Σ^* d'alphabet fini Σ dans un monoïde A^* d'alphabet A, $C = \phi(\Sigma)$, $\mathcal{L}(C)$ le graphe des équations de C^* et le système d'équations $S = \phi^{-1} \circ \phi$.

(i) Pour tout circuit unitaire σ de $\mathcal{L}(C)$, les motifs produits par σ sont des équations de S.

(ii) Pour toute équation (u,u') de S, il existe un circuit unitaire unique σ de $\mathcal{L}(C)$ tel que les motifs produits par σ soient (u,u') et (u',u).

Ce résultat a été démontré en [Spe 75] dans le cas où la restriction de ϕ à Σ et C est bijective mais la démonstration reste valable (voir aussi [Spe 76a] et G. Lallement [Lal 79] (pages 110-115)). Voir un exemple en 2.3.

2. Les systèmes entiers d'équations de rang 2 sur $\Sigma = \{x,y,z\}$ et leurs caractéristiques

Théorème 2.1.- Tout système entier S de rang 2 sur $\Sigma = \{x,y,z\}$ est, aux permutations de Σ près, d'une des formes suivantes :
ou bien $S = c(P,Q)$ avec $P = ((uv)^p u,(wv)^q w)$ et $Q = (uvw,wvu)$ où $p > 0$, $q > 0$, pgcd $(p+1,q+1) = 1$ et $\chi = (u,v,w)$ est d'un des types suivants :

(1) $\chi = (x,1,z)$;

(2) $\chi = (x^i yx^j, x^{k-i-j}, z)$ avec $0\leq i$, $0\leq j$, $0<k$ et $i+j\leq k$;

(3) $\chi = (x^{i+j-k}, x^{k-j}yx^{k-i}, z)$ avec $0<i\leq k$, $0<j\leq k$, $k<i+j$ et $q<p$;

(4) $\chi = (x^{i+j-k}, x^{k-j}yx^{k-i}, w)$ avec $0<i\leq k$, $0<j\leq k$, $k<i+j$, $p = q+1$ et $w \in z\{x,z\}^* z$;

(5) $\chi = (x^i y, x^m, zx^{j'})$ avec $0<i$, $0<j'$ et $0\leq m$ ou $\chi = (yx^j, x^m, x^{i'}z)$ avec $0<i'$, $0<j$ et $0\leq m$;

ou bien $S = c(P)$ avec χ et P d'un des types suivants :

(6) $\chi = (x^i yx^j, 1, w)$ et $P = (x^i yx^j, w^{q+1})$ avec $w \in x^{i'}z\{x,z\}^* \cap \{x,z\}^* zx^{j'}$, w primitif, $i\geq 0$, $j\geq 0$, $i'\geq 0$, $j'\geq 0$, $ii' = jj' = 0$ et $q \geq 0$;

(7) $\chi = (x,1,1)$ et $P = (x,1)$;

(8) $\chi = (x^i y, zx^{j'}, x^r)$ et $P = (x^r(x^i y)^{p+1}, (zx^{j'})^{q+1}x^r)$ avec $i\geq 0$, $j'\geq 0$, $p>0$, $q>0$, $r>0$ et pgcd $(p+1,q+1) = 1$.

Preuve.- Soit S un système entier de rang 2 sur $\Sigma = \{x,y,z\}$. Il existe une solution $\phi : \Sigma^* \to A^*$ de S telle que $S = \phi^{-1} \circ \phi$ avec $A = \{a,b\}$.

(i) Dans le cas où $C = \phi(\Sigma)$ est la plus petite partie génératrice du sous-monoïde $C^* = \phi(\Sigma^*)$ de A^* c'est-à-dire lorsque $C = \mathring{C}$, le théorème de [Spe 78] (voir aussi [Spe 76b]) donne toutes les présentations des sous-monoïdes C^* de A^* tels que card $\mathring{C} = 3$ à une erreur près qui est corrigée en annexe. On trouve ainsi tous les types de {1,2,3,4,5,8} et le type (6) lorsque $i+j > 0$, $w \neq z$ et $q \neq 0$.

(ii) Dans le cas contraire, on peut supposer, à une permutation de Σ près, que $\{\phi(x),\phi(z)\}$ est un code de A^* et que $\phi(y) \in \{\phi(x),\phi(z)\}^*$.

Il existe donc un élément unique w de $\{x,z\}^*$ tel que $\phi(y) = \phi(w)$. Soit alors $P = (y,w)$. $\forall E = (e,e') \in S$, soit f [resp. f'] l'unique élément de $\{x,z\}^*$ obtenu en remplaçant chaque occurrence de y dans e [resp. e'] par w. Alors $(e,f) \in c(P)$ et $(e',f') \in c(P)$ et, comme $c(P) \subset S = \phi^{-1} \circ \phi$, $\phi(f) = \phi(e) = \phi(e') = \phi(f')$. La restriction de ϕ à $\{x,z\}^*$ étant injective, il en résulte que $f = f'$ et, par suite, que $(e,e') \in c(P)$. Ceci prouve que $S = c(P)$ lorsque P est de type 6 avec $i = j = 0$ ou avec $w = z$ et lorsque P est de type 7 (après permutation de x et y).

Définition 2.2.- (i) Le théorème 2.1 associe à tout système entier S de rang 2 sur $\Sigma = \{x,y,z\}$ un entier $\tau \in \{1,2,3,4,5,6,7,8\}$ appelé le type du système S et une équation P appelée l'équation générique du système S.

(ii) Si S est un système entier de rang 2 et de type $\tau \in \{1,2,3,4,5,6,7\}$ sur $\Sigma = \{x,y,z\}$ et si P est son équation générique, tout triplet $\chi = (u,v,w)$ de mots de Σ^* tel que

. il existe $p \geq 0$, $q \geq 0$ tels que $P = ((uv)^p u,(wv)^q w)$ avec pgcd $(p+1,q+1)$

. les mots uv et wv de Σ^* sont primitifs ;

. $\mathrm{alph}(uv) \neq \Sigma$ ou $\mathrm{alph}(wv) \neq \Sigma$;

. $|u|_x + |u|_y > 0$ ou $u = 1$;

. $|w|_x + |w|_z > 0$ ou $w = 1$;

. $\mathrm{alph}(u) \neq \mathrm{alph}(w)$

est appelé une caractéristique de l'équation P et du système S.

Si τ est le type de S, la caractéristique χ est aussi dite de type τ.

L'équation $Q = (uvw,wvu)$ est appelée l'équation équilibrée associée à χ.

(iii) Si S est un système entier de rang 2 et de type 8 et si $P = (x^r(x^iy)^{p+1}, (zx^j)^{q+1}x^r)$ est son équation générique avec $i \geq 0$, $j \geq 0$, $p > 0$, $q > 0$, $r > 0$ et pgcd $(p+1,q+1) = 1$, le triplet $\chi = (x^iy,zx^j,x^r)$ est aussi appelé caractéristique de l'équation P et du système S.

La caractéristique χ est alors dite de type 8.

Remarque.- (i) Le théorème 2.1 associe explicitement une caractéristique à chaque système entier de rang 2 sur $\Sigma = \{x,y,z\}$.

Cette caractéristique est unique sauf lorsque $\tau = 6$ (voir l'exemple 2.3 et la proposition 2.5).

(ii) Toute caractéristique $\chi = (u,v,w)$ de type $\tau \in \{1,2,3,4,5\}$ est aussi de type 6 car, lorsque $\inf(p,q) = 0$, l'équation $P = ((uv)^p u,(wv)^q w)$ est de type 6.

Exemple 2.3.- Si S est le système entier de rang 2 d'équation générique $P = (xxyx,zzxzz)$, l'application ϕ de Σ^* dans A^* où $A = \{a,b\}$ telle que $\phi(x) = a, \phi(y) = ba^3ba^4ba^3b$ et $\phi(z) = aaba$ est une solution de S telle que $S = \phi^{-1}o\phi$. Si $C = \phi(\Sigma)$, le graphe $\mathcal{L}(C)$ est :

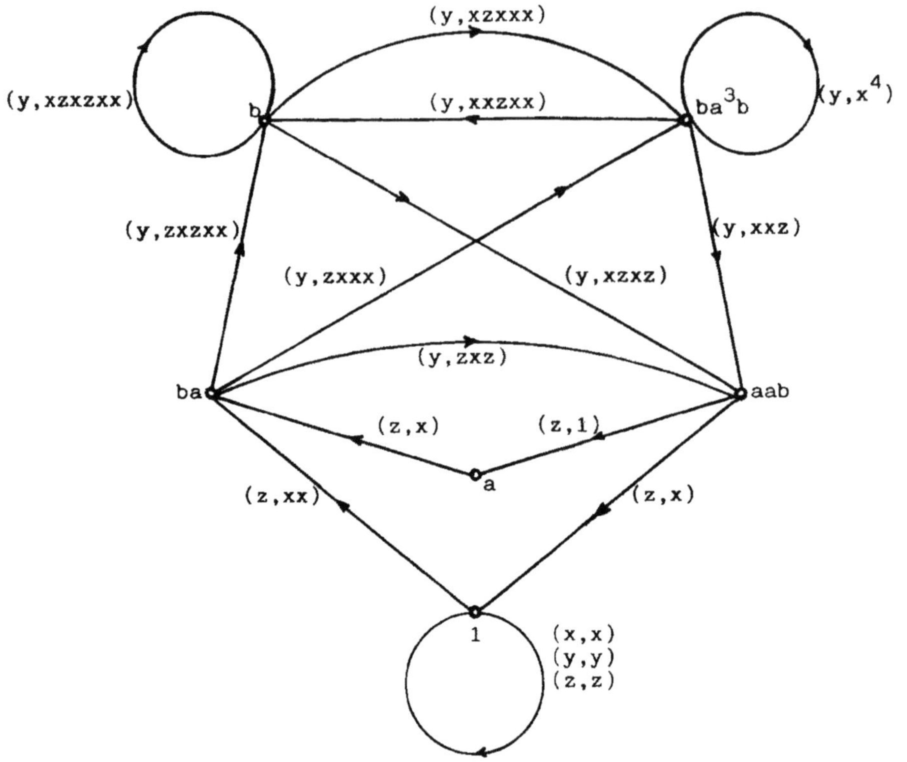

S admet les caractéristiques suivantes :
χ_1 = (xxyx,1,zzxzz), χ_2 = (x,xy,zzxzz), χ_3 = (xxyx,zxz,z) et χ_4 = (xxyx,x,zz) et contient les équations équilibrées associées Q_1 = (xxyx.zzxzz,zzxzz.xxyx), Q_2 = (x.xy.zzxzz,zzxzz.xy.x), Q_3 = (xxyx.zxz.z,z.zxz.xxyx) et Q_4 = (xxyx.x.zz,zz.x.xxyx).

L'équation Q_3 est produite par le chemin $\sigma_3 = ((1,ba)_{(z,xx)}, (ba,b)_{(y,zxzxx)}, (b,aab)_{(y,xzxz)}, (aab,1)_{(z,x)})$ de $\mathcal{L}(C)$.

<u>Définition 2.4.</u>- (i) Une <u>partie</u> C d'un monoïde libre A^* est dite <u>incontractable</u> si $C = \overset{\circ}{C}$ et si, pour toute partie D de A^* telle que les sous-monoïdes C^* et D^* soient isomorphes,

$$\sum_{x\in D} |x| \geq \sum_{x\in C} |x|$$

(ii) S étant un système d'équations sur Σ, une <u>solution</u> $\phi : \Sigma^* \to A^*$ de S est dite <u>incontractable</u> si la plus petite partie génératrice de $\phi(\Sigma^*)$ est une partie incontractable de A^*.

Toute solution incontractable de S est aussi une solution principale de S mais la réciproque est fausse d'après la remarque 2.6. (ii).

<u>Proposition 2.5.-</u> Tout système entier S de rang 2 sur $\Sigma = \{x,y,z\}$ admet, à un automorphisme de Σ^* près, une unique solution incontractable $\phi : \Sigma^* \to \{a,b\}^*$ telle que $S = \phi^{-1} \circ \phi$.

Si $\chi = (u,v,w)$ est une caractéristique de S et si $P = ((uv)^p u,(wv)^q w)$ est l'équation générique de S avec $p \geq 0$, $q \geq 0$ et pgcd $(p+1,q+1) = 1$ alors la partie $C = \phi(\Sigma)$ est

si $\tau = 1$ et $\chi = (x,1,z)$ alors $C = \{a^{q+1},b,a^{p+1}\}$ avec $p>0$ et $q>0$;

si $\tau = 2$ et $\chi = (x^i yx^j,x^{k-i-j},z)$ avec $0 \leq i$, $0 \leq j$, $0<k$ et $i+j \leq k$ alors

$C = \{a,(ba^k)^q b,\ a^i(ba^k)^p ba^j\}$ avec $p>0$ et $q>0$;

si $\tau = 3$ et $\chi = (x^{i+j-k},x^{k-j}yx^{k-i},z)$ avec $0<i<k$, $0<j<k$ et $k<i+j$ alors

$C = \{a,(ba^k)^q b,\ a^i(ba^k)^{p-q-1} ba^j\}$ avec $p>q>0$;

si $\tau = 4$ et $\chi = (x^{i+j-k},x^{k-j}yx^{k-i},w)$ avec $0<i \leq k$, $0<j \leq k$, $k<i+j$ et

$w = zx^{k_1}z \ldots \ldots zx^{k_n}z$ où $n \geq 1$, $k_1 \geq 0, \ldots, k_n \geq 0$ alors

$C = \{a,(b'a^k)^q b',\ a^i ba^j\}$ avec $b' = ba^{k_1+i+j} b \ldots \ldots ba^{k_n+i+j} b$, $q>0$ et $p = q+1$;

si $\tau = 5$ et $\chi = (x^i y,x^m,zx^{j'})$ avec $0<i$, $0<j'$ et $0 \leq m$ alors

$C = \{a,(ba^{i+j'})^q ba^{j'},\ a^i b(a^{i+j'+m}b)^p\}$ avec $p>0$ et $q>0$;

si $\tau = 6$ et $\chi = (x^i yx^j,1,w)$ où w est un mot primitif de la forme

$w = x^{i'} zx^{k_1}z \ldots \ldots zx^{k_n}zx^{j'}$ avec $0 \leq i$, $0 \leq j$, $0 \leq i'$, $0 \leq j'$, $n \geq 1$,

$0 \leq k_1, \ldots, 0 \leq k_n$ et $ii' = jj' = 0$ alors

$C = \{a,[a^{i'} ba^{k_1+i+j} b \ldots \ldots ba^{k_n+i+j} ba^{j'}]^q,\ a^i ba^j\}$ avec $p = 0$ et $q>0$;

si $\tau = 7$ et $\chi = (x,1,1)$ alors $C = \{1,a,b\}$.

Si $\tau = 8$, $\chi = (x^i y,zx^{j'},x^r)$ et $P = (x^r(x^i y)^{p+1},\ (zx^{j'})^{q+1}x^r)$ avec $i \geq 0$,

$j' \geq 0$, $r>0$, $p>0$, $q>0$ et pgcd $(p+1,q+1) = 1$ alors

$C = \{a,(ba^{i+j'+r})^q ba^{j'+r},a^{i+r}b(a^{i+j'+r}b)^p\}$.

L.G. Budkina et Al.A. Markov [BM 73] ont caractérisé les parties incontractables de cardinalité 3 dans le monoïde libre $\{a,b\}^*$ et leur résultat a été retrouvé en [Spe 76b] par une méthode totalement indépendante. Dans tous les autres cas, $A = \{a,b\}$ est la plus petite partie génératrice de C^* et le résultat est trivial.

<u>Remarque 2.6.-</u> (i) L'unicité de la solution incontractable pour tout système entier de rang 2 sur $\Sigma = \{x,y,z\}$ est remarquable. En effet, si S est un système sur Σ qui admet 2 solutions ϕ et ψ telles que les monoïdes $\phi(\Sigma^*)$ et $\psi(\Sigma^*)$ ne soient pas isomorphes, S admet aussi 2 solutions incontractables non équivalentes.

Il en est ainsi même pour des systèmes non entiers de rang 2 sur $\{x,y,z\}$ comme par exemple un système réduit à l'équation équilibrée $Q = (uvw,wvu)$ associée à une caractéristique $\chi = (u,v,w)$ avec $u \neq 1$ et $w \neq 1$.

Il en est de même aussi pour les solutions de rang 1 d'un système entier de rang 2 sur $\{x,y,z\}$ d'après l'exemple qui suit.

Exemple : Si S est le système entier de rang 2 sur $\Sigma = \{x,y,z\}$ et d'équation générique $P = (xxyx,zzxzz)$ de l'exemple 2.3 , S admet les applications $\phi,\psi : \Sigma^* \to \{a\}^*$ telles que

$\phi(x) = a^2$, $\phi(y) = 1$ et $\phi(z) = a$

$\psi(x) = 1$, $\psi(y) = a^4$ et $\psi(z) = a$

comme solutions incontractables de rang 1 et non équivalentes.

(ii) Toute solution principale n'est pas incontractable comme on peut le voir sur l'exemple suivant.

Exemple : Si S est le système entier d'équation générique $P = (xyx,zz)$, l'application $\phi_o : \Sigma^* \to \{a,b\}^*$ telle que $\phi_o(x) = a$, $\phi_o(y) = baab$ et $\phi_o(z) = aba$ est l'unique solution incontractable de rang 2 de S d'après la proposition 2.6 mais, $\forall\ r \geq 0$, l'application ϕ_r : $\Sigma^* \to \{a,b\}^*$ telle que

$\phi_r(x) = (ab)^r a$, $\phi_r(y) = baab$ et $\phi_r(z) = (ab)^{r+1}a$ est une solution principale de S.

Proposition 2.7.- Tout système entier S de rang 2 et de type $\tau \neq 6$ sur $\{x,y,z\}$, admet une unique caractéristique.

Preuve. (i) Si S est de type 8 et si

$P = (e,e') = (x^r(x^iy)^{p+1}, (zx^{j'})^{q+1}x^r)$ est son équation générique avec $r>0$, $i\geq 0$, $j'\geq 0$, $p>0$, $q>0$ et

pgdc $(p+1,q+1)$, $\{x^{i+r}yx^iy,zx^{j'}z\} \subset Fg(\{e,e'\})$ et

$\{yx^iy,zx^{j'}zx^{j'+r}\} \subset Fd(\{e,e'\})$, ce qui est incompatible avec toute caractéristique de type $\tau \neq 8$ et avec toute caractéristique de type 8 distincte de $\chi = (x^ij,zx^{j'},x^r)$.

(ii) Si S est de type 7, son équation générique $P = (x,1)$ ne peut admettre que la caractéristique $\chi = (x,1,1)$.

(iii) Si S est de type $\tau \in \{1,2,3,4,5,6\}$, soient $\chi = (u,v,w)$ une caractéristique de S, $P = ((uv)^pu,(wv)^qw)$ l'équation générique de S avec $p\geq 0$, $q\geq 0$ et pgcd $(p+1,q+1)$ et $\chi' = (u',v',w')$ une caractéristique de type $\tau' \neq 8$ telle que

$(u'v')^{p'}u' = (uv)^pu$ avec $p'>0$

$(w'v')^{q'}w' = (wv)^qw$ avec $q'>0$.

Si $|u'|_y > 1$, $|u'| > |uv|$ et, comme uv est primitif et que $u' \in Fb((uv)^pu)$, il existe $r>0$ tel que $u' = (uv)^ru$ d'où $(uv)^{p-r} = (u'v')^{p'}$ et, comme les mots uv et u'v' sont primitifs $uv= u'v'$ ce qui est incompatible avec $r>0$.

Si $|u'|_y = 1$, $u' \in Fb((uv)^pu)$ implique $u' = u$ d'après la forme des caractéristiques données en 2.1. Si $p>0$, $(uv)^p = (u'v')^{p'}$ et, par suite, $uv = u'v'$ par primitivité. Il en résulte que $v = v'$ d'où

$(wv)^{q+1} = (w'v')^{q'+1}$ et $wv = w'v'$ par primitivité. Ceci implique $w' = w$ et $\chi' = \chi$.

Si $|u'|_y = 0$, $|(uv)^p u|_y > 1$ implique $|v'|_y > 1$ ce qui est contraire à la primitivité de u'v'.

Si $\tau \neq 6$, S admet donc une unique caractéristique.

3.- La dimension des sytèmes entiers d'équations sur $\Sigma = \{x,y,z\}$

Définition 3.1.- (i) Deux systèmes d'équations sur un même alphabet sont dits équivalents s'ils admettent les mêmes solutions.

(ii) Si Sol(S) est l'ensemble des solutions d'un système d'équations S sur l'alphabet Σ,

$$e(S) = \bigcap_{\phi \in Sol(S)} \phi^{-1} \circ \phi$$

est appelé la clôture équationnelle de S.

Les systèmes S et e(S) sont alors équivalents et, par suite, deux systèmes S et T sont équivalents si, et seulement si, e(S) = e(T).

(iii) Un système S tel que, pour toute partie stricte T de S, $e(T) \neq e(S)$ est dit indépendant et les équations de S sont alors dites indépendantes.

(iv) Le plus petit cardinal d'une partie indépendante de S équivalente à S est appelé la dimension du système d'équations S et est noté dim(S).

La résolution de la conjecture d'Ehrenfeucht par M.H. Albert et J. Lawrence [AL 85] et par K.S. Guba [Gub 85] (voir aussi [Per 85] et [Sal 85]) implique que, si l'alphabet Σ est fini, alors tout système d'équations sur Σ est équivalent à un sous-système fini (voir aussi [Kar84]).

Un algorithme déterminant un tel sous-système fini lorsque le système S est entier est donné en [Spe 85].

$c(S) \subseteq e(S)$ car, pour tout solution ϕ du système S, $\phi^{-1} \circ \phi$ est une congruence de Σ^* et, par suite, $c(S) \subseteq \phi^{-1} \circ \phi$.

Lemme 3.2.- Pour tout système entier S de rang 2 sur $\Sigma = \{x,y,z\}$ d'équation générique P, S = e(P).

Preuve.- Si S est de type $\tau \in \{6,7,8\}$ alors S = c(P) d'après le théorème 2.1 et S = e(P) d'après 3.1.

Si $\tau \in \{1,2,3,4,5\}$, S admetune caractéristique unique $\chi = (u,v,w)$ d'après la proposition 2.7 et $P = ((uv)^p u, (wv)^q w)$ avec $p>0$, $q>0$ et pgcd $(p+1,q+1) = 1$.

Si Q = (uvw,wvu), S = c(P,Q) d'après le théorème 2.1. Comme

$(uv)^{p+1}w \to (wv)^q wvw = wv(wv)^q w \to w(vu)^{p+1}$

est une suite de $\{P\}$-transitions élémentaires,
$R = ((uv)^{p+1}w, w(vu)^{p+1}) \in c(P,Q) = S$. Pour toute solution $\phi : \Sigma^* \to A^*$ de S, $[\phi(uv)]^{p+1}\phi(w) = \phi(w)[\phi(vu)]^{p+1}$ et il existe donc un couple unique (h_1,h_2) de mots de A^* et des entiers $m>0$ et $n\geq 0$ tels que $[\phi(uv)]^{p+1} = (h_1h_2)^m$, $[\phi(vu)]^{p+1} = (h_2h_1)^m$ et $\phi(w) = (h_1h_2)^n h_1$ avec h_1h_2 primitif. Comme chaque mot admet une unique racine primitive, il existe $r \in N$ tel que $m = (p+1)r$. Alors $\phi(uvw) = (h_1h_2)^{r+n}h_1 = \phi(wvu)$ et $Q \in \phi^{-1} \circ \phi$. Il en résulte que $Q \in e(P)$ d'où $S = c(P,Q) = e(P,Q) = e(P)$.

<u>Lemme 3.3.</u>- Pour tout système entier S de rang 1 sur $\Sigma = \{x,y,z\}$, il existe des équations indépendantes P_1 et P_2 telles que $S = e(P_1,P_2)$.

Preuve.- (i) Le système entier S admet une solution $\phi : \Sigma^* \to a^*$ telle que $S = \phi^{-1} \circ \phi$ et celle ci est définie par $(i,j,k) = (|\phi(x)|,|\phi(y)|,|\phi(z)|) \neq (0,0,0)$.

Si $ijk \neq 0$, il existe des entiers non nuls p_1, q_1 et p_2, q_2 tels que $P_1 = (x^{p_1},y^{q_1}) \in S$ avec $\text{pgcd}(p_1,q_1) = 1$ et $ip_1 = jq_1$ et $P_2 = (x^{p_2},z^{q_2}) \in S$ avec $\text{pgcd}(p_2,q_2) = 1$ et $ip_2 = kq_2$.

Si $ij \neq 0$ et $k = 0$, il existe encore $p_1 > 0$ et $q_1 > 0$ tels que $P_1 = (x^{p_1},y^{q_1}) \in S$ avec $\text{pgcd}(p_1,q_1) = 1$ et $ip_1 = jq_1$ et, en outre, $P_2 = (z,1) \in S$.

Si $i \neq 0$ et $j = k = 0$, $P_1 = (y,1) \in S$ et $P_2 = (z,1) \in S$.

Dans tous les cas les équations P_1 et P_2 sont de rang 2 et indépendantes.

(ii) Lorsque $ijk \neq 0$, pour toute solution $\psi : \Sigma^* \to A^*$ de $\{P_1,P_2\}$, $[\psi(x)]^{p_1} = [\psi(y)]^{q_1}$ et $[\psi(x)]^{p_2} = [\psi(z)]^{q_2}$ et, comme chaque mot admet une unique racine primitive, il existe $a \in A^+$ et des entiers i', j', k' tels que $\psi(x) = a^{i'}$, $\psi(y) = a^{j'}$ et $\psi(z) = a^{k'}$ avec $p_1i' = p_2j'$ et $p_2i' = q_2k$ et ceci est encore vrai dans les cas où $ijk = 0$.

Le rang de ψ est donc égal à 0 ou 1 et, lorsqu'il est égal à 1, (i,j,k) et (i',j',k') sont proportionnels ce qui prouve que ψ est aussi une solution de S et que $S = e(P_1,P_2)$.

<u>Théorème 3.4.</u>- Pour tout système entier S d'équations sur $\Sigma = \{x,y,z\}$, $\dim(S) + \text{rang}(S) \leq 3$.

Preuve.- L'unique système entier de rang 3 sur Σ est $S = \{(e,e)\}$; $e \in \Sigma^*\}$ et admet les mêmes solutions que $\emptyset$ d'où $\dim(S) = 0$.

Pour tout système entier S de rang 2 sur Σ, $S = e(P)$ où P est l'équation générique de S d'après le lemme 3.2 et, par suite, $\dim(S) = 1$.

Si S est un système entier de rang 1 sur Σ, il existe des équations indépendantes P_1 et P_2 telles que $S = e(P_1,P_2)$ d'après le lemme 3.3 et, par suite, $\dim(S) \leq 2$. Dans ce cas $\dim(S) \neq 0$ mais $\dim(S) = 1$ est possible comme on peut le voir avec l'exemple suivant : si $P_1 = (x,1)$, $P_2 = (y,1)$, et $R = (xy,1)$ alors $e(P_1,P_2) = e(R)$.

L'unique système entier de rang 0 est $S = \Sigma^* x \Sigma^*$ et admet une unique solution ϕ telle que $\phi(x) = \phi(y) = \phi(z) = 1$ et alors $S = e(P)$ avec $P=(xyz,1)$ d'où $\dim(S) = 1$.

Remarque.- La dimension d'un système d'équations S pourrait aussi être, par définition, le plus grand cardinal d'une partie indépendante de S équivalente à S.

Avec cette nouvelle définition nous conjecturons l'équivalent suivant du théorème 3.4 :

$\dim(S) + \text{rang}(S) = 3.$

REFERENCES

[AL 85] M.H. Albert and J. Lawrence, A proof of Ehrenfeucht's conjecture, T.C.S. 41 (1985), 121-123.

[BPPR 79] J. Berstel, D. Perrin, J.F. Perrot and A. Restivo, Sur le théorème du défaut, J. Algebra 60 (1979), 169-180.

[BM 73] L.G. Budkina and Al.A. Markov, On F-semigroups with three generators, Mat. Zamtki 14 (1973) 267-277 (en russe), Math. Notes 14 (1974) 711-717.

[Eil 74] S. Eilenberg, Automata, Languages and Machines, Vol. 1, Academic Press (1974).

[Gub 85] V.S. Guba (1985).

[Kar 84] J. Karhumäki, The Ehrenfeucht conjecture ; a compactness claim for finitely generated free monoids, T.C.S. 29 (1984), 285-308.

[Lal 79] G. Lallement, Semigroups and Combinatorial Applications, Wiley (1979).

[Len 72] A. Lentin, Equations dans le monoïde libre, Gauthier-Villars-Mouton (1972).

[LS 67] A. Lentin et M.P. Schützenberger, A combinatorial problem in the theory of free monoids, Combinatorial Mathematics, North Carolina Press, Chapel Hill (1967), 112-144.

[Lot 83] M. Lothaire, Combinatorics on words, Addison Wesley (1983).

[Pec 84] J.P. Pécuchet, Solutions principales et rang d'un système d'équations avec constantes dans le monoïde libre, Discrete Math. 48 (1984), 253-274.

[Per 85] On the solution of Ehrenfeucht's conjecture, Bull. of EATCS 27 (1985), 68-70.

[Sal 85] The Ehrenfeucht conjecture : A proof of language Theorists, Bull. of EATCS 27 (9185), 71-82.

[Spe 75] J.C. Spehner, Quelques constructions et algorithmes relatifs aux sous-monoïdes d'un monoïde libre, Semigroup Forum, 9 (1975), 334-353.

[Spe 76a] J.C. Spehner, On external conjugation of submonoids of a free monoid, Algebraic Theory of semigroups, Szeged (1976), vol. 20, Noth-Holland, 545-576.

[Spe 76b] J.C. Spehner, Quelques problèmes d'extension, de conjugaison et de présentation des sous-monoïdes d'un monoïde libre, Thèse Paris (1976).

[Spe 78] J.C. Spehner, Les présentations des sous-monoïdes de rang 3 d'un monoïde libre, Lecture Notes in Math. 855, Semigroups (1978), 116-155.

[Spe 85] J.C. Spehner, Systèmes entiers d'équations sur un alphabet fini et conjecture d'Ehrenfeucht, T.C.S. 39 (1985), 171-188.

Jean-Claude SPEHNER

Université de Haute Alsace
4, rue des Frères Lumière
68093 MULHOUSE Cédex
FRANCE

Annexe : Les présentations des sous-monoïdes A^* de monoïde libre avec card $\mathring{A} = 3$

Le théorème de [Spe 78] qui donne toutes les présentations des sous-monoïdes A^* de monoîde libre avec card $\mathring{A} = 3$ comporte une erreur. La forme de présentation (VI) y était donnée avec $\tau = \{(\alpha^t\beta^r, \gamma^s\alpha^t)\}$ au lieu de $\tau = \{(\alpha^t(\alpha^i\beta)^r, (\gamma\alpha^j)^s\alpha^t\}$.
Le théorème rectifié est :

Théorème. - Tout sous-monoïde non libre A^* de monoïde libre avec card $\mathring{A} = 3$ admet une présentation (Σ,τ) où, à une permutation de $\Sigma = \{\alpha,\beta,\gamma\}$ près, τ est de l'une des formes suivantes :

(I) $\tau = \{(\alpha\beta,\beta\alpha),(\beta\gamma,\gamma\beta),(\alpha\gamma,\gamma\alpha),(\alpha^{r_1},\beta^{s_1}\gamma^{t_1}),(\beta^{r_2},\alpha^{s_2}\gamma^{t_2}),$
$(\gamma^{r_3},\alpha^{s_3}\beta^{t_3})\}$ avec, pour tout i de $\{1,2,3\}$, $r_i > 1$ et p.g.c.d. $(r_i,s_i,t_i) = 1$;

(II) $\tau = \{(\alpha\gamma,\gamma\alpha), (\alpha^r,\gamma^s)\}$
avec $r > 1$, $s > 1$ et p.g.c.d. $(r,s) = 1$;

(III) $\tau = \{(\alpha^r\gamma,\beta\omega)\}$
avec $r \geqslant 1$ et ω dans $\{\alpha,\beta\}^+$;

(IV) $\tau = \{(\gamma\alpha^r,\omega\beta)\}$
avec $r \geqslant 1$ et ω dans $\{\alpha,\beta\}^+$;

(V) $\tau = \{\alpha^i\gamma\alpha^j,\beta\omega\beta)\}$
avec $i > 1$, $j > 1$ et ω dans $\{\alpha,\beta\}^*$;

(VI) $\tau = \{(\alpha^t(\alpha^i\beta)^r,(\gamma\alpha^j)^s\alpha^t)\}$
avec $r \geqslant 1$, $s \geqslant 1$, $t \geqslant 1$, p.g.c.d. $(r,s) = 1$, $i \geqslant 0$ et $j \geqslant 0$;

(VII) $\tau = \{(\alpha^i\beta\alpha^{i'}\gamma,\gamma\alpha^{j'}\beta\alpha^j), ([\alpha^i\beta\alpha^{i'}]^p\alpha^i\beta\alpha^j,[\gamma\alpha^{i'-j}]^q\gamma)\}$
avec $i+i' = j+j' > 1$, $i' > j$, $p > 1$, $q > 1$ et
p.g.c.d. $(p+1,q+1) = 1$;

(VIII) $\tau = \{(\alpha^i\beta\alpha^{i'}\gamma,\gamma\alpha^{j'}\beta\alpha^j), ([\alpha^i\beta\alpha^{i'}]^p\alpha^{j-i'},[\gamma\alpha^{j'}\beta\alpha^{i'}]^q\gamma)\}$
avec $i+i' = j+j' \geqslant 1$, $j > i'$, $p > q \geqslant 1$ et
p.g.c.d. $(p+1,q+1) = 1$;

(IX) $\tau = \{(\beta\alpha^{i+j+k}\gamma,\alpha^k\gamma\alpha^j\beta\alpha^i), ([\beta\alpha^{i+j}]^p\beta\alpha^i,[\alpha^k\gamma\alpha^j]^q\alpha^k\gamma)\}$
avec $i \geqslant 1$, $j \geqslant 0$, $k \geqslant 1$, $p \geqslant 1$, $q \geqslant 1$ et
p.g.c.d. $(p+1,q+1) = 1$;

(X) $\tau = \{(\alpha^i\beta\alpha^{i'}\omega,\omega\alpha^{j'}\beta\alpha^j), ([\alpha^i\beta\alpha^{i'}]^{p+1}\alpha^{j-i'},[\omega\alpha^{j'}\beta\alpha^{i'}]^p\omega)\}$
avec $i+i' = j+j' \geqslant 1$, $i' < j$, $p \geqslant 1$ et ω dans $\gamma\{\alpha,\gamma\}^*\gamma$.

Nous nous limitons aux additifs et rectificatifs nécessaires pour justifier le théorème et nous gardons les notations de [Spe 78] : $\Sigma = \{\alpha,\beta,\gamma\}$, $A = \phi(\Sigma) = \{a,b,c\}$ avec $a = \phi(\alpha)$, $b = \phi(\beta)$ et $c = \phi(\gamma)$.

1.- Additif à "Cas où $\{a,b,c\}^*$ admet un "bloc" (page 148).-

Proposition 5.8.- Si $A = \{a,b,c\}$ est tel que $(ca^j)^r a^t = a^t (a^i b)^s$ avec $i \geq 0$, $j \geq 0$, $i+j > 0$, $r > 1$, $s > 1$, pgcd $(r,s) = 1$ et $t \geq 1$, $A = \overset{\circ}{A}$ et A^* admet la présentation

$(\{\alpha,\beta,\gamma\}, \{((\gamma\alpha^j)^r \alpha^t, \alpha^t(\alpha^i\gamma)^s)\})$.

Preuve. (i) Comme a^t est un facteur de conjugaison de $((ca^j)^r, (a^i b)^s)$, il existe des mots e et f de X^+ et des entiers $m > 0$, $n > 0$ et $k \geq 0$ tels que $ca^j = (ef)^m$, $a^i b = (fe)^n$ et $a^t = (ef)^k e$ avec ef primitif et $mr = ns = $ ppcm (m,n).

Comme $ef \neq fe$, $\max(i,j) \geq t$, implique $k = 0$. Si $t > 1$, $k \leq 1$ d'après le théorème 3.2. Si $t > 1$ et $k = 1$, il existe un mot g de X^+ tel que $f = (gee)^{t-1} g$, $a = ege$ et $ca^j = (ef)^m$ [resp. $a^i b = (fe)^n$] implique $eg = ge$ ce qui est contradictoire avec $ef \neq fe$. Il en résulte que $k = 0$.

Soient $c' = ca^j$, $A' = \{a,b,c'\}$, $\Sigma' = \{\alpha, \beta, \gamma'\}$ et les homomorphismes $\phi : \Sigma^* \to A^*$, $\phi' : \Sigma'^* \to A'^*$ et $\pi : \Sigma'^* \to \Sigma^*$ tels que $\phi(\alpha) = a$, $\phi(\beta) = b$, $\phi(\gamma) = c$, $\phi'(\alpha) = a$, $\phi'(\beta) = b$, $\phi'(\gamma') = c'$, $\pi(\alpha) = \alpha$, $\pi(\beta) = \beta$ et $\pi(\gamma') = \gamma\alpha^j$.

(ii) Lorsque $(i+j)|a| > |f|$, il existe des mots a_1, a_2 et a_3 de X^+ tels que $a = a_1 a_2 = a_2 a_3$ et $f = a^i a_3 a^{j-1} = a^{i-1} a_1 a^j$. Si $C = \{a, a_1, a_3\}$ et si $C' = \{a', a'f', f'a'\}$ où $a'f' \neq f'a'$, il existe un isomorphisme θ de C^* sur C'^* tel que $\theta(a) = a'$, $\theta(a_1) = a'f'$ et $\theta(a_3) = f'a'$ d'après le lemme 3.7 et, comme $A \subset C^*$, les monoïdes A^* et $\theta(A^*)$ sont isomorphes. Or $\theta(f) = a'^i f' a'^j$ avec $(i+j)|\theta(a)| < |\theta(f)|$. Il suffit donc d'étudier le cas où $(i+j)|a| < |f|$.

(iii) Si $(i+j)|a| < (f)$, il existe $g \in X^+$ tel que $f = a^i g a^j$, $b = ga^{j+t}(a^i g a^{j+t})^{n-1}$ et $c = (a^{i+t} g a^j)^{m-1} a^{i+t} g$ avec $ag \neq ga$ puisque $ef \neq fe$. Alors $A = \overset{\circ}{A}$ et $L(A) \subset \{a,g\}^+$ d'après la remarque 3.6.

Soient $L_o = \{a^s \,;\, 0 < s \leq r + \min(i,j)\}$, $L_1 = F\gamma(c) \cap F\delta(b) \setminus L_o$ et $L_2 = F\gamma(b) \cap F\delta(c)$. $\forall\, x \in F\beta(b)$ [resp. $x \in F\beta(c)$], x admet un unique A-successeur y et $y \in F\beta(b)$ [resp. A-précédesseur y et $y \in F\beta(c)$] et, par suite, $x \notin L(A)$ d'où $L(A) \subset L_o \cup L_1 \cup L_2$. Il en résulte aussi que, $\forall\, x \in L_o \cup L_1$ [resp. $x \in L_2$], tous les A-successeurs et A-prédécesseurs de x appartiennent à L_2 [resp. $L_o \cup L_1$].

$\forall\, x \in L_1$, $x' = x \in F\gamma(c') \cap F\delta(b)$ et, $\forall\, x \in L_2$, $x' = xa^j \in F\gamma(b) \cap F\delta(c)$. En outre $1' = 1$. Si (z,λ) est un A-motif produit par un A-couple (x,y) vérifiant $x,y \in L_1 \cup L_2 \cup \{1\}$, (x',y') est un A'-couple qui produit un A'-motif (z',λ') tel que $z' = z$ lorsque $z \neq \gamma$ et $z' = \gamma'$ sinon et $\pi(\lambda') = \lambda$ lorsque $x' = x$ et $\alpha^j \pi(\lambda') = \lambda$ sinon.

Pour tout relateur indéductible R de A^*, le circuit unitaire $\sigma = (1, x_1, \ldots, x_p, 1)$ de $\mathcal{L}(A)$ qui produit R ne contient aucun sommet de

L_o d'après la remarque 1.9. $\sigma' = (1,x'_1,\dots,x'_p,1)$ est donc un circuit unitaire de $\mathcal{L}(A')$ qui produit un relateur R' de A'^* tel que $\pi(R') = R$. Ceci prouve que $\omega = \gamma\alpha^j$ est un bloc pour A^* relativement à γ.

(iv) Lorsque i = 0 et j > 0, A'^* admet la présentation $(\{\alpha,\beta,\gamma'\}, \{(\gamma'^r\alpha^t,\alpha^t\beta^s)\})$ d'après le lemme 3.7 lorsque t = 1 et d'après la proposition 5.3 lorsque t > 1. D'après le lemme 5.2 , il en résulte que A^* admet alors la présentation $(\Sigma, \{((\gamma\alpha^j)^r\alpha^t,\alpha^t\beta^s)\})$.

En échangeant b et c et en passant à l'image en miroir, il en résulte que, lorsque i > 0 et j = 0, A^* admet la présentation $(\Sigma,\{(\gamma^r\alpha^t,\alpha^t(\alpha^i\beta)^s)\})$.

Lorsque i > 0 et j > 0, A'^* admet la présentation $(\Sigma', \{(\gamma'^r\alpha^t,\alpha^t(\alpha^i\beta)^s)\})$ d'après ce qui précède et, d'après le lemme 5.2 , il en résulte que A^* admet la présentation $(\Sigma,\{((\gamma\alpha^j)^r\alpha^t,\alpha^t(\alpha^i\beta)^s)\})$.

Exemple.- Si X = {x,y} et si A = {a,b,c} avec a = x, b = yxxxyxx et $c = xxy(xxxy)^2$, $c' = (xxyx)^3$ et, si A' = {a,b,c'}, les relateurs indéductibles $R = (\alpha(\alpha\beta)^3,(\gamma\alpha)^2\alpha)$ de A^* et $R' = (\alpha(\alpha\beta)^3,\gamma'^2\alpha)$ de A'^* sont tels que $\pi(R') = R$.

2.- Rectificatif du paragraphe intitulé "Cas où A^* admet une présentation de forme (VI)" (page 151)

Le paragraphe suivant lui est substitué.

Si $\Sigma = \{z_1,z_2,z_3\}$ et si $R_1 = (z_1^t(z_1^iz_2)^r, (z_3z_1^j)^sz_1^t)$ avec $i \geqslant 0$, $j \geqslant 0$, $r > 0$, $s > 0$, $t > 0$ et pgcd (r,s) = 1, tout relateur irréductible de A^* est de la forme $(z_1^t\omega_1,(z_3z_1^j)^{ns}z_1\omega_2)$ [resp. $(\omega'_2z_1(z_1^iz_2)^{nr},\omega'_1z_1^t)$] d'après le lemme 3.7 et les propositions 5.3 et 5.8. Le relateur R_1 doit donc vérifier la condition (r) et, après élimination des cas particuliers r = 1 et s = 1 déjà traitées, il reste uniquement le cas où a^t est un facteur de conjugaison de $((ca^j)^s, (a^ib)^r)$ avec j > 0 et, dans ce cas, a_o^{t+1} est un facteur de conjugaison de $((c_oa_o^{j-1})^s, (a_o^ib_o)^r)$ et A_o^* admet aussi une présentation de la forme (VI).

3.- Additif au paragraphe intitulé "Cas où A^* admet une présentation de la forme (IX)" (page 153)

Le cas suivant est ajouté aux 3 cas trouvés.

(4) a^kb est un facteur de conjugaison de (ca^i,ca^i) et j = 0 et alors, comme $(ca^i)^{p+1} = (a^kb)^{q+1}$ avec i > 0, $(c_oa_o^{i+1})^{p-1}a_o = a_o(a_o^kb_o)^{q+1}$ et A_o^* admet une présentation de la forme (VI).

A NEW INTERPRETATION OF FREE ORTHODOX AND GENERALIZED INVERSE *-SEMIGROUPS

Mária B. Szendrei

József Attila University
Bolyai Institute

H-6720 Szeged, Aradi vértanuk tere 1, Hungary

INTRODUCTION

The structure of free generalized inverse *-semigroups was described by Scheiblich in [9] by making use of the structure theorem of generalized inverse semigroups due to Yamada [11]. The author solved the word problem of free orthodox *-semigroups and gave a representation of free orthodox *-semigroups in [10]. The only thing common in these descriptions is that both are based on Scheiblich's well-known description [8] (cf. also [7] VIII.1) of free inverse semigroups. The aim of this note is to unify these results by presenting a free orthodox *-semigroup and a free generalized inverse *-semigroup as a subsemigroup in a semidirect product of a free band by a free group and of a free normal band by a free group, respectively. These presentations are analogous to the P-representation of a free inverse semigroup since the latter is just a presentation of a free inverse semigroup as a subsemigroup in a semidirect product of a free semilattice by a free group.

1. PRELIMINARIES

The notion of a regular *-semigroup was introduced by Nordahl and Scheiblich in [5] as follows. A semigroup with involution $(S;\cdot,^*)$ is called, for brevity, a **-semigroup*. Throughout the paper, *-semigroups are considered as algebras with two operations, a multiplication and a *-operation. The notions of subsemigroup, homomorphism and congruence are used in this sense. For example, by a subsemigroup of a *-semigroup we mean a subset closed with respect to both the multiplication and the

*-operation. By a *regular *-semigroup* we mean a *-semigroup satisfying the identity

$$xx^*x = x.$$

They proved that a regular *-semigroup is orthodox (that is, the idempotents form a subband) if and only if it satisfies the identity

$$(xx^*yy^*zz^*)^2 = xx^*yy^*zz^*.$$

A regular *-semigroup which is orthodox is termed an *orthodox *-semigroup*. Thus, the class of orthodox *-semigroups forms a variety.

It is well known (cf. [7] XII.1) that a regular *-semigroup is an inverse semigroup if and only if it satisfies the identity

$$xx^*x^*x = x^*xxx^*.$$

Adair [1] proved that an orthodox *-semigroup is a generalized inverse semigroup (that is, the idempotents form a normal band) if and only if it satisfies the identity

$$y(xx^*x^*x)z = y(x^*xxx^*)z.$$

A regular *-semigroup which is a generalized inverse semigroup is called a *generalized inverse *-semigroup*. So generalized inverse *-semigroups also constitute a variety.

A regular *-semigroup which is idempotent is termed a **-band*. Since the band of idempotents of an orthodox *-semigroup is closed under the *-operation it forms a *-band. Note that the lattice of all varieties of *-bands was described by Adair [2]. In the sequel, we denote the variety of all *-bands, that of all normal *-bands and that of all *-semilattices by $\mathcal{B}$, $\mathcal{NB}$ and $\mathcal{S}$, respectively. If $\mathcal{V}$ is one of these varieties then denote by $\mathcal{OV}$ the class of all orthodox *-semigroups which have a band of idempotents belonging to $\mathcal{V}$. Clearly, $\mathcal{OB}$, $\mathcal{ONB}$ and $\mathcal{OS}$ are the variety of all orthodox *-semigroups, that of all generalized inverse *-semigroups and that of all inverse semigroups, respectively.

Given a variety $\mathcal{V}$ of *-semigroups, the free object in $\mathcal{V}$ on a non-empty set X will be denoted by $F\mathcal{V}_X$. It is well known that $F\mathcal{V}_X$ is isomorphic to $F_X/\varepsilon_{\mathcal{V}}$ where F_X is the free *-semigroup on the set X and $\varepsilon_{\mathcal{V}}$ is the fully invariant congruence on F_X corresponding to the variety $\mathcal{V}$. For the construction of the free *-semigroup F_X we refer to [7] I. 10. 5. Throughout the paper the involution will be denoted by * except in case of groups, when the *-operation is the inversion, we make use of the usual sign $^{-1}$.

We say that the *word problem for* $F\mathcal{V}_X$ *is solvable* if there exists an algorithm for deciding whether or not $u \varepsilon_V v$ holds for any words u,v in F_X.

The reader is supposed to be familiar with both Scheiblich's and Munn's solution of the word problem for free inverse semigroups and the representations of free inverse semigroups obtained from these solutions (cf. [7] VIII. 1 and 3). The terminology and notations of [7] VIII. 1 and 3 will be used without any reference.

The description of the free generalized inverse *-semigroup on a set X due to Scheiblich [9] can be formulated in the following way. Consider the set

$$S_X = \{(x,A,g,y) \in (X \cup X^*) \times (E \setminus \{\{1\}\}) \times G_X \times (X \cup X^*) : \\ x, g \in A \text{ and } y \in g^{-1}A\}.$$

Define a multiplication and a *-operation on S_X such that, for any $(x,A,g,y),(x',B,h,y') \in S_X$, we have

$$(x,A,g,y)(x',B,h,y') = (x,A \cup gB,gh,y')$$

and

$$(x,A,g,y)^* = (y,g^{-1}A,g^{-1},x).$$

Result 1.1: The pair (S_X,f), *where* $f\colon X \to S_X$ *assigns* $(x,\hat{x},x,x^{-1})$ *to* x *for every* x *in* X, *is a free generalized inverse *-semigroup on* X.

The solution of the word problem for free orthodox *-semigroups published in [10] is based on a one-to-one correspondence between the elements of free *-semigroups and the spanning walks on birooted word trees. The notion of the Cayley graph of free groups (cf. Margolis and Meakin [4]) combines the advantages of Scheiblich's and Munn's approach to the solution of the word problem for free inverse semigroups. Now we restate the results of [10] needed in the sequel by making use of this notion.

Let X be a non-empty set. The Cayley graph of the free group G_X is the graph $\mathcal{G}_X = (G_X,E_X)$ with set of vertices G_X and with set of edges

$$E_X = \{\langle g,h\rangle \in G_X \times G_X : \text{there exists } x \in X \text{ with } gx = h\}.$$

If $\langle g,h\rangle \in E_X$ then we call g and h the *ends* of the edge $\langle g,h\rangle$. We label each edge $\langle g,h\rangle \in E_X$ with the unique $x \in X$ satisfying $gx = h$. Then each finite connected subgraph in $\mathcal{G}_X$ is a word tree and, conversely, each rooted word tree is isomorphic to a unique finite connected subgraph in $\mathcal{G}_X$ with a prescribed distinguished element. On the other hand, the set of vertices of each finite connected subgraph in $\mathcal{G}_X$ is an element of $\mathcal{X}$ and, conversely, each element of $\mathcal{X}$ is the set of vertices of a unique finite connected subgraph in G_X.

If Γ is a walk on a finite subgraph of G_X then we will simply say that Γ is a walk on G_X. Let W_X be the set of all walks on G_X. We define a partial multiplication $\circ$ and an involution ^{op} on W_X and an action of G_X on W_X as follows. For any $\Gamma = (\gamma_0,\gamma_1,\ldots,\gamma_m)$ and $\Delta = (\delta_0,\delta_1,\ldots,\delta_n)$ in W_X and for any $g\in G_X$, let

$$\Gamma\circ\Delta = \begin{cases} (\gamma_0,\gamma_1,\ldots,\gamma_m,\delta_1,\ldots,\delta_n) & \text{if } \gamma_m = \delta_0, \\ \text{undefined} & \text{otherwise,} \end{cases}$$

$$\Gamma^{op} = (\gamma_m,\gamma_{m-1},\ldots,\gamma_0)$$

and

$$g\Gamma = (g\gamma_0,g\gamma_1,\ldots,g\gamma_m).$$

Let

$$B_X = \{(\alpha,T,\beta): T \text{ is a finite connected subgraph in } G_X \text{ and } \alpha,\beta\in V(T)\}.$$

For any walk Γ in W_X, denote by $G_X(\Gamma)$ the finite connected subgraph in G_X spanned by Γ. Define a mapping $\sigma\colon W_X \to B_X$ by $\Gamma\sigma = (\alpha,G_X(\Gamma),\beta)$ for every (α,β)-walk Γ. In harmony with the preceding operations on W_X, we can define a partial multiplication $\circ$ and an involution ^{op} on B_X and an action of G_X on B_X in such a way that σ becomes a homomorphism with respect to these operations. Namely, if $(\alpha,T,\beta),(\alpha',T',\beta')\in B_X$ and $g\in G_X$ then

$$(\alpha,T,\beta)\circ(\alpha',T',\beta') = \begin{cases} (\alpha,T\cup T',\beta') & \text{if } \beta = \alpha', \\ \text{undefined} & \text{otherwise,} \end{cases}$$

$$(\alpha,T,\beta)^{op} = (\beta,T,\alpha)$$

and

$$g(\alpha,T,\beta) = (g\alpha,gT,g\beta)$$

where $T\cup T'$ is the subgraph $(V(T)\cup V(T'),E(T)\cup E(T'))$ and $gT = (\{gh\colon h\in V(T)\}, \{\langle gh_1,gh_2\rangle : \langle h_1,h_2\rangle\in E(T)\})$. Here $E(T)$ is used to denote the set of edges of the graph T.

The reason of these definitions is in the following correspondence between the words in F_X and certain walks in W_X. To every word u in F_X, assign the walk $\Gamma(u) = (1,r(x_1),r(x_1x_2),\ldots,r(x_1x_2\ldots x_n))$ provided $u = x_1x_2\ldots x_n$ ($x_i\in X\cup X^*$ for $i = 1,2,\ldots,n$). Clearly, the mapping $F_X \to W_X$, $u \mapsto \Gamma(u)$ is injective. Moreover, for any two words $u,v\in F_X$, we have

$$\Gamma(uv) = \Gamma(u)\circ r(u)\Gamma(v)$$

and

$$\Gamma(u^*) = (r(u))^{-1}(\Gamma(u))^{op}.$$

Consider the set W_X^1 of all (1,g)-walks ($g\in G_X$) on G_X and define operations $\cdot$ and $*$ as follows. For every $\Gamma,\Delta\in W_X^1$ where Γ is a (1,g)-walk, let

$$\Gamma\cdot\Delta = \Gamma o g\Delta$$

and

$$\Gamma^* = g^{-1}\Gamma^{op}.$$

Result 1.2: The mapping ω: $F_X \to W_X = (W_X^1;\cdot,*)$ *defined by* $u\omega = \Gamma(u)$ ($u\in F_X$) *is an isomorphism.*

By making use of the homomorphism σ, the operations $\cdot$ and $*$ on W_X^1 naturally induce operations $\cdot$ and $*$ on the set $B_X^1 = \{(T,\beta):(1,T,\beta)\in B_X\}$.

In this terminology, free inverse semigroups can be described as follows.

Result 1.3 (cf. [7] VIII. 3.8): *The mapping* β: $F_X \to B_X = (B_X^1;\cdot,*)$ *defined by* $u\beta = (G_X(\Gamma(u)),r(u))$ *is an onto homomorphism and its kernel is* $\varepsilon_{\mathcal{OS}}$. *Consequently, the pair* (B_X,f), *where* f: $X \to B_X$ *maps* $((\{1,x\},\{\langle 1,x\rangle\}),x)$ *to* x *for each* x *in* X, *is a free inverse semigroup on* X.

Denote the free semigroup on a non-empty set X by $\bar{F}_X$. In order to make difference between varieties of bands and $*$-bands, denote the variety of all bands, normal bands and semilattices by $\bar{\mathcal{B}}$, $\overline{\mathcal{NB}}$ and $\bar{\mathcal{S}}$, respectively. The fully invariant congruences on $\bar{F}_X$ corresponding to the varieties $\bar{\mathcal{B}}$, $\overline{\mathcal{NB}}$ and $\bar{\mathcal{S}}$ are denoted by $\varepsilon_{\bar{\mathcal{B}}}$, $\varepsilon_{\overline{\mathcal{NB}}}$ and $\varepsilon_{\bar{\mathcal{S}}}$, respectively.

Consider the walks in W_X as words in the alphabet G_X, the set of vertices of the Cayley graph, that is, as elements in $\bar{F}_{G_X}$. Then we can restate the solution of the word problem for free orthodox $*$-semigroups in the following way.

Result 1.4 ([10] Theorem 3.4): *For each pair of words* u,v *in* F_X, *we have* $u\,\varepsilon_{\mathcal{OB}}v$ *if and only if* $\Gamma(u)\varepsilon_{\bar{\mathcal{B}}}\Gamma(v)$ *in* $\bar{F}_{G_X}$.

For the description of the fully invariant congruence $\varepsilon_{\bar{\mathcal{B}}}$ and for the notations involved we refer to [3] IV. 4. Furthermore, denote the empty word in the alphabet $\{0,1\}$ by $\emptyset$. Put $P_0 = \{\emptyset\}$ and $\Sigma_k^0 = \Sigma_k\cup P_0$ and define $w(\emptyset)$ to be w for any word w. It is usual to call the first and the last letter of a word w the head and the tail of w, respectively, and to denote them by h(w) and t(w) (cf. [6]). Clearly, if $|C(w)| = k$ then $h(w) = \bar{w}(0^k)$ and $t(w) = \bar{w}(1^k)$.

The following statement characterizes those words in $\bar{F}_{G_X}$ which belong to the $\varepsilon_{\bar{\mathcal{B}}}$-class of a walk in W_X^1.

Result 1.5 ([10] Proposition 3.6): *For any sequence* Δ *of vertices of* G_X, *that is, for any* $\Delta\in\bar{F}_{G_X}$, *there exists a word* u *in* F_X *with* $\Gamma(u)\varepsilon_{\bar{\mathcal{B}}}\Delta$ *if and only if the following conditions are fulfilled with* $k = |C(\Delta)|$:

(i) $h(\Delta) = 1$;

(ii) $C(\Delta(\alpha)) \in X$ *for every* α *in* Σ^0_{k-1}; *and*

(iii) *both pairs of vertices* $\bar{\Delta}(\alpha 0)$, $t(\Delta(\alpha 0))$ *and* $\bar{\Delta}(\alpha 1)$, $h(\Delta(\alpha 1))$ *are adjacent in* G_X *for every* α *in* Σ^0_{k-2}.

The descriptions of the fully invariant congruences $\varepsilon_{\overline{S}}$ and $\varepsilon_{\overline{NB}}$ are well known. Since they are not formulated in [3] explicitely we draw up them in order to make later references easier.

Result 1.6 ([3] IV. proof of Theorem 5.6 and Exercise 14, or [6] II. Lemma 2.5 and Proposition 3.10)*: For a non-empty set* X *and for any* $u,v \in \overline{F}_X$, *we have*

(i) $u\ \varepsilon_{\overline{S}}\ v$ *if and only if* $C(u) = C(v)$;

(ii) $u\ \varepsilon_{\overline{NB}}\ v$ *if and only if* $C(u) = C(v)$, $h(u) = h(v)$ *and* $t(u) = t(v)$.

We will need the notion of a semidirect product of a *-band by a group. Let H be a group and B a *-band. We say that H *acts on* B *on the left* if an antihomomorphism α of H is given into the group of automorphisms of the *-band B. For any $h \in H$ and $b \in B$, we denote $b(h\alpha)$ simply by hb.

Suppose H is a group acting on the left on a *-band B. Define a multiplication and a unary operation * on B×H as follows. For every (b,h), $(c,k) \in B \times H$, let

$$(b,h)(c,k) = (b \cdot hc, hk)$$

and

$$(b,h)^* = (h^{-1}b^*, h^{-1}).$$

A straightforward calculation shows that $(B\times H;\cdot,^*)$ is an orthodox *-semigroup with band of idempotents isomorphic to B. $(B\times H;\cdot,^*)$ is termed the *semidirect product* of B by H and is denoted by B∗H.

2. NEW INTERPRETATION

Results 1.2 and 1.4 suggest that $FO\mathcal{B}_X$ can be described by means of a semidirect product of $F\overline{\mathcal{B}}_{G_X}$, the free band on the set G_X by the free group G_X. First of all, we introduce the construction needed.

Let $\overline{V}$ be a variety of bands. We say that $\overline{V}$ is *self-dual* if $B \in \overline{V}$ implies that the dual of B also belongs to $\overline{V}$. For example, $\overline{S}$, $\overline{NB}$ and $\overline{\mathcal{B}}$ are self-dual varieties of bands.

Let $\overline{V}$ be a self-dual variety of bands. For every non-empty set X, the dual $F\overline{V}_X^d$ of the free object $F\overline{V}_X$ belongs to $\overline{V}$. Therefore the identity mapping on X can be uniquely extended to an isomorphism α of $F\overline{V}_X$ into $F\overline{V}_X^d$. Clearly, α can be considered as an antiautomorphism of $F\overline{V}_X$ and α^2 is the identity automorphism of $F\overline{V}_X$. Thus α is an involution. In the sequel we will denote it by $*$ and the $*$-band $(F\overline{V}_X;\cdot,{}^*)$ will be denoted by $F\overline{V}_X^*$.

Let $\overline{V}$ be a variety of bands. If H is a group and A is a left H-system (cf. [3] VII. 2) then the left action of H on A can be trivially extended to a left action of H on $F\overline{V}_A$. If, moreover, $\overline{V}$ is self-dual then H acts on the $*$-band $F\overline{V}_A^*$ and one can consider the semidirect product $F\overline{V}_A^* * H$.

Now we intend to interpret $F\mathcal{OB}_X$ as a subsemigroup in the semidirect product $F\overline{B}_{G_X}^* * G_X$ where the definition of the semidirect product is based on the left G_X-system on itself.

Lemma 2.1: The mapping $\varphi\colon W_X \to F\overline{B}_{G_X}^* * G_X$ *defined by* $\Gamma\varphi = (\Gamma\varepsilon_{\overline{B}}, g)$ *provided* Γ *is a* $(1,g)$*-walk is a homomorphism.*

Proof: First we verify that φ is a homomorphism with respect to the multiplication. Assume that Γ is a $(1,g)$-walk and Δ is a $(1,h)$-walk. Then $\Gamma\cdot\Delta = \Gamma o g\Delta$ is clearly a $(1,gh)$-walk. So all we have to check is that $(\Gamma o g\Delta)\varepsilon_{\overline{B}} = (\Gamma\varepsilon_{\overline{B}})(g(\Delta\varepsilon_{\overline{B}}))$. Taking into consideration the definition of the action this equality holds if $\Gamma o g\Delta \; \varepsilon_{\overline{B}} \; (\Gamma)(g\Delta)$ where the latter means the product of the words Γ and $g\Delta$ in $\overline{F}_{G_X}$. However, this relation is valid since $\Gamma o g\Delta$ considered as a word in $\overline{F}_{G_X}$ is obtained from $(\Gamma)(g\Delta)$ by omitting the first letter of $g\Delta$ which is the same as the last letter in Γ.

Now we turn to proving that φ is a homomorphism with respect to the $*$-operation. Let Γ be a $(1,g)$-walk. Then $\Gamma^* = g^{-1}\Gamma^{op}$ is a $(1,g^{-1})$-walk. Moreover, we clearly have $(\Gamma\varepsilon_{\overline{B}})^* = \Gamma^{op}\varepsilon_{\overline{B}}$. Thus $(\Gamma\varphi)^* = (g^{-1}(\Gamma\varepsilon_{\overline{B}})^*, g^{-1}) = (g^{-1}(\Gamma^{op}\varepsilon_{\overline{B}}), g^{-1}) = ((g^{-1}\Gamma^{op})\varepsilon_{\overline{B}}, g^{-1}) = (\Gamma^*\varepsilon_{\overline{B}}, g^{-1}) = \Gamma^*\varphi$ which was to be proved.

Since ω is an isomorphism of F_X onto W_X by Result 1.2, the product $\kappa = \omega\varphi\colon F_X \to F\overline{B}_{G_X}^* * G_X$ is a homomorphism. Observe that, for every $u \in F_X$, the second component of $u\kappa$ is $r(u)$. Since $r(u) = r(v)$ necessarily holds when $u\varepsilon_{\mathcal{OB}}v$, Result 1.4 implies that the kernel of κ is just $\varepsilon_{\mathcal{OB}}$. Its range $F_X\kappa$ can be described by Result 1.5 as follows. For brevity, let us say that $\Delta \in \overline{F}_{G_X}$ is *connected* if it satisfies the conditions (ii) and (iii) in Result 1.5. Since $\Delta\varepsilon_{\overline{B}}\Delta'$ implies $\overline{\Delta}(\alpha) = \overline{\Delta'}(\alpha)$ for each $\alpha \in \Sigma_k$ and $C(\Delta(\alpha)) = C(\Delta'(\alpha))$ for each $\alpha \in \Sigma_{k-1}^0$ where $k = |C(\Delta)| = |C(\Delta')|$ we can talk in an unambiguous way about the connectedness of an element in $F\overline{B}^*_{G_X}$ and can make use of the notations h(b) and t(b) for any $b \in F\overline{B}^*_{G_X}$. Hence we infer by Result 1.5 and Lemma 2.1 that, for $(b,g) \in F\overline{B}^*_{G_X} * G_X$, we have $(b,g) \in F_X\kappa$ if and only if b is connected, $h(b) = 1$ and $t(b) = g$. Thus we obtained the following representation of free orthodox *-semigroups.

Let X be a non-empty set. Consider the subsemigroup

$$O_X = \{(b,g)\colon b \text{ is connected, } h(b) = 1,\ t(b) = g\}$$

in the semidirect product $F\overline{B}^*_{G_X} * G_X$.

Theorem 2.2: The pair (O_X, f), *where* $f\colon X \to O_X$ *assigns* $((1,x)\varepsilon_{\overline{B}}, x)$ *to* x *for every* x *in* X, *is a free orthodox* *-*semigroup on* X.

This representation of free orthodox *-semigroups is analogous to the P-representation of free inverse semigroups if we consider $P(E, G_X; X)$ as a subsemigroup in the semidirect product $F\overline{S}^*_{G_X} * G_X$ of the free semilattice $F\overline{S}^*_{G_X}$ by the free group G_X. (Note that the *-operation on $F\overline{S}^*_{G_X}$ is identical.) However, the analogous representation for free generalized inverse *-semigroups is false. In order to show it notice that Results 1.1 and 1.2 imply the following solution of the word problem for free generalized inverse *-semigroups.

Proposition 2.3: For each pair of words u,v *in* F_X, *we have* $u\ \varepsilon_{ONB}\ v$ *if and only if* $G_X(\Gamma(u)) = G_X(\Gamma(v))$, $r(u) = r(v)$ *and* $g_1 = h_1$, $g_{n-1} = h_{m-1}$ *where* $\Gamma(u) = (1, g_1, \ldots, g_{n-1}, r(u))$ *and* $\Gamma(v) = (1, h_1, \ldots, h_{m-1}, r(v))$.

Thus $\Gamma(u)\ \varepsilon_{\overline{NB}}\ \Gamma(v)$ does not ensure $u\ \varepsilon_{ONB}\ v$. Notice that $\Gamma(u)\varepsilon_{\overline{NB}}\Gamma(v)$ is equivalent just to $u\ \varepsilon_{OS}\ v$ (cf. Result 1.3). This shows that if we want to represent a free generalized inverse *-semigroup as a subsemigroup in a semidirect product of a free normal band by a free group then we have to modify the previous method.

The main point is to notice that the condition in Proposition 2.3 says that $G_X(\Gamma(u)) = G_X(\Gamma(v))$, $r(u) = r(v)$ and the first and, respectively, the last edges are equal in the walks $\Gamma(u)$ and $\Gamma(v)$. This gives the idea that we should consider the walks in W_X as series of edges, that is, as words in $\overline{F}_{E_X}$. More precisely, to any walk Γ in W_X, we assign a word $\hat{\Gamma}$ in $\overline{F}_{E_X}$ as follows. If $\Gamma = (\gamma_0, \gamma_1)$ is a walk of length 1 then either $e = \langle\gamma_0, \gamma_1\rangle$ or $e = \langle\gamma_1, \gamma_0\rangle$ is an edge in E_X. Let $\hat{\Gamma} = e$. If Γ is a walk of length at least 2 then Γ can be uniquely written in the form $\Gamma = \Gamma_1 \circ (\gamma, \delta)$ where either $e = \langle\gamma, \delta\rangle$ or $e = \langle\delta, \gamma\rangle$ is an edge in E_X. Then define $\hat{\Gamma} = \hat{\Gamma}_1 e$. For example, if $\Gamma = (1, x, 1, x^{-1}, 1, x)$ then $\hat{\Gamma} = \langle 1,x\rangle\langle 1,x\rangle \langle x^{-1},1\rangle\langle x^{-1},1\rangle\langle 1,x\rangle \in \overline{F}_{E_X}$. Note that the mapping $W_X \to \overline{F}_{E_X}$, $\Gamma \mapsto \hat{\Gamma}$ is not injective since, for example, $\widehat{(1,x)} = \langle 1,x\rangle = \widehat{(x,1)}$. However, if we restrict it to W_X^1 then it is easily seen to be injective.

For simplicity, denote $\widehat{\Gamma(u)}$ by $\hat{\Gamma}(u)$ for any $u \in F_X$. We can unify the solutions of the word problems for free orthodox *-semigroup, free generalized inverse *-semigroups and free inverse semigroups in the following way.

*Theorem 2.4: Let V be one of the varieties of *-bands S, NB and B. For each pair of words u,v in F_X, we have $u\ \varepsilon_{0V}\ v$ if and only if $r(u) = r(v)$ and $\hat{\Gamma}(u)\ \varepsilon_{\bar{V}}\ \hat{\Gamma}(v)$ in $\bar{F}_{E_X}$.*

Proof: For $V = S$ the statement immediately follows from Results 1.3 and 1.6(i) if we observe that $C(\Gamma(u))$ is just the set of edges in $G_X(\Gamma(u))$ and, for every walk Γ, $C(\Gamma)$ and $G_X(\Gamma)$ uniquely determine each other. The same observation together with that preceding the introduction of the mapping $\Gamma \mapsto \hat{\Gamma}$ shows by Result 1.6(ii) that our statement is valid in case $V = NB$, too.

Now let $V = B$. We have to verify that, for every $u,v \in F_X$, the relation $\Gamma(u)\varepsilon_{\bar{B}}\ \Gamma(v)$ holds in $\bar{F}_{G_X}$ if and only if $r(u) = r(v)$ and $\hat{\Gamma}(u)\ \varepsilon_{\bar{B}}\ \hat{\Gamma}(v)$ in $\bar{F}_{E_X}$. Since $r(u) = t(\Gamma(u))$ and $r(v) = t(\Gamma(v))$ it suffices to show that

(*) for any $\Gamma, \Delta \in W_X$, we have $\Gamma\ \varepsilon_{\bar{B}}\ \Delta$ in $\bar{F}_{G_X}$ if and only if $h(\Gamma) = h(\Delta)$, $t(\Gamma) = t(\Delta)$ and $\hat{\Gamma}\ \varepsilon_{\bar{B}}\ \hat{\Delta}$ in $\bar{F}_{E_X}$.

The following lemma finds connection between $\bar{\Gamma}(0)$ and $\bar{\hat{\Gamma}}(0)$ and, respectively, between $\Gamma(0)$ and $\hat{\Gamma}(0)$ for any walk Γ in W_X.

Lemma 2.5: Let $\Gamma \in W_X$.

(i) *$\bar{\Gamma}(0)$ is an extreme vertex in $G_X(\Gamma)$ and $\bar{\Gamma}(0)$ is one of the ends of the edge $\bar{\hat{\Gamma}}(0)$.*

(ii) *The other end of $\bar{\hat{\Gamma}}(0)$ is $t(\Gamma(0))$.*

(iii) $\widehat{\Gamma(0)} = \hat{\Gamma}(0)$.

Proof: Let $\Gamma = (\gamma_0,\gamma_1,\ldots,\gamma_m)$ and suppose that $\bar{\Gamma}(0) = \gamma_k$ and $\Gamma(0) = (\gamma_0,\gamma_1,\ldots,\gamma_{k-1})$, $k \le m$. Since $\gamma_k \notin C(\Gamma(0))$ we infer that the edge e with ends γ_{k-1} and γ_k does not belong to $C(\widehat{\Gamma(0)})$. If $f \in C(\hat{\Gamma})$ and $f \neq e$ then $f \in C(\widehat{\Gamma(0)})$ since $(\gamma_0,\gamma_1,\ldots,\gamma_k)$ spans $G_X(\Gamma)$. On the one hand, this implies that the only edge in $G_X(\Gamma)$ having γ_k as an end is e. On the other hand, we obtain that e is the edge making its first appearence last, and thus $e = \bar{\hat{\Gamma}}(0)$. Hence the statements of the lemma immediately follow.

Now we turn to proving the "only if" part of (*). First observe that $\Gamma\ \varepsilon_{\bar{B}}\ \Delta$ implies $h(\Gamma) = h(\Delta)$, $t(\Gamma) = t(\Delta)$ and $C(\Gamma) = C(\Delta)$. We will prove by induction on $n = |C(\Gamma)| = |C(\Delta)|$ that, for every $\Gamma,\Delta \in W_X$, the relation $\Gamma \,\varepsilon_{\overline{B}}\, \Delta$ implies $\hat{\Gamma} \,\varepsilon_{\overline{B}}\, \hat{\Delta}$. If $n = 2$ then $G_X(\Gamma) = G_X(\Delta)$ consists of a single edge e and hence $\hat{\Gamma} \,\varepsilon_{\overline{B}}\, e \,\varepsilon_{\overline{B}}\, \hat{\Delta}$. Assume that, for any $\Gamma',\Delta' \in W_X$, $\Gamma' \,\varepsilon_{\overline{B}}\, \Delta'$ implies $\hat{\Gamma}' \,\varepsilon_{\overline{B}}\, \hat{\Delta}'$ provided $2 \leq |C(\Gamma')| = |C(\Delta')| < n$. Suppose that $\Gamma,\Delta \in W_X$ with $\Gamma \,\varepsilon_{\overline{B}}\, \Delta$ and $|C(\Gamma)| = |C(\Delta)| = n$. Then $\bar{\Gamma}(i) = \bar{\Delta}(i)$ and $\Gamma(i) \,\varepsilon_{\overline{B}}\, \Delta(i)$ for $i = 0,1$. By Lemma 2.5(i) and its dual, the equality implies that

(1) $\qquad \bar{\hat{\Gamma}}(i) = \bar{\hat{\Delta}}(i)$ for $i = 0,1$.

Since $|C(\Gamma(i))| = |C(\Delta(i))| < n$ for $i = 0,1$, the induction hypothesis ensures the relation $\widehat{\Gamma(i)} \,\varepsilon_{\overline{B}}\, \widehat{\Delta(i)}$, $i = 0,1$. Hence we infer by Lemma 2.5(iii) and its dual that $\hat{\Gamma}(i) \,\varepsilon_{\overline{B}}\, \hat{\Delta}(i)$ for $i = 0,1$. By combining this relation with (1) we deduce that $\hat{\Gamma} \,\varepsilon_{\overline{B}}\, \hat{\Delta}$ which was to be proved.

The "if" part of (*) will be proved by induction on $n = |C(\hat{\Gamma})| = |C(\hat{\Delta})|$. If $n = 1$ then $G_X(\Gamma) = G_X(\Delta)$ has only two vertices and, since $h(\Gamma) = h(\Delta)$ and $t(\Gamma) = t(\Delta)$, the relation $\Gamma \,\varepsilon_{\overline{B}}\, \Delta$ immediately follows. Assume that, for every $\Gamma',\Delta' \in W_X$, the relation $\Gamma' \,\varepsilon_{\overline{B}}\, \Delta'$ holds provided $h(\Gamma') = h(\Delta')$, $t(\Gamma') = t(\Delta')$, $\hat{\Gamma}' \,\varepsilon_{\overline{B}}\, \hat{\Delta}'$ and $1 \leq |C(\hat{\Gamma}')| = |C(\hat{\Delta}')| < n$. Let $\Gamma,\Delta \in W_X$ such that $h(\Gamma) = h(\Delta)$, $t(\Gamma) = t(\Delta)$, $\hat{\Gamma} \,\varepsilon_{\overline{B}}\, \hat{\Delta}$ and $|C(\hat{\Gamma})| = |C(\hat{\Delta})| = n$. The relation $\hat{\Gamma} \,\varepsilon_{\overline{B}}\, \hat{\Delta}$ implies $\bar{\hat{\Gamma}}(i) = \bar{\hat{\Delta}}(i)$ and

(2) $\qquad \hat{\Gamma}(i) \,\varepsilon_{\overline{B}}\, \hat{\Delta}(i) \quad$ for $\quad i = 0,1$.

By Lemma 2.5(i) and its dual, we obtain from the equality that, for $i = 0,1$, each of $\bar{\Gamma}(i)$ and $\bar{\Delta}(i)$ is an extreme vertex which is an end of the edge $e_i = \bar{\hat{\Gamma}}(i) = \bar{\hat{\Delta}}(i)$. Since, by assumption, $G_X(\Gamma) = G_X(\Delta)$ has at least two edges, at most one of the ends of an edge is an extreme vertex. Hence

(3) $\qquad \bar{\Gamma}(i) = \bar{\Delta}(i) \qquad (i = 0,1)$.

This implies by Lemma 2.5(ii) and its dual that $t(\Gamma(0)) = t(\Delta(0))$ and $h(\Gamma(1)) = h(\Delta(1))$. Moreover, we clearly have $h(\Gamma(0)) = h(\Gamma) = h(\Delta) = h(\Delta(0))$ and, dually, $t(\Gamma(1)) = t(\Delta(1))$. By making use of Lemma 2.5(iii) and its dual, we see that $|C(\widehat{\Gamma(i)})| = |C(\widehat{\Delta(i)})| < n$ and, by (2), the relation $\widehat{\Gamma(i)} \,\varepsilon_{\overline{B}}\, \widehat{\Delta(i)}$ holds for $i = 0,1$. Then the induction hypothesis implies that $\Gamma(i) \,\varepsilon_{\overline{B}}\, \Delta(i)$ $(i = 0,1)$. By combining this relation with (3) we obtain that $\Gamma \,\varepsilon_{\overline{B}}\, \Delta$. The proof of the theorem is complete.

Similarly to the way as we applied Result 1.4 to interpret FOB_X as a subsemigroup in the semidirect product $F\overline{B}^*_{G_X} * G_X$, now we will make use of Theorem 2.4 to obtain a representation of FOV_X with V one of S, NB and B as a subsemigroup in the semidirect product $F\overline{V}^*_{E_X} * G_X$ where the definition of this semidirect product is based on the left G_X-system E_X determined by

$$h\langle\gamma,\delta\rangle = \langle h\gamma,h\delta\rangle$$

for every $h\in G_X$ and $\langle\gamma,\delta\rangle\in E_X$.

Before formulating the analogue of Lemma 2.1 we notice several simple properties of the mapping $W_X \to \bar{F}_{E_X}$, $\Gamma \mapsto \hat{\Gamma}$ which can be easily verified by definitions. In the second property the left action of G_X on $\bar{F}_{E_X}$ is that naturally determined by the left G_X-system E_X.

Lemma 2.6: For every $\Gamma,\Delta\in W_X$, *we have*

(i) $\widehat{\Gamma o \Delta} = \hat{\Gamma}\hat{\Delta}$ *provided* $\Gamma o \Delta$ *is defined in* W_X;

(ii) $\widehat{g\Gamma} = g\hat{\Gamma}$;

(iii) *if* $\hat{\Gamma} = e_1e_2\ldots e_n$ *then* $\widehat{\Gamma^{op}} = e_ne_{n-1}\ldots e_1$.

Let V be one of the varieties of $*$-bands S, NB and B.

Lemma 2.7: The mapping $\psi_V\colon W_X \to F\bar{V}^{*}_{E_X}*G_X$ *defined by* $\Gamma\psi_V = (\hat{\Gamma}\varepsilon_{\bar{V}},g)$ *for every* $(1,g)$*-walk* Γ *is a homomorphism.*

Proof: First we show that ψ_V is a homomorphism with respect to the multiplication. Let Γ be a $(1,g)$-walk and Δ a $(1,h)$-walk in W_X. Since $\Gamma\cdot\Delta = \Gamma o g\Delta$ is a $(1,gh)$-walk we have to prove only that $\widehat{\Gamma o g\Delta}\varepsilon_{\bar{V}} = (\hat{\Gamma}\varepsilon_{\bar{V}})(g(\hat{\Delta}\varepsilon_{\bar{V}}))$. However, this equality is valid since, by Lemma 2.6(i) and (ii), we have $\widehat{\Gamma o g\Delta} = (\hat{\Gamma})(g\hat{\Delta})$. As far as the $*$-operation is concerned, Lemma 2.6(iii) ensures that $\widehat{\Gamma^{op}}\varepsilon_{\bar{V}} = (\hat{\Gamma}\varepsilon_{\bar{V}})^*$. Thus we obtain by Lemma 2.6(ii) that $(\Gamma\psi_V)^* = (g^{-1}(\hat{\Gamma}\varepsilon_{\bar{V}})^*,g^{-1}) = (g^{-1}(\widehat{\Gamma^{op}}\varepsilon_{\bar{V}}),g^{-1}) = (\widehat{g^{-1}\Gamma^{op}}\varepsilon_{\bar{V}}, g^{-1}) = (\widehat{\Gamma^*}\varepsilon_{\bar{V}},g^{-1}) = \Gamma^*\psi_V$. The proof is complete.

Consider the product $\lambda_V = \omega\psi_V\colon F_X \to F\bar{V}^{*}_{E_X}*G_X$ of the isomorphism ω (cf. Result 1.2) and the homomorphism ψ_V. Clearly, λ_V is a homomorphism. Theorem 2.4 ensures that its kernel is ε_{0V}. Now we determine its range.

Let $w\in\bar{F}_{E_X}$ and $g,h\in G_X$. We say that w is an *S-connected* (g,h)*-word* if $C(w)$ is the set of edges of a finite connected subgraph T in G_X with $g,h\in V(T)$. If w has the properties that $C(w)$ is the set of edges of a finite connected subgraph in G_X and the vertices g and h are one of the ends of $h(w)$ and $t(w)$, respectively, then we term w an *NB-connected* (g,h)*-word*. Finally, let us call w a *B-connected* (g,h)*-word* if the following conditions are satisfied where $k = |C(w)|$: (i) for every α in Σ^0_{k-1}, $C(w(\alpha))$ is the set of edges of a finite connected subgraph in G_X, (ii) for every α in Σ^0_{k-2}, both pairs of edges $\bar{w}(\alpha 0),t(w(\alpha 0))$ and $\bar{w}(\alpha 1)$, $h(w(\alpha 1))$ have a common end, (iii) the vertices g and h are one of the ends of the edges $h(w)$ and $t(w)$, respectively. Observe that if V is one of the varieties of $*$-bands S, NB and B, $w,w'\in\bar{F}_{E_X}$ with $w\ \varepsilon_{\bar{V}}\ w'$, $g,h\in G_X$ and w is a V-connected (g,h)-word then w' is also a V-connected (g,h)-word. Therefore we can call the $\varepsilon_{\bar{V}}$-class of a V-connected (g,h)-word a *connected* (g,h)*-element* in $F\overline{V}_{E_X}$.

Lemma 2.8: Let V *be one of the varieties of *-bands* S, NB *and* B. *Let* $w \in \overline{F}_{E_X}$ *and* $g,h \in G_X$. *There exists a* (g,h)*-walk* Γ *on* G_X *with* $\hat{\Gamma} \varepsilon_{\overline{V}} w$ *if and only if* w *is a* V*-connected* (g,h)*-word.*

Proof: First let $V = S$ or $V = NB$. Necessity can be easily verified by checking that if Γ is a (g,h)-walk on G_X then Γ is a V-connected (g,h)-word. In order to prove sufficiency, consider a V-connected (g,h)-word w in $\overline{F}_{E_X}$. Since $C(w)$ is the set of edges of a finite connected subgraph T in G_X therefore, for each pair of vertices g',h' in $V(T)$, there exists a spanning (g',h')-walk $\Gamma_{g',h'}$ on T. Clearly, we have $\Gamma_{g,h} \varepsilon_{\overline{S}} w$ and, if $g''[h'']$ is the end of $h(w)[t(w)]$ different from $g[h]$ then the walk $\Gamma =$ $= (g,g'') \circ \Gamma_{g'',h''} \circ (h'',h)$ has the property that $\Gamma \varepsilon_{\overline{NB}} w$. This completes the proof in cases $V = S$ and $V = NB$.

Turn to the case $V = B$. First we prove by induction on $n = |C(\Gamma)|$ that if Γ is a (g,h)-walk on G_X then $\hat{\Gamma}$ is a B-connected (g,h)-word. If $|C(\Gamma)| = 2$ then $G_X(\Gamma)$ consists of a single edge e. Clearly, g and h are ends of e. Then $\hat{\Gamma} \varepsilon_{\overline{B}} e$ and thus $\hat{\Gamma}$ is trivially a B-connected (g,h)-word. Assume that, for any $g',h' \in G_X$ and any (g',h')-walk Γ' on G_X with $|C(\Gamma')| < n$ $(n > 2)$, $\hat{\Gamma'}$ is a B-connected (g',h')-word. Let $g,h \in G_X$ and Γ a (g,h)-walk on G_X with $|C(\Gamma)| = n$. Clearly, $C(\hat{\Gamma})$ is the set of edges of the finite connected subgraph $G_X(\Gamma)$ in G_X. Moreover, Lemma 2.5(ii) and (iii) imply that the edges $\bar{\hat{\Gamma}}(0)$ and $t(\hat{\Gamma}(0))$ have a common end. Dually we obtain that $\bar{\hat{\Gamma}}(1)$ and $h(\hat{\Gamma}(1))$ also have a common end. The rest of the properties (i)-(iii) in the definition of a B-connected (g,h)-word can be easily deduced by applying the induction hypothesis for the walks $\Gamma(0)$ and $\Gamma(1)$, taking into consideration Lemma 2.5(iii). Now we turn to proving the converse statement. We will prove by induction on $n = |C(w)|$ that, for every $g,h \in G_X$ and every B-connected (g,h)-word w in $\overline{F}_{E_X}$, there exists a (g,h)-walk Γ with $\hat{\Gamma} \varepsilon_{\overline{B}} w$. If $n = 1$ then $w \varepsilon_{\overline{B}} e$ for an edge $e = \langle\gamma,\delta\rangle$. Since $g,h \in \{\gamma,\delta\}$, one of the walks $(\gamma,\delta),(\delta,\gamma)$, (γ,δ,γ) and (δ,γ,δ) has the property required. Assume that, for any $g',h' \in G_X$ and any B-connected (g',h')-word w' in $\overline{F}_{E_X}$ with $|C(w')| < n$ $(n > 1)$, there exists a (g',h')-walk Γ' with $\hat{\Gamma'} \varepsilon_{\overline{B}} w'$. Let w be a B-connected (g,h)-word with $|C(w)| = n$. Then $|C(w(0))| = |C(w(1))| < n$. Denote the common ends of the pairs of edges $\bar{w}(0), t(w(0))$ and $\bar{w}(1)$, $h(w(1))$ by g_1 and h_1, respectively. Furthermore, denote the end of $\bar{w}(0)[\bar{w}(1)]$ different from $g_1[h_1]$ by $g_2[h_2]$. Since w is a B-connected (g,h)-word it is clear by the definition that $w(0)$ is a $\mathcal{B}$-connected (g,g_1)-word and $w(1)$ is a $\mathcal{B}$-connected (h_1,h)-word. Thus, by the induction hypothesis, we infer that there exist a (g,g_1)-walk Γ_0 and a (h_1,h)-walk Γ_1 with $\widehat{\Gamma_0}\ \varepsilon_{\overline{\mathcal{B}}}\ w(0)$ and $\widehat{\Gamma_1}\ \varepsilon_{\overline{\mathcal{B}}}\ w(1)$. Consider the walk $\Gamma = \Gamma_0 \circ (g_1,g_2) \circ \Pi \circ (h_2,h_1) \circ \Gamma_1$ where Π is the (g_2,h_2)-path on G_X. Since $\widehat{\Gamma_0 \circ (g_1,g_2)}\ \varepsilon_{\overline{\mathcal{B}}}\ w(0)\bar{w}(0)$, $\widehat{(h_2,h_1)\circ\Gamma_1}\ \varepsilon_{\overline{\mathcal{B}}}\ \bar{w}(1)w(1)$ and $C(w) = C(w(0)\bar{w}(0)) = C(\bar{w}(1)w(1))$ we obtain that $\Gamma_0 \circ (g_1,g_2)$ and $(h_2,h_1)\circ\Gamma_1$ span the same finite connected subgraph T in G_X. In particular, this implies that $g_2,h_2 \in V(T)$ and hence Π is a walk on T. Therefore $\bar{\hat{\Gamma}}(0)$ is the edge with ends g_1,g_2 and $\hat{\Gamma}(0) = \widehat{\Gamma_0}$. Consequently, $\bar{\hat{\Gamma}}(0) = \bar{w}(0)$ and $\hat{\Gamma}(0)\ \varepsilon_{\overline{\mathcal{B}}}\ w(0)$. Similarly, we can deduce that $\bar{\hat{\Gamma}}(1) = \bar{w}(1)$ and $\hat{\Gamma}(1)\ \varepsilon_{\overline{\mathcal{B}}}\ w(1)$. Thus $\hat{\Gamma}\ \varepsilon_{\overline{\mathcal{B}}}\ w$ follows which completes the proof.

Lemmas 2.7 and 2.8 ensure that the range of $\psi_{\mathcal{Y}}$ and, consequently, the range of $\lambda_{\mathcal{Y}}$ consists of those pairs (b,g) for which b is a connected (1,g)-element in $F\overline{\mathcal{Y}}^*_{E_X}$. Thus we obtained the following unified interpretation of $FO\mathcal{Y}_X$ for $\mathcal{Y} = \mathcal{S}, \mathcal{NB}, \mathcal{B}$.

Let $\mathcal{Y}$ be one of the varieties of *-bands $\mathcal{S}$, $\mathcal{NB}$ and $\mathcal{B}$. For any non-empty set X, consider the subsemigroup

$$U\mathcal{Y}_X = \{(b,g)\colon b \text{ is a connected } (1,g)\text{-element}\}$$

in the semidirect product $F\overline{\mathcal{Y}}^*_{E_X} * G_X$.

Theorem 2.9: The pair $(U\mathcal{Y}_X, f)$, *where* $f\colon X \to U\mathcal{Y}_X$ *assigns* $(\langle 1,x\rangle\varepsilon_{\overline{\mathcal{Y}}}, x)$ *to* x *for every* x *in* X, *is a free object in the variety* $O\mathcal{Y}$.

The unified solution of the word problem for free inverse, generalized inverse and orthodox *-semigroups (cf. Theorem 2.4) immediately raises the question whether it remains true for any variety of *-bands $\mathcal{Y}$. In a forthcoming paper a more general question will be answered in the positive and, consequently, a number of relatively free orthodox *-semigroups will be described analogously to Theorem 2.9.

REFERENCES

[1] Adair, C.L., Varieties of * orthodox semigroups, Ph.D.Thesis, University of South Carolina (1979).

[2] Adair, C.L., Bands with an involution, J. Algebra 75(1982), 297-314.

[3] Howie, J.M., An introduction to semigroup theory, Academic Press, London, 1976.

[4] Margolis, S.W. and J.C. Meakin, E-unitary inverse monoids and the Cayley graph of a group presentation, preprint.

[5] Nordahl, T.E. and H.E. Scheiblich, Regular * semigroups, Semigroup Forum 16(1978), 369-377.

[6] Petrich, M., Lectures in semigroups, Akademie-Verlag, Berlin, 1977.

[7] Petrich, M., Inverse semigroups, Wiley & Sons, New York, 1984.

[8] Scheiblich, H.E., Free inverse semigroups, Proc. Amer. Math. Soc. 38(1973), 1-7.

[9] Scheiblich, H.E., Generalized inverse semigroups with involution, Rocky Mountain J. Math. 12(1982), 205-211.

[10] Szendrei, M.B., Free *-orthodox semigroups, Simon Stevin 59(1985), 175-201.

[11] Yamada, M., Regular semigroups whose idempotents satisfy permutation identities, Pacific J. Math. 21(1967), 371-392.

VARIETIES OF COMPLETELY REGULAR SEMIGROUPS: THEIR INJECTIVES.

P.G. Trotter,
Department of Mathematics,
University of Tasmania,
Hobart, Tasmania,
Australia.

An object I in a category of semigroups C is a *C-injective* if and only if for any objects $S,T \in C$ such that S is a subsemigroup of T, any morphism $\phi: S \to I$ extends to a morphism $\psi: T \to I$. In [8], Schein showed that if each object of C is properly embeddable in a congruence-free object in C then C has only trivial injectives. If C is a variety of completely regular semigroups, this property applies only for some varieties C of groups. We will see that non-group varieties of completely regular semigroups are rich in non-trivial injectives (as are some varieties of groups).

In some varieties of completely regular semigroups all of the injectives are known; in particular see Gerhard [2] for any variety of bands, Schein [8] for the variety of all commutative semilattices of groups, García and Larrión [1] and Kovacs and Newman [5] for any variety of groups, and Trotter [9] for any variety of completely simple semigroups. The aim here is to determine the injectives in any variety of completely regular semigroups.

1. Preliminaries

If S and T are semigroups such that $\phi: T \to S$ and $\theta: S \to T$ are homomorphisms and $\theta\phi$ is the identity map on S then S is a *retract* of T *under the retraction* ϕ. In a variety V, a direct product of V-injectives is a V-injective, as is every retract of a V-injective. A V-injective is an *absolute retract*; that is, it is a retract of any extension of itself in V.

Let S be a *completely regular semigroup* (that is, S is a union of its subgroups). Let E(S) denote the set of idempotents of S. Then S is a semilattice of completely simple semigroups S_α; $\alpha \in \Gamma$, where S_α is a $\mathcal{D}$-class of S and $\Gamma = S/\mathcal{D} \cong E(S)$. Furthermore S is a *normal band of groups* if and only if there exists homomorphisms $\psi_{\alpha,\beta}: S_\alpha \to S_\beta$ for each $\alpha,\beta \in \Gamma$ where $\alpha \geq \beta$, such that $\psi_{\alpha,\alpha}$ is the identity map, $\psi_{\alpha,\beta}\psi_{\beta,\gamma} = \psi_{\alpha,\gamma}$ if $\beta \geq \gamma$, and $xy = x\psi_{\alpha,\alpha\delta}y\psi_{\delta,\alpha\delta}$ for each $x \in S_\alpha, y \in S_\delta$. In this case call $\psi_{\alpha,\beta}$ a *structure map* of S and write

$$S = [\Gamma; S_\alpha, \psi_{\alpha,\beta}].$$

We may define a *natural partial order* $\geq$ on the normal band of groups S as follows: for $a \in S_\alpha$, $b \in S_\beta$ then $a \geq b$ if and only if $\alpha \geq \beta$ and $a\psi_{\alpha,\beta} = b$. A subset H of S is *compatible* if and only if for each $a \in H \cap S_\alpha$, $b \in H \cap S_\beta$ then $a\psi_{\alpha,\alpha\beta} = b\psi_{\beta,\alpha\beta}$. Define S to be *complete* if each compatible subset H of S has a least upper bound $\vee H$ in S. A complete normal band of groups is *infinitely distributive* if and only if for each pair of compatible subsets H and K, $(\vee H)(\vee K) = \vee(HK)$.

The following notation will be used to describe varieties.

CR - the variety of all completely regular semigroups.
CS - the variety of all completely simple semigroups.
B - the variety of all bands.
SL - the variety of all semilattices.
NB - the variety of all normal bands.
LNB, RNB - respectively the variety of all left normal bands, or all right normal bands.
RB - the variety of all rectangular bands.
O - the variety of all orthodox completely regular semigroups.
G - the variety of all groups.
$L(V)$ - the lattice of subvarieties of a subvariety V of CR.
UG - the variety of all bands of groups S such that $S/H \in U \in L(B)$.
$[U,V]$ - the interval $\{W \in L(CR);\ U \subseteq W \subseteq V\}$, for $U,V \in L(CR)$.

<u>THEOREM 1.1</u> [2]. Suppose $V \in L(B)$. Then I is a V-injective if and only if
(i) $V \subseteq RB$ and $I \in V$;
(ii) $SL \subseteq V \subsetneq NB$ and I is a complete infinitely distributive semilattice, or
(iii) $V \in [SL, NB]$, $I \in V$ and I is complete and infinitely distributive with retractions for structure maps.

Now suppose $V \in L(CR)$. The *index* of V is the least positive integer n such that $(xx^{-1}yy^{-1})^n = xy(xy)^{-1}$ is a law in $V \cap CS$; if there is no such integer the index is infinite (of course x^{-1} denotes the inverse of x in the same H-classes as x). A group $G \in V \cap G$ is a V-injective group is and only if G is a $V \cap G$-injective such that

(i) if V has infinite index then $|G| = 1$, and

(ii) if V has index n then G has no non-identity element of order dividing n.

THEOREM 1.2 [9]. Suppose $V \in L(CS)$. Then I is a V-injective if and only if I is a direct product of a rectangular band in V by a V-injective group.

2. The injectives

Since the results of this section are to appear elsewhere, the proofs are abbreviated.

The section begins with three propositions that considerably restrict the possibilities for $I \in V \in L(CR)$ to be a V-injective.

PROPOSITION 2.1. Suppose $V \in L(CR)$ and I is a V-injective. Then

(i) The H-classes and D-classes of I are respectively $V \cap G$-injectives and $V \cap CS$-injectives,

(ii) $I \in O \cap NBG$,

(iii) E(I) is both a $V \cap B$-injective and a V-injective.

Proof. (i) follows directly from definitions. By [9; Theorem 1] a $V \cap CS$-injective is orthodox, so $I \in O$ by (i) and [6; IV.3.2]. It follows that E(I) is a $V \cap B$-injective. But then by Theorem 1.1, $E(I) \in NB$; consequently $I \in NBG$ (see [6; Exercise IV.4.9.1]). Since $I \in O \cap NBG$ then E(I) is a retract of I under $H^{\#}$ so E(I) is a V-injective.

PROPOSITION 2.2. Let $V \in L(CR)$ and I be a V-injective. Then the structure maps of I are retractions.

Proof. Let $I = [\Gamma; I_\alpha, \psi_{\alpha,\beta}]$. Assume $|\Gamma| > 1$, otherwise the result is trivial. For $\alpha > \beta$ in Γ let $e \in E(I_\alpha)$ and $S = \{e\} \cup I_\beta$ be a subsemigroup of I. Select $C \in CS$ such that there is an isomorphism $\theta: C \to I_\beta$, $e \in C$ and $e\theta = e\psi_{\alpha,\beta}$. There exists $T = C \cup I_\beta \in V \cap NBG$ with non-identity structure map θ. So T embeds S. Since I is a V-injective there is a homomorphism $\psi: T \to I$ that extends the natural embedding $\phi: S \to I$. It can be routinely checked that $\theta\psi$ is an isomorphism and $(\theta\psi)^{-1}\psi\psi_{\alpha,\beta}$ is the identity map on I_β. Hence $\psi_{\alpha,\beta}$ is a retraction.

PROPOSITION 2.3. Suppose there exists $V \in \mathcal{V} \in \mathcal{L}(\mathcal{CR})$ such that $\mathcal{R}$ is not a congruence on V. If I is a $\mathcal{V}$-injective then $I \in \mathcal{RNBG}$.

Proof. There are $\mathcal{R}$-classes R_1, R_2 of V such that R_1R_2 is not in an $\mathcal{R}$-class. Let S be the $\mathcal{D}$-class containing R_1R_2 and $T = R_1 \cup R_2 \cup S$ be a subsemigroup of V. Then $S, T \in \mathcal{V}$. There exists $r, s \in R_1$ such that rR_2 lies in an $\mathcal{R}$-class R of S and $sR_2 \cap R$ is empty. Suppose $I \notin \mathcal{RNBG}$; so I contains a two element left zero subsemigroup $\{a,b\}$. Since $S \in \mathcal{CS}$ there is a homomorphism $\phi: S \to I$ such that $x\phi = a$ or b according as $x \in R$ or $x \notin R$. But $\mathcal{R}$ is a congruence on I so for any extension homomorphism $\psi: T \to I$ of ϕ we get $(ry)\psi \; \mathcal{R} \; (sy)\psi$ for any $y \in R_2$. Thus $a \; \mathcal{R} \; b$, which is a contradiction.

Attention will now be focused on $\mathcal{V}$-injectives that are semilattices of groups.

PROPOSITION 2.4. Let $I = [\Gamma; I_\alpha, \psi_{\alpha,\beta}] \in \mathcal{SLG}$ be a $\mathcal{V}$-injective for $\mathcal{V} \in [\mathcal{SL}, \mathcal{CR}]$. Then I is complete and infinitely distributive and I_α is a $\mathcal{V}$-injective group for each $\alpha \in \Gamma$.

Proof. By [7; 1.10, 1.15 and 1.33] I is embedded in a complete infinitely distributive semigroup $C(I) \in \mathcal{SLG}$ and the $\mathcal{H}$-classes of $C(I)$ are subdirect products of $\mathcal{H}$-classes of I. So $C(I) \in \mathcal{V} \cap \mathcal{SLG}$. Since I is a $\mathcal{V}$-injective embedded in $C(I)$ then I is a retract of $C(I)$; as such it is a straightforward exercise to check that I is complete. By Proposition 2.1 and Theorem 1.1, $E(I)$ is infinitely distributive so by [7; 1.13], I is infinitely distributive. By Proposition 2.1 and Theorem 1.2, I_α is a $\mathcal{V}$-injective group.

We further restrict our attention so as to determine the $\mathcal{V}$-injectives in $\mathcal{SLG}$ with just two $\mathcal{H}$-classes where $\mathcal{V} \in [\mathcal{SL}, \mathcal{CR}]$. Any $\mathcal{V}$-injective must contain a zero.

Proposition 2.5. Let $\mathcal{V} \in [\mathcal{SL}, \mathcal{CR}]$ and G be a $\mathcal{V}$-injective group. Then G^0 is a $\mathcal{V}$-injective.

Proof. The author's proof of this is rather long; an outline only is presented here.

Suppose $S, T \in \mathcal{V}$ and S is a subsemigroup of T. Let $\Lambda = S/\mathcal{D}$, $\Omega = T/\mathcal{D}$ and S_α, T_α be $\mathcal{D}$-classes for $\alpha \in \Lambda$, $\beta \in \Omega$. Let $\phi: S \to G^0$ be a homomorphism; we must find an extension homomorphism $\psi: T \to G^0$.

Let τ be the least inverse semigroup congruence on T and $\tau' = \tau \cap (S \times S)$. Using [4] it can be shown that $\phi \circ \phi^{-1} \supseteq \tau'$ so $\phi = (\tau')^{\#}\phi_1$ for some homomorphism $\phi_1: S/\tau' \to G^0$. Since S/τ' can be identified with a subsemigroup of T/τ it suffices to find a homomorphism $\psi_1: T/\tau \to G^0$ extending ϕ. For then $\psi = \tau^{\#}\psi_1$.

Hence assume $S = [\Lambda; S_\alpha, \phi_{\alpha,\beta}]$ and $T = [\Omega; I_\alpha, \psi_{\alpha,\beta}]$ are semilattices of groups. Define

$$\Delta = \{\alpha \in \Lambda;\ S_\alpha\phi = \{0\}\}, \quad S_\Delta = \cup\{S_\alpha;\ \alpha \in \Delta\}.$$

Then S_Δ is an ideal of S and $S\backslash S_\Delta$ is a subsemigroup. Furthermore it is easy to check that

(1) $\qquad h\phi = k\phi$ if $h,k \in H$, H is a compatible subset of $S\backslash S_\Delta$.

Using [7; 1.33] it can be shown that T is embeddable in a complete infinitely distributive semigroup $W \in \mathcal{V} \cap \mathcal{SLG}$ with retractions for structure maps. It is easy to see that the completion V of S in W is infinitely distributive. Since V is an infinitely distributive completion of S it can be shown, using (1), that there is a homomorphism $\theta: V \to G^0$ that extends ϕ. If a homomorphism $\theta': W \to G^0$ can be found to extend θ then we can select ψ to be the restriction of θ' to T.

Hence also assume T is complete and infinitely distributive with retractions for structure maps and S is complete and infinitely distributive. Since $T \in \mathcal{SLG}$ then $E(T)S^1$ is an inverse subsemigroup of T. With $\varepsilon = \wedge\Lambda$ and $\alpha \in \Omega$ then $\alpha' = \vee\{\gamma \in \Lambda \cup \{\varepsilon\};\ \gamma \leqslant \alpha\} \in \Lambda$. Using (1) it can be checked that there is a homomorphism $\theta: E(T)S^1 \to G^0$ defined by $a\theta = a\psi_{\alpha,\alpha'}\phi$ for $a \in T_\alpha$, that extends ϕ.

So we may assume T is as above and $S = [\Lambda; S_\alpha, \phi_{\alpha,\beta}] \in \mathcal{V} \cap \mathcal{SLG}$ where $E(S) = E(T)$; that is $\Omega = \Lambda$. With $T_\Delta = \cup\{T_\alpha;\ \alpha \in \Omega\}$ then $T_\Delta \cup S$ is an inverse subsemigroup of T with ideal T_Δ and there is a homomorphism $\theta: T_\Delta \cup S \to G^0$ extending ϕ such that $T_\Delta\theta = \{0\}$.

So we assume $T_\Delta = S_\Delta$. Let

$$A = \{t \in T;\ t \geqslant a \ \text{ for some } a \in S\backslash S_\Delta\} \cup S.$$

Then A is an inverse subsemigroup of T and using (1) it can be verified that there is a homomorphism $\theta: A \to G^0$ extending ϕ, given by $t\theta = a\phi$ whenever $t \geqslant a \in S\backslash S_\Delta$ and $t\theta = 0$ otherwise.

We may further assume $S = A$. Let $\eta = \vee\Lambda$ and assume

$S_\eta\phi \subseteq G$; otherwise $S_\eta\phi = 0$ and we may choose ψ to be the trivial map. Since G is a V-injective group there is a homomorphism $\psi_\eta: T_\eta \to G^0$ extending $\phi|_{S_\eta}$. Using the fact that $E(S) = E(T)$, $S = A$ and $\psi_{\eta,\alpha}$ is a retraction, it can be shown that there is a homomorphism $\psi_\alpha: T_\alpha \to G^0$ given by $\psi_{\eta,\alpha}\psi_\alpha = \psi_\eta$ that extends $\phi|_{S_\alpha}$ for each $\alpha \in \Omega\backslash\Delta$. For $\alpha \in \Delta$ put $T\psi_\alpha = \{0\}$. It can now be routinely checked that $\psi: T \to G^0$ given by $\psi|_{T_\alpha} = \psi_\alpha$ is a homomorphism extending ϕ.

<u>COROLLARY 2.6.</u> Suppose $V \in [SL, CR]$ and $I \in SLG$. Then I is a V-injective if and only if I is a retract of a direct product $\Pi\{G_\alpha^0;\ \alpha \in \Gamma\}$ where G_α is a V-injective group for each $\alpha \in \Gamma$.

<u>Proof.</u> Sufficiency is by Proposition 2.5. So suppose $I = [\Gamma; I_\alpha, \psi_{\alpha,\beta}] \in SLG$ is a V-injective. For $\alpha \in \Gamma$ let $\bar{\alpha}$ denote the ideal of Γ generated by α. The inverse limit $I_{\bar{\alpha}}$ of the ideal $\{I_\beta;\ \beta \leqslant \alpha\}$ of I is the subsemigroup of $\Pi\{I_\beta;\ \beta \leqslant \alpha\}$ consisting of the elements p such that if $p(\beta) \in I_\beta$, $p(\gamma) \in I_\gamma$ are components of p for $\beta \geqslant \gamma$ then $p(\beta)\psi_{\alpha,\beta} = p(\gamma)$ (see [3; §21]). For $\alpha,\beta \in \Gamma$, $\alpha \geqslant \beta$, let $\psi_{\bar{\alpha}\bar{\beta}}: I_{\bar{\alpha}} \to I_{\bar{\beta}}$ be the projection map. Let $\bar{\Gamma}$ be the semilattice under intersection of principal ideals of Γ. It can be shown that $I \cong \bar{I} = [\bar{\Gamma}; I_{\bar{\alpha}}, \psi_{\bar{\alpha}\bar{\beta}}]$. It is not hard to see that $\bar{I}$ is isomorphic to a subsemigroup of $\Pi\{I_\alpha^0;\ \alpha \in \Gamma\} \in V \cap SLG$. Note by Proposition 2.4 that I_α is a V-injective group. As a V-injective, I is an absolute retract and hence I is a retract of $\Pi\{I_\alpha^0;\ \alpha \in \Gamma\}$.

<u>REMARK 2.7.</u> Subsequent to the conference the author has shown that *for* $V \in [SL, CR]$ *and* $I = [\Gamma; I_\alpha, \psi_{\alpha,\beta}] \in V \cap SLG$ *then* I *is a* V*-injective if and only if* I *is complete and infinitely distributive,* $\psi_{\alpha,\beta}$ *is a retraction and the kernel of* $\psi_{\alpha,\beta}$ *is a* V*-injective group for each* $\alpha,\beta \in \Gamma$ *where* $\alpha \geqslant \beta$.

<u>THEOREM 2.8.</u> Suppose $V \in [SL, CR]$. Then I is a V-injective if and only if

(i) $I \in V \cap O \cap NBG$, E(I) is a $V \cap B$-injective and $eIe \in V \cap SLG$ is a V-injective for some maximal idempotent e of I,

(ii) either $I \in SLG$ or $V \cap B \in \{LNB, RNB, NB\}$, and

(iii) if L or R is not a congruence on some member of V then $I \in LNBG$ or $I \in RNBG$ respectively.

<u>Proof.</u> Say I is a V-injective. By Proposition 2.1, $I \in V \cap O \cap NBG$ and E(I) is a $V \cap B$-injective. By Theorem 1.1, E(I) has a maximal

element e. Clearly $eIe \in V \cap SLG$ and the map $\theta: I \to eIe$ given by $a\theta = eae$ is a retraction; so eIe is a V-injective. The remaining conditions are by Theorem 1.1 and Proposition 2.3.

Now suppose I satisfies (i), (ii) and (iii), S is a subsemigroup of T, $S,T \in V$, and $\phi: S \to I$ is a homomorphism. Since eIe is a V-injective then there is a homomorphism $\psi_H: T \to eIe$ extending $\phi\theta$, where θ is given above.

Assume $I \notin SLG$, otherwise $I = eIe$ is a V-injective. By (iii) we may assume L (or R) is a congruence on each member of V. Let L_T and L_S respectively denote the L-relations on T and S; L_S is the restriction of L_T to S since $T \in CR$. We may regard S/L_S as being a subsemigroup of T/L_T. Since $I \in O \cap NBG$ then

(2) $$afb \; D \; ab \Rightarrow afb = ab; \quad a,b \in I, \; f \in E(I).$$

Furthermore if $a \; L \; b$ then $ea^{-1}a = eb^{-1}b$. So there is a homomorphism

$$\phi_L: S/L_S \to E(I); \quad sL_S^{\#}\phi_L = e((s^{-1}s)\phi).$$

But E(I) is a $V \cap B$-injective so there is a homomorphism $\psi_L: T/L_T \to E(I)$ extending ϕ_L; then $L_T^{\#}\psi_L$ extends $L_S^{\#}\phi_L$. By (iii) either $I \in RNBG$ or $I \in NBG\backslash RNBG$. So suppose $I \in RNBG$; then e is a left identity of I and

$$s\phi = s\phi((s^{-1}s)\phi) = e(s\phi)ee((s^{-1}s)\phi) = (s\phi\theta)sL_S^{\#}\phi_L; \; s \in S.$$

So define $\psi: T \to I$ by $t\psi = t\psi_H(tL_T^{\#}\psi_L)$. Since $tL_T^{\#}\psi_L \in E(I)$, then by (2)

$$\begin{aligned}(tr)\psi &= (t\psi_H(r\psi_H)(tL_T^{\#}\psi_L))(rL_T^{\#}\psi_L) = t\psi_H(tL_T^{\#}\psi_L)(r\psi_H)(tL_T^{\#}\psi_L)(rL_T^{\#}\psi_L) \\ &= t\psi_H((tL_T^{\#}\psi_L)(r\psi_H)(rL_T^{\#}\psi_L)) = t\psi(r\psi).\end{aligned}$$

The homomorphism ψ extends ϕ so I is a V-injective.

A dual result applies when $I \in LNBG$. If $I \notin RNBG \cup LNBG$ then L_T and R_T are both congruences on T and $s\phi = sR_S^{\#}\phi_R(s\phi\theta)(sL_S^{\#}\phi_L)$. There is an extension homomorphism $\psi: T \to I$ given by $t\psi = tR_T^{\#}\psi_R(t\psi_H)(tL_T^{\#}\psi_L)$. Again I is a V-injective.

<u>REMARK 2.9.</u> In light of Remark 2.7 it can be shown that condition (i) of the Theorem can be replaced by (i)′ $I = [\Gamma; I_\alpha, \psi_{\alpha,\beta}] \in V \cap O \cap NBG$

is complete and infinitely distributive, the structure maps are retractions and the kernel of $\psi_{\alpha,\beta}|_{H_\alpha}$ *is a V-injective group for each* $\alpha,\beta \in \Gamma$, $\alpha \geqslant \beta$, *and any H-class* H_α *of* I_α.

REFERENCES

1. García, O.C. and F. Larrión, Injectivity in varieties of groups, Algebra Universalis 14 (1982), 280-286.

2. J.A. Gerhard, Injectives in equational classes of idempotent semigroups, Semigroup Forum 9 (1974), 36-53.

3. G. Gratzer, Universal algebra, Van Nostrand, Princeton (1968).

4. P.R. Jones, The least inverse and orthodox congruences on a completely regular semigroup, Semigroup Forum 27 (1983), 390-392.

5. L.G. Kovacs and M.F. Newman, Injectives in varieties of groups, Algebra Universalis 14 (1982), 398-400.

6. M. Petrich, Introduction to semigroups, Merill, Columbus (1973).

7. B.M. Schein, Completions, translation hulls and ideal extensions of inverse semigroups, Czechoslovak Math. J. 23 (1973), 575-610.

8. B.M. Schein, Injectives in certain classes of semigroups, Semigroup Forum 9 (1974), 159-171.

9. P.G. Trotter, Injectives in varieties of completely simple semigroups, Semigroup Forum 33 (1986), 47-55.

GENERALIZED SEMIALGEBRAS OVER SEMIRINGS

Respectfully dedicated to
Professor G. Pickert on his 70th birthday

Hanns Joachim Weinert
Technische Universität Clausthal
D-3392 Clausthal-Zellerfeld, Germany

For an associative ring S, the classical concept of an algebra A over S forces S to be commutative in nearly all cases of some interest. E. g., polynomial rings or matrix rings over a non-commutative ring S with identity are not classical algebras over S. For this reason, H. Zassenhaus [21] and G. Pickert [13] have introduced a more general concept of an algebra A over S. If A has an infinite basis over S, both concepts can be generalized in another direction, which may be illustrated by rings of formal power series. At third, certain semirings A constructed from semirings S have become important tools e. g. in the theory of automata and formal languages, in particular again those in which also infinite sums occur. The purpose of this paper is to deal with all these different concepts in a unique way as special cases of generalized semialgebras over semirings. Since we have to describe all the material we want to combine, some parts of this paper have the character of a survey article.

§ 1 Introduction

Let S be an associative ring with identity and $({}_SA,+)$ a unitary (left) S-module with a basis U, which means that each $a \in A$ has a unique presentation

(1.1) $a = \sum_{u \in U} \alpha_u u$ with $\alpha_u \in S$, where almost all α_u equal the zero ω of S.

If $(A,+,\cdot)$ is a not necessarily associative ring, $({}_SA,+,\cdot)$ is usually called an algebra over S iff it satisfies

(1.2) $\sigma(ab) = (\sigma a)b = a(\sigma b)$ for all $\sigma \in S$ and $a,b \in A$.

Such an algebra is associative iff $(uv)w = u(vw)$ holds for all elements u,v,w of a basis U of $({}_SA,+)$, which is in particular important to construct various associative rings as algebras $({}_SA,+,\cdot)$. One purpose of this paper is to replace the ring S by an additively commutative semiring S. If the latter (cf. § 2) has an identity and an absorbing zero ω, the corresponding concepts of a unitary S-semimodule $({}_SA,+)$ with a basis U and of a *semialgebra* $({}_SA,+,\cdot)$ *over a semiring* S will prove meaningful. They can be treated similarly as in the ring-theoretical case.

But in both cases, (1.2) obviously implies

(1.3) $(\alpha\beta)(ab) = (\beta\alpha)(ab)$ for all $\alpha,\beta \in S$ and $a,b \in A$.

Now assume that there is some product ab in A such that for some basis U of $({}_SA,+)$ some of the coefficients in the presentation $ab = \Sigma\sigma_u u$, say $\sigma_v \in S$, is right cancellable in $(S,\cdot)$, for instance the identity. Then (1.3) implies $\alpha\beta\sigma_v v = \beta\alpha\sigma_v v$ for some $v \in U$ and hence $\alpha\beta = \beta\alpha$ for all $\alpha,\beta \in S$. So e.g. matrix (semi)rings, polynomial (semi)rings, or semigroup (semi)-rings (cf. [19], § 3) over a (semi)ring S are (semi)algebras over S only if $(S,\cdot)$ is commutative, whereas their definitions are completely independent on the latter. In fact, it is even somewhat tedious to construct non-trivial (semi)algebras over a (semi)ring S such that $(S,\cdot)$ is not commutative. A complete answer to this question is contained in Prop. 4.10.

These disadvantages can be avoided in generalizing (1.2) by

(1.4) $(\alpha u)(\beta v) = (\alpha\beta)(uv)$ for all $\alpha,\beta \in S$ and $u,v \in U$,

where U is one suitable basis of $({}_SA,+)$. In the ring case, this was already done 1937 in [21], p.67 (using assumptions equivalent to (1.4), cf. Remark 4.5), and ten years later continued in [13]. The resulting concept is called a *generalized algebra* $({}_SA,+,\cdot)$ *over a ring* S in the following ("Algebra im weiteren Sinne" in [13]). Apart from the fact that (1.4) need not hold for elements u,v of another basis of $({}_SA,+)$, this concept has the same properties as the classical one defined by (1.2), and it can be introduced and handled with no more effort. Nearly the same is true for algebras and generalized algebras $({}_SA,+,\cdot)$ over a ring S which is not assumed to have an identity, as considered for the first time in [13].

We will not show all this explicitely in this paper, because we give in § 4 a corresponding theory for *generalized semialgebras* $({}_SA,+,\cdot)$ *over a semiring* S, which includes all cases mentioned above. That section, the central part of the paper, is prepared by some basic definitions and statements in § 2. In particular we show that S-(semi)modules $({}_SA,+)$ over a (semi)ring S with basis U exist iff S has a right identity and a right absorbing zero (cf. Prop. 2.6) and that they are uniquely determined up to S-semimodule isomorphisms by the cardinality $|U|$ of U (Cor. 2.7). In particular, we may represent them by certain S-(semi)modules $({}_SS\langle U\rangle,+)$ described in Fact 2.3.

The intermediate section § 3 is not needed to go on and devoted to the following. The concept of an S-module $({}_SA,+)$ with a basis U over a ring

S, also called a "vector space over a ring" (cf. [3], [14], § 64), was frequently used in earlier papers. If S has an identity, it is just a free unitary S-module according to its universal property with respect to U. We shall see that a corresponding statement (Thm. 3.3) holds for S-semimodules over a semiring S with an identity, which fails to be true for (semi)modules over (semi)rings S with a basis if S has no identity (Fact 3.6). On the other hand, we can use the results obtained so far to show that, for each (semi)ring S and each cardinality |U| of U, there exist free S-(semi)modules on U in the class of all S-(semi)modules. (Thm. 3.8 and Remark 3.9).

Generalized semialgebras $({}_SA,+,\cdot)$ over a semiring S as investigated in § 4 may be roughly described as S-semimodules $({}_SA,+)$ with a basis U, hence with elements uniquely presented by (1.1), and a distributive multiplication satisfying (1.4). Each such multiplication is determined and can be defined by the products $uv \in A$ for all $u,v \in U$ of a suitable basis U (Thms. 4.2 and 4.4), and one obtains numerous semirings and rings in this way (Thm. 4.6, Cor. 4.7, Expl. 4.8). Further results concern e. g. the embedding of S as a subsemiring of $({}_SA,+,\cdot)$, and generalized semialgebras over generalized semialgebras. With respect to the following we note that each of these generalized semialgebras $({}_SA,+,\cdot)$ is isomorphic to one, say $({}_SS\langle U\rangle,+,\cdot)$, with an underlying S-semimodule $({}_SS\langle U\rangle,+)$.

In the theories of automata, languages, machines and graphs various semirings have become useful (cf. e. g. [2], [7], [8], [10], [15], [16], [20]). Most of them are (generalized) semialgebras as described above, or again generalizations of the latter with elements uniquely presented by

(1.5) $\quad a = \sum_{u\in U} \alpha_u u \quad \text{with} \quad \alpha_u \in S$

for some semiring S and some infinite set U. Now, for each semiring with a right identity and an absorbing zero and each U there is essentially one S-semimodule $({}_SS\langle\langle U\rangle\rangle,+)$ containing U with elements given by (1.5) (Fact 2.2). So we define in § 5 a corresponding concept $({}_SS\langle\langle U\rangle\rangle,+,\cdot)$, also called a generalized semialgebra over S. We consider at first a special case to which all results of § 4 transfer directly, since all infinite sums needed in S are formal infinite ones (Thm. 5.2). It contains all examples where the elements are called "formal power series", from classical rings to several semirings used for applications as mentioned above, e. g. those where $(U,\cdot)$ is a free monoid.

Further investigations of generalized semialgebras $({}_SS\langle\langle U\rangle\rangle,+,\cdot)$ need an algebraic treatment of effective infinite sums, in particular semimodules and semirings in which certain infinite sums are defined. Caused by different purposes, various concepts are scattered in the literatur in this context (cf. e. g. [2], [5], [7], [9], [10], [12], [15]). This material is rather extensive and non-uniform, such that we cannot give a complete survey in this paper, which we hope to do elsewhere. So we present what we need in a self-contained way in § 6 and continue our considerations on generalized semialgebras $({}_S S\langle\langle U\rangle\rangle,+,\cdot)$ in § 7. Assuming that S is a partial complete semiring according to Def. 6.7 we give necessary and sufficient conditions such that the results of § 4 transfer to generalized semialgebras $({}_S S\langle\langle U\rangle\rangle,+,\cdot)$, which become then also partial complete in a suitable meaning. In particular we obtain that, for an S-semimodule $({}_S S\langle\langle U\rangle\rangle,+)$ over a complete semiring S, each choice of the products $uv \in S\langle\langle U\rangle\rangle$ for all $u,v \in U$ provides a complete generalized semialgebra $({}_S S\langle\langle U\rangle\rangle,+,\cdot)$. Moreover, the question whether its multiplication is associative is answered by criterions corresponding to those given in the case of § 4.

§ 2 Semimodules over semirings

Let $S = (S,+,\cdot)$ be a (universal) algebra such that $(S,+)$ and $(S,\cdot)$ are semigroups, connected by ring-like distributivity, i. e. $\alpha(\beta+\gamma) = \alpha\beta+\alpha\gamma$ and $(\beta+\gamma)\alpha = \beta\alpha+\gamma\alpha$ hold for all $\alpha,\beta,\gamma \in S$. Such an algebra is usually called a *semiring*, but troughout this paper *all semirings are assumed to have commutative addition.* (The bibliography in [18] contains about 20 papers on semirings published before 1960.) By the above definition, concepts as *subsemirings* and *homomorphisms* of semirings are clear and refer merely to the two binary operations. If a semiring S has an additive neutral, say ω, it is called the *zero* of S. An element $\alpha \in S$ is called *left absorbing* iff $\alpha\sigma = \alpha$ holds for all $\sigma \in S$, which is shortened by $\alpha S = \alpha$. It is well known that for a semiring S with a zero ω the latter need neither be left nor right absorbing. We speak about an *absorbing zero* ω iff $\omega S = S\omega = \omega$ holds, and note that each semiring S, with or without a zero, can be embedded into a semiring $S \cup \{\omega'\}$ with $\omega' \notin S$ as absorbing zero. Clearly, a *left identity* ε_ℓ (an *identity* ε) of a semiring S means such an element of its semigroup $(S,\cdot)$. We also recall that each semiring as considered in this paper can be embedded into one with an identity (cf. [4], in particular for counter examples for semirings S such that $(S,+)$ is not commutative as excluded here by the above assumption) according to Lemma 2.1 given below.

An additively written commutative semigroup $(M,+)$, which may have a zero (neutral) o or not, is called a *semimodule*. We speak about a *(left)* S-*semimodule* $({}_SM,+)$ iff a semiring S works as a (left) operator domain on M such that

(2.1) $\sigma(a+b) = \sigma a+\sigma b$, $(\sigma+\varrho)a = \sigma a+\varrho a$ and $(\sigma\varrho)a = \sigma(\varrho a)$

are satisfied for all $\sigma,\varrho \in S$ and $a,b \in M$. An S-semimodule $({}_SM,+)$ is called *unitary* iff S has an identity ε and $\varepsilon a = a$ holds for all $a \in M$. As far as we know, these concepts have been used for the first time independently in [6] and [17].

Note that each semiring S may be considered e. g. as a left S-semimodule $({}_SS,+)$ with respect to the multiplication of S. Further, each semimodule $(M,+)$ is in a natural way a (left or right) unitary $\mathbb{N}$-semimodule over the semiring $(\mathbb{N},+,\cdot)$ of positive integers by

(2.2) $\nu a = a\nu = \sum_{i=1}^{\nu} a$ for all $\nu \in \mathbb{N}$, $a \in M$.

If $(M,+)$ has a zero o, the same applies to the semiring $(\mathbb{N}_0,+,\cdot)$ of non-negative integers by (2.2) and $0a = a0 = o$. Clearly, each (left) S-module over a ring S is an S-semimodule, but there are various examples of S-semimodules $({}_SM,+)$ over a ring S which are not S-modules, even those where $({}_SM,+)$ has no zero.

An S-*semimodule-homomorphism* $f: ({}_SA,+) \to ({}_SM,+)$ is defined by

$$f(a+b) = f(a)+f(b) \text{ and } f(\sigma a) = \sigma f(a) \text{ for all } a,b \in A \text{ and } \sigma \in S,$$

and f is said to *respect the zeroes* iff both S-semimodules have those elements and $f(o_A) = o_M$ holds. The latter is always the case if f is an isomorphism and one of the S-semimodules has a zero.

Dealing with S-semimodules in this paper, the rules (2.1) and their obvious consequences e. g. for more than two summands will be used without any comment. In particular, considering a semiring S with a zero ω as a semimodule $(S,+)$ over $\mathbb{N}_0$ according to (2.2) and $0\alpha = \alpha 0 = \omega$, we formulate the above announced

LEMMA 2.1. Let S be a semiring with an absorbing zero ω. Then the set $T = \mathbb{N}_0 \times S = \{(\nu,\alpha) \mid \nu \in \mathbb{N}_0, \alpha \in S\}$ is a semiring $(T,+,\cdot)$ with respect to componentwise addition and the multiplication

(2.3) $(\nu,\alpha)\cdot(\mu,\beta) = (\nu\cdot\mu, \nu\beta+\alpha\mu+\alpha\cdot\beta)$.

The element $(0,\omega)$ is the absorbing zero of T and $(1,\omega)$ its identity, and

$\alpha \to (0,\alpha)$ _for all_ $\alpha \in S$ _provides a monomorphism of_ $(S,+,\cdot)$ _into_ $(T,+,\cdot)$.

A corresponding statement holds if one merely assumes $S\omega = \omega$. Then the zero $(0,\omega)$ of T is also right absorbing and $(1,\omega)$ a right identity of T. The S-semimodules considered in the following are basic for this paper.

Fact 2.2 a) Let S be a semiring, $U \neq \emptyset$ a set and $A = S\langle\langle U\rangle\rangle$ the set of all mappings $a:U \to S$ denoted by $a(u) = \alpha_u \in S$. Defining

(2.4) $\quad (a+b)(u) = a(u)+b(u) \quad$ and $\quad (\sigma a)(u) = \sigma(a(u))$

for all $a,b \in A$, $u \in U$ and $\sigma \in S$, we obtain an S-semimodule $({}_SA,+)$. It is unitary iff S has an identity. If S has a zero ω, the mapping o defined by $o(u) = \omega$ is the zero of $({}_SA,+)$. Then

(2.5) $\quad \sigma o = o \quad$ for all $\sigma \in S \quad$ iff $\quad S\omega = \omega \quad$ and

(2.6) $\quad \omega a = o \quad$ for all $a \in A \quad$ iff $\quad \omega S = \omega$

are satisfied, and $(A,+)$ is a module iff S is a ring, hence $({}_SA,+)$ an S-module in this case.

b) If S has a zero ω, we use the, say at first formal, notation

(2.7) $\quad a = \sum_{u \in U} \alpha_u u \quad$ for each $a = (a(u) = \alpha_u \mid u \in U) \in A = S\langle\langle U\rangle\rangle$.

Then $a+b = \Sigma\alpha_u u + \Sigma\beta_u u = \Sigma(\alpha_u+\beta_u)u$ and $\sigma a = \sigma(\Sigma\alpha_u u) = \Sigma(\sigma\alpha_u)u$ hold by (2.4). Now we interpret $\alpha_u u$ as the mapping defined by

(2.8) $\quad (\alpha_u u)(u) = \alpha_u \quad$ and $\quad (\alpha_u u)(v) = \omega$ for all $v \neq u$ in U,

hence as an element of A. Then, provided that U is finite, (2.7) is a finite sum as defined in $(A,+)$. Otherwise we can define the infinite sum $\Sigma\alpha_u u$ in (2.7) by $a \in A$ according to the considerations in § 6 (Prop. 6.5).

c) We can go a step further and interpret each $u \in U$ as an element of $A = S\langle\langle U\rangle\rangle$ iff S has a right identity ε_r and, for $|U| \geq 2$, $S\omega = \omega$ holds. Then we consider u as the mapping

(2.9) $\quad u(u) = \varepsilon_r \quad$ and $\quad u(v) = \omega$ for all $v \neq u$ in U.

Proof Since a) and b) are clear, we only show c). Assume that one $u \in U$ is interpreted as a mapping $u \in A$, say $u(u) = \xi$ and $u(v) = \tau$, the latter for some $v \neq u$ if U contains such an element. By (2.8) and (2.4) we obtain $\sigma = (\sigma u)(u) = \sigma(u(u)) = \sigma\xi$ and $\omega = (\sigma u)(v) = \sigma(u(v)) = \sigma\tau$ for all $\sigma \in S$. Hence ξ is a right identity of S and $\omega = \sigma\tau$ for all $\sigma \in S$ implies $S\omega = \omega$ by $\sigma\omega = \sigma(\omega\tau) = (\sigma\omega)\tau = \omega$. Conversely, (2.9) defines u as an element of A such that (2.8) holds. As an immediate consequence of Fact 2.2 we state:

Fact 2.3 Let S be a semiring with a zero satisfying $S\omega = \omega$ and $U \neq \emptyset$ a set. Then the subset $A = S\langle U\rangle$ of the S-semimodule $({}_S S\langle\langle U\rangle\rangle, +)$ which consists of the mappings

$$(2.10) \qquad a = \sum_{u \in U} \alpha_u u \ , \text{ almost all } \alpha_u \in S \text{ equal } \omega$$

forms an S-subsemimodule of $({}_S A, +)$. It has the zero $o = \Sigma\omega u$, satisfies $\sigma o = o$ for all $\sigma \in S$ by (2.5), and all statements on $S\langle\langle U\rangle\rangle$ given in Fact 2.2 a) transfer to $S\langle U\rangle$. If U is infinite, each sum (2.10) is a formal infinite one, as usual defined as the sum of its non-zero summands (if there are those) or as o (otherwise). Moreover, U may be considered as contained in $A = S\langle U\rangle$ by (2.9) iff S has a right identity ε_r.

Remark 2.4 Note that $S\langle\langle U\rangle\rangle$ may be considered as the direct product and $S\langle U\rangle$ as the direct sum of $|U|$ copies of $({}_S S, +)$. For a semiring S with an identity and an absorbing zero, the elements of $S\langle\langle U\rangle\rangle$ are called "S-subsets of U" in [2], p. 126, and for $S = \mathbb{N}_0$ also "multisets" (cf. e. g. [12]). If $(U, \cdot)$ is a monoid, in particular the free monoid generated by a set X, certain semirings defined on $S\langle\langle U\rangle\rangle$ and $S\langle U\rangle$ are used e. g. in [2], [7], [8] and [16]. They will turn out to be examples of generalized semialgebras as considered in this paper.

DEFINITION 2.5 Let $({}_S A, +)$ be an S-semimodule with a zero o over a semiring S with a zero ω. A subset $U \subseteq A$ is called a *basis* of $({}_S A, +)$ iff $\omega u = o$ holds for all $u \in U$ and each element $a \in A$ has a unique presentation

$$(2.11) \qquad a = \sum_{u \in U} \alpha_u u \ , \ \alpha_u \in S \ , \text{ almost all } \alpha_u = \omega,$$

interpreted as formal infinite sums if the cardinal number $|U|$ of U is not finite.

Clearly, each S-semimodule $({}_S A, +)$ with a basis U is unitary iff S has an identity, and an S-module iff S is a ring. On the other hand, the cardinality $|U|$ of U need not be an invariant of $({}_S A, +)$. There are even (non-commutative) rings S with identity and S-modules $({}_S A, +)$ with a basis U and another one, say V, such that $|U| \neq |V|$ holds (cf. [3], [14], § 64 and [1]).

PROPOSITION 2.6 Let S be a semiring with a zero ω.

a) If there exists one S-semimodule $({}_S A, +)$ which has a basis U, then S has a right identity ε_r and, provided that $|U| \geq 2$, $S\omega = \omega$ holds.

b) Conversely, let S satisfy these conditions. Then, for each cardinal number $c \neq 0$, there exists an S-semimodule $({}_S A, +)$ with a basis U satis-

fying $|U| = \mathfrak{c}$, e. g. the S-semimodule $A = S\langle U\rangle$ described in Fact 2.3.

Proof a) Fix one element $v \in U$. Then $v = \Sigma\tau_u u$ holds by (2.11) and hence $\sigma v = \Sigma\sigma\tau_u u$ for all $\sigma \in S$. This yields $\sigma = \sigma\tau_v$ and $\omega = \sigma\tau_u$ for each $u \neq v$. So τ_v is a right identity of S and, if $|U| \geq 2$ holds, there is at least one element $\tau \in S$ satisfying $\omega = \sigma\tau$ for all $\sigma \in S$, which implies $S\omega = \omega$ as shown above in the proof of Fact 2.2 c).

b) Let U be a set satisfying $|U| = \mathfrak{c}$. The existence of a right identity ε_r of S allows to consider U as a subset of $S\langle U\rangle$ by (2.9). Hence U is a basis of $({}_S S\langle U\rangle,+)$ according to (2.10). Note that the case $|U| = 1$ needs in fact no assumption concerning a zero of S.

Applying Prop. 2.6 to rings S and S-modules, the property of S occuring in a) and b) reduces to the existence of a right identity. This statement is due to [13]. Note also that a unique right identity ε_r of a ring S is an identity (cf. [14], § 23), whereas semirings satisfying $\omega S = S\omega = \omega$ with exactly one right identity exist.

COROLLARY 2.7 Let $({}_S A,+)$ be an S-semimodule with a basis U and fix one right identity ε_r of S. Then the set $\varepsilon_r U = \{\varepsilon_r u \mid u \in U\}$ is also a basis of $({}_S A,+)$ such that each $a \in A$ has the same family $(\alpha_u \mid u \in U)$ as coefficients according to

$$(2.12)\qquad a = \sum_{u\in U} \alpha_u u = \sum_{u\in U} (\alpha_u \varepsilon_r) u = \sum_{\varepsilon_r u \in \varepsilon_r U} \alpha_u (\varepsilon_r u).$$

Moreover, for each S-semimodule $({}_S M,+)$ with a basis V satisfying $|V| = |U|$ and each bijection $\varphi: U \to V$, there is a unique S-semimodule isomorphism $f: ({}_S A,+) \to ({}_S M,+)$ which extends $\varphi': \varepsilon_r U \to \varepsilon_r V$ defined by $\varphi'(\varepsilon_r u) = \varepsilon_r \varphi(u)$. In particular, each S-semimodule with a basis U is isomorphic to $({}_S S\langle U\rangle,+)$.

Proof The first statement is clear by (2.12). Thus φ' is a bijection of the basis $\varepsilon_r U$ of $({}_S A,+)$ onto the basis $\varepsilon_r V$ of $({}_S M,+)$. Hence f defined by $f(\Sigma\alpha_u(\varepsilon_r u)) = \Sigma\alpha_u \varphi'(\varepsilon_r u)$ is an S-semimodule isomorphism of $({}_S A,+)$ onto $({}_S M,+)$. By $f(\varepsilon_r u) = f(\varepsilon_r(\varepsilon_r u)) = \varepsilon_r \varphi'(\varepsilon_r u) = \varphi'(\varepsilon_r u)$ it extends φ', and as an S-semimodule homomorphism, f is uniquely determined by this property.

Note that a bijection $\varphi: U \to V$ as in Cor. 2.7 need not be extendable to an S-semimodule homomorphism $f: ({}_S A,+) \to ({}_S M,+)$ which is clearly the case if S has an identity. This applies to S-semimodules $({}_S A,+)$ over a semiring S as well as to S-modules over a ring S. We show it for a ring S in the following example, which also illustrates some other statements:

Example 2.8 For $S=\{\omega,\varepsilon_r,\eta_r,\tau\}$, let $(S,+)$ be Klein's four-group and $(S,\cdot)$ given by the right hand table. It is well known that $(S,+,\cdot)$ is a ring, the smallest one with a right identity which is not a two sided one. The direct sum $A=\{(\alpha_1,\alpha_2)\mid\alpha_i\in S\}$ of two copies of $({}_SS,+)$ is an S-module (cf. Remark 2.4). Clearly, the elements $u=(\varepsilon_r,\omega)$ and $v=(\varepsilon_r,\eta_r)$ form a basis V of $({}_SA,+)$.

$\cdot$	ω	ε_r	η_r	τ
ω	ω	ω	ω	ω
ε_r	ω	ε_r	ε_r	ω
η_r	ω	η_r	η_r	ω
τ	ω	τ	τ	ω

Whereas $\varepsilon_r u=u$ holds, the unique presentation (2.11) of v reads $v=\tau u+\eta_r v$ (cf. the proof of Prop. 2.6). Near by hand, the elements $u=(\varepsilon_r,\omega)$ and $w=(\omega,\varepsilon_r)$ from a basis U of $({}_SA,+)$, which even satisfies $\varepsilon_r u=u$ and $\varepsilon_r w=w$. Now, for $({}_SM,+)=({}_SA,+)$, consider the bijection $\varphi:U\to V$ given by $\varphi(u)=u$ and $\varphi(w)=v$. This bijection cannot be extended to an S-(semi)-module homomorphism f because of

$$\varepsilon_r f(w)=\varepsilon_r v=(\varepsilon_r,\varepsilon_r) \quad\text{and}\quad f(\varepsilon_r w)=f(w)=v=(\varepsilon_r,\eta_r).$$

§ 3 S-semimodules with universal properties

DEFINITION 3.1 Let S be a semiring, $({}_SF,+)$ an S-semimodule, $U\neq\emptyset$ a set, $\psi:U\to F$ a mapping and K a class of S-semimodules which contains at least one S-semimodule $({}_SM,+)$ such that $|M|\geq 2$ holds. Then we call $({}_SF,+)$ *universal with respect to* ψ *for the class* K iff the following conditions are satisfied:

(i) The subset $\psi(U)\subseteq F$ generates $({}_SF,+)$ which means that each S-subsemimodule of $({}_SF,+)$ containing $\psi(U)$ coincides with $({}_SF,+)$.

(ii) For each S-semimodule $({}_SM,+)\in K$ and each mapping $\varphi:U\to M$ there is an S-semimodule homomorphism $f:({}_SF,+)\to({}_SM,+)$ which satisfies $f\circ\psi=\varphi$.

In particular, $({}_SF,+)$ is called *freely generated by* $\psi(U)$ *in the class* K (briefly: *free on* $\psi(U)$ *in* K) if additionally $({}_SF,+)\in K$ holds. Clearly, we use the same concepts for S-modules over a ring S.

Considering each S-semimodule $({}_SM,+)$ as a universal algebra with the binary operation + and one unary operation $a\to\alpha a$ for each $\alpha\in S$, the following statement is well known, but also easily checked in a direct way.

LEMMA 3.2 a) If $({}_SF,+)$ is universal with respect to ψ for K, the mapping ψ is injective.

b) The homomorphism f in Def. 3.1 is uniquely determined by $f\circ\psi=\varphi$, and it is surjective iff $\varphi(U)$ generates $({}_SM,+)$.

c) If $({}_SF_1,+)$ is free in $\mathbb{K}$ on $\psi_1(U)$ and $({}_SF_2,+)$ is free in $\mathbb{K}$ on $\psi_2(U)$, then there exists an S-semimodule isomorphism $f:({}_SF_1,+) \to ({}_SF_2,+)$, in particular a unique one which satisfies $f \circ \psi_1 = \psi_2$.

The following theorem corresponds to the well known one on free unitary S-modules over a ring S, and we refer to Prop. 2.6 for the existence of $({}_SA,+)$.

THEOREM 3.3 Let S be a semiring with an identity ε and a zero ω satisfying $S\omega = \omega$, $({}_SA,+)$ an S-semimodule with a basis U, and $\psi:U \to A$ the identical injection. Then $({}_SA,+)$ is universal with respect to ψ for the class $\mathbb{K}$ of all unitary S-semimodules $({}_SM,+)$ with a zero o_M satisfying

(3.1) $\omega m = o_M$ for all $m \in M$,

which implies $\sigma o_M = o_M$ for all $\sigma \in S$ due to $S\omega = \omega$. In particular, $({}_SA,+)$ is contained in $\mathbb{K}$ and hence free on U in $\mathbb{K}$ iff the zero ω of S is absorbing.

Proof The condition (i) of Def. 3.1. is satisfied since $\psi(U) = U$ generates $({}_SA,+)$ by (2.11). For (ii), let $\varphi:U \to M$ be a mapping into an S-semimodule $({}_SM,+) \in \mathbb{K}$. Then

(3.2) $f(\sum_{u\in U} \alpha_u u) = \sum_{u\in U} \alpha_u \varphi(u)$ for all $\sum_{u\in U} \alpha_u u \in A$

defines a mapping $f:A \to M$, since $\Sigma\alpha_u u$ is a formal infinite sum, which implies the same for $\Sigma\alpha_u\varphi(u)$ due to (3.1). Clearly, f is an S-semimodule homomorphism of $({}_SA,+)$ into $({}_SM,+)$ and $f \circ \psi = \varphi$ holds by (3.2) because of $f(u) = f(\varepsilon u) = \varepsilon\varphi(u) = \varphi(u)$. Moreover, $\sigma o_M = \sigma(\omega m) = (\sigma\omega)m = \omega m = o_M$ holds, and the last statement follows from (2.6) sind $({}_SA,+)$ is isomorphic to $({}_SS\langle U\rangle,+)$ by Cor. 2.7.

Note that the (unique) S-semimodule homomorphism f which extends φ respects the zeroes. This was not intended in Def. 3.1 and seems to be merely a consequence of our assumptions on the class $\mathbb{K}$ in Thm. 3.3 concerning the zero o_M for each S-semimodule $({}_SM,+) \in \mathbb{K}$. However, these assumptions are necessary:

SUPPLEMENT 3.4 Let S be a semiring as in Thm. 3.3, $({}_SA,+)$ an S-semimodule with a basis U such that $|U| \geq 2$, and $\psi:U \to A$ the identical injection. Assume that (ii) holds for a unitary S-semimodule $({}_SM,+)$. Then $({}_SM,+)$ has a zero o_M and satisfies (3.1).

Proof For arbitrary elements $m,k \in M$, define $\varphi: U \to M$ such that $\varphi(u) = m$ and $\varphi(v) = k$ hold for some $u,v \in U$. By (ii) let f be an S-semimodule homomorphism $f: ({}_SA,+) \to ({}_SM,+)$ which extends φ. Then $\omega u = o_A$ implies $f(\omega u+\varepsilon v) = f(\varepsilon v)$ and hence $\omega m+k = k$ for all $m,k \in M$. This shows that ωm is the zero o_M of $(M,+)$ for each $m \in M$ and (3.1).

In the context of Thm. 3.3 and Suppl. 3.4 we state:

Fact 3.5 There are semirings S with an identity ε and a zero ω satisfying $S\omega = \omega$, but not $\omega S = \omega$. For such a semiring S, each S-semimodule $({}_SA,+)$ with a basis U is not contained in the class K of Thm. 3.3, since $({}_SA,+)$ satisfies $So_A = o_A$, but not $\omega A = o_A$ due to (2.5) and (2.6). An example of such a semiring S is given by the tables

+	ω	ε	α	β
ω	ω	ε	α	β
ε	ε	ε	α	α
α	α	α	α	α
β	β	α	α	β

$\cdot$	ω	ε	α	β
ω	ω	ω	β	β
ε	ω	ε	α	β
α	ω	α	α	β
β	ω	β	β	β

We further remark that, for semirings S with an identity and an absorbing zero, S-semimodules as introduced in [11] (they are called "S-modules" there) are just the S-semimodules contained in the class K of Thm. 3.3 in our notion. With this restrictions on the basic concepts it was stated in [11], § 2.4, that "S-modules" $({}_SS\langle U\rangle,+)$ are "free S-modules", which corresponds to the last statement of Thm. 3.3.

Finally, the assumption in Thm. 3.3 that S has an identity is essential for the statements on $({}_SA,+)$ even if one deals with rings:

Fact 3.6 Let S be a (semi)ring with a, say absorbing, zero ω which has a right identity but not a two-sided one, and $({}_SA,+)$ an S-(semi)module with a basis U. Then there exists no class K of S-(semi)modules such that $({}_SA,+)$ is free on U in K.

Proof We go by contradiction and assume that such a class K would exist. Then $({}_SA,+)$ is contained in K. By the proof of Prop. 2.6 a), each $v \in U$ has a presentation $v = \Sigma\tau_u u$ where $\tau_v = \varepsilon_r$ is a right identity of S. Since S has no identity, there are elements $\sigma,\varrho \in S$ satisfying $\varepsilon_r\sigma = \varrho \neq \sigma$. We define a mapping $\varphi: U \to A$ by $\varphi(v) = \sigma v$ and $\varphi(u) = o_A$ for all other $u \in U$ (if there are those). As assumed, there is an S-semimodule endomorphism f of $({}_SA,+)$ which extends φ. Then $f(v) = \Sigma\tau_u f(u)$ implies $\sigma v = \varepsilon_r \sigma v + o_A$, hence

$\sigma v = \varrho v$, contradicting $\sigma \neq \varrho$.

In the second part of this section we prove the existence of free S-semimodules in the class of all S-semimodules for any semiring S. For this purpose let $S' = S \cup \{\omega'\}$ be the semiring obtained from S by adjoining an absorbing zero $\omega' \notin S$ (regardless whether S has a zero or not), and $S'' = \mathbb{N}_0 \times S'$ the semiring obtained from S' by Lemma 2.1 (again regardless whether S and hence S' has an identity or not). Recall that $\varepsilon'' = (1,\omega')$ is the identity and $\omega'' = (0,\omega')$ the absorbing zero of S". It is straightforward to check that $S''\setminus\{\omega''\}$ is a subsemiring of S". We further use for the announced theorem that each S-semimodule can be embedded into a unitary S"-semimodule:

Fact 3.7 Let S be a semiring, S' and S" the semirings introduced above, and $({}_S M,+)$ an S-semimodule. Adjoin a zero $o' \notin M$ to $(M,+)$ by $o'+x = x+o' = x$ for all $x \in M' = M \cup \{o'\}$. Then $(M',+)$ becomes an S'-semimodule defining

(3.3) $\omega' x = o'$ for all $x \in M'$ and $\alpha' o' = o'$ for all $\alpha' \in S'$.

As a second step, applicable to each semiring S' with an absorbing zero ω', each S'-semimodule $({}_{S'}M',+)$ with a zero o' satisfying (3.3), and the semiring $S'' = \mathbb{N}_0 \times S'$ of Lemma 2.1, define

(3.4) $\alpha'' x = (\nu,\alpha')x = \nu x + \alpha' x$ for all $\alpha'' = (\nu,\alpha') \in S''$ and $x \in M'$,

where νx is the natural operation of $\mathbb{N}_0$ on $(M',+)$, cf. (2.2). Then $({}_{S''}M',+)$ is a unitary S"-semimodule with o' as zero, satisfying

(3.5) $\omega'' x = o' = \alpha'' o'$ for all $x \in M'$ and $\alpha'' \in S''$.

Further, if S' is obtained from S and M' from M as assumed above,

(3.6) $\alpha'' x = o'$ for $\alpha'' \in S''$ and $x \in M'$ implies $\alpha'' = \omega''$ or $x = o'$.

Hence $(M,+)$ is a unitary $(S''\setminus\{\omega''\})$-semimodule, and the S-subsemimodule of $({}_S M,+)$ generated by a subset $Y \subseteq M$ consists of the formal infinite sums (if Y is infinite) or finite sums (otherwise)

(3.7) $\sum_{y\in Y} \alpha''_y y \neq o'$ for all $\alpha''_y \in S''$ and $y \in Y$.

Proof It is not hard to check that $({}_{S'}M',+)$ is an S'-semimodule by (3.3). Somewhat more tedious, but also straightforward is to prove that (3.4) combines $({}_{S'}M',+)$ and $({}_{\mathbb{N}_0}M',+)$ to an S"-semimodule $({}_{S''}M',+)$. The latter is clearly unitary and satisfies (3.5) due to the definitions. For (3.6), $\alpha'' x = \nu x + \alpha' x = o'$ implies $\nu x = \alpha' x = o'$ and, for $x \neq o'$, as well $\nu = 0$ as $\alpha' = \omega'$ since $({}_S M,+)$ is an S-semimodule. Finally, the S-subsemimodule of $({}_S M,+)$ generated by $Y \subseteq M$ consists of all finite sums

$\Sigma\nu_i x_i$, $\Sigma\alpha_j y_j$ and $\Sigma\nu_i x_i + \Sigma\alpha_j y_j$ for $\nu_i \in \mathbb{N}$, $\alpha_j \in S$, $x_i, y_j \in Y$

which coincide by (3.4) with the sums (3.7) in $({}_{S''}M',+)$.

<u>THEOREM 3.8</u> *Let S be a semiring, $U \neq \emptyset$ a set, S" the semiring introduced above, $({}_{S''}A,+)$ the S"-semimodule with U as a basis, $o''_A = \Sigma\omega''u$ its zero, and $\psi: U \to A = S''\langle U\rangle$ the identical injection. Then $A\setminus\{o''_A\} = B$ is an S-subsemimodule $({}_S B,+)$ of $({}_{S''}A,+)$, and B is free on $\psi(U) = U$ in the class $\mathbb{K}$ of all S-semimodules.*

<u>Proof</u> Since S" is a semiring with an absorbing zero ω'' and an identity ε'', for each set $U \neq \emptyset$ the S"-semimodule $({}_{S''}A,+) = ({}_{S''}S''\langle U\rangle,+)$ with U as a basis exists. From $\nu+\mu = 0 \Rightarrow \nu = \mu = 0$ in $\mathbb{N}_0$ and $\alpha'+\beta' = \omega' \Rightarrow \alpha' = \beta' = \omega'$ in S' it follows that $B = A\setminus\{o''_A\}$ is an S-subsemimodule of $({}_{S''}A,+)$, and $\psi(U) = U$ clearly generates this S-semimodule $({}_S B,+)$, which is (i) of Def. 3.1. For (ii), let $\varphi: U \to M$ be a mapping into an S-semimodule $({}_S M,+)$. Then φ is also a mapping into the unitary S"-semimodule $({}_{S''}M',+)$ obtained from $({}_S M,+)$ by Fact 3.7. Since $({}_{S''}M',+)$ has a zero o' and satisfies (3.1) by (3.5), we can apply Thm. 3.3 and obtain an S"-semimodule-homomorphism f of $({}_{S''}A,+)$ into $({}_{S''}M',+)$ which extends φ. Clearly, f is also an S-semimodule homomorphism, and we know $f(o''_A) = o'$. According to (3.2),

$$f(\sum_{u\in U} \alpha''_u u) = o' \quad \text{implies} \quad \sum_{u\in U} \alpha''_u \varphi(u) = o'$$

hence $\alpha''_u\varphi(u) = o'$ for all $u \in U$. By $\varphi(u) \neq o'$ and (3.6), this yields $\Sigma\alpha''_u u = o''_A$. So we can restrict f to $B = A\setminus\{o''_A\}$ and obtain an S-semimodule homomorphism f of $({}_S B,+)$ into $({}_S M,+)$ which extends φ proving (ii).

<u>Remark 3.9</u> A simplified version of Thm. 3.8 holds if one only considers semirings S with an absorbing zero ω and S-semimodules $({}_S M,+)$ with a zero o_M satisfying (3.1). Then, for the semiring $T = \mathbb{N}_0 \times S$ according to Lemma 2.1, the unitary T-semimodule $({}_T A,+)$ with a basis U is an S-semimodule which is free on U in the class $\mathbb{K}$ of all S-semimodules as described above. This statement, in turn, corresponds to one on S-modules over arbitrary rings S. One only has to replace the semiring $T = \mathbb{N}_0 \times S$ by the well known Dorroh ring extension $T = \mathbb{Z} \times S$ of S (cf. e.g. [14], § 35), where $\mathbb{Z}$ denotes the ring of integers.

<u>§ 4 Generalized semialgebras $(S\langle U\rangle,+,\cdot)$ over a semiring S</u>

In the following we restrict ourselves to semirings S with an absorbing zero ω. We further need that S has a right identity in order that, for each cardinality of U, S-semimodules $({}_S A,+)$ with a basis U exist (Prop. 2.6). Each of the latter is isomorphic to $({}_S S\langle U\rangle,+)$ (Cor. 2.7,

Fact 2.3), and we use (2.11) to denote its elements. In the important case that S has an identity, these S-semimodules are unitary and free on U in the class $\mathbb{K}$ described in Thm. 3.3.

DEFINITION 4.1 Let S be a semiring with an absorbing zero ω and at least one right identity. Then $({}_SA,+,\cdot)$ is called a *generalized semialgebra over* S (or a *generalized* S-*semialgebra*) iff

a) $({}_SA,+)$ is an S-semimodule with a zero o,
b) the binary operations of $(A,+,\cdot)$ satisfy

(4.1) $a(b+c) = ab+ac$ and $(b+c)a = ba+ca$ for all $a,b,c \in A$, and

c) there exists a basis U of $({}_SA,+)$ which satisfies

(4.2) $(\alpha u)(\beta v) = (\alpha\beta)(uv)$ for all $\alpha,\beta \in S$ and $u,v \in U$.

In particular, $({}_SA,+,\cdot)$ is called a *semialgebra over* S iff it satisfies (1.2), which clearly implies (4.2). Finally, a (generalized) semialgebra $({}_SA,+,\cdot)$ is called a (*generalized*) *algebra over* S iff $(A,+,\cdot)$ is a (not necessarily associative) ring, which is just the case iff S is assumed to be a ring.

We shall see that each of the basic statements on the classical concept of algebras over a ring S corresponds to one on generalized semialgebras over a semiring S. There are some minor simplifications if one turns to semialgebras over semirings, whereas in both cases the assumption that S is an (associative) ring only yields that $({}_SA,+)$ is an S-module, but no further simplifications in our context. These specializations will be mentioned in the following only sporadically.

Moreover, the main interest will be with the case that the multiplication of a generalized semialgebra $({}_SA,+,\cdot)$ is associative and hence $(A,+,\cdot)$ a semiring according to our definition in § 1. But even to obtain corresponding criterions (Thm. 4.6, Cor. 4.7) some statements on generalized S-semialgebras as defined in Def. 4.1 are indispendable. At first, we obtain as an immediate consequence of (4.1) and (4.2):

THEOREM 4.2 Let $({}_SA,+,\cdot)$ be a generalized S-semialgebra and U a basis of $({}_SA,+)$ satisfying (4.2). Then the multiplication of A is uniquely determined by the products

(4.3) $uv = \sum_{w\in U} \gamma_{u,v}^{w}\, w$ for all $u,v \in U$ according to

(4.4) $ab = (\sum_{u\in U} \alpha_u u)(\sum_{v\in U} \beta_v v) = \sum_{u,v\in U} (\alpha_u\beta_v)(uv)$ or

(4.5) $\quad ab = \sum_{w \in U} \left(\sum_{u,v \in U} \alpha_u \beta_v \gamma_{u,v}^{w} \right) w$ *for all* $a,b \in A$.

The elements $\gamma_{u,v}^{w} \in S$ *occuring on the right side of* (4.3) *are called the* ***structure constants*** *of* $({}_SA,+,\cdot)$ *with respect to* U. *Clearly,*

(4.6) $\quad$ *for each* $(u,v) \in U \times U$ *almost all* $\gamma_{u,v}^{w}$ *equal* ω.

Note that the zero $o = \Sigma\omega u$ *of* $({}_SA,+)$ *is absorbing in* $({}_SA,+,\cdot)$.

Remark 4.3 According to (2.12), U is a basis of $({}_SA,+)$ iff the same holds for $\varepsilon_r U = \{\varepsilon_r u \mid u \in U\}$ and a right identity ε_r of S. One also checks that (4.2) for U implies (4.2) for $\varepsilon_r U$. (The converse is also true, but harder to prove.) Thus, for an S-semimodule $({}_SA,+)$ or a generalized S-semialgebra $({}_SA,+,\cdot)$, without loss of generality we may assume that

(4.7) $\quad \varepsilon_r u = u$ for all $u \in U$ and a fixed right identity ε_r of S

holds for each basis U under consideration. Clearly, (4.7) yields $\varepsilon_r \gamma_{u,v}^{w} = \gamma_{u,v}^{w}$ for all structure constants by (4.2) and (4.3), and all this is self-evident if S has an identity.

THEOREM 4.4 *Let* S *be a semiring as in Def. 4.1 and* $({}_SA,+)$ *an S-semimodule with a basis* U. *According to Remark 4.3, we may assume that* U *satisfies (4.7). Then each multiplication of the elements of* U *subjected to* $uv = \varepsilon_r A$ *(equivalently: defined by (4.3) for any structure constants* $(\gamma_{u,v}^{w} \in \varepsilon_r S \mid (u,v,w) \in U \times U \times U)$ *satisfying (4.6)) can be extended to a multiplication on* A *by (4.4) or (4.5) such that* $({}_SA,+,\cdot)$ *is a generalized S-semialgebra and* U *satisfies* (4.2).

The proof itself is straightforward, but we want to point out the effect of (4.7). Consider an arbitrary basis U of $({}_SA,+)$ and a multiplication (4.3), say $u \cdot v$ for all $u,v \in U$. Now, in the pattern of Def. 4.1 and Thm. 4.2, define a multiplication on A by (4.4) or (4.5), say $a \circledast b$ for all $a,b \in A$. Using $u \cdot v$ on the right side of (4.4), one easily checks that (4.4) and (4.5) define the same multiplication $a \circledast b$ on A, which also satisfies (4.1). Nevertheless, if S has no identity, the multiplication $\circledast$ on A need not extend the multiplication $u \cdot v$ for $u,v \in U$ used to define it - not even if one deals with generalized algebras over rings. E. g., let S be the ring of Expl. 2.8, $({}_SA,+)$ an S-semimodule with a basis U, and define $u \cdot v = \tau v \neq o$ for all $u,v \in U$. Then, by (4.4) or (4.5), we get $a \circledast b = o$ for all $a,b \in A$. A similar consideration was given in [13] and caused some restrictions to the unitary case in that paper, whereas (4.7) allows to include rings and semirings S without an identity simultaneously. In this context we mention:

Remark 4.5 In Def. 4.1, the assumption c) is equivalent to

(4.2') $\alpha(ab) = (\alpha a)b$ and $u(\beta b) = \varepsilon_r\beta(ub)$

for all $\alpha,\beta \in S$, all $a,b \in A$ and all $u \in U$ for a basis U of $({}_SA,+)$ which satisfies (4.7). The implication (4.2') ⇒ (4.2) is immediate, and the converse follows from (4.5). Clearly, it is superflouos to note ε_r in (4.2') if it is the identity of S. In this form (4.2') was used in [21] to introduce generalized algebras over a ring with identity. Note also that $\alpha(ab) = (\alpha a)b$ is the first equation of (1.2).

THEOREM 4.6 *Let $({}_SA,+,\cdot)$ be a generalized semialgebra over a semiring S with a basis U satisfying (4.2) and (4.7) and structure constants $\gamma_{u,v}^{w} \in \varepsilon_r S$. Then the multiplication is associative and hence $(A,+,\cdot)$ a semiring iff*

(4.8) $$\sum_{x\in U} \gamma_{u,v}^{x}\delta\gamma_{x,w}^{y} = \sum_{x\in U} \varepsilon_r\delta\gamma_{v,w}^{x}\gamma_{u,x}^{y}$$

holds for all $u,v,w,y \in U$, each $\delta \in S$ and the fixed right identity ε_r of S, which in turn is equivalent to

(4.9) $(uv)(\delta w) = u(v(\delta w))$ *for all* $u,v,w \in U$ *and* $\delta \in S$.

Proof For elements $a = \Sigma\alpha_u u$, $b = \Sigma\beta_v v$ and $d = \Sigma\delta_w w$ of A we get

$$(ab)d = \sum_y\Big(\sum_{x,w}\Big(\sum_{u,v}\alpha_u\beta_v\gamma_{u,v}^{x}\Big)\delta_w\gamma_{x,w}^{y}\Big)y \quad \text{and}$$

$$a(bd) = \sum_y\Big(\sum_{u,x}\alpha_u\Big(\sum_{v,w}\beta_v\delta_w\gamma_{v,w}^{x}\Big)\gamma_{u,x}^{y}\Big)y$$

by (4.5) and (4.6). So $(A,\cdot)$ is associative iff

(4.10) $$\sum_{u,v,w,x}\alpha_u\beta_v\gamma_{u,v}^{x}\delta_w\gamma_{x,w}^{y} = \sum_{u,v,w,x}\alpha_u\beta_v\delta_w\gamma_{v,w}^{x}\gamma_{u,x}^{y}$$

holds for each $y \in U$ and all $\alpha_u,\beta_v,\delta_w \in S$ such that almost all equal ω. The latter implies each of the equations (4.8): we choose $\alpha_u = \beta_v = \delta_w = \omega$ for all $u,v,w \in U$ always with the exception of one triple (u,v,w) for which we put $\alpha_u = \beta_v = \varepsilon_r$ and $\delta_w = \delta$, where of course $\varepsilon_r\gamma_{u,v}^{w} = \gamma_{u,v}^{w}$ is used. Conversely, if all equations (4.8) hold, we clearly get each equation of (4.10) as a linear combination of the former. The equivalence of (4.8) and (4.9) corresponds to the specializations $a = \varepsilon_r u$, $b = \varepsilon_r v$ and $d = \delta w$.

COROLLARY 4.7 *Let $({}_SA,+,\cdot)$ be a generalized S-semialgebra with a basis U which satisfies (4.2), (4.7) and*

(4.11) $(uv)w = u(vw)$ *for all* $u,v,w \in U$.

Then $({}_S A,+,\cdot)$ is a semiring if all structure constants $\gamma_{u,v}^{w}$ are contained a) in the centre of $(S,\cdot)$, or b) in $\{\omega,\varepsilon_r\}$, or c) in $\{\omega,\varepsilon_r,-\varepsilon_r\}$, the latter clearly if S is a ring.

Proof By (4.9), (4.11) is equivalent to the condition obtained from (4.8) in choosing $\delta=\varepsilon_r$. The latter implies again (4.8) if a), b) or c) holds, because of $\varepsilon_r\delta\gamma_{u,v}^{x}=\gamma_{u,v}^{x}\delta$ for all $\delta\in S$ in all three cases.

Now various examples of semirings and rings with associative multiplication can be obtained (and even defined) directly from these considerations. We restrict ourselves to the following ones:

Example 4.8 Let S be a (semi)ring with an absorbing zero ω and a right identity ε_r.

a) For each semigroup $(U,\cdot)$, the *semigroup (semi)ring of* $(U,\cdot)$ *over* S is defined as a generalized (semi)algebra $({}_S S\langle U\rangle,+,\cdot)$ with $\varepsilon_r u=u$ for each $u\in U$, where the multiplication of $(U,\cdot)$ is used to define the products (4.3) (cf. Thm. 4.4 and Cor. 4.7 b)). Choosing $(U,\cdot)$ as the free (commutative) semigroup with identity on a set $X=\{x_1,x_2,\ldots\}$, we obtain *polynomial (semi)rings* $S[x_1,x_2,\ldots]$ *in the (commutative or) non-commutative indeterminates* $x_1,x_2,\ldots$ *over* S (cf. also Cor. 4.14).

b) If the semigroup $(U,\cdot)$ has an absorbing element 0, one mostly considers the *contracted semigroup (semi)ring of* $(U,\cdot)$ *over* S. It is obtained in the same way as in a) replacing for (4.3) the products $0u=u0=0\in U$ for all $u\in U$ by $0u=u0=0=o=\Sigma\omega u$. Choosing $(U,\cdot)$ as a semigroup of matrix units, say $\{0,e_{\nu,\mu}\mid \nu,\mu=1,\ldots,n\}$ and $e_{\nu,\mu}\,e_{\kappa,\lambda}=\delta_{\mu,\kappa}\,e_{\nu,\lambda}$, the *full matrix (semi)ring* $M_n(S)$ is obtained as a generalized (semi)algebra over S.

Remark 4.9 For many applications of generalized S-(semi)algebras the (semi)ring S will in fact have an identity. On the other hand, let S be a (semi)ring which does not even have a right identity (but an absorbing zero ω). Then, for each set U, the S-(semi)module $({}_S S\langle U\rangle,+)$ according to Fact 2.3 exists yet U is not a subset of $S\langle U\rangle$. But we can embed S into the semiring $\mathbb{N}_0\times S=T$ of Lemma 2.1 and consider $S\langle U\rangle$ as an S-sub(semi)-module of the T-semimodule $({}_T T\langle U\rangle,+)$ which contains U as a basis. Now each generalized T-semialgebra $({}_T T\langle U\rangle,+,\cdot)$ induces a multiplication on $({}_S S\langle U\rangle,+)$ according to (4.5), where the $\gamma_{u,v}^{w}\in T$ work as right operators on the products $\alpha_u\beta_v\in S$. In particular, if $(T\langle U\rangle,+,\cdot)$ is associative, the same holds for $(S\langle U\rangle,+,\cdot)$. If S is a ring, one may prefer to use the Dorroh ring extension $T=\mathbb{Z}\times S$ (cf. Remark 3.9) instead of the semiring $\mathbb{N}_0\times S$ in order that $(T\langle U\rangle,+,\cdot)$ is also a generalized T-algebra, but this is in fact superfluous.

PROPOSITION 4.10 *Let $({}_SA,+,\cdot)$ be a generalized semialgebra over a semiring S with a basis U satisfying (4.2) and (4.7) and structure constants $\gamma_{u,v}^{w}$. Then $({}_SA,+,\cdot)$ is a semialgebra over S iff*

(4.12) $\sigma\alpha\gamma_{u,v}^{w} = \alpha\sigma\gamma_{u,v}^{w}$ *holds for all* $u,v,w \in U$ *and* $\sigma,\alpha \in S$.

In this case, each basis V of $({}_SA,+)$ satisfies (4.2) such that $\varepsilon_r V$ can be used instead of U, and $(A,\cdot)$ is associative iff (4.11) holds for some basis. Clearly, (4.12) is satisfied if $(S,\cdot)$ is commutative.

Proof A generalized S-semialgebra $({}_SA,+,\cdot)$ is an S-semialgebra iff (1.2) holds, or according to Remark 4.5

(4.13) $(\sigma a)b = a(\sigma b)$ for all $\sigma \in S$ and $a,b \in A$.

By (4.5), (4.13) is equivalent to $\sigma\alpha_u\beta_v\gamma_{u,v}^{w} = \alpha_u\sigma\beta_v\gamma_{u,v}^{w}$ for all $\sigma,\alpha_u,\beta_v \in S$ and $u,v,w \in U$. The latter implies (4.12) by $\beta_v = \varepsilon_r$, and one obtains the converse by $\sigma\alpha\beta\gamma_{u,v}^{w} = \beta\sigma\alpha\gamma_{u,v}^{w} = \beta\alpha\sigma\gamma_{u,v}^{w} = \alpha\sigma\beta\gamma_{u,v}^{w}$ applying three times (4.12). The other statements are obvious.

A corresponding criterion such that a generalized algebra $({}_SA,+,\cdot)$ is an algebra by the aid of ideals of the ring S was given in [13]. We now turn to other questions, again for generalized semialgebras $({}_SA,+,\cdot)$. The first one concerns the existence of one- or two-sided identities of A, where the main interest is of course with sufficient conditions, in particular those which follow from the multiplication (4.3) of the elements of a basis U.

LEMMA 4.11 *Let $({}_SA,+,\cdot)$ be a generalized semialgebra over a semiring S with a basis U satisfying (4.2) and (4.7). Then we have:*

a) *An element $c \in A$ is a right identity of A iff $uc = u$ holds for all $u \in U$.*
b) *If $c \in A$ is a left identity of A, then S has an identity and $cu = u$ holds for all $u \in U$. The converse is true if c itself is an element of U, or if $({}_SA,+,\cdot)$ is an S-semialgebra.*
c) *If $c \in A$ is an identity of A, then S has an identity and $cu = uc = u$ holds for all $u \in U$. The converse is true if one of the conditions of b) is satisfied, or if $(A,\cdot)$ is associative.*

Proof Part a) is obvious by (4.1). For b) assume that $c \in A$ is a left identity of A. Now $ab = a(\varepsilon_r b)$ holds for all $a,b \in A$ by (4.5). Hence $a = c$ yields $b = \varepsilon_r b$ for all $b \in A$, in particular $\beta u = \varepsilon_r(\beta u) = (\varepsilon_r\beta)u$ for all $\beta \in S$ and some $u \in U$. This shows that the right identity ε_r of S is a two-sided one and hence the first implication of b). For the converse, note that $cu = u$ for all $u \in U$ implies by (4.1) that c is a left identity of A iff $c(\beta u) = \beta(cu)$ holds for all $\beta \in S$ and all $u \in U$. The latter follows for $c \in U$ from (4.2) by $(\varepsilon c)(\beta u) = (\varepsilon\beta)(cu)$, and without further assumptions from (1.2) if $({}_SA,+,\cdot)$ is an S-semialgebra.

For c) it remains to show: if S has an identity and $(A,\cdot)$ is associative, then $cu = uc = u$ for all $u \in U$ implies that c is a left identity of A, or that $c(\beta u) = \beta(cu)$ holds as discussed above. Now the first assumption yields $\beta(uc) = u(\beta c)$ by (4.2'). From this and the other assumptions we obtain $c(\beta u) = c(\beta(uc)) = c(u(\beta c)) = (cu)(\beta c) = u(\beta c) = \beta(uc) = \beta(cu)$.

We could not decide whether the converse of b) holds if one assumes that $(A,\cdot)$ is associative as in c), not even in the case that S is a ring. Apart from this gap, our sufficient conditions in b) and c) such that c is a (left) identity cannot be weakened considerably according to the following simple examples:

Example 4.12 a) Let S be a ring with a right identity ε_r which is not two-sided and $({}_SA,+,\cdot)$ any polynomial ring of Expl 4.8 a), for simplicity $S[x]$ with the basis $U = \{\varepsilon_r x^\nu \mid \nu \in \mathbb{N}_0\}$. Then $c = \varepsilon_r x^0 \in U$ satisfies $cu = uc = u$ for all $u \in U$, but $c\alpha x^\nu = \varepsilon_r x^0 \alpha x^\nu = \varepsilon_r \alpha x^\nu$ shows that c is not a left identity of $S[x]$, though $(S[x],\cdot)$ is associative (cf. Cor. 4.14).

b) Let S be a ring with an identity ε and elements $\alpha,\beta \in S$ satisfying $\alpha\beta = \varepsilon$ and $\beta\alpha \neq \varepsilon$, e. g. the semigroup ring of the bicyclic semigroup over $\mathbb{Z}$. Define a generalized algebra $({}_SA,+,\cdot)$ with a basis $U = \{u\}$ by the multiplication $uu = \beta u$ according to Thm. 4.4. Then the element $c = \alpha u \in A\setminus U$ satisfies $cu = uc = \alpha\beta u = u$, but $c(\beta\alpha u) = \alpha\beta\alpha\beta u \neq \beta\alpha u$ shows that c is not a left identity of A. With respect to Lemma 4.11 c) $(A,\cdot)$ cannot be associative To check the latter by Thm. 4.6, we get that (4.8) for $({}_SA,+,\cdot)$ is equivalent to $\beta\delta\beta = \delta\beta\beta$ for each $\delta \in S$, which fails to be true e. g. for $\delta = \alpha^2$ by our assumptions on α and β.

Let $({}_SA,+,\cdot)$ and $({}_SB,+,\cdot)$ be generalized S-semialgebras. Then a homomorphism $f:(A,+,\cdot) \to (B,+,\cdot)$ with respect to both binary operations which is also an S-semimodule homomorphism $f:({}_SA,+) \to ({}_SB,+)$ should be called a *generalized S-semialgebra homomorphism*, but we believe that *S-semialgebra homomorphism* will cause no confusion. We now want to embed $(S,+,\cdot)$ by a semiring monomorphism g into $({}_SA,+,\cdot)$. There are cases with various possibilities to do this, but of course we are only interested in those embeddings g for which all products $g(\sigma)a$ in A coincide with σa obtained in $({}_SA,+)$, i. e.

(4.14) $\quad g(\sigma)a = \sigma a$ for all $\sigma \in S$ and $a \in A$.

Note that S has a right identity ε_r and may hence be considered as a generalized S-semialgebra $({}_SS,+,\cdot)$ with $\{\varepsilon_r\}$ as a basis. Then (4.14) implies $g(\sigma\delta) = g(\sigma)g(\delta) = \sigma g(\delta)$, hence we may consider $g: ({}_SS,+,\cdot) \to ({}_SA,+,\cdot)$ as an S-semialgebra monomorphism.

PROPOSITION 4.13 *Let $({}_SA,+,\cdot)$ be a generalized semialgebra over a semiring S with a basis U satisfying (4.2) and (4.7). Then, depending on the given assumptions, there are the following monomorphic embeddings g of $(S,+,\cdot)$ into $({}_SA,+,\cdot)$ which satisfy (4.14):*

a) *If A has a left identity $c \in A$, define $g(\sigma) = \sigma c$.*

b) *If some $v \in U$ satisfies $vu = u$ for all $u \in U$, define $g(\sigma) = \sigma v$.*

Proof Assume $\delta,\sigma \in S$. Then, in both cases, $g(\delta+\sigma) = g(\delta) + g(\sigma)$ is clear and g is injective since $\delta c = \sigma c$ implies $\delta v = \sigma v$ for some $v \in U$ in case a). Further, $g(\delta\sigma) = g(\delta)g(\sigma)$ follows from $(\delta\sigma)c = \delta(\sigma c) = \delta(c(\sigma c)) = (\delta c)(\sigma c)$ in the first case and from $(\delta\sigma)v = (\delta\sigma)(vv) = (\delta v)(\sigma v)$ by (4.2) in the second one, for which only $v^2 = v \in U$ was used so far. Finally, (4.14) is clear for a), and it follows for b) from (4.1) and (4.2) by

$$g(\sigma)(\beta u) = (\sigma v)(\beta u) = (\sigma\beta)(vu) = \sigma(\beta u).$$

COROLLARY 4.14 *Let the generalized semialgebra $({}_SA,+,\cdot)$ be the semigroup semiring of a semigroup $(U,\cdot)$ over S (cf. Expl. 4.8 a)). Assume that $v \in U$ is a left (or a two-sided) identity of $(U,\cdot)$. Then $g(\sigma) = \sigma v$ defines an embedding of $(S,+,\cdot)$ into $({}_SA,+,\cdot)$ such that (4.14) holds. But v need not be a left (or a two-sided) identity of A, since the latter holds iff S has an identity.*

We conclude this section with the consideration of generalized semialgebras over generalized semialgebras over a semiring S. If S is assumed to be a ring, the resulting statements on generalized algebras are essentially those given in [13]:

THEOREM 4.15 *Let $({}_SA,+,\cdot)$ be a generalized semialgebra over a semiring S with a basis U satisfying (4.2) and (4.7) such that $(A,+,\cdot)$ is a semiring with a right identity e_r. Consider a generalized semialgebra $({}_A\mathfrak{B},+,\cdot)$ over A with a basis $\mathfrak{V}$ satisfying (4.2) and (4.7), i.e.*

$$(a\mathfrak{u})(b\mathfrak{v}) = (ab)(\mathfrak{u}\mathfrak{v}) \quad \text{and} \quad e_r\mathfrak{u} = \mathfrak{u} \quad \text{for all } a,b \in A \text{ and } \mathfrak{u},\mathfrak{v} \in \mathfrak{V}.$$

Then $(\mathfrak{B},+)$ is an S-semimodule $({}_S\mathfrak{B},+)$ according to

(4.15) $\quad \alpha\mathfrak{b} = \alpha(\sum_{\mathfrak{v}\in\mathfrak{V}} b_{\mathfrak{v}}\mathfrak{v}) = \sum_{\mathfrak{v}\in\mathfrak{V}} (\alpha b_{\mathfrak{v}})\mathfrak{v}$ *for all* $\alpha \in S$ *and* $\mathfrak{b} = \Sigma b_{\mathfrak{v}}\mathfrak{v} \in \mathfrak{B}$

such that $\mathfrak{U}\mathfrak{V} = \{u\mathfrak{v} \mid u \in U,\ \mathfrak{v} \in \mathfrak{V}\}$ *is a basis of* $({}_S\mathfrak{B},+)$ *which satisfies* (4.2) *and* (4.7) *for* $({}_S\mathfrak{B},+,\cdot)$. *Hence* $({}_S\mathfrak{B},+,\cdot)$ *is also a generalized semialgebra over the semiring* S. *Further, if* $\gamma_{u,v}^{w} \in \varepsilon_r S$ *and* $c_{\mathfrak{u},\mathfrak{v}}^{\mathfrak{w}} = \sum_{y\in U} \delta_{\mathfrak{u},\mathfrak{v}}^{y,\mathfrak{w}} y \in e_r A$ *are the structure constants for both steps,*

$$(4.16)\qquad \Gamma_{u\mathfrak{u},v\mathfrak{v}}^{w\mathfrak{w}} = \sum_{x,y\in U} \gamma_{u,v}^{x}\,\delta_{\mathfrak{u},\mathfrak{v}}^{y,\mathfrak{w}}\,\gamma_{x,y}^{w}$$

are the structure constants of $({}_S\mathfrak{B},+,\cdot)$ *with respect to the basis* $U\mathfrak{V}$.

In particular, $({}_S\mathfrak{B},+,\cdot)$ *is a semialgebra over* S *if* $({}_SA,+,\cdot)$ *is one, whereas* $({}_A\mathfrak{B},+,\cdot)$ *may be merely a generalized semialgebra over* A.

Proof One easily checks that $({}_A\mathfrak{B},+)$ is also an S-semimodule $({}_S\mathfrak{B},+)$ with respect to (4.15) and that both are connected by

$$(4.17)\qquad \alpha(a\mathfrak{b}) = (\alpha a)\mathfrak{b} \quad \text{for all} \quad \alpha\in S,\ a\in A \quad \text{and} \quad \mathfrak{b}\in\mathfrak{B}.$$

Now

$$(4.18)\qquad \mathfrak{b} = \sum_{\mathfrak{v}} b_{\mathfrak{v}}\mathfrak{v} = \sum_{\mathfrak{v}}\Big(\sum_{u}\beta_{u,\mathfrak{v}}u\Big)\mathfrak{v} = \sum_{\mathfrak{v}}\Big(\sum_{u}(\beta_{u,\mathfrak{v}}u)\mathfrak{v}\Big) = \sum_{\mathfrak{v}}\Big(\sum_{u}\beta_{u,\mathfrak{v}}(u\mathfrak{v})\Big)$$

shows that $U\mathfrak{V}$ is a basis of $({}_S\mathfrak{B},+)$. By (4.17) one obtains $\varepsilon_r(u\mathfrak{v}) = u\mathfrak{v}$ and that (4.2) for both steps yields $(\alpha(u\mathfrak{u}))(\beta(v\mathfrak{v})) = (\alpha\beta)((u\mathfrak{u})(v\mathfrak{v}))$ for $({}_S\mathfrak{B},+,\cdot)$ with respect to the multiplication of $({}_A\mathfrak{B},+,\cdot)$. Since the latter satisfies (4.1), $({}_S\mathfrak{B},+,\cdot)$ is a generalized semialgebra over S according to Def. 4.1. Applying Thm. 4.2 to $({}_S\mathfrak{B},+,\cdot)$, this multiplication is according to

$$(4.19)\qquad \Big(\sum_{u\mathfrak{u}\in U\mathfrak{V}}\alpha_{u,\mathfrak{u}}u\mathfrak{u}\Big)\Big(\sum_{v\mathfrak{v}\in U\mathfrak{V}}\beta_{v,\mathfrak{v}}v\mathfrak{v}\Big) = \sum_{w\mathfrak{w}\in U\mathfrak{U}}\Big(\sum_{u\mathfrak{u},v\mathfrak{v}\in U\mathfrak{U}}\alpha_{u,\mathfrak{u}}\beta_{v,\mathfrak{v}}\Gamma_{u\mathfrak{u},v\mathfrak{v}}^{w\mathfrak{w}}\Big)w\mathfrak{w}$$

determined by the structure constants $\Gamma_{u\mathfrak{u},v\mathfrak{v}}^{w\mathfrak{w}}$, for which one obtains (4.16) in calculating the products $(u\mathfrak{u})(v\mathfrak{v})$. The last implication follows from Prop. 4.10, since (4.12) for the structure constants $\gamma_{u,v}^{w}$ of $({}_SA,+,\cdot)$ yields the corresponding statement for the structure constants (4.16) of $({}_S\mathfrak{B},+,\cdot)$. But $({}_A\mathfrak{B},+,\cdot)$ need not be an A-semialgebra in this case:

Example 4.16 Let S be a commutative (semi)ring, say with an identity, and $({}_SA,+,\cdot)$ a matrix (semi)ring $M_n(S)$. Then the polynomial (semi)ring $\mathfrak{B} = A[x]$ is merely a generalized (semi)algebra $({}_A\mathfrak{B},+,\cdot)$ over A, whereas $({}_S\mathfrak{B},+,\cdot)$ as considered in Thm. 4.15 and $({}_SA,+,\cdot)$ are (semi)algebras since S was assumed to be commutative.

§ 5 Generalized semialgebras $(S\langle\langle U\rangle\rangle,+,\cdot)$ over a semiring S

Let S be again a semiring with an absorbing zero ω and a fixed right identity ε_r, clearly the identity of S if there is one. For a generalized S-semialgebra as considered in § 4, we may always identify $({}_SA,+,\cdot)$ with $({}_SS\langle U\rangle,+,\cdot)$, where $({}_SS\langle U\rangle,+)$ is the S-semimodule of Fact 2.3. In the case that U is infinite, certain semirings are used in the applications mentioned in § 1 which are generalizations of the above concept to the S-semimodules $({}_SS\langle\langle U\rangle\rangle,+)$ of Fact 2.2. To deal with generalizations of this kind, we assume that U is contained in $S\langle\langle U\rangle\rangle$ by (2.9), which implies (4.7). Now, corresponding to (4.3), consider products $uv = \Sigma\gamma_{u,v}^{w}w$ defined in $S\langle\langle U\rangle\rangle$ for all $u,v \in U$ with structure constants $\gamma_{u,v}^{w} = \varepsilon_r\gamma_{u,v}^{w}$, clearly not subjected to (4.6). These products can be extended to a multiplication on $({}_SS\langle\langle U\rangle\rangle,+)$ by (4.5) provided that

(5.1) $\quad \sum_{u,v\in U} \alpha_u\beta_v\gamma_{u,v}^{w}$ is defined in S for all $w \in U$ and $\alpha_u,\beta_v \in S$.

Then $({}_SS\langle\langle U\rangle\rangle,+,\cdot)$ obviously satisfies (4.2), but one already needs further assumptions to show (4.1). So we generalize Def. 4.1 as follows:

<u>DEFINITION 5.1</u> For S and $({}_SS\langle\langle U\rangle\rangle,+)$ as above, $({}_SS\langle\langle U\rangle\rangle,+,\cdot)$ is called a *generalized semialgebra over* S if the multiplication is defined by (4.5) and (5.1), and if (4.1) holds. In particular, we talk about a *semialgebra* $({}_SS\langle\langle U\rangle\rangle,+,\cdot)$ or a *(generalized) algebra* $({}_SS\langle\langle U\rangle\rangle,+,\cdot)$ *over* S according to the use of these terms in Def. 4.1.

For a general treatment of those generalized S-semialgebras one clearly needs some assumptions and statements on infinite sums in S and in $S\langle\langle U\rangle\rangle$ (cf. § 6 and § 7). In this section we only sketch two extreme cases which are often used in applications. The first one can in fact be handled without any preparations as given in § 6:

<u>THEOREM 5.2</u> <u>Let S be a semiring with an absorbing zero ω and a right identity ε_r and define a multiplication on $({}_SS\langle\langle U\rangle\rangle,+)$ for an (infinite) set U by (4.5) with structure constants $\gamma_{u,v}^{w} \in \varepsilon_rS$ satisfying</u>

(5.2) <u>for each $w \in U$ almost all $\gamma_{u,v}^{w}$ equal ω</u>

<u>and hence (5.1). Then $({}_SS\langle\langle U\rangle\rangle,+,\cdot)$ is a generalized S-semialgebra, and all statements on generalized S-semialgebras $({}_SS\langle U\rangle,+,\cdot)$ given in § 4 transfer to $({}_SS\langle\langle U\rangle\rangle,+,\cdot)$.</u>

Due to (5.2) the distributivity (4.1) is clear, and also all other statements can be proved in the pattern of § 4. But the near by hand argument that "all infinite sums needed in S have only a finite number of non-zero summands" has to be interpreted in a suitable way according to the following. Let us e. g. consider the proof of Thm. 4.6 in order to transfer it to $({}_S S\langle\langle U\rangle\rangle,+,\cdot)$. Then one needs, in a simplified version, that for certain $\sigma_{i,j} \in S$ sums of the form

(5.3) $\quad \sum_{i\in I}(\sum_{j\in J}\sigma_{i,j})$ can be transformed to $\sum_{j\in J}(\sum_{i\in I}\sigma_{i,j})$.

If one only assumes that each sum occuring on the left side of (5.3) is formal infinite, the same need not be true for the right side, and even if the latter is the case, both sides need not be equal. (E. g. for $S = \mathbb{Z}$ and $I = J = \mathbb{N}$ let all non-zero $\sigma_{i,j}$ be $\sigma_{i,i} = i$ and $\sigma_{i,i+1} = -i$, or $\sigma_{i,i} = 1$ and $\sigma_{i,i+1} = -1$.) In fact, however, one can check that in the discussed situation almost all $\sigma_{i,j}$ occuring in (5.3) equal ω. So the above argument has to be applied to iterated sums and corresponding expressions, and one has to check that this interpretation is sufficient to transfer all proofs given in § 4.

In this paper we shall show that each $({}_S S\langle\langle U\rangle\rangle,+,\cdot)$ as defined in Thm. 5.2 is a partial complete generalized S-semialgebra (cf. Def. 7.1 and Thm. 7.4). So all assertions of Thm. 5.2 will follow from the more general considerations in § 7. This applies in particular to the following statement which corresponds to Thm. 4.15 and Thm. 7.8.

THEOREM 5.3 Let A be a semiring with a right identity which is a generalized S-semialgebra $A = ({}_S S\langle\langle U\rangle\rangle,+,\cdot)$ and $\mathcal{B} = ({}_A A\langle\langle \mathfrak{V}\rangle\rangle,+,\cdot)$ a generalized A-semialgebra, both according to Thm. 5.2. Then $(\mathcal{B},+) = (S\langle\langle U\rangle\rangle\langle\langle \mathfrak{V}\rangle\rangle,+)$ is an S-semimodule by (4.15), essentially the S-semimodule $({}_S S\langle\langle U\mathfrak{V}\rangle\rangle,+)$ on the set $U\mathfrak{V} = \{u\mathfrak{v} \mid u \in U,\ \mathfrak{v} \in \mathfrak{V}\}$, and $\mathcal{B} = ({}_A A\langle\langle \mathfrak{V}\rangle\rangle,+,\cdot)$ is the generalized S-semialgebra $({}_S S\langle\langle U\mathfrak{V}\rangle\rangle,+,\cdot)$ with structure constants given by (4.16), which also satisfy (5.2). If $({}_S S\langle\langle U\rangle\rangle,+,\cdot)$ is a semialgebra over S, the same holds for $({}_S S\langle\langle U\mathfrak{V}\rangle\rangle,+,\cdot)$.

Considering Thm. 5.3 as a consequence of Thm. 7.8, one only has to check that the structure constants (4.16) satisfy (5.2), i. e. that for each $w\mathfrak{w} \in U\mathfrak{V}$ only a finite number of the $\Gamma^{w\mathfrak{w}}_{u\mathfrak{u},v\mathfrak{v}}$ can differ from ω.

Remark 5.4 An important class of semirings can be obtained by Thm. 5.2 as follows. Let $(U,\cdot)$ be a semigroup such that each $w \in U$ admits only a finite number of factorizations $w = uv$ in $(U,\cdot)$ and define structure constants for $({}_S S\langle\langle U\rangle\rangle,+)$ by $uv = \Sigma\gamma^w_{u,v}w$, i. e. $\gamma^w_{u,v} = \varepsilon_r$ for $uv = w$ and $\gamma^w_{u,v} = \omega$ otherwise (cf. Expl. 4.8). This yields (5.2) and associativity according to Cor. 4.7. Monoids $(U,\cdot)$ of this kind are called locally finite in

[2], p. 170, and for the free (commutative) monoid $(U,\cdot)$ on an alphabet $X = \{x_1, x_2, \dots\}$ one obtains (semi)rings $S\langle\langle U\rangle\rangle = S[[x_1, x_2, \dots]]$ in the (commutative or) non-commutative indeterminates $x_1, x_2, \dots$ over S in this way.

In the second case one assumes that S is a complete (or $\mathfrak{c}$-complete) semiring which means that all infinite sums $\sum_{i\in I} \alpha_i$ (or all where the index sets I are restricted by $|I| \leqq \mathfrak{c}$) are defined in S and subjected to rules corresponding to those for finite sums. Independent on different definitions, these concepts seem to have the same meaning throughout the literature (cf. Def. 6.7 and Remark 6.10), and everyone will suspect that a theorem corresponding to Thm. 5.2 holds for S-semimodules $({}_S S\langle\langle U\rangle\rangle, +)$ over those semirings and structure constants which may be chosen arbitrarily. We need not formulate such a theorem here, since it is contained in the second part of Thm. 7.3 and following considerations of § 7.

To apply the latter, a semiring S (in particular $\mathbb{N}_0$, cf. [2], [9], [20]) is occasionally embedded into a complete one, say $T = S\cup\{\infty\}$, by adjoining an "infinite element" ∞ and defining $\sum_{i\in I} \alpha_i = \infty$ in T for all infinite sums which are not formal infinite. To prepare a corresponding statement in § 6 we note:

<u>Fact 5.5</u> Let $(S,+,\cdot)$ be a semiring with an absorbing zero ω and $T = S\cup\{\infty\}$ for an element $\infty \notin S$. Extend the operations of S to T by

$$\xi + \infty = \infty + \xi = \infty \quad \text{for all} \quad \xi \in T$$

$$\xi \cdot \infty = \infty \cdot \xi = \infty \quad \text{for all} \quad \xi \in T\setminus\{\omega\} \quad \text{and} \quad \omega \cdot \infty = \infty \cdot \omega = \omega .$$

Then $(T,+,\cdot)$ is a semiring iff $(S\setminus\{\omega\},+,\cdot)$ is a subsemiring of $(S,+,\cdot)$, and in this case ω is the absorbing zero of T.

<u>Proof</u> Clearly, $S\setminus\{\omega\}$ is a subsemiring iff no element $\alpha \neq \omega$ of S has an additive inverse $-\alpha$ and S has no zero divisors. Both conditions are necessary. If $\alpha+\beta = \omega$ holds for $\alpha \neq \omega$, we get $\infty(\alpha+\beta) = \omega$ and $\infty\alpha+\infty\beta = \infty$, whereas $\alpha\beta = \omega$ for $\alpha \neq \omega \neq \beta$ yields $\omega = \infty(\alpha\beta) \neq (\infty\alpha)\beta = \infty$. For the converse it is convenient to adjoin at first ∞ to $S\setminus\{\omega\}$ as an element which is absorbing for both operations, a procedure obviously applicable to each semiring. Then T is obtained from $(S\setminus\{\omega\})\cup\{\infty\}$ by adjoining an absorbing zero (cf. § 2).

§ 6 Partial complete S-semimodules and semirings

Depending on different applications, various concepts are defined in the literature to deal with infinite sums in an algebraic manner (cf. § 1). We cannot compare all those concepts explicitely in this section, but we try to sketch some common ideas before we restrict ourselves to cases in which all finite sums are defined. The basic concept used more or less in all these considerations may be described as follows:

DEFINITION 6.1 Let $A \neq \emptyset$ be a set and $\mathfrak{S} \neq \emptyset$ a class of families $(a_i \mid i \in I)$ in A (more precisely: with elements $a_i \in A$), indexed by sets I which may be restricted with respect to their cardinality $|I|$. Let Σ be a mapping of $\mathfrak{S}$ into A denoted by

$$(6.1) \qquad \Sigma(a_i \mid i \in I) = \sum_{i \in I} a_i = a \in A.$$

We call such a triple $(A,\Sigma,\mathfrak{S})$ a *partial Σ-algebra* (in the universal meaning of "algebra"), the families $(a_i \mid i \in I)$ *summable* in A, and (6.1) the *sum* of $(a_i \mid i \in I)$. If $\mathfrak{S}$ consists of all families $(a_i \mid i \in I)$ in A (or of all satisfying $|I| \leq \mathfrak{c}$ for some infinite cardinal number $\mathfrak{c}$), $(A,\Sigma,\mathfrak{S})$ is called *complete* (or *$\mathfrak{c}$-complete*).

With a fixed triple $(A,\Sigma,\mathfrak{S})$ in mind, most authors merely write (A,Σ) and replace $(a_i \mid i \in I) \in \mathfrak{S}$ by a phrase like " $\sum_{i \in I} a_i$ exists", and we often proceed similarly to simplify formulations. In this way we list the following five assumptions on a partial Σ-algebra $(A,\Sigma,\mathfrak{S})$, already called axioms with respect to their later use.

(E) Equivalent families axiom Let $(a_i \mid i \in I)$ and $(b_k \mid k \in K)$ be *equivalent families* in A, defined by the existence of a bijection $\varphi : I \to K$ which satisfies $b_{\varphi(i)} = a_i$ for all $i \in I$. Then $\sum_{i \in I} a_i$ exists iff $\sum_{k \in K} b_k$ does, and in this case they are equal.

With respect to the various possibilities to denote "the same" family $(a_i \mid i \in I)$ by different index sets, some assumption of this kind (which determines what "the same" or "equivalent" means) is indispensable in the context of Def. 6.1. Since (E) involves a generalized kind of commutativity for Σ, it is rather strong, and there are near by hand applications of Def. 6.1 for which (E) has to be weakened.

(U) Unary sum axiom If $I = \{i\}$ holds for a family $(a_i \mid i \in I)$ in A, then $\sum_{i \in I} a_i$ exists and equals a_i.

The following axioms concern existence and equality in

(6.2) $$\sum_{i\in I} a_i = \sum_{j\in J} \left(\sum_{i\in I_j} a_i \right) \quad \text{for} \quad I = \bigcup_{j\in J} I_j,$$

for each family $(a_i \mid i \in I)$ in A and each family $(I_j \mid j \in J)$ of pairwise disjoint subsets I_j of I without restriction on the set of indices $j \in J$ such that $I_j = \emptyset$ holds. To avoid confusion we talk about *generalized partitions* of I in this context. If one formulates the following axioms using partitions ($I_j \neq \emptyset$ for all $j \in J$), one obtains considerably weaker assumptions.

<u>(GP) Generalized partition axiom</u> If $\sum_{i\in I} a_i$ exists, each sum occuring on the right side of (6.2) does and equality holds.

<u>(GP') Strong converse of (GP)</u> If each sum on the right side of (6.2) exists, also $\sum_{i\in I} a_i$ does and equality holds.

<u>(GP'_F) Weak converse of (GP)</u> The same as (GP') for finite sets J.

<u>Fact 6.2</u> a) Let $(A,\Sigma,\mathcal{S})$ be a partial Σ-algebra satisfying (U) and (GP). Then (E) holds, and each subfamily of a summable family is summable, in particular the empty family. Denoting the sum of the latter by $o \in A$, the sum of each family $(a_k \mid k \in K)$ satisfying $a_k = o$ exists and equals o. Moreover, for each summable family $(a_i \mid i \in I)$ in A such that $I \cap K = \emptyset$ holds,

(6.3) $$\sum_{i\in I\cup K} a_i \quad \text{exists and equals} \quad \sum_{i\in I} a_i.$$

b) If $(A,\Sigma,\mathcal{S})$ also satisfies (GP'), for each family $(a_{i,j} \mid (i,j) \in I \times J)$ in A

(6.4) $$\sum_{i\in I} \left(\sum_{j\in J} a_{i,j} \right) = \sum_{j\in J} \left(\sum_{i\in I} a_{i,j} \right)$$

holds in the meaning that one side exists if the other does.

<u>Proof</u> a) To show (E), let $(a_i \mid i \in I)$, $(b_k \mid k \in K)$ and φ be as in (E). We assume that $\sum_{k\in K} b_k$ exists and use the partition $K = \bigcup_{i\in I} K_i$ with $K_i = \{\varphi(i)\}$. Then $\sum_{k\in K} b_k = \sum_{i\in I} \left(\sum_{k\in K_i} b_k \right) = \sum_{i\in I} b_{\varphi(i)} = \sum_{i\in I} a_i$ holds, where existence and equality follow by (GP) and (U) from the left to the right. The other statements are either clear or proved similarly.

b) Each side of (6.4) implies by (GP') that also $\sum_{(i,j)\in I\times J} a_{i,j}$ exists, which equals the other side by (GP).

In this context we introduce for later use:

<u>(WDS) Weak double sum axiom</u> Let $(a_{i,j} \mid (i,j) \in I \times J)$ be a family in A such that the left side of (6.4) exists and also $\sum_{i \in I} a_{i,j}$ for each $j \in J$. Then $\sum_{j \in J} (\sum_{i \in I} a_{i,j})$ exists and (6.4) holds.

<u>Fact 6.3</u> a) Let $(A,\Sigma,\mathfrak{S})$ be a partial Σ-algebra which satisfies (U), (GP) and (GP'_F) and assume that all finite families $(a_i \mid i \in I)$ in A are summable. Then we obtain a semimodule (A,+) defining a_1+a_2 by (6.1) according to $a_1+a_2 = \sum_{i \in \{1,2\}} a_i$ for all $a_1,a_2 \in A$, which implies

$$(6.5) \qquad a_1+\ldots+a_\nu = \sum_{i \in I} a_i \quad \text{for all families} \quad (a_i \mid i \in \{1,\ldots,\nu\}) \text{ in } A,$$

where $a_1+\ldots+a_\nu$ is the sum in the semimodule (A,+). By Fact 6.2, (A,+) has a zero o, and each family $(a_i \mid i \in I)$ in A such that $a_i = o$ holds for almost all $i \in I$ is summable in $(A,\Sigma,\mathfrak{S})$ and their sum corresponds to the formal infinite sum defined in (A,+).

b) If $(A,\Sigma,\mathfrak{S})$ as above satisfies also (GP') and $(a_i \mid i \in I)$ is summable in A, then $\sum_{i \in I} a_i = o$ implies $a_i = o$ for all $i \in I$.

<u>Proof</u> We only show the statement b), which can be reduced by (GP) to $x+y = o \Rightarrow x = o$ for $x,y \in A$. For all $\nu \in \mathbb{N}$, define $z_\nu = x$ if $2 \nmid \nu$ and $z_\nu = y$ if $2 \mid \nu$. Then from our statement about formal infinite sums, (GP') and (GP) we obtain

$$o = \sum_{\nu \in 2\mathbb{N}} o = \sum_{\nu \in 2\mathbb{N}} (z_{\nu-1}+z_\nu) = \sum_{\nu \in \mathbb{N}} z_\nu = z_1 + \sum_{\nu \in 2\mathbb{N}} (z_\nu+z_{\nu+1}) = z_1+o.$$

Partial Σ-algebras $(A,\Sigma,\mathfrak{S})$ satisfying (U), (GP) and (GP'), but without the assumption that all finite families in A are summable, have been considered as "positive partial monoids (A,Σ)" in [12]. Replacing generalized partitions $I = \cup I_j$ by injective functions $f : I \to J$, a concept called "Σ-monoid" was introduced in [5]. It coincides with partial Σ-algebras $(A,\Sigma,\mathfrak{S})$ satisfying (U), (GP) and (GP'_F) for which all finite families in A are summable as considered in Fact 6.3 a). This concept is basic for us in the following. But, as usual in papers dealing with semirings, we define it the other way around, i.e. starting with a semimodule (A,+):

<u>DEFINITION 6.4</u> A *partial complete semimodule* $(A,+,\Sigma)$ is defined as a semimodule (A,+) with a zero o, which is at the same time a partial Σ-algebra $(A,\Sigma,\mathfrak{S})$ satisfying (U), (GP) and (GP'_F) such that (A,Σ) extends (A,+) according to (6.5). Starting with an S-semimodule $({}_SA,+)$ for a se-

miring S, we call $({}_S A,+,\Sigma)$ a *partial complete S-semimodule* provided that

(Op) $\sigma(\sum_{i\in I} a_i) = \sum_{i\in I} \sigma a_i$ holds for all $(a_i | i \in I)$ in A and $\sigma \in S$ in the meaning that $\sum_{i\in I} \sigma a_i$ exists whenever $\sum_{i\in I} a_i$ does.

A *complete semimodule* $(A,+,\Sigma)$ (or a *complete S-semimodule* $({}_S A,+,\Sigma)$) is defined in the same way with respect to a complete Σ-algebra $(A,\Sigma,\$)$, and likewise with $\mathfrak{c}$-*complete*. In these cases, (GP) implies (GP') and hence (GP_F') and (WDS), which depends for $\mathfrak{c}$-completeness on our strong definition of this concept in Def. 6.1.

Partial complete S-semimodules $({}_S A,+,\Sigma)$ and $({}_S M,+,\Sigma)$ are called *isomorphic* if there is an S-semimodule isomorphism $f: ({}_S A,+) \to ({}_S M,+)$ which also satisfies

(6.6) $f(\sum_{i\in I} a_i) = \sum_{i\in I} f(a_i)$ in the meaning that one side exists iff the other does.

There are several possibilities to define homomorphisms for partial complete S-semimodules which we do not need here.

Note that for each partial complete semimodule $(A,+,\Sigma)$ at least all formal infinite sums are defined, and that the latter may be all infinite sums which exist. So each semimodule $(A,+)$ with a zero o can be considered as a partial complete semimodule $(A,+,\Sigma)$, and we shall do so henceforth. We further remark that each partial complete semimodule $(A,+,\Sigma)$ may be considered as a partial complete $\mathbb{N}_0$-semimodule $({}_{\mathbb{N}_0} A,+,\cdot)$ by (2.2).

We give an example of a partial complete S-semimodule which is a module. Let $({}_{\mathbb{R}}\mathbb{R},+)$ be the $\mathbb{R}$-module of the real numbers and define Σ by

$\sum_{\nu=1}^{\infty} a_\nu$ for all families $(a_\nu | \nu \in \mathbb{N})$ such that this series is absolutely convergent. In the same way, the non-negative real numbers $\mathbb{P}_0$ form a partial complete $\mathbb{P}_0$-semimodule. Note that the latter also satisfies (GP'), whereas even (WDS) fails to be true for $({}_{\mathbb{R}}\mathbb{R},+,\Sigma)$, and that both are partial complete semirings (cf. Def. 6.7).

The following statement defines infinite sums on $S\langle\langle U\rangle\rangle$ by those given on S. It is basic in our context and, by the way, it subordinates the infinite sums (2.7) already used for the elements of $S\langle\langle U\rangle\rangle$ as a "formal notation" to the concepts of this section:

PROPOSITION 6.5 Let $(S,+,\cdot)$ be a semiring with a right identity and a right absorbing zero ω, and $A = S\langle\langle U\rangle\rangle$ the S-semimodule of Fact 2.2 a) consisting of all mappings $a:U \to S$ for some set $U \neq \emptyset$. Assume that $({}_SS,+)$ is a partial complete S-semimodule $({}_SS,+,\Sigma)$ with a mapping $\Sigma = \Sigma_S : \mathcal{S}_S \to S$ according to (6.1). Then $({}_SA,+)$ becomes a partial complete S-semimodule $({}_SA,+,\Sigma)$ where $\Sigma = \Sigma_A : \mathcal{S}_A \to A$ is defined as follows. For each family $(a_i \mid i \in I)$ in A,

(6.7) $\quad \sum_{i\in I} a_i$ exists iff $\sum_{i\in I} a_i(u)$ exists for each $u \in U$, and it is defined as the mapping $\left(\sum_{i\in I} a_i\right)(u) = \sum_{i\in I} a_i(u)$.

If $({}_SS,+,\Sigma)$ satisfies (GP') or (WDS), the same holds for $({}_SS\langle\langle U\rangle\rangle,+,\Sigma)$. Finally, $({}_SS\langle\langle U\rangle\rangle,+,\Sigma)$ is a complete S-semimodule iff $({}_SS,+,\Sigma)$ is one, and likewise with $\mathfrak{c}$-complete provided that $\mathfrak{c} \geq |U|$ holds.

Proof According to Fact 2.2 c), we select one right identity ε_r of S and consider each $u \in U$ as the mapping $u:U \to S$ given by $u(u) = \varepsilon_r$ and $u(v) = \omega$ for all $v \neq u$ in U, which yields $(\alpha u)(u) = \alpha$ and $(\alpha u)(v) = \omega$ for all $\alpha \in S$. Let $a \in S\langle\langle U\rangle\rangle$ be given by $a(u) = \alpha_u$ and apply (6.7) to the family $(a_u = \alpha_u u \mid u \in U)$ in A. Since at least all formal infinite sums exist in $({}_SS,+,\Sigma)$, for each $v \in U$ we get

$$\sum_{u\in U} a_u(v) = \alpha_v, \text{ hence } \sum_{u\in U} a_u = \sum_{u\in U} \alpha_u u \text{ exists in } ({}_SA,+,\Sigma)$$

and is defined by (6.7) as the mapping $a \in S\langle\langle U\rangle\rangle$ given by $a(v) = \alpha_v$ for all $v \in U$. This shows that the notation (2.7) is always a special case of the sum $\Sigma = \Sigma_A$ defined by (6.7), in particular if U is infinite (cf. Fact 2.2 b)). Moreover, denoting the elements of a family $(a_i \mid i \in I)$ in A by $a_i = \sum_{u\in U} \alpha_{i,u} u$, (6.7) can now be reformulated:

(6.8) $\quad \sum_{i\in I} a_i = \sum_{i\in I}\left(\sum_{u\in U} \alpha_{i,u} u\right) = \sum_{u\in U}\left(\sum_{i\in I} \alpha_{i,u}\right) u$, where the left hand sums exist iff $\sum_{i\in I} \alpha_{i,u}$ does for each $u \in U$.

One easily checks that each of the conditions (GP), (GP_F') and (Op) as well as (GP') or (WDS) transfers from $({}_SS,+,\Sigma)$ to $({}_SA,+,\Sigma) = ({}_SS\langle\langle U\rangle\rangle,+,\Sigma)$ (which in fact also holds conversely). The statements concerning completeness follow from (6.7) or (6.8), where the restriction $\mathfrak{c} \geq |U|$ is only added to avoid that the sums $\sum_{u\in U} \alpha_u u$, by $\mathfrak{c} < |U|$ excluded from (6.1), have again to be considered as a formal notation for the mappings $a:U \to S$.

Remark 6.6 In later applications we use $({}_S S\langle\langle U\rangle\rangle, +, \Sigma)$ with the above interpretation of U as a subset of $S\langle\langle U\rangle\rangle$, which yields $\varepsilon_r u = u$ for $u \in U$ and a fixed right identity ε_r of S. Note further that Prop. 6.5 also includes partial complete S-semimodules $({}_S S\langle U\rangle, +, \Sigma)$ for finite sets U. Finally, the assumption (Op) for the partial complete S-semimodule $({}_S S, +, \Sigma)$ is just the infinite left distributive law (D_ℓ) and hence satisfied if one starts with a partial complete semiring $(S, +, \Sigma, \cdot)$ according to the following considerations.

DEFINITION 6.7 A *partial complete semiring* $(S, +, \Sigma, \cdot)$ is defined as a semiring $(S, +, \cdot)$ which is also a partial complete semimodule $(S, +, \Sigma)$ and satisfies the *infinite distributive law*:

(D) for all families $(\alpha_i \mid i \in I)$ and $(\beta_j \mid j \in J)$ in S, $(\sum_{i\in I} \alpha_i)(\sum_{j\in J} \beta_j) = \sum_{(i,j)\in I\times J} \alpha_i\beta_j$ holds in the meaning that the right hand sum exists whenever both sums on the left side do.

Corresponding to Def. 6.4, *complete* ($\mathfrak{c}$-*complete*) *semirings* $(S, +, \cdot, \Sigma)$ are defined, which satisfy (GP') as a consequence of (GP).

Remark 6.8 In the context of Def. 6.7, (D) implies the *infinite left* and *right distributive laws* (D_ℓ) and (D_r), defined by $|I| = 1$ or $|J| = 1$ in (D), respectively. The converse holds if $(S, +, \Sigma)$ satisfies (GP'), but not in general (cf. Expl. 6.13).

Remark 6.9 Partial complete semirings as defined above coincide with "Σ-semirings" in [5], § 1. For semirings S with an absorbing zero and an identity, other concepts called "partial complete semiring" occur in the literature. In [10], Def. 3.1, additionally (GP') is included. The same concept is intended in [9], p. 63, but the definition given there yields in fact that each infinite sum exists in S. On the other hand, in [15], p. 195, only (U), (GP) and (D) is assumed, which is (apart from the restriction to countably infinite sums) weaker than our concept. It has the disadvantage that there may be families in S such that $\alpha_0 + \sum_{\nu\in\mathbb{N}} \alpha_\nu$ exists, whereas $\sum_{\nu\in\mathbb{N}_0} \alpha_\nu$ does not (cf. Expl. 6.12).

Remark 6.10 Each of the three different concepts called "partial complete semiring" compared in Remark 6.9 leads to the same concept of a complete (or $\mathfrak{c}$-complete) semiring, since in this case (GP) implies (GP') and (GP'_F). It coincides with complete semirings as defined in [2], p.125.

One easily checks that the semiring $(T,+,\cdot)$ of Fact 5.5 becomes a complete semiring $(T,+,\Sigma,\cdot)$ by defining $\sum_{i\in I}\alpha_i=\infty$ for all (not formal) infinite sums.

Fact 6.11 Each semiring $(S,+,\cdot)$ with an absorbing zero ω can be considered as a partial complete semiring $(S,+,\Sigma,\cdot)$ if one restricts Σ to finite and formal infinite sums. It satisfies (GP') iff $\alpha+\beta=\omega$ implies $\alpha=\omega$ for all $\alpha,\beta\in S$, which is easily obtained from Fact 6.3. b). Otherwise $(S,+,\Sigma,\cdot)$ does not even satisfy (WDS) according to the second example given in the context of (5.3).

The following examples are constructed to prove statements in the above remarks and will not deserve interest beyond this purpose.

Example 6.12 Let $(T,+,\cdot)$ be the semiring obtained from $(\mathbb{N}_0,+,\cdot)$ by Fact 5.5. Apart from formal infinite sums with the usual meaning, an infinite family $(\alpha_i\mid i\in I)$ is defined to be summable iff $2\mid\alpha_i$ holds for all $i\in I$, clearly including $2\mid\infty$. For those families we define $\sum_{i\in I}\alpha_i=\infty$.

If we restrict all indexing sets I to be countable, $(T,+,\Sigma,\cdot)$ is a "partial complete semiring" in the meaning of [15], i. e. (U), (GP) and (D) are satisfied. But e. g. for $\alpha_\nu=2^\nu$

$$\alpha_0+\sum_{\nu\in\mathbb{N}}\alpha_\nu=1+\infty=\infty \text{ exists, whereas } \sum_{\nu\in\mathbb{N}_0}\alpha_\nu \text{ does not,}$$

which also disproves (GP'_F).

Example 6.13 For the set $\mathbb{N}_0$, let $S=(\mathbb{N}_0,\oplus,\cdot)$ be the semiring with the usual multiplication and $\alpha\oplus\beta=\max(\alpha,\beta)$, and $(T,\oplus,\cdot)$ the semiring on $T=\mathbb{N}_0\cup\{\infty\}$ as in Fact 5.5. Define a family $(\alpha_i\mid i\in I)$ in T to be summable iff there exists a natural number n such that each $\alpha\in\mathbb{N}$ occurs at most n times in $(\alpha_i\mid i\in I)$, and their sum by

$$\sum_{i\in I}\alpha_i=\sup\{\alpha_i\mid i\in I\} \text{ with respect to } 0<1<2<\ldots<\infty.$$

Then $(T,\oplus,\Sigma,\cdot)$ is a semiring which satisfies (U), (GP), (GP'_F), (D_ℓ) and (D_r), but not (D). A counter example for the latter is $\sum_{\nu\in\mathbb{N}}\nu=\sum_{\mu\in\mathbb{N}}\mu=\infty$, whereas $\sum_{(\nu,\mu)\in\mathbb{N}\times\mathbb{N}}\nu\mu$ does not exist.

§ 7 Partial complete generalized semialgebras

DEFINITION 7.1 Let $S = (S,+,\Sigma,\cdot)$ be a partial complete semiring with an absorbing zero ω and a right identity. Then $({}_SA,+,\Sigma,\cdot)$ is called a *partial complete generalized semialgebra over* S iff the following statements hold:

a) $({}_SA,+,\Sigma)$ is the partial complete S-semimodule $({}_SS\langle\langle U\rangle\rangle,+,\Sigma)$ obtained by Prop. 6.5 from the partial complete S-semimodule $({}_SS,+,\Sigma)$ of $(S,+,\Sigma,\cdot)$ for some $U \neq \emptyset$. Hence $\varepsilon_r u = u$ holds for each $u \in U$ and a selected right identity ε_r of S (cf. Remark 6.6).

b) $(A,+,\Sigma,\cdot)$ satisfies the infinite distributive law (D), i.e. $(\sum_{i\in I} a_i)(\sum_{j\in J} b_j) = \sum_{(i,j)\in I\times J} a_i b_j$ provided that the left hand sums exist.

c) (4.2) holds for all $u,v \in U$ and $\alpha,\beta \in S$.

The terms *partial complete semialgebra over* S and *partial complete (generalized) algebra over* S are used according to Def. 4.1.

A *complete* (*𝔠-complete*) *generalized semialgebra* $({}_SS\langle\langle U\rangle\rangle,+,\Sigma,\cdot)$ *over* S is defined in the same way, assuming that the semiring $(S,+,\Sigma,\cdot)$ and hence the S-semimodule $({}_SS\langle\langle U\rangle\rangle,+,\Sigma)$ are complete (𝔠-complete for some $\mathfrak{c} \geqq |U|$).

THEOREM 7.2 Let $({}_SS\langle\langle U\rangle\rangle,+,\Sigma,\cdot)$ be a partial complete generalized S-semialgebra. Then its multiplication is uniquely determined by the products

(7.1) $$uv = \sum_{w\in U} \gamma_{u,v}^{w}\, w \quad \text{for all} \quad u,v \in U \quad \text{according to}$$

(7.2) $$ab = (\sum_{u\in U} \alpha_u u)(\sum_{v\in U} \beta_v v) = \sum_{u,v\in U} (\alpha_u\beta_v)(uv) \quad \text{or}$$

(7.3) $$ab = \sum_{w\in U} (\sum_{u,v\in U} \alpha_u\beta_v\gamma_{u,v}^{w})w \quad \text{for all} \quad a,b \in S\langle\langle U\rangle\rangle.$$

The elements $\gamma_{u,v}^{w}$, again called the *structure constants* of $({}_SS\langle\langle U\rangle\rangle,+,\Sigma,\cdot)$, are contained in $\varepsilon_r S$ and clearly not subjected to (4.6). Moreover,

(7.4) $$\sum_{u,v\in U} \alpha_u\beta_v\gamma_{u,v}^{w} \quad \text{exists for each } w \in U \quad \text{and all} \quad \alpha_u,\beta_v \in S,$$

hence $({}_SS\langle\langle U\rangle\rangle,+,\Sigma,\cdot)$ satisfies Def. 5.1.

Proof From b) and c) in Def. 7.1 we obtain (7.2) for all $a,b \in S\langle\langle U\rangle\rangle$. The latter yields step by step

$$ab = \sum_{u,v\in U} \alpha_u\beta_v(\sum_{w\in U}\gamma_{u,v}^{w}\, w) = \sum_{u,v\in U}\ \sum_{w\in U} (\alpha_u\beta_v\gamma_{u,v}^{w}\, w) = \sum_{w\in U} (\sum_{u,v\in U} \alpha_u\beta_v\gamma_{u,v}^{w})\, w$$

including the existence of all sums by (7.1), (Op) and (6.8), which also proves (7.4).

THEOREM 7.3 *Let $({}_S S\langle\langle U\rangle\rangle,+,\Sigma)$ be a partial complete S-semimodule obtained from a partial complete semiring $(S,+,\Sigma,\cdot)$ as in Def. 7.1 a) and $(\gamma_{u,v}^{w} \in \varepsilon_r S \mid (u,v,w) \in U \times U \times U)$ a family satisfying (7.4). Then (7.3) defines a multiplication on ${}_S S\langle\langle U\rangle\rangle$ such that $({}_S S\langle\langle U\rangle\rangle,+,\cdot)$ is a generalized S-semialgebra according to Def. 5.1, and $({}_S S\langle\langle U\rangle\rangle,+,\Sigma,\cdot)$ is a partial complete generalized S-semialgebra provided that $({}_S S,+,\Sigma)$ satisfies (WDS).*

In particular, for a complete semiring $(S,+,\Sigma,\cdot)$ each choice of structure constants $\gamma_{u,v}^{w} \in \varepsilon_r S$ yields in this way a complete generalized S-semialgebra $({}_S S\langle\langle U\rangle\rangle,+,\Sigma,\cdot)$, and the same holds with $\mathfrak{c}$-complete for some $\mathfrak{c} \geq |U|$.

Proof As already noted in § 5, the generalized version (7.3) of (4.5) implies (7.1) and (4.2), and (4.1) can be checked now using (GP'_F) and (GP) for $(S,+,\Sigma,\cdot)$. Next we show that the supplementary assumption of (WDS) implies (D) for $({}_S S\langle\langle U\rangle\rangle,+,\Sigma,\cdot)$ and assume that the families

(7.5) $\qquad (a_i = \sum_{u\in U} \alpha_{i,u} u \mid i \in I) \quad$ and $\quad (b_j = \sum_{v\in U} \beta_{j,v} v \mid j \in J)$

are summable. Then we get (always including the existence of all occuring sums) by (6.8) and (7.3)

$$(\sum_{i\in I} a_i)(\sum_{j\in J} b_j) = (\sum_{u\in U}(\sum_{i\in I}\alpha_{i,u})u)(\sum_{v\in U}(\sum_{j\in J}\beta_{j,v})v)$$

$$= \sum_{w\in U}(\sum_{u,v\in U}((\sum_{i\in I}\alpha_{i,u})(\sum_{j\in J}\beta_{j,v})\gamma_{u,v}^{w}))w.$$

As (D) is assumed for $(S,+,\Sigma,\cdot)$, the latter equals

$$= \sum_{w\in U}(\sum_{u,v\in U}(\sum_{(i,j)\in I\times J}\alpha_{i,u}\beta_{j,v}\gamma_{u,v}^{w}))w.$$

Since all sums $\sum_{u,v\in U}\alpha_{i,u}\beta_{j,v}\gamma_{u,v}^{w}$ exist by (7.4), we can apply (WDS):

$$= \sum_{w\in U}(\sum_{(i,j)\in I\times J}(\sum_{u,v\in U}\alpha_{i,u}\beta_{j,v}\gamma_{u,v}^{w}))w.$$

Again by (6.8) and (7.3) we obtain as intended

$$= \sum_{(i,j)\in I\times J}(\sum_{w\in U}(\sum_{u,v\in U}\alpha_{i,u}\beta_{j,v}\gamma_{u,v}^{w})w) = \sum_{(i,j)\in I\times J} a_i b_j .$$

The remaining statements follow since a complete ($\mathfrak{c}$-complete) semiring $(S,+,\Sigma,\cdot)$ satisfies (GP') and hence (WDS) (cf. Fact 6.2 b)).

By the last part of Thm. 7.3, the second case of § 5 is included in the following considerations on partial complete generalized semialgebras

$({}_S S\langle\langle U\rangle\rangle,+,\Sigma,\cdot)$. To include also the first case we use Fact 6.11 and state:

THEOREM 7.4. Let $({}_S S\langle\langle U\rangle\rangle,+,\cdot)$ be a generalized S-semialgebra as defined in Thm. 5.2 and consider the semiring $(S,+,\cdot)$ as a partial complete semiring $(S,+,\Sigma,\cdot)$ with respect to formal infinite sums. Then $({}_S S\langle\langle U\rangle\rangle,+,\Sigma,\cdot)$ is a partial complete S-semialgebra according to Def. 7.1.

Proof We consider $({}_S S\langle\langle U\rangle\rangle,+)$ as the partial complete S-semimodule $({}_S S\langle\langle U\rangle\rangle,+,\Sigma)$ obtained from $(S,+,\Sigma,\cdot)$ by Prop. 6.5. Correspondingly we reproduce $({}_S S\langle\langle U\rangle\rangle,+,\cdot)$ in defining a multiplication on $({}_S S\langle\langle U\rangle\rangle,+,\Sigma)$ by (7.3) with the same structure constants $\gamma_{u,v}^{w}$, which satisfy (7.4) in $(S,+,\Sigma,\cdot)$ by (5.2). According to Thm. 7.3, it remains to prove (D) for $({}_S S\langle\langle U\rangle\rangle,+,\Sigma,\cdot)$. But we cannot apply Thm. 7.3 for this purpose, because $(S,+,\Sigma,\cdot)$ need not satisfy (WDS) (cf. Fact 6.11). Now (WDS) was used in the above proof for (D) only for one step. It can be replaced there for formal infinite sums in $(S,+,\cdot)$ by the fact that only a finite number of all summands $\alpha_{i,u}\beta_{j,v}\gamma_{u,v}^{w}$ can differ from ω. This follows from (5.2) and since for each u and each v almost all $\alpha_{i,u}$ and almost all $\beta_{i,v}$ equal ω, according to the assumption that the families (7.5) are summable in $({}_S S\langle\langle U\rangle\rangle,+,\Sigma)$ by (6.8).

Remark 7.5 In Def. 7.1, the assumption of (4.2) in c) is equivalent to

(4.2') $\quad \alpha(ab) = (\alpha a)b \quad$ and $\quad u(\beta b) = \varepsilon_r\beta(ub)$

for all $\alpha,\beta \in S$, all $a,b \in S\langle\langle U\rangle\rangle$ and all $u \in U$. As in Remark 4.5, (4.2') implies (4.2) without any further knowledge about $({}_S S\langle\langle U\rangle\rangle,+,\Sigma,\cdot)$. The converse follows from (7.3) (we may use Thm. 7.2 in this direction) and (D_ℓ) for $(S,+,\Sigma,\cdot)$ needed to obtain the left equation in (4.2').

THEOREM 7.6 Let $({}_S S\langle\langle U\rangle\rangle,+,\Sigma,\cdot)$ be a partial complete generalized S-semialgebra. Then the multiplication is associative iff the structure constants $\gamma_{u,v}^{w} \in \varepsilon_r S$ satisfy (4.8), which in turn is equivalent to (4.9). These conditions are implied by (4.11) and one of the assumptions a), b) or c) given in Cor. 4.7.

Proof By (7.3), the equivalence of (4.8) and (4.9) is obvious, and (4.9) holds if $(S\langle\langle U\rangle\rangle,\cdot)$ is associative. We show the converse of the latter applying four times (D) for $({}_S S\langle\langle U\rangle\rangle,+,\Sigma,\cdot)$ to $a = \Sigma\alpha_u u$, $b = \Sigma\beta_v v$ and $d = \Sigma\delta_w w$. Then we get

$$(ab)d = \Big(\sum_{u,v\in U}(\alpha_u u)(\beta_v v)\Big)d = \sum_{u,v,w\in U}((\alpha_u\beta_v)(uv))(\delta_w w) \quad \text{and}$$

$$a(bd) = a\Big(\sum_{v,w\in U}(\beta_v v)(\delta_w w)\Big) = \sum_{u,v,w\in U}(\alpha_u u)((\beta_v v)(\delta_w w)).$$

Now we compare the summands using several times the left and once the right side of (4.2'). Then $((\alpha_u\beta_v)(uv))(\delta_w w) = (\alpha_u\beta_v)((uv)(\delta_w w))$ equals $(\alpha_u u)((\beta_v v)(\delta_w w)) = \alpha_u(u(\beta_v(v(\delta_w w)))) = \alpha_u(\varepsilon_r\beta_v(u(v(\delta_w w)))) = (\alpha_u\beta_v)(u(v(\delta_w w)))$ due to (4.9). The rest follows as in § 4, using now (D_ℓ) for $(S,+,\Sigma,\cdot)$.

PROPOSITION 7.7 *Let* $A = ({}_SA,+,\Sigma,\cdot) = ({}_SS\langle\langle U\rangle\rangle,+,\Sigma,\cdot)$ *be a partial complete generalized S-semialgebra.*

i) *A is a partial complete S-semialgebra iff its structure constants satisfy* (4.12) *given in Prop. 4.10.*

ii) *A satisfies the statements of Lemma 4.11 concerning the existence of (one-sided) identities of* A.

iii) $(S,+,\Sigma,\cdot)$ *can be embedded into* $({}_SA,+,\Sigma,\cdot)$ *according to a) or b) of Prop 4.13, and in both cases the monomorphism* g *satisfies also (cf. (6.6))*

(7.6) $\quad g(\sum_{i\in I}\sigma_i) = \sum_{i\in I} g(\sigma_i)$ *iff one of the families* $(\sigma_i \mid i\in I)$ *and* $(g(\sigma_i)\mid i\in I)$ *is summable.*

The proofs are essentially the same as for the cited statements in § 4, now based on (7.3) and, now and then, on some of the infinite distributive laws satisfied for $({}_SA,+,\Sigma,\cdot)$ as well as for $(S,+,\Sigma,\cdot)$.

THEOREM 7.8 *Let* $({}_SA,+,\Sigma,\cdot) = ({}_SS\langle\langle U\rangle\rangle,+,\Sigma,\cdot)$ *be a partial complete generalized S-semialgebra such that* $(A,+,\Sigma,\cdot)$ *is a partial complete semiring with a right identity* e_r, *and* $({}_A\mathfrak{B},+,\Sigma,\cdot) = ({}_AA\langle\langle\mathfrak{V}\rangle\rangle,+,\Sigma,\cdot)$ *a partial complete generalized A-semialgebra. Then* $(\mathfrak{B},+,\Sigma) = (S\langle\langle U\rangle\rangle\langle\langle\mathfrak{V}\rangle\rangle,+,\Sigma)$ *is an S-semimodule by* (4.15), *essentially the partial complete S-semimodule* $({}_SS\langle\langle U\mathfrak{V}\rangle\rangle,+,\Sigma)$ *on the set* $U\mathfrak{V} = \{u\mathfrak{v} \mid u\in U,\ \mathfrak{v}\in\mathfrak{V}\}$, *and* $\mathfrak{B} = ({}_AA\langle\langle\mathfrak{V}\rangle\rangle,+,\Sigma,\cdot)$ *is the partial complete generalized S-semialgebra* $({}_SS\langle\langle U\mathfrak{V}\rangle\rangle,+,\Sigma,\cdot)$. *The structure constants of the latter are given by* (4.16), *and it is a partial complete semialgebra over* S *if* $({}_SS\langle\langle U\rangle\rangle,+,\Sigma,\cdot)$ *is one. Finally,* $({}_SS\langle\langle U\mathfrak{V}\rangle\rangle,+,\Sigma,\cdot)$ *is complete* ($\mathfrak{c}$-*complete for some* $\mathfrak{c}\geq|U|$) *iff the same holds for* $(S,+,\Sigma,\cdot)$.

Proof Based on (6.8), one checks that $({}_A\mathfrak{B},+,\Sigma)$ is also a partial complete S-semimodule $({}_SS\langle\langle U\rangle\rangle\langle\langle\mathfrak{V}\rangle\rangle,+,\Sigma)$ by (4.15), and (4.17). Its elements

(7.7) $\quad \mathfrak{b} = \sum_{\mathfrak{v}\in\mathfrak{V}}(\sum_{u\in U}\beta_{u,\mathfrak{v}}u)\mathfrak{v} = \sum_{\mathfrak{v}\in\mathfrak{V}}(\sum_{u\in U}\beta_{u,\mathfrak{v}}u\mathfrak{v})$, or $\mathfrak{b} : \mathfrak{V}\to S^U$,

correspond bijectively to the elements

(7.8) $\quad \mathfrak{b}' = \sum_{(u,\mathfrak{v})\in U\times\mathfrak{V}}\beta_{u,\mathfrak{v}}(u\mathfrak{v}) = \sum_{(u,\mathfrak{v})\in U\times\mathfrak{V}}\beta_{u,\mathfrak{v}}(u,\mathfrak{v})$, or $\mathfrak{b}' : U\times\mathfrak{V}\to S$, of the partial complete S-semimodule $({}_S S\langle\langle U\mathfrak{V}\rangle\rangle, +, \Sigma)$, where the set $U \times \mathfrak{V}$ and the subset $U\mathfrak{V}$ of $\mathfrak{B}$ can be identified. Moreover, this bijection provides that both partial complete S-semimodules are isomorphic according to Def. 6.4. The question as to whether both underlying sets $(S^U)^{\mathfrak{V}}$ and $S^{U\times\mathfrak{V}}$ are equal is the same as that for $\mathfrak{V} \times (U \times S)$ and $(U \times \mathfrak{V}) \times S$. So we may use the multiplication of $({}_A\mathfrak{B}, +, \Sigma, \cdot)$ which satisfies (D) for $({}_S S\langle\langle U\mathfrak{V}\rangle\rangle, +, \Sigma)$ and check $\varepsilon_r u v = uv$ and (4.2) for $({}_S S\langle\langle U\mathfrak{V}\rangle\rangle, +, \Sigma, \cdot)$. Hence the latter is a partial complete generalized S-semialgebra, to which we may apply Thm. 7.2 to calculate (4.16) for its structure constants. The next assertion follows from Prop. 7.7 i), and the last one in applying the last statement of Prop. 6.5 twice.

REFERENCES

[1] Cohn, P.M., Some remarks on the invariant basis property, Topology 5 (1966), 215 - 228.

[2] Eilenberg, S., Automata, Languages and Machines, Vol. A, Academic Press, 1974.

[3] Everett, C.J., Vector Spaces over Rings, Bull. Amer. Math. Soc. 48 (1942), 312 - 316.

[4] Griepentrog, R.D. and Weinert, H.J., Embedding semirings into semirings with identity (Proc. Conf. Szeged, 1976), Coll. Math. Soc. J. Bolyai 20 (1979), 225 - 245.

[5] Higgs, D., Axiomatic infinite sums - an algebraic approach to integration theory, Contemp. Math. 2 (1980), 205 - 212.

[6] Iizuka, K., On the Jacobson radical of a semiring, Tôhoku Math. J. (2) 11 (1959), 409 - 421.

[7] Kuich, W. and Salomaa, A., Semirings, Automata, Languages, Springer-Verlag, 1986.

[8] Lallement, G., Semigroups and Combinatiorial Applications, John Wiley & Sons, 1979.

[9] Mahr, B., Semirings and transitive closure, TU Berlin, FB 20, Bericht-Nr. 82-5 (1982).

[10] Mahr, B., Iteration and summability in semirings, Ann. Discrete Math. 19 (1984), 229 - 256.

[11] Main, M.G. and Benson, D.B., Functional behavior of nondeterministic programs, Lecture Notes in Comp. Sci. 158 (1983), 290 - 301.

[12] Manes, E.G. and Benson, D.B., The inverse semigroup of a sum-ordered semiring, Semigroup Forum 31 (1985), 129 - 152.

[13] Pickert, G., Bemerkungen zum Algebrenbegriff, Math. Ann. 120 (1947 - 1949), 158 - 164.

[14] Rédei, L., Algebra, Geest & Portig, 1959 (Engl. Ed. 1967).

[15] Rote, G., A Systolic Array Algorithm for the Algebraic Path Problem, Computing 34 (1985), 191 - 219.

[16] Salomaa, A. and Soittola, M., Automata-Theoretic Aspects of Formal Power Series, Springer-Verlag, 1978.

[17] Steinfeld, O., Über die Struktursätze der Semiringe, Acta Math. Acad. Sci. Hungar. 10 (1959), 149 - 155.

[18] Weinert, H.J., Über Halbringe und Halbkörper I, Acta Math. Acad. Sci. Hungar. 13 (1962), 365 - 378.

[19] Weinert, H.J., On 0-simple semirings, semigroup semirings, and two kinds of division semirings, Semigroup Forum 28 (1984), 313 - 333.

[20] Wongseelashote, A., Semirings and path spaces, Discrete Math. 26 (1979), 55 - 78.

[21] Zassenhaus, H., Lehrbuch der Gruppentheorie, Teubner-Verlag, 1937 (Engl. Ed. Chelsea, 1949).

LECTURE NOTES IN MATHEMATICS

Edited by A. Dold and B. Eckmann

Some general remarks on the publication of proceedings of congresses and symposia

Lecture Notes aim to report new developments - quickly, informally and at a high level. The following describes criteria and procedures which apply to proceedings volumes. The editors of a volume are strongly advised to inform contributors about these points at an early stage.

§1. One (or more) expert participant(s) of the meeting should act as the responsible editor(s) of the proceedings. They select the papers which are suitable (cf. §§ 2, 3) for inclusion in the proceedings, and have them individually refereed (as for a journal). It should not be assumed that the published proceedings must reflect conference events faithfully and in their entirety. Contributions to the meeting which are not included in the proceedings can be listed by title. The series editors will normally not interfere with the editing of a particular proceedings volume - except in fairly obvious cases, or on technical matters, such as described in §§ 2, 3. The names of the responsible editors appear on the title page of the volume.

§2. The proceedings should be reasonably homogeneous (concerned with a limited area). For instance, the proceedings of a congress on "Analysis" or "Mathematics in Wonderland" would normally not be sufficiently homogeneous.

One or two longer survey articles on recent developments in the field are often very useful additions to such proceedings - even if they do not correspond to actual lectures at the congress. An extensive introduction on the subject of the congress would be desirable.

§3. The contributions should be of a high mathematical standard and of current interest. Research articles should present new material and not duplicate other papers already published or due to be published. They should contain sufficient information and motivation and they should present proofs, or at least outlines of such, in sufficient detail to enable an expert to complete them. Thus resumes and mere announcements of papers appearing elsewhere cannot be included, although more detailed versions of a contribution may well be published in other places later.

Surveys, if included, should cover a sufficiently broad topic, and should in general not simply review the author's own recent research. In the case of surveys, exceptionally, proofs of results may not be necessary.

"Mathematical Reviews" and "Zentralblatt für Mathematik" require that papers in proceedings volumes carry an explicit statement that they are in final form and that no similar paper has been or is being submitted elsewhere, if these papers are to be considered for a review. Normally, papers that satisfy the criteria of the Lecture Notes in Mathematics series also satisfy this

.../...

requirement, but we would strongly recommend that the contributing authors be asked to give this guarantee explicitly at the beginning or end of their paper. There will occasionally be cases where this does not apply but where, for special reasons, the paper is still acceptable for LNM.

§4. Proceedings should appear soon after the meeeting. The publisher should, therefore, receive the complete manuscript within nine months of the date of the meeting at the latest.

§5. Plans or proposals for proceedings volumes should be sent to one of the editors of the series or to Springer-Verlag Heidelberg. They should give sufficient information on the conference or symposium, and on the proposed proceedings. In particular, they should contain a list of the expected contributions with their prospective length. Abstracts or early versions (drafts) of some of the contributions are very helpful.

§6. Lecture Notes are printed by photo-offset from camera-ready typed copy provided by the editors. For this purpose Springer-Verlag provides editors with technical instructions for the preparation of manuscripts and these should be distributed to all contributing authors. Springer-Verlag can also, on request, supply stationery on which the prescribed typing area is outlined. Some homogeneity in the presentation of the contributions is desirable.

Careful preparation of manuscripts will help keep production time short and ensure a satisfactory appearance of the finished book. The actual production of a Lecture Notes volume normally takes 6 -8 weeks.

Manuscripts should be at least 100 pages long. The final version should include a table of contents and as far as applicable a subject index.

§7. Editors receive a total of 50 free copies of their volume for distribution to the contributing authors, but no royalties. (Unfortunately, no reprints of individual contributions can be supplied.) They are entitled to purchase further copies of their book for their personal use at a discount of 33.3 %, other Springer mathematics books at a discount of 20 % directly from Springer-Verlag. Contributing authors may purchase the volume in which their article appears at a discount of 33.3 %.

Commitment to publish is made by letter of intent rather than by signing a formal contract. Springer-Verlag secures the copyright for each volume.

LECTURE NOTES

ESSENTIALS FOR THE PREPARATION OF CAMERA-READY MANUSCRIPTS

The preparation of manuscripts which are to be reproduced by photo-offset requires special care. Manuscripts which are submitted in technically unsuitable form will be returned to the author for retyping. There is normally no possibility of carrying out further corrections after a manuscript is given to production. Hence it is crucial that the following instructions be adhered to closely. If in doubt, please send us 1 - 2 sample pages for examination.

Typing area. On request, Springer-Verlag will supply special paper with the typing area outlined.

The CORRECT TYPING AREA is 18 x 26 1/2 cm (7,5 x 11 inches).

Make sure the TYPING AREA IS COMPLETELY FILLED. Set the margins so that they precisely match the outline and type right from the top to the bottom line. (Note that the page-number will lie outside this area). Lines of text should not end more than three spaces inside or outside the right margin (see example on page 4).

Type on one side of the paper only.

Type. Use an electric typewriter if at all possible. CLEAN THE TYPE before use and always use a BLACK ribbon (a carbon ribbon is best).

Choose a type size large enough to stand reduction to 75%.

Word Processors. Authors using word-processing or computer-typesetting facilities should follow these instructions with obvious modifications. Please note with respect to your printout that
i) the characters should be sharp and sufficiently black;
ii) if the size of your characters is significantly larger or smaller than normal typescript characters, you should adapt the length and breadth of the text area proportionally keeping the proportions 1:0.68.
iii) it is not necessary to use Springer's special typing paper. Any white paper of reasonable quality is acceptable.
IF IN DOUBT, PLEASE SEND US 1-2 SAMPLE PAGES FOR EXAMINATION. We will be glad to give advice.

Spacing and Headings (Monographs). Use ONE-AND-A-HALF line spacing in the text. Please leave sufficient space for the title to stand out clearly and do NOT use a new page for the beginning of subdivisions of chapters. Leave THREE LINES blank above and TWO below headings of such subdivisions.

Spacing and Headings (Proceedings). Use ONE-AND-A-HALF line spacing in the text. Start each paper on a NEW PAGE and leave sufficient space for the title to stand out clearly. However, do NOT use a new page for the beginning of subdivisions of a paper. Leave THREE LINES blank above and TWO below headings of such subdivisions. Make sure headings of equal importance are in the same form.

The first page of each contribution should be prepared in the same way. Therefore, we recommend that the editor prepares a sample page and passes it on to the authors together with these ESSENTIALS. Please take

.../...

the following as an example.

MATHEMATICAL STRUCTURE IN QUANTUM FIELD THEORY

John E. Robert
Fachbereich Physik, Universität Osnabrück
Postfach 44 69, D-4500 Osnabrück

Please leave THREE LINES blank below heading and address of the author. THEN START THE ACTUAL TEXT OF YOUR CONTRIBUTION.

Footnotes. These should be avoided. If they cannot be avoided, place them at the foot of the page, separated from the text by a line 4 cm long, and type them in SINGLE LINE SPACING to finish exactly on the outline.

Symbols. Anything which cannot be typed may be entered by hand in BLACK AND ONLY BLACK ink. (A fine-tipped rapidograph is suitable for this purpose; a good black ball-point will do, but a pencil will not). Do not draw straight lines by hand without a ruler (not even in fractions).

Equations and Computer Programs. Equations and computer programs should begin four spaces inside the left margin. Should the equations be numbered, then each number should be in brackets at the right-hand edge of the typing area.

Pagination. Number pages in the upper right-hand corner in LIGHT BLUE OR GREEN PENCIL ONLY. The final page numbers will be inserted by the printer.

There should normally be NO BLANK PAGES in the manuscript (between chapters or between contributions) unless the book is divided into Part A, Part B for example, which should then begin on a right-hand page.

It is much safer to number pages AFTER the text has been typed and corrected. Page 1 (Arabic) should be THE FIRST PAGE OF THE ACTUAL TEXT. The Roman pagination (table of contents, preface, abstract, acknowledgements, brief introductions, etc.) will be done by Springer-Verlag.

Corrections. When corrections have to be made, cut the new text to fit and PASTE it over the old. White correction fluid may also be used.

Never make corrections or insertions in the text by hand.

If the typescript has to be marked for any reason, e.g. for TEMPORARY page numbers or to mark corrections for the typist, this can be done VERY FAINTLY with BLUE or GREEN PENCIL but NO OTHER COLOR: these colors do not appear after reproduction.

Table of Contents. It is advisable to type the table of contents later, copying the titles from the text and inserting page numbers.

Literature References. These should be placed at the end of each paper or chapter, or at the end of the work, as desired. Type them with single line spacing and start each reference on a new line.
Please ensure that all references are COMPLETE and PRECISE.

MIX
Papier aus verantwortungsvollen Quellen
Paper from responsible sources
FSC® C105338

If you have any concerns about our products,
you can contact us on
ProductSafety@springernature.com

In case Publisher is established outside the EU,
the EU authorized representative is:
Springer Nature Customer Service Center GmbH
Europaplatz 3, 69115 Heidelberg, Germany

Printed by Libri Plureos GmbH
in Hamburg, Germany